# Archaeological Ceramics: Interpreting Artefacts

# Archaeological Ceramics: Interpreting Artefacts

Edited by Christian Green

MURPHY & MOORE
www.murphy-moorepublishing.com

Murphy & Moore Publishing,
1 Rockefeller Plaza,
New York City, NY 10020, USA



ISBN: 978-1-63987-055-4

**Cataloging-in-Publication Data**

Archaeological ceramics : interpreting artefacts / edited by Christian Green.
p. cm.
Includes bibliographical references and index.
ISBN 978-1-63987-055-4
1. Pottery--Analysis. 2. Archaeology--Methodology. I. Green, Christian.
CC79.5.P6 A73 2021
930.102 85--dc23

For information on all Murphy & Moore Publications
visit our website at www.murphy-moorepublishing.com

MURPHY & MOORE

# Contents

# Preface

The world is advancing at a fast pace like never before. Therefore, the need is to keep up with the latest developments. This book was an idea that came to fruition when the specialists in the area realized the need to coordinate together and document essential themes in the subject. That's when I was requested to be the editor. Editing this book has been an honour as it brings together diverse authors researching on different streams of the field. The book collates essential materials contributed by veterans in the area which can be utilized by students and researchers alike.

An inorganic non-metallic solid made up of either metal or non-metal compounds that has been shaped and then hardened by heating to high temperatures is referred to as a ceramic. Earthenware, porcelain, and brick are some of the common examples of ceramics. In archeology, ceramic artifacts have an important role in understanding the culture, technology, and behavior of peoples of the past. They are among the most common artifacts that are found at an archaeological site, especially in the form of small fragments of broken pottery called sherds. This book brings forth some of the most innovative concepts and elucidates the unexplored aspects of archeological ceramics. Its aim is to present researches that have transformed this discipline and aided its advancement. This book will serve as a reference to a broad spectrum of readers.

Each chapter is a sole-standing publication that reflects each author´s interpretation. Thus, the book displays a multi-facetted picture of our current understanding of application, resources and aspects of the field. I would like to thank the contributors of this book and my family for their endless support.

**Editor**

1

# Luminescence dating of ceramic building materials: Application to the study of early medieval churches in north-western France and south-eastern England

**S. Blain[a,e], I.K. Bailiff[b], P. Guibert[a], C. Sapin[c], D. Prigent[d] and A. Bouvier[a]**

[a]IRAMAT-CRP2A - UMR 5060, CNRS - Université de Bordeaux 3, Pessac, France, blain.sophie@gmail.com, guibert@u-bordeaux3.fr, bouvier_armel@yahoo.fr

[b]Luminescence Laboratory, Dept of Archaeology, University of Durham, South Road, Durham, DH1 3LE, England, ian.bailiff@durham.ac.uk

[c]Laboratoire Artehis - UMR 5594, CNRS -Université de Bourgogne, Dijon, France, sapin.christian@wanadoo.fr

[d]Service archéologique départemental de Maine-et-Loire, France, d.prigent@cg49.fr

[e]FRS-FNRS, Centre Européen d'Archéométrie, Université de Liège, Belgium

**Abstract** - This study focuses on early medieval buildings in England and France, specifically on those displaying the presence of brick in their masonry. The purpose was to determine whether medieval builders reused salvaged Roman building materials or if the bricks used were contemporary to the building under construction. Thus, luminescence dating was applied to bricks sampled from 11 standing Anglo-Saxon or Carolingian churches. The luminescence dates show that the two types of practice were in use in both countries. Where contemporary brickmaking appears to be the case, the study also provided a more precise chronology of the buildings.

## 1. Introduction

Brickmaking, introduced by the Romans to Britain and Gaul, seems to have ceased following their departure and the subsequent arrival of Germanic settlers in the 5th century AD, to reappear only at the end of the 12th century with the introduction of continental monastic institutions in England and France. In spite of this apparent cessation in brickmaking, early medieval builders continued to use ceramic building material (CBM), which is generally assumed to be reused Roman material. The hypothesis of the reuse of Roman CBM in early medieval buildings was proposed in the 18th century in England (Morant 1768, 298; Davey 1961; Jope 1964, 113). However, doubts regarding this assumption were raised by the discovery of decorative floor tiles from 10th - 12th century archaeological contexts (Betts *et al.* 1991, 37; Betts 1996) in England (Westminster, York, Peterborough, Coventry, St Albans, and Canterbury) and also by the preliminary results of a thermoluminescence (TL) dating study performed during the developmental stage of the method on wall bricks from the famous Anglo-Saxon church of All Saints, Brixworth, Northamptonshire. The results obtained from the analysis of four bricks indicated Roman, Saxon, and medieval manufacture (Everson and Parsons 1979, 406). However, it was subsequently discovered that the technique used in the original TL work was potentially susceptible to producing underestimates of the age (Everson and Parsons 1979), and recent re-testing of the same samples by the Durham luminescence laboratory has indicated that the bricks are of Roman date (Bailiff 2013). Hence, uncertainties concerning early medieval brick manufacture in England have persisted.

In the Carolingian Empire, indirect evidence of brick production during the early Middle Ages exists in the form of a letter from Einhard, friend, counsellor, and biographer of Charlemagne, who ordered the making of 260 large "*lateres*" (Norton 1983, 36; Greenhalgh 1989). Another testimony was provided by Dudo of St Quentin, biographer of the Dukes of Normandy, who mentions the building site of the sanctuary of la Trinité, at Fécamp (granted by Richard I), the techniques used at the site, and particularly the manufacture of CBM: "*lateribus artificialiter cumpotis*" (Lair 1865, 290-1; Renoux 1991, 473-5). However, there was no indication of whether these were wall bricks or roof tiles. In the 1950s, Forsyth (1953, 25) suggested the existence of Carolingian bricks in the masonry of the Collegiate Church of St Martin in Angers. Between 1969 and 1972, on the 9th-10th century site of La Médecinerie at Saran (Loiret), a dozen kilns were discovered, revealing a production site of CBM, especially for antefixa and modillions (Debal 1969; Chapelot 1970). Despite such evidence, de Bouärd (1975, 55-6) declared that on the basis of the limited amount of CBM used in early medieval buildings in northern France, these materials were likely to have been salvaged from abandoned Roman sites.

The focus of this study is a group of buildings in north-western France and south-eastern England which displays analogous building methods and techniques, such as the use of brick as voussoirs and/or laid in horizontal rows in the masonry, similar to the style and practice of Roman or Byzantine architecture. Bricks are a potentially important component since they can be used to date the monument if their manufacture is contemporary with the construction. A number of religious buildings in Pays de Loire and Normandy in France, and in Kent and Essex in England was selected on the basis of the use of CBM. In particular, the selected buildings display an organised and systematic use of CBM, indicating a purposeful selection of CBM, which was more likely to have been made for the building and consequently contemporary with it.

The buildings included in the study comprise: the abbey Church of Saint-Philbert-de-Grandlieu (with a well-established Carolingian origin); the Collegiate Church of Saint-Martin in Angers; and the underground Church of Notre-Dame-sous-Terre, Mont-Saint-Michel, which contains the earliest surviving monastic origins of the church. The remaining buildings from Normandy included the parish rural churches of Vieux-Pont-en-Auge, Rugles, and Condé-sur-Risle. In England, the selected buildings included the Anglo-Saxon parish churches of St Martin's, Canterbury, which is the earliest English church in this study, St Margaret's, Darenth, and St Margaret's, Lower Halstow, both in Kent, Holy Trinity, Colchester, and the Norman parish Church of St Martin's, Chipping Ongar, in Essex.

Within this group of buildings included in this study, some provide architectural markers for dating, as they represent key-sites in the history of architecture, such as St Philbert-de-Grandlieu, Notre-Dame-sous-Terre, or St Martin in Angers, acting as comparisons for other buildings displaying similar architectural features. However, not all such buildings may be contemporary, and the situation is usually more complex. The dates for the study group have been the subject of debate, fluctuating between the 7th century and the early 11th century. The dating of these key-sites is therefore essential for establishing the extent of the influence within the group between the different areas.

## 2 Method

### *a) Principle*

The luminescence dating method determines the date of the last firing of the ceramic material being tested (i.e., brick), and a key luminescent mineral used for measurements in this study is mineral quartz (Aitken 1985). Thus, according to this method, the date of manufacture is determined. The time elapsed from the last firing of the material to analysis in the laboratory can be expressed as follows: T (yrs) = P (Gy) / D (mGy/yr), where T is the age of the brick; P is the paleodose, the total dose absorbed and stored by the grains within the ceramic. Two techniques can be employed to determine the paleodose depending on the chosen grain size: the polymineral fine grain technique, a mixture of quartz, feldspars and other alumino-silicates within a range between 3 and 12 μm, or the coarse grain technique, which involves only quartz in a range between 53 to 90 or 90 to 150 μm for the laboratory at Durham, and between 80 to 200 μm for the laboratory at Bordeaux.

Finally, D is the annual dose rate provided by the ceramic sample itself and its environment, i.e., the energy released from ionising radiation by naturally occurring radioactive minerals in brick and surrounding materials, and, to a small extent, cosmic rays. Therefore, the $\alpha$ and $\beta$ particle contributions are determined from the ceramic (only $\beta$ when the coarse grain technique is used), and $\gamma$ and cosmic rays from the environment.

### *b) Sampling*

The samples were cut using a diamond faced core drill of 20 to 50 mm diameter, designed for dry or wet cutting. The brick cavities were back-filled with lime-based mortar, and the surface impregnated with brick dust. The choice of the bricks to be sampled is usually made in conjunction with archaeologists or after close examination of the masonry structures, to make sure the chosen bricks are not later insertions. Only a couple of bricks are sampled when the aim is to verify whether the bricks are indeed reused Roman materials within an early medieval wall. Between four to fifteen bricks can be sampled when the aim is to date a structure or define the phasing of a building precisely.

### *c) Technique*

Two types of measurement procedures were used to determine the paleodose: thermoluminescence (TL) in Bordeaux and optically stimulated luminescence (OSL) in Durham. The main difference between both methods lies only in the the experimental means by which the luminescence was generated, and therefore they are expected to produce equivalent dating results.

After petrological observations on thin sections of the samples, the determination of the paleodose was either made with coarse grains of quartz (notes to Table 1) in both laboratories if the fabric was rather coarse, or with polymineral fine grains (3-12 μm) in Bordeaux only when the fabric appeared to be particularly fine (diameter of the grains less than 40 μm). The crystalline samples were obtained using standard extraction procedures, including crushing and sieving followed by acid etching treatments (Blain 2011). In Durham, by OSL, the single-aliquot regenerative-dose protocol (SAR) was chosen to determine the paleodose at it requires less material. In Bordeaux, by TL, the additive dose and regeneration technique was chosen, as SAR is not suitable for TL.

The annual dose was determined using direct and indirect measurement techniques (Blain 2011). In situ dosimetry was performed with thermoluminescent dosemeter capsules to determine the dose rate at the sample location due to gamma and cosmic radiation; high resolution gamma spectrometry was used to measure the activities of the lithogenic radionuclides in the brick fabric; and, in addition, for the English samples, beta

**Table 1.** Individual dating results by TL or OSL for the bricks from the studied sites: P: paleodose value (Gy), D. annual dose rate (mGy/an) and dates. Fg refers to the fine grain technique (3 - 12 μm) and CG to the coarse grain technique (53 - 90, 90 - 150 or 80 - 200 μm). In the case of reused materials, the uncertainty in the luminescence date (of manufacture) is higher because of uncertainty in the gamma component of the annual dose rate with each displacement. However, it was possible to estimate a date of manufacture for these materials with a corrected dose rate. The model involved in this correction is based on the assumption of a provenance from a Roman structure made of mortar and bricks of the same composition as the analysed brick sample. Since this was a working hypothesis, uncertainties of 20 - 25% were assigned to the calculation of the gamma dose rate. Typically the effect of this correction on the age falls within a range from 2.6 to 6.7%.

| Site | Sample | Technique (fadingcorr.model) | P ± s.d. (Gy) | | $D_{overall}$ ± 1σ (mGy/yr) | | Date ± 1σ (random) | |
|---|---|---|---|---|---|---|---|---|
| | | | Bdx | Dur | Bdx | Dur | Bdx | Dur |
| St Martin, Canterbury | Dur345-1 | CG | | 6.43 ± 0.40 | | 3.40 ± 0.10 | | 119 ± 122 (62) |
| | Dur345-2 | CG | | 5.98 ± 0.48 | | 3.55 ± 0.10 | | 323 ± 117 (58) |
| St Margaret, Lower Halstow | Dur344-2 | CG | | 5.59 ± 0.77 | | 3.50 ± 0.13 | | 411 ± 125 (87) |
| St Margaret, Darenth | Dur342-1 | CG | | 6.33 ± 0.31 | | 3.78 ± 0.15 | | 333 ± 118 (71) |
| Holy Trinity, Colchester | Dur343-2 | CG | | 6.65 ± 0.63 | | 3.55 ± 0.15 | | 136 ± 139 (93) |
| St Martin of Tours, Chipping Ongar | Dur363-1A | CG | | 2.98 ± 0.15 | | 3.20 ± 0.08 | | 1077 ± 57 (25) |
| | Dur363-1B | CG | | 3.08 ± 0.14 | | 3.19 ± 0.08 | | 1043 ± 59 (25) |
| | Dur363-2A | CG | | 2.86 ± 0.12 | | 2.91 ± 0.07 | | 1025 ± 61 (25) |
| | Dur363-3 | CG | | 3.08 ± 0.08 | | 3.06 ± 0.07 | | 1005 ± 58 (23) |
| St Philbert-de-Grandlieu | Bdx10219 | CG | 7.69 ± 0.58 | | 4.71 ± 0.18 | | 405 ± 137(123) | |
| | Bdx10220 | CG | 5.85 ± 0.57 | | 5.02 ± 0.22 | | 841 ± 121 (113) | |
| | Bdx10221 | CG | 3.52 ± 0.56 | | 4.29 ± 0.19 | | 1184 ± 134 (130) | |
| | Bdx10222 | CG | 4.61 ± 0.34 | | 4.21 ± 0.32 | | 910 ± 115 (107) | |
| | Bdx10223 | CG | 8.49 ± 0.70 | | 5.06 ± 0.22 | | 327 ± 153 (139) | |
| | Bdx10224 | CG | 4.55 ± 0.39 | | 5.15 ± 0.22 | | 1075 ± 84 (76) | |
| | Bdx10225 | CG | 6.53 ± 0.57 | | 4.16 ± 0.16 | | 377 ± 149 (135) | |
| | Bdx10226 | CG | 6.32 ± 0.58 | | 5.00 ± 0.19 | | 741 ± 125 (116) | |
| | Bdx10227 | CG | 5.41 ± 0.49 | | 4.56 ± 0.19 | | 820 ± 115 | |

*(Continued)*

**Table 1** - *continued*

| Site | Sample | Technique $_{(fadingcorr.model)}$ | P ± s.d. (Gy) | | $D_{overall}$ ± 1σ (mGy/yr) | | Date ± 1σ $_{(random)}$ | |
|---|---|---|---|---|---|---|---|---|
| | | | Bdx | Dur | Bdx | Dur | Bdx | Dur |
| | Bdx10228 | CG | 5.65 ± 0.54 | | 4.61 ± 0.18 | | 779 ± 126 (118) | |
| | Bdx10229 | CG | 6.22 ± 0.55 | | 4.60 ± 0.18 | | 654 ± 130 | |
| | Bdx10230 | CG | 5.55 ± 0.48 | | 4.12 ± 0.17 | | 661 ± 128 | |
| St Martin, Angers | Bdx9456 | Fg | 10.91 ± 0.47 | | 10.11 ± 0.53 | | 928 ± 65 (47) | |
| | Bdx9457 | Fg | 10.67 ± 0.62 | | 9.64 ± 0.44 | | 900 ± 79 (64) | |
| | Bdx9459 | Fg | 11.66 ± 0.53 | | 9.86 ± 0.46 | | 825 ± 79 (59) | |
| | Bdx9461 | Fg | 10.76 ± 0.51 | | 9.25 ± 0.46 | | 844 ± 82 (63) | |
| | Bdx9462 | Fg | 10.99 ± 0.60 | | 9.71 ± 0.42 | | 875 ± 81 (63) | |
| | Bdx9472 | Fg | 12.38 ± 0.64 | | 9.74 ± 0.42 | | 735 ± 83 (66) | |
| Notre-Dame-sous-Terre, Mont-Saint-Michel | Bdx8853 | CG | 4.26 ± 0.50 | | 4.02 ± 0.15 | | 947 ± 131 (125) | |
| | Bdx8854 | CG | 4.26 ± 0.39 | | 4.45 ± 0.16 | | 1047 ± 93 (86) | |
| | Bdx8856 | CG | 4.62 ± 0.40 | | 4.37 ± 0.16 | | 949 ± 100 (92) | |
| | Bdx8866 | CG | 4.14 ± 0.31 | | 3.88 ± 0.15 | | 941 ± 89 (79) | |
| | Bdx8861 | CG | 4.29 ± 0.29 | | 4.05 ± 0.18 | | 947 ± 85 (76) | |
| | Bdx8862 | CG | 4.08 ± 0.16 | | 3.91 ± 0.15 | | 962 ± 56 (41) | |
| | Bdx8858 | Fg | 5.90 ± 0.62 | | 4.98 ± 0.20 | | 916 ± 86 | |
| | | CG | 4.24 ± 0.39 | | 4.10 ± 0.15 | | | |
| | Bdx8859 | CG | 4.99 ± 0.69 | | 4.26 ± 0.16 | | 833 ± 168 (162) | |
| | Bdx8851 | CG | 4.89 ± 0.33 | | 4.46 ± 0.17 | | 995 ± 75 (68) | |
| | Bdx8857 | CG | 5.71 ± 0.40 | | 5.46 ± 0.21 | | 959 ± 82 (73) | |

(*Continued*)

**Table 1** - *continued*

| Site | Sample | Technique (fadingcorr.model) | P ± s.d. (Gy) | | $D_{overall}$ ± 1σ (mGy/yr) | | Date ± 1σ (random) | |
|---|---|---|---|---|---|---|---|---|
| | | | Bdx | Dur | Bdx | Dur | Bdx | Dur |
| | Bdx8864 | CG | 4.23 ± 0.30 | | 4.45 ± 0.17 | | 1054 ± 75 (67) | |
| | Bdx8865 | CG | 4.83 ± 0.36 | | 4.72 ± 0.18 | | 985 ± 85 (76) | |
| | Bdx8863 | CG | 4.58 ± 0.28 | | 4.35 ± 0.17 | | 954 ± 75 (64) | |
| | Bdx8860 | CG | 5.90 ± 0.32 | | 4.57 ± 0.17 | | 717 ± 84 (70) | |
| St Aubin, Vieux-Pont-en-Auge | Bdx9624 | Fg | 5.38 ± 0.48 | | 3.87 ± 0.13 | | 616 ± 159 | |
| | Bdx9625 | Fg | 8.01 ± 0.78 | | 4.25 ± 0.16 | | 122 ± 238 | |
| | Bdx9626 | Fg | 7.09 ± 0.86 | | 3.35 ± 0.15 | | − 111 ± 326 | |
| | Bdx9628 | Fg | 9.11 ± 1.13 | | 4.33 ± 0.17 | | 6 ± 282 | |
| | Bdx11987 | Fg | 10.14 ± 0.94 | | 3.62 ± 0.18 | | − 100 ± 327 | |
| | Bdx11989 | Fg | 4.69 ± 0.26 | | - | | − 170 ± 247 | |
| Notre-Dame-Outre-l'Eau, Rugles | Bdx11979/ Dur346-3 | CG | 6.26 ± 0.71 | 5.54 ± 0.19 | 2.51 ± 0.16 | 2.51 ± 0.12 | − 487 ± 354 | − 162 ± 179 |
| | Bdx11982/ Dur346-1 | CG | 6.15 ± 0.91 | 5.27 ± 0.26 | 2.61 ± 0.16 | 2.76 ± 0.13 | − 350 ± 441 | 117 ± 154 |
| | Bdx11983/ Dur346-2 | CG | | 3.96 ± 0.34 | | | - | 239 ± 139 |
| St Martin, Conde -sur-Risle | Bdx11984/ Dur347-1 | CG | 7.77 ± 0.56 | 8.70 ± 0.56 | 5.09 ± 0.24 | 4.91 ± 0.18 | 482 ± 212 | 244 ± 136 |
| | Bdx11985/ Dur347-2 | CG | - | 4.68 ± 0.46 | | | - | 225 ± 136 |

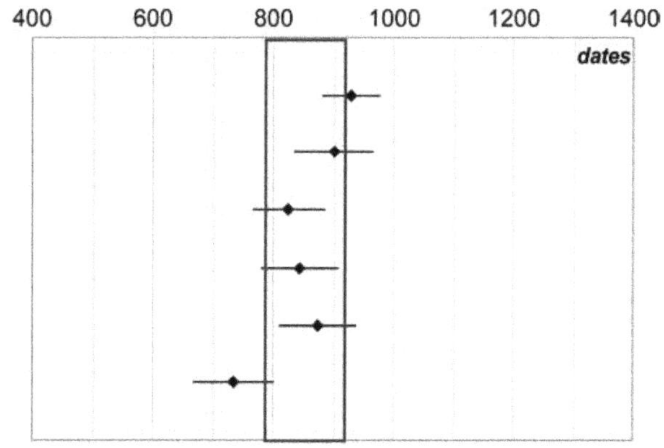


**Figure 1.** Individual dating results and overall dating (square zone) of the bricks from St Martin, Angers.

thermoluminescence dosimetry was used to measure the beta dose rate directly.

## 3 Results and discussion

The dating results obtained are listed in Table 1. The luminescence dates obtained for both the Anglo-Saxon and French Norman churches consistently confirm the archaeological assessment of reuse of Roman CBM. These include the English buildings of: St Martin in Canterbury (2 bricks sampled; 119 ± 122 and 323 ± 117 AD), St Margaret of Antioch in Lower Halstow (one brick: 411 ± 125 AD), St Margaret's Darenth (one brick: 333 ± 118 AD), and Holy Trinity in Colchester (one brick: 136 ± 139 AD), and the French buildings of St Aubin in Vieux-Pont-en-Auge (6 bricks sampled: from −170 ± 247 to 122 ± 238), Notre-Dame in Rugles (3 bricks analysed: from −487 ± 354 to 239 ± 139), and St Martin in Condé-sur-Risle (2 bricks: 225 ± 136 and 482 ± 212 AD).

The results for intercomparison samples show large error limits of the Bordeaux dates, likely due to the technique and protocol used: the coarse grains technique, which might involve a greater dispersion of the luminescence signals, and the multiple aliquot protocol, which is always less accurate than SAR, the latter being a direct comparison of the natural signal with the regenerated ones on the same aliquot. However, the Bordeaux and Durham dates overlap and are therefore relatively consistent with each other (Blain *et al.* 2010).

For most of these churches, the archaeological evidence which suggests reused material in the masonry include traces of *opus signinum* (typical pink Roman mortar) still adhering to the surface of some bricks, and the use of roof tiles such as *imbrices* and *tegulae* diverted from their primary function as used within walls. Moreover, most of these churches were located near Roman remains, as is the case for the church of Rugles, built over hypocaust remains. The church at Darenth was built 500 m from a Roman villa, and in Colchester there was an extensive Roman settlement with some remains surviving in the vicinity of the church of Holy Trinity.

In cases where CBM appears to be contemporary with the early medieval building, the dating results assist in ordering the chronology of the key-sites and in comparing it with the established historical interpretation. In the following discussion, the average of a series of OSL dates was obtained by calculating the weighted mean of the individual data.

The results for six bricks from the crossing tower of St Martin's, Angers, are homogeneous as a group (from 735 ± 83 to 928 ± 65 AD) and enable the lower parts of the phase to be re-dated to the 9$^{th}$ century (851 ± 60 AD; Fig. 1). This result has implications not only concerning previous historical-political knowledge, but also on the development of architectural style, showing that the church was first built two hundred years before the long supposed architectural activity of the famous count of Anjou, Fulk Nerra, whose contribution was probably limited to repair work only. This revision in dating is potentially of significance to the Anjou province in the early Middle Ages, since it confers to the church greater significance in the chronology of the history of art, by making it a precursor in building and artistic techniques.

The influence of Anjou spread to Normandy, particularly in the 10$^{th}$ and 11$^{th}$ centuries, where the architecture of vernacular buildings tends to imitate those seen in Anjou, most notably in the use of CBM in the masonry. This could well be the case for other buildings included in the study, such as the parish churches of Vieux-Pont-en-Auge, Rugles, or Condé-sur-Risle, and especially the Church of Notre-Dame-sous-Terre, Mont-Saint-Michel, where the luminescence dates showed that it was built after St Martin of Angers, for instance. The TL dates obtained on 14 bricks from Notre-Dame-sous-Terre (Fig. 2), used in conjunction with the archaeological interpretation, allow the chronological gap separating the two building phases to be better defined, and enable a clear hypothesis to be proposed (among the many controversial claims made for this particular site). The construction of the surrounding wall can be dated to the mid-10$^{th}$ century (952 ± 47 AD; weighted mean, range 832-1049), and the subsequent median wall and eastern sanctuaries, archaeologically defined as later adjunctions, were built later, at the end of the 10$^{th}$ century (986 ± 48; weighted mean, range 952-994). These results are interesting because they suggest a very short gap between the two phases, contrary to the

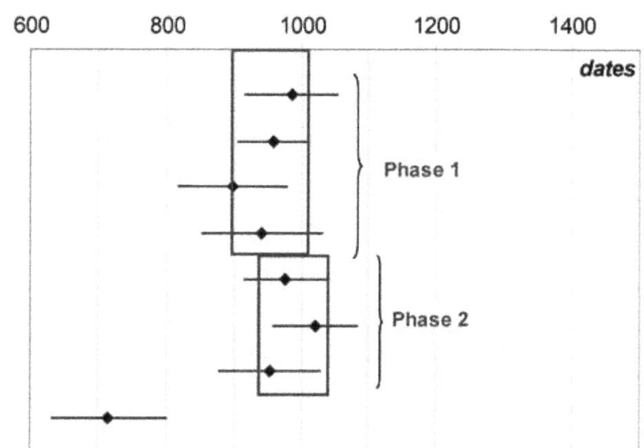


**Figure 2.** Dating results for each masonry structure and each building phase (square zones) for the Church of Notre-Dame-sous-Terre, Mont-Saint-Michel.

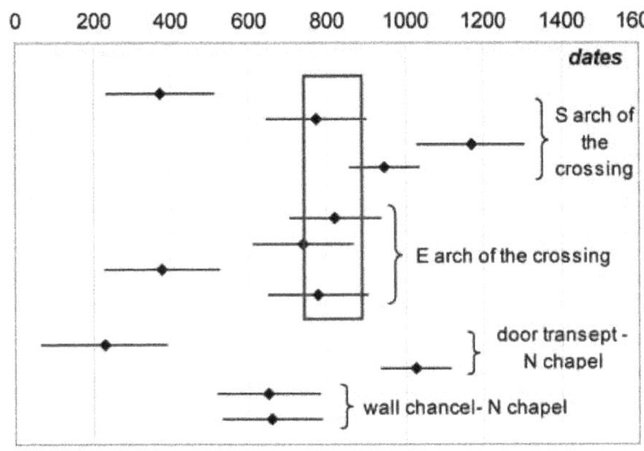


**Figure 3.** Individual dating results on the 12 bricks sampled at St Philbert-de-Grandlieu: the square zone indicates the dating obtained by $^{14}C$ on charcoals from mortar of the arches of the transept crossing.

archaeological assumption of de Boüard (1961), who put forth the hypothesis of a gap of one century between them.

The analysis of 12 ceramic samples from the St Philbert-de-Grandlieu abbey church was less successful due to technical problems arising from the complexity of the brick fabric (Guibert *et al.* 2009), and it was not possible to refine the dating of the construction and phasing. Although the dispersion in the dates is significant, the results indicate brickmaking during the period ca. $8^{th}$-$10^{th}$ century, a chronology which is consistent with the archaeological assessment of the construction of the original church. There is also an indication of reused or displaced material and repair work (Fig. 3).

This study also provided the opportunity to examine the links between Normandy and England. An example of Anglo-French relations is the Chipping Ongar church, which is assigned to the Norman style of architecture and dedicated to Saint Martin of Tours, whose cult was particularly fervent in medieval France. The CBM does not compare with known English medieval brick types, yet they are of medieval manufacture according to the OSL results (1038 ± 32 AD; weighted mean, range 1005-1077; Fig. 4), indicating an origin of the brickmaking prior to or near the Conquest. The variations in size, the large number of fragmentary CBM, also used in the structural features of the church, and the sporadic use of other bricks suggest the use of imported materials. Since Chipping Ongar is one of the few churches in England whose bricks are used in a continuous and regular way throughout the walls of the building, a feature which is more usual in France, it may be significant that the ownership of the church can be linked with the Continent. All this evidence reinforces the idea of strong ties with the Continent, and it is possible that the bricks are an example of early, post-Roman CBM in England, or of imported material from the Continent. These results warrant a wider investigation of buildings in Essex displaying similar unidentified bricks (Ryan 1996), and also provide supporting arguments (Ryan 1996; Rodwell 1998; Andrews 2008) that the previously held view of the pioneering role of Cistercians in the technological reintroduction of brickmaking in England are not correct, and that the first medieval bricks pre-date Cistercian or Flemish imports.

## 4 Conclusion

Surviving medieval buildings, particularly those of the early post-Roman period, are likely to have been reused, restored, modified, and transformed; this aspect raises a series of complex archaeological questions and problems when attempting to reconstruct their history. Defining different building phases is often difficult, and providing dates for such phases of construction enables a better understanding of the evolution of the building. This study has shown that the luminescence dating techniques can be successfully applied to the study of medieval buildings containing brick. When applied to test archaeological assumptions concerning the dating and phasing of buildings, the TL or OSL results have enhanced our understanding of the significance of the buildings in the cases of the Collegiate Church of Saint-Martin in Angers, the Church of Notre-Dame-sous-Terre, Mont-Saint-Michel, and the parish Church of St Martin of Tours, Chipping Ongar. Although the study of the bricks from St Philbert-de-Grandlieu proved to be problematic, producing a highly dispersed group of dates, attributed to heterogeneous composition of the ceramic fabric, they have provided valuable methodological information regarding radiochemically complex ceramics (Guibert *et al.* 2009).

When applied to address questions relating to the reuse of building materials as assessed by archaeological examination, this type of study could confirm or, on the contrary, invalidate archaeological assumptions related to the origin of the material (Blain 2011; Bailiff *et al.* 2010). Indeed, Roman bricks were actually reused in the walls of Anglo-Saxon churches (St Martin in Canterbury, St Margaret of Antioch, Lower Halstow, St Margaret, Darenth, and Holy Trinity, Colchester) and in pre-Romanesque masonries in France (St Aubin in Vieux-Pont-en-Auge, St Martin in Condé-sur-Risle, and Notre-Dame-Outre-l'Eau, Rugles), but, in contrast to what was long assumed, in some cases such as Chipping Ongar, the bricks were not of Roman origin but were in fact medieval productions.

Finally, the study of CBM as indicators of past material culture contributes to the understanding of past human behaviour. The architectural history of a building reflects local practices, personalities and events, and therefore constitutes a graphic witness of local history. It indirectly

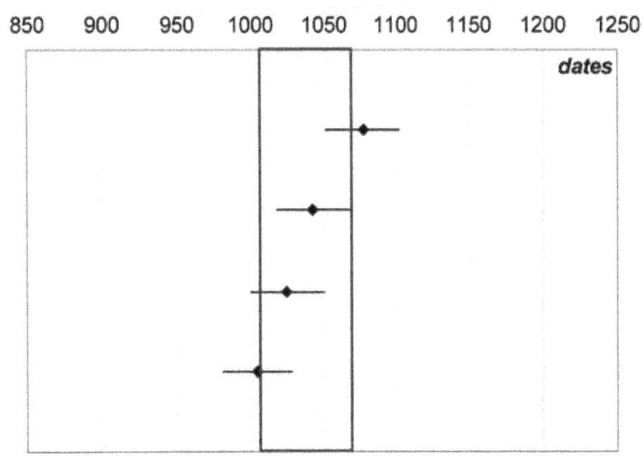


**Figure 4.** Individual dating results and overall dating (square zone) of the bricks from the Church of St Martin of Tours, Chipping Ongar.

presents information on economic vicissitudes, religious development and the culture of a people, their technical skills and aesthetic choices. Architectural creations are influenced by traditions and fashions, and are also dependent on politico-economic contexts, which determine material constraints or the commissioning requirements. The study of the origin and mode of use of building materials in the early Middle Ages therefore provides information on the cultural behaviour of early medieval people, together with the politico-economic choices they made. This study thus sheds light on the social or ritual practices, the economic organisation and activity, the development and social implications of technology and aesthetics, and the ideologies and beliefs prevalent at the time.

## References

Aitken M.J., 1985, *Thermoluminescence Dating*, Academic Press, London.

Andrews D., 2008, The Influence of the Low Countries on Medieval Brick Building in Eastern England. In *Medieval Brick Architecture in Flanders and northern Europe: The Question of the Cistercian Origin*, (eds. T. Coomans and H. van Royen), 53-72, Jaarboek Abdijmuseum Ten Duinen 1138, Novi Monasterii, Academia Press, Koksijde.

Bailiff I.K., 2013, Luminescence dating of brick from Brixworth Church: re-testing by the Durham Laboratory. In *The Anglo-Saxon Church of All Saints, Brixworth, Northamptonshire: Survey, Excavation and Analysis 1972-2010*, (eds. D. Parsons and D.S. Sutherland), 243-250, Oxbow, Oxford.

Bailiff I.K., Blain S., Graves C.P., Gurling T., Semple S., 2010, Uses and recycling of brick in medieval and Tudor English buildings: insights from the application of luminescence dating and new avenues for further research. *The Archaeological Journal*, 167, 165-196.

Betts I.M., 1996, Glazed Eleventh-Century Wall Tiles from London. *Medieval Ceramics*, 20, 19-24.

Betts I., Crowley N. and Keily J., 1991, Recent Work on Medieval and Later Building Materials in London. *Medieval Ceramics*, 15, 37-41.

Blain S., 2011, *Les terres cuites architecturales des églises du haut Moyen Age du NW de la France et du SE de l'Angleterre. Application de la datation par luminescence à l'archéologie du bâti*, BAR International Studies 2189, Archaeopress, Oxford.

Blain S., Bailiff I.K., Guibert P., Bouvier A. and Baylé M., 2010, An intercomparison study of luminescence dating protocols and techniques applied to medieval brick samples from Normandy (France). *Quaternary Geochronology*, 5, 311-316.

Boüard M. (de), 1961, L'église Notre-Dame-Sous-Terre au Mont-Saint-Michel: essai de datation. *Journal des Savants*, 1, 10-27.

Boüard M. (de), 1975, *Manuel d'archéologie médiévale: de la fouille à l'histoire*, Collection Regards sur l'Histoire, Paris.

Chapelot J., 1970, L'atelier céramique carolingien de Saran. Les fouilles de 1969 à 1970. *Bulletin de la société archéologique et historique de l'Orléanais*, 6, 49-72.

Davey N., 1961, *History of Building Materials*, Phoenix House, London.

Debal J., 1969, La découverte du site de la Médecinerie à Saran [Loiret]. *Bulletin de la société archéologique et historique de l'Orléanais*, 5, 311-338.

Everson P. and Parsons D., 1979, Brixworth Church - Are the Bricks Really Roman? In *Roman Brick and Tile Studies in Manufacture, Distribution and Use in the Western Empire*, (ed. A. McWhirr), 405-411, British Archaeological Reports International Series 68. Archaeopress, Oxford.

Forsyth G.H., 1953, *The Church of St Martin at Angers*, Princeton University Press, London.

Greenhalgh M., 1989, *The Survival of Roman Antiquities in the Middle Ages*, Duckworth, London.

Guibert P., Bailiff I.K., Blain S., Gueli A.M., Martini M., Sibilia E., Stella G. and Troja S., 2009, Luminescence dating of architectural ceramics from an early medieval abbey: the St-Philbert intercomparison (Loire Atlantique, France). *Radiation Measurements*, 44, 488-493.

Jope E.M., 1952-53, Late Saxon Pits under Oxford Castle Mound: Excavations in 1952. *Oxoniensia*, 17-18, 405-411.

Lair, J. (ed.), 1865, *Dudon de Saint-Quentin, De moribus et actis primorum Normanniae ducum, M.S.A.N.*, XXIII, Le Blanc-Hardel, Caen.

Morant P., 1768, *The History and Antiquities of the County of Essex, Vol. II*, Printed for T. Osborne, J. Whiston, S. Baker, L. Davis and C. Reymers, B. White, London.

Norton C.N, 1983, *Twelfth and thirteenth century decorated tile pavements in France and related material in England*, Unpublished PhD thesis, University of Cambridge, UK.

Renoux A., 1991, *Fécamp, du palais ducal au palais de Dieu: bilan historique et archéologique des recherches menées sur le site du château des ducs de Normandie: IIe siècle A.C.-XVIIIe siècle P.C.*, Editions du CNRS, Paris.

Rodwell W., 1998, "Bradwell-iusta-Coggeshall". *Essex Archaeology and History*, 29, 59-114.

Ryan P., 1996, *Brick in Essex from the Roman Conquest to the Reformation*, P. Ryan, Chelmsford.

## ACKNOWLEDGEMENTS

This research was supported by the CNRS, Department of Human and Social Sciences, within the framework of the GdRE (*Groupe de Recherche Européen*) "*Terres Cuites Architecturales et nouvelles méthodes de datation*", the University of Durham (UK), the University of Bordeaux 3 (France), the Conseil Régional d'Aquitaine, and the European EGIDE project: Programme Alliance.

2

# Assyrian palace ware definition and *chaîne opératoire*: Preliminary results from Nineveh, Nimrud, and Aššur

**Alice M.W. Hunt**
Center for Applied Isotope Studies, University of Georgia, ahunt@uga.edu

**Abstract** - Palace ware, an 8th–7th century BC drabware, is used as an indicator of Neo-Assyrian imperial elite contact and occupation, despite the lack of an established definition. Definitional criteria must incorporate formal characteristics, such as shape, and manufacture behaviours, such as raw material processing, in order to facilitate the distinction between imported or Assyrian produced vessels and imitations from the imperial provinces. This paper presents eleven preliminary definitional criteria for palace ware based on petrographic analysis and scanning electron microscopy of 24 sherds, radiography of 9 vessels, and typological analysis of 75 vessels from the Neo-Assyrian imperial heartland: Aššur, Nineveh, and Nimrud.

## 1. Introduction

Palace ware, a late 8th–7th century BC drabware, was first identified as a 'ware' by Rawson (1954) during Mallowan's excavation of Nimrud, and has since become synonymous with the Neo-Assyrian empire and elite (cf. Ohtsu 1991; Hausleiter 2008). Archaeologists working in peripheral regions of the Neo-Assyrian Empire consider the presence of Assyrian palace ware (APW) evidence of Assyrian contact and occupation (cf. Engstrom 2004; Na'aman & Thareani-Sussely 2006; Singer-Avitz 2007). Despite its importance, APW has not been comprehensively studied or defined since Rawson's research. This paper provides preliminary definitional criteria for APW as a 'ware' from the Assyrian imperial heartland, Aššur, Nineveh, and Nimrud, facilitating a better understanding of palace ware and Neo-Assyrian ceramic technology in general, and providing concrete tools for assessing whether APW from the imperial provinces is imported, produced locally by an Assyrian trained potter, or an imitation of APW forms using local technology.

Every technological choice of the potter is imprinted on the completed vessel. Therefore, when defining a 'ware', it is important to understand the sequence of behaviours which created it, in addition to formal characteristics, such as shape and decoration. Rawson (1954) published a series of technical observations based on macroscopic examination, summarised as follows: "sloppy", "wheel-made" pottery, probably "thrown in one piece from the hump". This paper uses archaeometry to refine these observations and define central polity palace ware typologically and technologically. Petrological and radiographic methods are used to determine the *chaîne opératoire* of palace ware, from raw material acquisition and processing through firing and decoration. Metric and formal analyses establish the range of APW forms and shapes in the Neo-Assyrian heartland and provide a quantitative typology.

## 2. Materials and methods

The entire assemblage of complete and repaired palace ware vessels from Aššur, Nineveh, and Nimrud/Kalḫu housed in the Vorderasiatisches and British Museums, and UCL Institute of Archaeology collections were analysed (Table 1). Aššur APW, excavated by the *Deutsche Morgenländische Gesellschaft* (DMG) between 1903 and 1913 (cf. Andrae 1909), comes from 'elite' burials dated to the late 8th–7th centuries BC (Haller 1954). Contextual information for Nineveh APW housed in the British Museum and UCL Institute of Archaeology is not available. Nimrud APW, excavated by Mallowan between 1949 and 1957, housed in the British Museum and UCL Institute of Archaeology, comes from palatial and administrative contexts dated to the 8th–7th centuries BC: specifically the NW Palace, Nabu temple, Governor's Palace, '49 Palace, and Fort Shalmaneser (Mallowan 1966).

The typological analysis of all complete palace ware vessels from Aššur, Nineveh, and Nimrud is based on the following attributes: rim type and diameter; neck/collar length and diameter; shoulder angle and diameter; minimum and maximum diameter; capacity, measured as the sum of successive truncated conic sections or frustums; body height; base type and diameter; wall thickness;

**Table 1.** Table of palace ware analysed from Aššur, Nimrud, and Nineveh.

| Excavation | Museum Reg. No | Typology Ref. No. | Object location | Form | Scientific analyses |
|---|---|---|---|---|---|
| Aššur | VA Ass.505 | AA | Vorderasiatisches Museum | jar | |
| Aššur | VA Ass.176 | AB | Vorderasiatisches Museum | cup | |
| Aššur | VA Ass.2 | AC | Vorderasiatisches Museum | jar | |
| Aššur | VA Ass.435 | AD | Vorderasiatisches Museum | jar | |
| Aššur | VA Ass.379 | AE | Vorderasiatisches Museum | jar | |
| Aššur | VA Ass.143 | AF | Vorderasiatisches Museum | cup | |
| Aššur | VA Ass.163 | AG | Vorderasiatisches Museum | jar | |
| Aššur | VA Ass.126 | AH | Vorderasiatisches Museum | jar | |
| Aššur | VA Ass.561 | AI | Vorderasiatisches Museum | jar | |
| Aššur | VA Ass.174 | AJ | Vorderasiatisches Museum | jar | |
| Aššur | VA Ass.166 | AK | Vorderasiatisches Museum | jar | |
| Aššur | VA Ass.171 | AL | Vorderasiatisches Museum | cup | |
| Aššur | VA Ass.542 | AM | Vorderasiatisches Museum | cup | |
| Aššur | VA Ass.169 | AN | Vorderasiatisches Museum | jar | |
| Aššur | VA Ass.433 | AO | Vorderasiatisches Museum | jar | |
| Aššur | VA Ass.840 | AP | Vorderasiatisches Museum | bowl | |
| Aššur | VA Ass.1424 | AQ | Vorderasiatisches Museum | bowl | |
| Aššur | VA Ass.849 | AR | Vorderasiatisches Museum | bowl | |
| Aššur | VA Ass.1585 | AS | Vorderasiatisches Museum | bowl | |
| Aššur | VA Ass.1481 | AT | Vorderasiatisches Museum | bowl | |
| Aššur | VA Ass.1397 | AU | Vorderasiatisches Museum | NA | |
| Aššur | VA Ass.848 | AV | Vorderasiatisches Museum | bowl | |
| Aššur | VA 842 | AW | Vorderasiatisches Museum | bowl | |
| Aššur | VA Ass.1483 | AX | Vorderasiatisches Museum | bowl | |
| Nimrud | 1992-3-2-66 | KA | British Museum | bowl | |
| Nimrud | 1992-3-2-64 | KB | British Museum | bowl | |
| Nimrud | 1992-3-2-62 | KC | British Museum | bowl | |
| Nimrud | 1992-3-2-239 | KD | British Museum | bowl | |
| Nimrud | 1992-3-2-237 | KE | British Museum | bowl | |
| Nimrud | 1994-11-5-148 | KF | British Museum | bowl | |
| Nimrud | 1992-3-2-238 | KG | British Museum | bowl | |
| Nimrud | 1992-3-2-143 | KH | British Museum | cup | |
| Nimrud | 1992-3-2-186 | KI | British Museum | N/A | |
| Nimrud | 1932-12-12-37 | KJ | British Museum | cup | |

*(Continued)*

**Table 1.** - *continued*

| Excavation | Museum Reg. No | Typology Ref. No. | Object location | Form | Scientific analyses |
|---|---|---|---|---|---|
| Nimrud | 1992-3-2-187 | KK | British Museum | cup | |
| Nimrud | 1992-3-2-191 | KL | British Museum | cup | |
| Nimrud | 1992-3-2-192 | KM | British Museum | cup | |
| Nimrud | 1992-3-2-190 | KN | British Museum | cup | |
| Nimrud | 1992-3-2-524 | KO | British Museum | cup | |
| Nimrud | 1992-3-2-182 | KP | British Museum | cup | |
| Nimrud | 1992-3-2-184 | KQ | British Museum | cup | |
| Nimrud | 1992-3-2-525 | KR | British Museum | cup | |
| Nimrud | 1992-3-2-188 | KS | British Museum | cup | |
| Nimrud | 1992-3-2-526 | KT | British Museum | cup | |
| Nimrud | 1992-3-2-477 | KU | British Museum | cup | |
| Nimrud | 1992-3-2-522a | KV | British Museum | cup | |
| Nimrud | 1992-3-2-522b | KW | British Museum | cup | |
| Nimrud | 1992-3-2-484 | KX | British Museum | cup | |
| Nimrud | 1992-3-2-494 rim | KY | British Museum | N/A | |
| Nimrud | 1992-3-2-494base | KZ | British Museum | bowl | |
| Nimrud | 1992-3-2-522.1 | | British Museum | N/A | X |
| Nimrud | 1992-3-2-522.2 | | British Museum | N/A | X |
| Nimrud | 1992-3-2-522.3 | | British Museum | N/A | X |
| Nimrud | 1992-3-2-522.4 | | British Museum | N/A | X |
| Nimrud | 1992-3-2-522.5 | | British Museum | N/A | X |
| Nimrud | 1992-3-2-522.6 | | British Museum | N/A | X |
| Nimrud | 1992-3-2-522.7 | | British Museum | N/A | X |
| Nimrud | 1992-3-2-522.8 | | British Museum | N/A | X |
| Nimrud | 1992-3-2-522.9 | | British Museum | N/A | X |
| Nimrud | 1992-3-2-522.10 | | British Museum | N/A | X |
| Nimrud | 1992-3-2-522.11 | | British Museum | N/A | X |
| Nimrud | 1992-3-2-522.12 | | British Museum | N/A | X |
| Nineveh | 1932-1212-859 | NA | British Museum | jar | |
| Nineveh | 1992-3-2-188 | NB | British Museum | cup | |
| Nineveh | 1932-12-12-37 | NC | British Museum | cup | |
| Nineveh | 1992-3-2-143 | ND | British Museum | cup | |
| Nineveh | 1992-3-2-182 | NE | British Museum | cup | |

(*Continued*)

**Table 1.** - *continued*

| Excavation | Museum Reg. No | Typology Ref. No. | Object location | Form | Scientific analyses |
|---|---|---|---|---|---|
| Nineveh | G 1441 | NF | UCL Institute of Archaeology | cup | |
| Nineveh | ND 883 | NG | UCL Institute of Archaeology | bowl | |
| Nineveh | ND 1312 | NH | UCL Institute of Archaeology | cup | |
| Nineveh | ND 3066 | NI | UCL Institute of Archaeology | cup | |
| Nineveh | ND 6662 | NJ | UCL Institute of Archaeology | cup | |
| Nineveh | 3/1030-122 | | UC Berkeley Excavation | N/A | X |
| Nineveh | 3/1316-22 | | UC Berkeley Excavation | N/A | X |
| Nineveh | 1058 | | UC Berkeley Excavation | N/A | X |
| Nineveh | 3/1307-15 | | UC Berkeley Excavation | N/A | X |
| Nineveh | 1129 | | UC Berkeley Excavation | N/A | X |
| Nineveh | 3/1278-17 | | UC Berkeley Excavation | N/A | X |
| Nineveh | 1129(2) | | UC Berkeley Excavation | N/A | X |
| Nineveh | 3/1390-50 | | UC Berkeley Excavation | N/A | X |
| Nineveh | 3/1031-65 | | UC Berkeley Excavation | N/A | X |
| Nineveh | 3/1059-2 | | UC Berkeley Excavation | N/A | X |
| Nineveh | 3/1065-3 | | UC Berkeley Excavation | N/A | X |
| Nineveh | 1219 | | UC Berkeley Excavation | N/A | X |

weight-height ratio (complete vessels only); interior and exterior fabric colour; and decorative element type and height from base. Attribute pairs were plotted on a Cartesian coordinate system using SPSS to detect latent patterns, and multiresponse permutation procedure (MRPP) analyses were run to confirm the validity of the observed grouping behaviour. MRPP tests the significance of sample groups by comparing intragroup and intergroup Euclidean distance, and was selected as an analytical tool because it does not require conditions, such as homogeneity of variance, which cannot be met by an assemblage of archaeological artefacts Mielke and Berry 2001.

Scientific analyses of APW were conducted on diagnostic and body sherds from the UC Berkeley excavation of Nineveh (1987–1990), area MG22, and Mallowan's excavation of the NW palace at Nimrud. Petrological analyses were conducted using a Leica DMRX petrographic microscope and Hitachi S3700N scanning electron microscope (SEM). SEM micrographs were taken by C.R. Cartwright at 15 kV and are published here courtesy of the Trustees of the British Museum. Petrographic descriptions use terminology and methods outlined by Whitbread (1995). Digital radiographs were collected using Varian Viva software (revision K3, build 45), a tungsten X-ray source (X-tek 160 kV gun), and Varian PaxScan 4030 detector, at 40 kV, 400 A, and 20 frames per image (approximately 10 minutes run time).

## 3. Results

### *Typology*

Every attribute of a ceramic vessel results from intentional human behaviour; however, not all of the relationships among attributes are typologically useful. Attributes can correlate naturally, for example wall thickness and weight: as wall thickness increases, so does the weight of the vessel. These intrinsic relationships between attributes display linear or horizontal asymptotic behaviour (the flattening of a curve with respect to the y-axis as it approaches infinity on the x-axis), easily distinguished from the non-continuous or cluster behaviour of intentional attribute relationships.

Attributes were considered definitional of palace ware when 90% or more of the assemblage consistently formed a cluster or non-continuous pattern. These definitional clusters are reported in Table 2 as value ranges, two endpoints, and a mode. Mode more accurately describes definitional attribute clusters than mean, because mode is a measure of frequency which is not affected by end behaviour. Outliers are defined as vessels falling outside

**Table 2.** Table summarising the metric characterisation of palace ware forms.

| | Bowls | Cups | Jars |
|---|---|---|---|
| Rim diameter | 8.5–13 cm<br>12 cm mode | 4–10 cm<br>7 cm mode | 6.5–11.5 cm<br>8.5 cm mode |
| Neck length | 0.2–1 cm<br>1 cm mode | 2.5–6.7 cm<br>3 cm mode | 2.5–7.5 cm<br>4.6 cm mode |
| Neck diameter (mouth) | 8–12 cm<br>11 cm mode | 4.5 -7.5 cm<br>5 cm mode | 5.5–8.5 cm<br>5.7 cm mode |
| Shoulder angle | – | 25–58°<br>45° mode | 28–57°<br>45° |
| Maximum diameter | 9–14 cm<br>12 cm mode | 4–10 cm<br>7 cm mode | 8.8–11 cm<br>10.5 cm mode |
| Body length | 1.6–3 cm<br>2.8 cm mode | 4.4–10 cm<br>8 cm mode | 8–14 cm<br>13.5 mode |
| Wall thickness | 0.2–0.4 cm<br>0.3 cm mode | 0.15–0.3 cm<br>0.2 cm mode | 0.2–0.4 cm<br>0.3 cm mode |
| Vessel capacity | 190–750 cc<br>478 cc mean | 300–800 cc<br>658 cc mean | 1250–1840 cc 2350–3049 cc<br>1450 cc mean 2700 cc mean |
| Total height | 2.5–5.5 cm<br>3 cm mode | 7.6–13.6 cm<br>12.4 cm mode | 10–20 cm<br>20 cm mode |
| Base diameter | 2–7.5 cm<br>4 cm mode | 0–2.5 cm<br>0.9 cm mode | 0–1.6 cm<br>1.5 cm mode |

the cluster clouds of more than one attribute and are not considered palace ware.

Five diagnostic typological criteria were detected for the entire palace ware assemblage: maximum diameter is 6–14 cm; capacity is discontinuous and forms three clusters around 500, 1500 and 3000 cc.; the wall thickness range is 0.15–0.4 cm with a mode of 0.2 cm; rim diameter ranges from 6 to 14 cm; and base diameter ranges from 0 to 8 cm. Three form clusters were also identified in the assemblage, differentiated by neck/collar length and capacity (Fig. 1): bowls, cups, and jars.

Palace ware bowls (Fig. 2) occur in two related forms. Type I bowls are prevalent in the bowl assemblage and have everted collars and rims extending from a sharp shoulder carination and thinned lips. Their bases are rounded to flat, with no visible basal element such as a disk, ring, or knob. Type I bowls are unequally biconical and the ratio of collar to body height is between 1:3 and 1:6. Type II bowls are equally biconical, with flat disk bases and everted rims with 'rounded' lips. Both type I and II bowls can be decorated with engraved circumferential lines; however, only type I bowls are dimpled.

Palace ware bowl capacity ranges between 200 and 700 cc. (500 cc. mode), regardless of type; however, a small group of thick walled outlier bowls have capacities between 967 and 1119 cc. (1000 cc. mode). These vessels (Fig. 2) are equally biconical, have stepped bases and bodies, 'bow tie' or triangular lips, thick walls (0.6–1 cm), and, while they may represent the 'thick' palace ware described by Rawson (1954), these bowls are probably 'typical' Neo-Assyrian tableware (Anastasio 2010).

Palace ware cups also occur in two shapes (Fig. 2). Type I is characterised by short, wide necks (mode 3 cm long and 6 cm wide) and slightly truncated body height (4.4–8.2 cm). The second cup shape (type II) has longer (mode 4.5 cm), narrower (mode 5 cm) necks and an elongated body (mode 8 cm). Despite these subtle differences, palace ware cups are remarkably consistent in shape. They are unequally biconical, with pronounced v-shaped bodies and carinated shoulders. Cup necks are 'trumpet' shaped, with everted, horizontally flattened rims and thinned or rolled lips. APW cup bases are typically flat, although pointed, knob, ring, and elevated ring bases are also represented in the

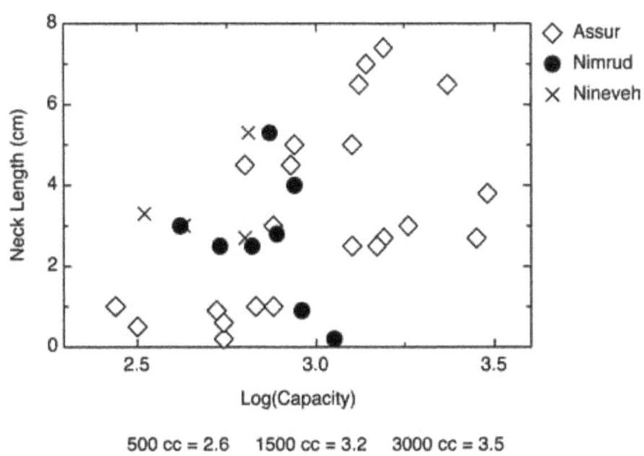


**Figure 1.** Scatter plot of palace ware vessel capacity against neck/collar length. Three distinct form clusters are visible: bowls, cups, and jars.

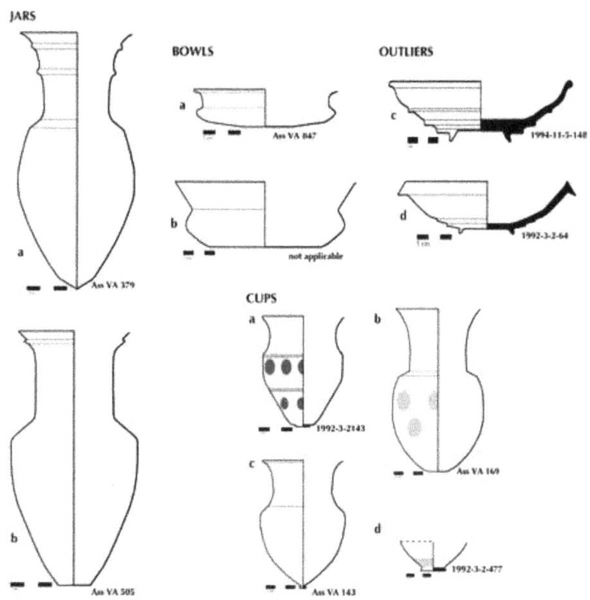


**Figure 2.** Neo-Assyrian palace ware typology. Note that due to the thinness of vessel walls, in order for vessel profiles to print, they could not be drawn to scale, except for the outliers. Bowls: (a) Type II; (b) Type I. Cups: (a) and (c) Type I; (b) Type II. Cup and Jar base shapes: (a) ring; (b) flat; (c) knob; (d) elevated ring; jar (a) pointed.

assemblage (Fig. 2). Palace ware cup capacity is slightly larger than that of bowls (550 cc. mode), although the cluster clouds for both vessel types are similar. Palace ware cups are decorated by dimples, circumferential engraved lines, appliqué ridges, and basal steps.

Palace ware jars are morphologically similar to cups, with increased body height (13.5cm mode) and total height (20 cm mode) to accommodate their larger capacities (1500 cc. and 3000 cc. mode) (Fig. 2). The majority of APW jars have capacity clusters around 1500 cc., triple the volume held by cups and bowls. Like cups, jars are unequally biconical, with pronounced v-shaped bodies and 'trumpet' necks. Most palace ware jar bases are pointed, however flat and rounded bases are also represented. Jars are decorated with engraved circumferential lines, appliqué, and dimples.

## *Chaîne opératoire*

*Raw material acquisition and processing*
Fabric characterisation of APW reveals an extremely fine-grained fabric, < 2% inclusions, and no inclusions larger than 0.05 mm. Petrological analysis of Neo-Assyrian tableware, selected for comparison because tableware does not contain the chaff temper common in most Assyrian ceramics, reveals the fine-grained nature of Assyrian fabrics in general (inclusions < 2 mm). Comparison of grain size distribution profiles for APW and non-APW vessels shows significant truncation, suggestive of levigation (Fig. 3).

Differentiation between alluvial and artificial levigation is difficult and subjective. The proximity of Aššur, Nimrud, and Nineveh to the Tigris suggests an alluvial origin for the matrix material, a suggestion supported by the rounded/well-rounded (almost spherical) mineral inclusions. Given the alluvial origin of the palace ware matrix and the energy of the Tigris river system, it is reasonable to assume that natural levigation took place. However, the marked difference in inclusion frequency between APW and 'typical' Neo-Assyrian tableware, and the sharp truncation of palace ware grain size distribution are probably indicative of human behaviour. Palace ware fabrics are intentionally and consistently finer than other Assyrian 'fine wares', either through mechanical processing of the raw material or exploitation of natural processes, such as seasonal acquisition or sediment beds out in the flood plain.

*Forming and shaping*
Scarring, parallel to the base, on the internal and external surface of palace ware vessels indicates rotary processing, leading Rawson to conclude that the vessels were wheel-made. Radiography of palace ware cups (types I and II) revealed diagonal orientation of voids and inclusions on the surface of the vessel and the subtle variation in wall thickness resulting from drawing the paste upwards using rotative kinetic energy (cf. Middleton 1995; 1997; Berg 2008), confirming that APW was wheel formed (Fig. 4). The lack of obvious joins also seems to confirms Rawson's assertion that palace ware was 'thrown in one piece'. Further work on a larger sample set including all three forms is needed to confirm this observation.

*Firing*
Determination of firing temperature is difficult without conducting experimental firing of the raw material in question. Colour, while a useful guideline for calcareous clays, can be altered by depositional and post-depositional environments. Quantifying the vitrification of a ceramic body can also be used to estimate firing temperature (cf. Tite *et al.* 1982; Wolf 2002). Palace ware fires buff (10YR) to green-grey (2.5YR, 5YR), indicating a firing temperature above 900 °C (Nicholson and Patterson 1989). Moderate vitrification visible in palace ware sherds also suggests a relatively high firing temperature (850–900°C) for an intermediate soak time (Freestone 1982) (Fig. 5). Firing temperature is important for the determination of the palace ware production cost, therefore, further work including experimental firing of raw material from the Tigris is planned.

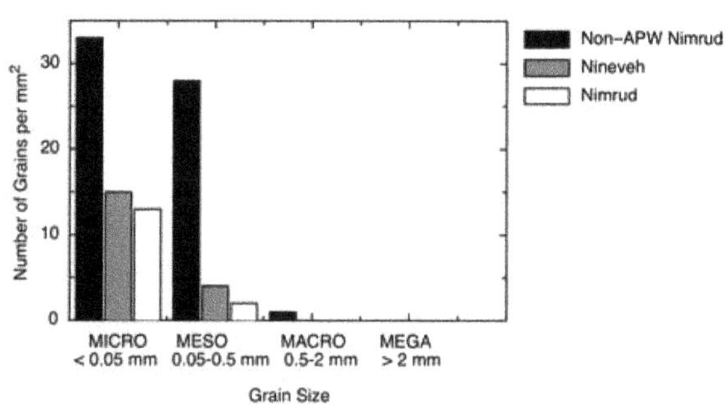


**Figure 3.** Grain size distribution plot for palace ware fabrics from Nimrud, Nineveh, and non-palace ware ceramic bowls from Nimrud. Note the pronounced truncation indicative of levigation.

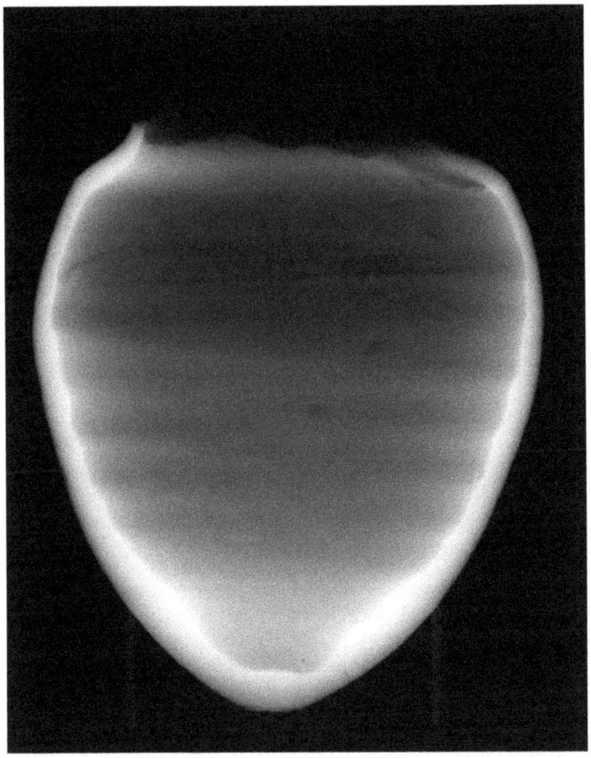

**Figure 4.** Radiograph of a palace ware cup from Nineveh (ND1312). Gradation between the base and shoulder reflects subtle variations in wall thickness introduced as the paste is pulled during formation on the wheel. Voids diagonal to the basal plane are also indicative of processing using rotary motion.

*Decoration*

The elegance of palace ware decoration, circumferential engraving, appliqué, and dimples, is deceptive in its simplicity. The perfect execution of these techniques, the symmetry and regularity of the engraving and consistent spacing of the dimples, reflects the mastery of the potter. Engraved lines and appliqué bands were created when the vessel was leather-hard, using rotary motion and probably a tool to ensure even spacing and placement. Dimpling methods are currently under investigation and will be the subject of a future publication. However, experiments replicating APW dimpling reveal the difficulty of creating dimples on a leather-hard surface without rupturing the vessel wall: deformation and dendritic cracking visible 'behind' the dimple are evidence that dimpling was conducted during the leather-hard stage. While dimples are unique in the Neo-Assyrian ceramic corpus and restricted to palace ware, engraving and appliqué are widely used. None of the APW decorative styles are unique to any particular form, although dimpling on bowls is less common.

## 4. Conclusions

Typological and scientific analyses identified 11 definitional criteria for palace ware in the Neo-Assyrian heartland:

- Three basic forms: bowl, cup and jar, described above and in Figure.
- Restricted vessel capacities: bowl and cups approximately 500 cc.; jars approximately 1500 or 3000 cc.
- Maximum diameter: 6–14 cm.
- Rim diameter: 6–14 cm.
- Base diameter: 0–8 cm.
- Wall thickness: 0.15–0.4 cm, with a mode of 0.2 cm.
- Fine-grained levigated fabric: < 2% inclusions; no inclusion larger than 0.05 mm.
- Wheel thrown formation.
- Fired colours: buff (10YR) to greenish-grey (2.5YR, 5YR).
- Estimated firing temperature: 850–900°C.
- Decoration applied when vessels were leather-hard, often using rotary motion.

## 5. Archaeological implications and further work

In light of these definitional criteria inferred from artefacts from the Neo-Assyrian central polity, it becomes possible to evaluate the movement of palace ware as object (trade) and as idea (local non-Assyrian manufacture), in contrast to the movement of Assyrian potters throughout the empire (local Assyrian manufactured APW). During the next phase of this research, *chaîne opératoire* and provenance of palace ware in the Neo-Assyrian imperial provinces of Syria, Turkey, and Israel will be evaluated to assess mechanisms of palace ware diffusion throughout the empire.

Another implication of this study is the potential use and social significance of palace ware. Palace ware is expensive and technically difficult to produce. At least in the Neo-Assyrian central polity, palace ware forms are not unique, occurring in Middle Assyrian assemblages as early as the $13^{th}$ century BC (cf. Duistermaat 2008) and, in the Neo-Assyrian period, as contemporary thicker forms (cf. Hausleiter 1999). Therefore, it is likely that the value of these vessels was related to the palace ware fabric itself and vessel capacity. The specific capacities of palace ware vessels, 500, 1500 and 3000 cc., suggest that these vessels

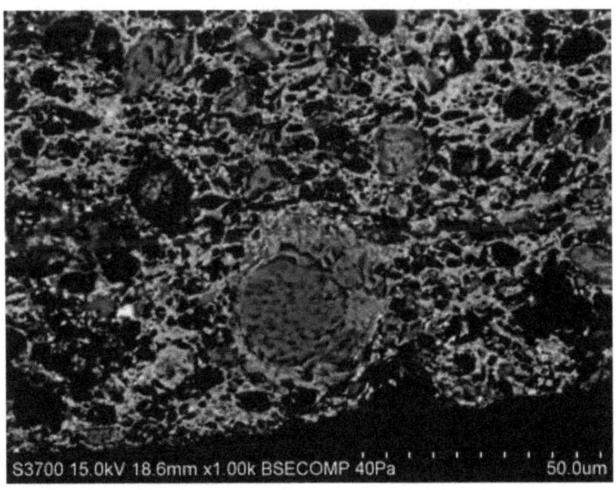


**Figure 5.** SEM-BSE micrograph of palace ware fabric from Nineveh (1129). The spongy or honeycomb texture of the matrix is indicative of moderate vitrification at high firing temperatures (850–900°C). Note the almost spherical quartz grain. Image reproduced courtesy of the Trustees of the British Museum. Image: C.R. Cartwright.

were used together, probably as a drinking set. Planned future work includes investigating the Neo-Assyrian drinking culture and rituals, and liquid measurements listed in the cuneiform texts to refine this hypothesis.

Experimental firing of raw material from the Tigris and raw material sources in Syria, Turkey, and Israel is also planned, to provide a quantitative baseline for firing temperature determinations and a better understanding of production cost, particularly with regard to fuel.

## References

Anastasio S., 2010, *Atlas of the Assyrian Pottery of the Iron Age*, Subartu 24, Brepols, Turnhout.

Andrae W., 1909, *Der Anu-Adad-Tempel in Assur*, J.C. Hinrichs, Leipzig.

Berg I., 2008, Looking through Pots: Recent Advances in Ceramics X-Radiography. *Journal of Archaeological Science*, 35, 1177-1188.

Duistermaat K., 2008, *The Pots and Potters of Assyria: Technology and Organization of Production, Ceramic Sequence, and Vessel Function at Late Bronze Age Tell Sabi Abyad, Syria*, Papers on archaeology of the Leiden Museum of Antiquities 4, Brepols, Turnhout.

Engstrom C., 2004, The Neo-Assyrians at Tell El-Hesi: A Petrographic Study of Imitation Assyrian Palace Ware. *Bulletin of the American Schools of Oriental Research*, 333, 69-81.

Freestone I.C., 1982, Applications and Potential of Electron Probe Micro-Analysis in Technological and Provenance Investigations of Ancient Ceramics. *Archaeometry*, 24, 99-116.

Haller A., 1954, *Die Gräber Und Grüfte Von Assur*, Gebr. Mann, Berlin.

Hausleiter A., 1999, Neo-Assyrian Pottery from Kalḫu/ Nimrud: With Special Reference to the Polish Excavations in The "Central Building". In *Iron Age Pottery in Northern Mesopotamia, Northern Syria and South-Eastern Anatolia*, (eds. A. Hausleiter and A. Reiche), 17-60, Ugarit-Verlag, Münster.

Hausleiter A., 2008, Nimrud in the Context of Neo-Assyrian Pottery Studies. In *New Light on Nimrud: Proceedings of the Nimrud Conference 11–13th March 2008*, (eds. J. Curtis, H. McCall, D. Collon and L. Al-Gailani Werr), 215-224, British Institute for the Study of Iraq in association with The British Museum, London.

Mallowan M.E.L., 1966, *Nimrud and Its Remains*, Collins, London.

Middleton A.P., 1995, Integrated Approaches to the Understanding of Early Ceramics: The Role of Radiography. In *The Cultural Ceramic Heritage European Ceramic Society Fourth Conference*, (ed. B. Fabbri), 63-74, Gruppo Editoriale Faenza, Faenza.

Middleton A., 1997, Ceramics. In *Radiography of Cultural Materials*, (eds. A. Middleton and J. Lang Cornwall), 76-95, Butterworth - Heinemann, Oxford.

Mielke P.W.Jr and Berry K.J., 2001, *Permutation Methods: A Distance Function Approach*, Springer, New York.

Na'aman N. and Thareani-Sussely Y., 2006, Dating the Appearance of Imitations of Assyrian Ware in Southern Palestine. *Tel Aviv*, 33, 61-82.

Nicholson P.T. and Patterson H.L., 1989, Ceramic Technology in Upper Egypt: A Study of Pottery Firing. *World Archaeology*, 21, 71-86.

Ohtsu T., 1991, Late Assyrian "Palace Ware" - Concerning Dimpled Goblet. In *Essays on Ancient Anatolian and Syrian Studies in the 2nd and 1st Millennium B.C.*, (ed. H.I.H. Prince Takahito Mikasa), 131-154, Bulletin of the Middle Eastern Culture Center in Japan 4, Harrassowitz, Wiesbaden.

Rawson P.S., 1954, Palace Ware from Nimrud: Technical Observations. *Iraq*, 16, 168-172.

Singer-Avitz L., 2007, On Pottery in Assyrian Style: A Rejoinder. *Tel Aviv*, 34, 182-203.

Tite M.S., Freestone I.C., Meeks N.D. and Bimson M., 1982, The Use of Scanning Electron Microscopy in the Technological Examination of Ancient Ceramics. In *Archaeological Ceramics: Papers Presented at a Seminar on Ceramics as Archaeological Material, Held at the National Bureau of Standards and the Smithsonian Institution, Washington, D.C., 29 September–1 October 1980*, (eds. J.S. Olin and A.D. Franklin), 109-120, Smithsonian Institution Press, Washington, DC.

Whitbread I.K., 1995, *Greek Transport Amphorae: A Petrological and Archaeological Study*, British School at Athens, Athens.

Wolf S., 2002, Estimation of the Production Parameters of Very Large Medieval Bricks from St. Urban, Switzerland. *Archaeometry*, 44, 37-65.

## ACKNOWLEDGEMENTS

Trustees of The British Museum and Dr. J. Curtis; Dr. C.R. Cartwright, The British Museum; Freie Universität Berlin, T. Sheh Hammad Project and Prof. H. Kühne and Dr. F.J. Kreppner; Dr. A. Middleton, The British Museum; Prof. C. Orton, UCL Institute of Archaeology; Dr. R. Sparks, UCL Institute of Archaeology; Prof. Th. Rehren, UCL Institute of Archaeology; UC Berkeley Excavation, Nineveh and Dr. S. Lumsden and Dr. E.B. Wilkinson; UCL Medical Physics and Prof. R. Speller, Dr. C. Reid and Dr. B. Price; Vorderasiatisches Museum and Stellvertretender Direktor Dr. R-B. Wartke.

# 3

# Archaeometric investigation of Punic lamps from Ibiza

**Bruno Fabbri[a], Sabrina Gualtieri[a] and Enrico Acquaro[b]**
[a]CNR, Institute of Science and Technology for Ceramics, Faenza, Italy, bruno.fabbri@istec.cnr.it, sabrina.gualtieri@istec.cnr.it
[b]University of Bologna, Ravenna, Italy

**Abstract** - Punic shell-shaped lamps are attested in the Island of Ibiza for a period of almost five centuries (6$^{th}$-2$^{nd}$ century BC). These two-spouted (*Bilicni*) lamps gradually evolved from an open to a closed shape, as shown by the six typologies identified on the island. The results of archaeometric investigations indicate that the lamps were locally manufactured using the same type of clays throughout this period; that is a calcium-rich fine clay. The absence of specific treatments for the clays and the low firing temperature confirm the scarce quality of the products, as macroscopically ascertained. This can be justified by the possible votive function of the lamps, considering the high number of artefacts without traces of combustion.

## 1. Introduction

### *1.1. Archaeological notes*

The data regarding the ceramic production of Punic lamps in the Island of Ibiza show an evolutionary and coherent trend of the shell-shaped typology during a period of five centuries (6$^{th}$-2$^{nd}$ century BC). Starting from an open shape, characterised by small interventions on the borders, the progressive transformations led to the production of a closed shape (Fig. 1). In addition, the 'Ibiza' lamps never show traces of pictorial decorations, which, in contrast, are present in the productions of other Punic contexts of the Mediterranean basin.

Only few cases of *monolicni* Punic lamps have been found on Ibiza Island. Most of the lamps belong to the two-spouted type (*bilicni*). As at other sites of the western Mediterranean, they are mainly present in sepulchral areas (necropoles) and secondarily in sanctuaries. According to a recent study on 175 Punic *bilicni* lamps from the Museum of Ibiza and Formentera (Savio 2005), the lamps attested at Ibiza originate from different sites on the island, although the provenance is unknown for approximately 15% of them. More than 90% of the findings whose provenance is known originate from the necropolis known as Puig des Molins, while the remaining 10% come from small rural settlements or necropoles.

The *bilicni* lamps can be subdivided into six typologies, from II to VII, with typology I including *monolicni* lamps.

Typology II is the most diffused (approximately 51% of the total), and belongs to the period between the 6$^{th}$ and 4$^{th}$ centuries BC; the lamps have two spouts, whose borders are well separated, and a large open basin with a well-defined central depression. The lamps of typology III (approximately 9%, 5$^{th}$-4$^{th}$ century BC) are very similar to the ones of typology II, from which they are distinguished due to the presence of a little foot.

Typologies IV and V are the best expressions of the evolutionary trend; the dimensions of the basin tend to become smaller and the spouts tend to close. They date back to the beginning and first half of the 4$^{th}$ century BC (typology IV), and to the end of the 4$^{th}$ – beginning of the 3$^{rd}$ century BC (typology V), and represent approximately 6% and 5% of the total assemblage, respectively.

Typology VI dates back to the 3$^{rd}$-2$^{nd}$ century BC and is the second in the order of abundance (approximately 26% of the total) after typology II. The lamps are made of two tubular spouts with a small basin. Finally, typology VII is represented only by five finds, i.e., about 3% of the total; these vessels present three openings (two spouts and a closed basin) and a circular foot.

The majority of the lamps do not show any traces of combustion, especially those of typologies V and VI. In the other cases, only few traces of combustion are present. Considering also their small dimensions (around ten centimetres or less), this indicates that they were not used for illumination purposes; they were probably produced only as symbolic votive objects.

### *1.2. Objectives*

The main objective of this work is to determine the clay raw materials and the details of the production technology of

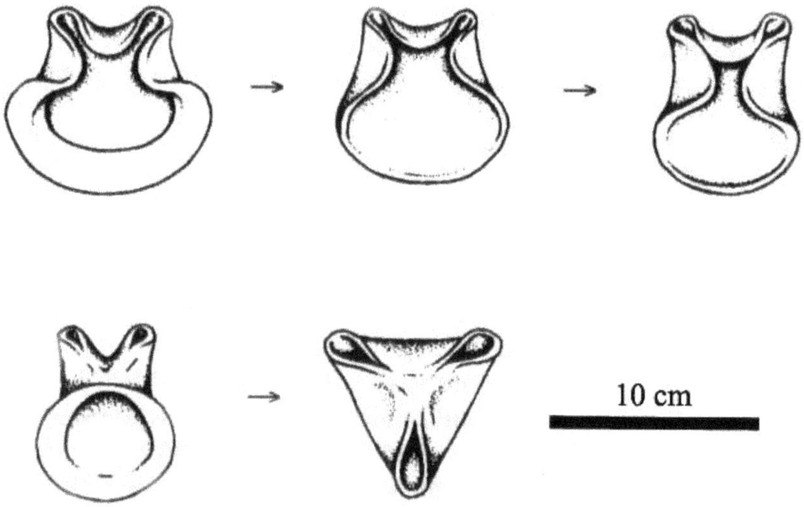


**Figure 1.** Proposal of evolutionary scheme of the Punic 'shell-lamps' (modified after Savio 2005).

**Table 1.** Description of the samples.

| Sample | Inv. Nr | Place | Typology | Macroscopic observation |
|---|---|---|---|---|
| 47 | 3964 | Puig des Molins – without context | II | Pink paste and burning traces |
| 53 | 4122 | Puig des Molins – hypogeum n° 14 | II | Pink paste |
| 70 | 5379 | Puig des Molins – Via Romana n° 40 | II | Orange paste, red engobe and burning traces |
| 85 | 10004/25 | Puig des Molins – Calle de Léon house 10-12, hypogeum 5 | II | Light brown paste |
| 87 | 14000 | Illa Plana- without context | II | |
| 103 | 4649 | Puig des Molins- without context | III | Pink paste and burning traces |
| 106 | 7768 | Puig des Molins- Excavation 5, sector D | III | Pink paste and burning traces |
| 150 | 4441 | Puig des Molins- hypogeum n° 11 | VI | Brown paste |
| 155 | 4897 | Puig des Molins- hypogeum n° 15 | VI | Pink paste |
| 156 | 7033 | Puig des Molins- hypogeum n° 2 | VI | Orange-pink paste |
| 169 | 14578 | without context | VI | Light pink paste |

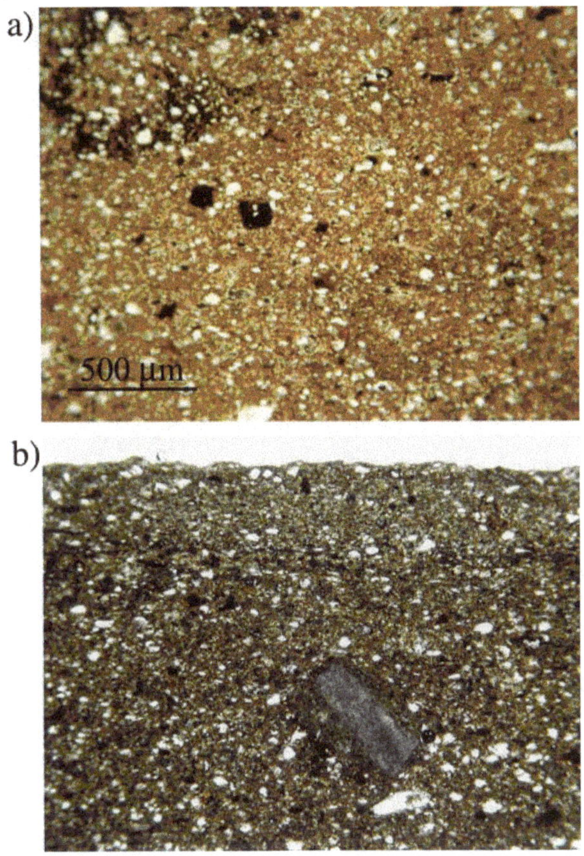


**Figure 2.** Typical appearance of the the clay body under plane-polarised light: (a) sample 47 (type II); (b) sample 155 (type III).

the lamps, and to verify the existence, in the evolutionary trend, of either a continuity in the use of raw materials and technology, or their eventual change during the period under consideration. In addition, it aims to establish which type of fuel was used for combustion, when such combustion actually took place.

This information could be of some archaeological and historical importance with regard to the customs prevalent in the society of Ibiza and its contacts with other Punic areas. In this context, the study of ceramic artefacts is particularly useful, because pottery has been produced everywhere and continuously over time.

## 2. Materials

A total of 11 samples of shell-shaped lamps have been opportunely selected from the different typological groups, for a better representation of the evolutionary trend and an improvement of the archaeological conclusions reached to date. In Table 1, a detailed description of the lamps is reported. Five fragments belong to type II, two samples represent type III, and four samples belong to type VI.

The above-mentioned *bilicni* lamps, held in the archaeological museum of Ibiza Island, are artefacts collected from tombs in the necropolis of Puig des Molins, with only one object originating from the sanctuary of Illa Plana. Practically these artefacts were never used for their typical function, as only some of them show just traces of combustion on the surface.

## 3. Analytical methods

The artefacts were characterised by optical microscopy in thin section, using a Leika POL11 microscope, in order to obtain mineralogical, petrographic, and microstructural information on the ceramic body. The chemical composition ($SiO_2$, $Al_2O_3$, $TiO_2$, $Fe_2O_3$, MnO, MgO CaO, $Na_2O$, $K_2O$, $P_2O_5$) was obtained by inductively coupled plasma atomic emission spectrometry (ICP-AES). At least 100 mg of powdered sample were melted with lithium borates at about 1000°C, and then dissolved into a nitric acid solution. The element concentrations were determined using a Varian Liberty 200 instrument. The measurements were

**Table 2.** Chemical composition of the analysed samples (wt%).

| Typology | Sample | $SiO_2$ | $Al_2O_3$ | $TiO_2$ | $Fe_2O_3$ | MnO | MgO | CaO | $Na_2O$ | $K_2O$ | $P_2O_5$ | LOI |
|---|---|---|---|---|---|---|---|---|---|---|---|---|
| II | 47 | 66.33 | 17.15 | 0.99 | 5.12 | 0.01 | 1.38 | 5.52 | 0.91 | 2.59 | 0.09 | 5.13 |
| | 53 | 64.98 | 16.47 | 0.90 | 4.87 | 0.03 | 1.43 | 6.94 | 1.02 | 3.35 | 0.12 | 3.58 |
| | 70 | 61.91 | 15.96 | 0.84 | 5.63 | 0.02 | 1.65 | 9.82 | 0.75 | 3.41 | 0.15 | 10.96 |
| | 85 | 68.52 | 15.72 | 0.96 | 5.52 | 0.03 | 1.67 | 2.89 | 0.85 | 3.83 | 0.11 | 5.38 |
| | 87 | 52.22 | 17.36 | 0.85 | 5.87 | 0.10 | 1.62 | 18.52 | 0.84 | 2.63 | 0.10 | 17.14 |
| III | 103 | 62.03 | 17.55 | 0.89 | 5.05 | 0.02 | 1.49 | 8.96 | 0.99 | 3.02 | 0.30 | 3.89 |
| | 106 | 58.11 | 17.60 | 0.80 | 5.33 | 0.02 | 1.66 | 12.10 | 0.92 | 3.46 | 0.27 | 5.81 |
| VI | 150 | 65.02 | 15.87 | 0.79 | 5.11 | 0.02 | 1.51 | 7.85 | 1.38 | 2.45 | 0.17 | 4.76 |
| | 155 | 60.61 | 15.79 | 0.81 | 4.27 | 0.01 | 1.47 | 13.05 | 1.19 | 2.80 | 2.27 | 6.22 |
| | 156 | 59.18 | 18.04 | 0.86 | 5.46 | 0.03 | 1.56 | 10.52 | 0.77 | 3.57 | 0.11 | 5.04 |
| | 169 | 61.60 | 14.90 | 0.83 | 4.17 | 0.01 | 1.14 | 13.79 | 1.25 | 2.29 | 0.07 | 8.11 |

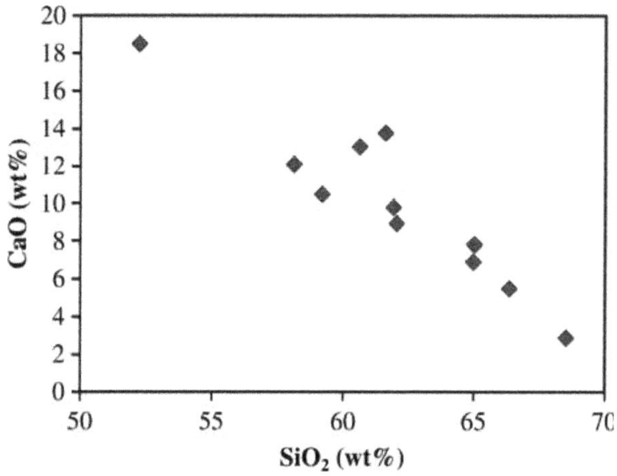


**Figure 3.** Scatterplot of CaO vs. $SiO_2$ showing an inverse correlation between the two oxides.

carried out using a single, sequential multi-element programme. The calculation of the loss on ignition (LOI) in the range 105–1000°C completed the chemical analysis.

The composition of the crystalline phases was obtained using X-ray powder diffraction; a Geigerflex instrument was used with the working conditions usually adopted in the field of ceramics, 4 to 64° 2θ scanning interval with a speed of 2° per minute.

In addition, five samples were characterised by means of gas chromatography (5894 Hewlett Packard) coupled with quadrupole mass spectroscopy (5971A) in order to verify the presence of organic matter inside the ceramic body, as residue of the fuel used. For this purpose, 2-3 mg of material were gently scraped from the surfaces where combustion traces were recognisable.

## 4. Results

The observations in thin section evidenced that all the examined ceramic samples are characterised by a very fine grain size. The abundance of non-plastic inclusions is around 15–20%, and the dimensions of the grains are less than 100 microns (Fig. 2). In some samples, a few grains of micrite (0.5–1 mm in size), partially decomposed due to the firing, are recognisable. The presence of carbonate grains, although infrequent, is attributable to natural variability rather than an intentional addition of carbonate temper by the potter. Taking into consideration these results, it is possible to indicate that the lamps were always produced using a quite homogeneous and fine paste. When large carbonate grains are present, the maximum firing temperature was around 800°C; a semi-isotropic matrix confirms the low firing temperature of these artefacts.

Despite the microstructural homogeneity, the chemical compositions are quite variable, in particular the calcium oxide contents (Table 2). Considering the high values of loss on ignition (from 2 to 11%, reaching 17% in sample 87), the obtained compositions were normalised excluding this parameter and the phosphorous percentage, which sometimes can be related to post-depositional phenomena. The high LOI percentages are mainly due to the presence of calcium carbonate in the body rather than to rehydration phenomena, as confirmed by the correlation between LOI and CaO values. The actual $P_2O_5$ values are very low, with the exception of one sample (n. 155) that has approximately 2.3% $P_2O_5$, very likely due to post-depositional processes.

Comparing the chemical data, the results indicate that the CaO content is actually the only parameter that differs from sample to sample. The CaO vs. $SiO_2$ diagram (Fig. 3) shows a very evident inverse correlation between these two oxides. This suggests the use of clays originating from the same type of geological sediment, probably characterised by large variations in the calcite content in different outcropping zones. In addition, it can be suggested that the different lamp typologies were made in the same area, so that they presumably have the same local provenance, i.e., Ibiza Island. Different production centres probably existed within the island and they used raw materials outcropping in their surroundings.

These hypotheses seem to be in agreement with the results of archaeometric studies on bodies of Punic amphorae found at Ibiza, which are locally produced and practically contemporary ($6^{th}$ to $3^{rd}$ century BC) to the samples analysed in the present paper (Iliopoulos *et al.* 2009). They are characterised by a very fine paste and a variable amount of carbonatic temper. These characteristics are compatible with the geolithologic features of Ibiza (IGME 1972; Cau 2003). Unfortunately, no data are available on contemporary pottery.

As regards the samples with the highest quantities of calcium, it has to be emphasised that most carbonate grains have very small sizes. This is in agreement with the hypothesis of a non-homogeneous source of raw material, rather than a different purification degree of the clay. In conclusion, it is likely that no attention was given to the preparation of the clay raw material.

The crystalline phase compositions of the five analysed samples are in accordance with the chemical results and the hypothesised firing temperature. Quartz is the prevalent phase, with subordinate calcite, illite, and plagioclase. In three samples (n. 85, 106, and 155), gehlenite is also present in small amounts, while traces of anatase were detected in samples n. 70, 85, and 87.

By considering the variation in the abundance of these mineral phases, it is possible to estimate a range of firing temperatures between 800 and 900°C (Cultrone *et al.* 2001). These firing temperatures are in good agreement with the conclusions reached by archaeometric studies on Punic pottery in several Mediterranean settlements (Amadori and Fabbri 1998a; 1998b; 1998c; Iliopoulos *et al.* 2009). Taking into account the age of the lamps and their phase composition, a minor increase in the temperature seems to characterise the more recent productions. In fact, in the samples belonging to type II (n. 70, 85 and 87), the maximum temperature can be estimated to have been around 850°C, while the other two samples (n. 106 and 155), belonging to the III and VI type, respectively, show minerals associated with slightly higher temperatures (850–900°C). The limited number of samples, however, does not support this hypothesis with sufficient certainty.

The gas chromatography analyses evidenced a very scarce quantity of residual organic compounds due to combustion.

Notwithstanding the small amount of sample, it was possible to recognise that the residues are made of acid fats, whose molecular structure is typical of animal fats. In particular, it is possible to exclude an attribution of the combustion residues to the use of olive oil.

## 5. Conclusions

The results indicate that the lamps were locally manufactured using the same type of clay during the entire period from the 6th to the 2nd century BC. This was a calcium-rich, fine, presumably local clay, with evident variations of carbonate content depending on the supply area. The absence of specific treatments of the clay denotes scarce attention to the quality of the products by the potters. This can be justified by the votive destination of the lamps, in agreement with the high number of artefacts without traces of combustion, and the minor such traces that are present in the remaining ones. The composition of the organic residues indicates the use of animal fats, instead of olive oil, as documented in many other archaeological sites in the western Mediterranean area.

In conclusion, there is an apparent contradiction between the typological changes undergone by these lamps and the stability and continuity in the raw materials and the processing technology used for their production. The contradiction is actually only apparent, as different factors affected these two aspects. The typological evolution could have been influenced by cultural and commercial contacts of the population of Ibiza with other Punic settlements in the Mediterranean. In contrast, technological changes are typically prompted by specific quality and/or technological needs. The votive use of the studied lamps did not involve any requirements for quality improvements, while the complex shape of the more recent lamps could have required clay raw materials that were well workable in a plastic state. Evidently, the local clays used were suitable for these purposes.

## References

Amadori M.L. and Fabbri B., 1998a, Indagini archeometriche su ceramica fenicia da mensa proveniente da Cartagine (VIII-VI sec. a.C.). In *Atti della II Giornata di Archeometria della Ceramica. Produzione e circolazione della ceramica fenicia e punica nel Mediterraneo: il contributo delle analisi archeometriche*, (eds. E. Acquaro and B. Fabbri), 43–55, University Press Bologna, Imola.

Amadori M.L. and Fabbri B., 1998b, Studio archeometrico di ceramica fenicia (VIII-VI secolo a.C.) proveniente da siti archeologici della Sardegna e Ischia. In *Atti della II Giornata di Archeometria della Ceramica. Produzione e circolazione della ceramica fenicia e punica nel Mediterraneo: il contributo delle analisi archeometriche*, (eds. E. Acquaro and B. Fabbri), 68–84, University Press Bologna, Imola.

Amadori M.L. and Fabbri B., 1998c, Produzione locale e importazioni di ceramiche fenicie da mensa (fine VIII-fine VI secolo a.C.) a Toscanos (Spagna Meridionale). In *Atti della II Giornata di Archeometria della Ceramica. Produzione e circolazione della ceramica fenicia e punica nel Mediterraneo: il contributo delle analisi archeometriche*, (eds. E. Acquaro and B. Fabbri), 85–94, University Press Bologna, Imola.

Cau M.A., 2003, *Cerámica tardoromana de cocina de las Islas Baleares: estudio arqueométrico*, British Archeological Reports International Series 1182, Archaeopress, Oxford.

Cultrone G., Rodríguez-Navarro C., Sebastian E., Cazalla O. and De La Torre M.J., 2001, Carbonate and silicate phase reactions during ceramic firing. *European Journal of Mineralogy*, 13, 621–634.

Iliopoulos I., Cau M.A. and Montana G., 2009, Le anfore fenicio-puniche prodotte nel Mediterrane occidentale: caratteristiche petrografiche degli impasti siciliani e spagnoli. In *Atti della X Giornata di Archeometria della Ceramica. Le classi ceramiche*, (eds. S. Gualtieri, B. Fabbri and G. Bandini), 157–162, Edipuglia, Bari.

Instituto Geológico y Minero de España (IGME), 1972, Menorca-Ibiza-Formentera. *Mapa geológico de España 1:200.000*, 49–65, Instituto Geológico y Minero de España, Madrid.

Savio G., 2005, *Le Lucerne fenicie e puniche del Museo Archeologico di Ibiza e Formentera*, Agorà Edizioni, Sarzana (La Spezia).

## ACKNOWLEDGEMENTS

The authors thank Prof. Maria Perla Colombini for the analyses of the organic residues.

# 4

# The ways of the lustre: Looking for the Tunisian connection

**Yona Waksman[a], Claudio Capelli[b], Trinitat Pradell[c] and Judit Molera[d]**
[a]CNRS UMR5138, Lyon, France, waksman@mom.fr
[b]DIP.TE.RIS, Università degli Studi di Genova, Italy, capelli@dipteris.unige.it
[c]Dpt. FEN. Universitat Politècnica de Catalunya, Barcelona, Spain, Trinitat.Pradell@upc.edu
[d]GRMAiA, EPS, Universitat de Vic, Spain, Judit.Molera@uvic.cat

**Abstract** - Recent excavations at the Fatimid and Zirid site of Sabra al-Mansuriya near Kairouan (Tunisia) provide the first evidence of lustreware production in medieval Ifriqiya, in the $10^{th}$–$11^{th}$ centuries AD. As the Fatimid dynasty moved from Ifriqiya to Egypt to establish its capital in Fustat (Cairo), technological connections with the Egyptian lustreware could be expected. Tunisian lustreware may also be the link in the transmission of the technique towards Muslim Spain. It represents a new piece of the puzzle of understanding the diffusion of lustre technology from the East to the West of the Mediterranean. The composition and microstructure of the bodies and glazes, and the micro- and nano-structure of the lustre layer are compared in a preliminary approach to the technological relationships between Tunisian, Egyptian, and Spanish lustrewares.

## 1. Introduction

The site of Sabra al-Mansuriya, founded by the Fatimids on the outskirts of Kairouan (Tunisia), was occupied according to written documents from 947/948 to 1057 AD. Recent excavations were carried out at the site by a French-Tunisian team directed by P. Cressier (CNRS UMR 5648, France) and M. Rammah (INP, Tunisia) (Cressier and Rammah 2004; 2006). Of particular interest to us are the remains of an artisanal area with a glass and a pottery kiln (Gragueb and Tréglia 2005; Thiriot forthcoming). Laboratory investigations of ceramics, including wasters, indicated a local manufacture of lustreware in Sabra al-Mansuriya (Fig. 1; Waksman *et al.* forthcoming; Capelli *et al.* 2011). The characterisation of its body and glaze is complemented in the present paper by a preliminary study of the properties of the lustre layer.

Although lustreware tiles dated back as early as the $9^{th}$ century AD ornate Kairouan's great mosque, they are likely to be imports from Mesopotamia (Bobin *et al.* 2003), and the excavations at Sabra al-Mansuriya provide the first evidence for lustreware production in medieval Ifriqiya. One may wonder which role it could have played in the transmission of the lustreware technique. As the Fatimid dynasty moved from Ifriqiya to Egypt in 972 to establish its capital in Fustat (Cairo), technological connections with the Egyptian lustreware could be expected. Tunisian lustreware may also be the link in the transmission of the technique towards Muslim Spain. Comparisons of technological characteristics of Tunisian and other lustreware productions (Bobin *et al.* 2003; 2005; Borgia *et al.* 2004; Mason 2004; Darque-Ceretti *et al.* 2005; Pradell *et al.* 2005; 2007; 2008a; 2008b; 2008c; Chabanne *et al.* 2006; 2008; Polvorinos del Rio *et al.* 2006; Smith *et al.* 2006; Molera *et al.* 2007; Roqué *et al.* 2007; Polvorinos del Rio and Castaing 2010), especially Egyptian and Spanish ones, were expected to shed some light on its role in the diffusion of the lustre technology in the Mediterranean regions.

## 2. Materials and methods

Chemical analyses of the bodies were carried out by WD-XRF at the Laboratoire de Céramologie (CNRS UMR5138, Lyon, France), allowing the determination of twenty-four elements. Oxygen was added by stoichiometry and the results normalised to 100 wt%; all the major and minor elements were determined (totals before normalisation are typically between 98 and 102%; Waksman 2011 for details of the analytical protocol and the calibration procedure). Seventeen elements (major and minor elements MgO, $Al_2O_3$, $SiO_2$, $K_2O$, CaO, $TiO_2$, $Fe_2O_3$, MnO; trace elements V, Cr, Ni, Zn, Rb, Sr, Zr, Ba, Ce) are generally used to classify ceramics into groups of similar chemical compositions. The latter may correspond to the production of the same workshop, under certain conditions mainly related to the geological environment (Picon 1993).

Petrographic (thin section) and SEM-EDS analyses of bodies and glazes were performed at the archaeometry laboratory of the DIP.TE.RIS (University of Genoa, Italy). Chemical microanalyses of glazes were carried out on at least

**Figure 1.** Example of lustreware manufactured at Sabra al-Mansuriya (sample TUN134; scale 5 cm).

3 selected points (spot mode) along a cross-section of the glaze from the base to the surface, avoiding alterations, interface, reaction zones, and mineral inclusions. The composition was normalised to 100 wt% after subtraction of $SnO_2$ values, attributable to the presence of cassiterite (the opacifier, added separately from other glaze ingredients).

Chemical analysis of the lustre layers has been performed by SEM-EDS at SCT-UB (University of Barcelona, Spain). Chemical analyses were carried out over the surface of the glazes (lustre decorations and plain glaze). The penetration depth of the microprobe is greater (about 2 μm) than the lustre layer thickness (probably between 0.1 and 1 μm), so that the analyses provide information on the overall composition. Oxygen was added by stoichiometry and the results normalised to 100 wt%, with all the major and minor elements being determined (in all cases, the totals before normalisation were between 94.2 and 99%, which is reasonable considering that we were working directly on the lustre and glaze surfaces). Although the chemical composition of the lustre layer itself was only indirectly assessed, the copper and silver determined correspond only to the lustre layer, and therefore their Cu/(Cu + Ag) ratio corresponds to their ratio in the lustre layer itself.

UV-Vis diffuse reflectance (DR) was measured directly from the surface of the lustre layers. A small circular spot of 5 mm diameter was used to collect the data. The data are presented as log(1/DR), which is equivalent to absorption for highly absorbing materials. The characteristic Surface Plasmon Resonance (SPR) absorption peaks associated with the metal nanoparticles appear in the UV-Vis spectra providing information concerning the type and size of the nanoparticles (Kreibig and Vollmer 1995).

## 3. Characterisation of Tunisian lustreware

As part of a larger assemblage from Sabra al-Mansuriya, eight samples of Tunisian lustreware were analysed by WD-XRF to determine the composition of the body. Two of them were further characterised by petrography and SEM-EDS analysis of the glazes (TUN3 and TUN8). Two others were investigated for the properties of their lustre layer (TUN134 and TUN118). The latter sample, a waster of lustreware, was not considered representative of the finished product and was thus not included here.

### *Bodies*

In terms of chemical composition, Tunisian lustreware produced in Sabra al-Mansuriya have calcareous clay bodies, containing high percentages of silica due to abundant quartz inclusions (petrographic features *infra*, and Table 1). Other elements possibly related to the sand fraction, such as zirconium, titanium, chromium, and nickel, are only present in small quantities. The clay proportion is low, as can be seen from the low aluminium concentrations. The iron, potassium, and rubidium concentrations are low as well, the latter falling under detection limits. Occasional high contents of sodium, strontium, and manganese are likely to derive from secondary phases developed during burial (Walter and Besnus 1989; Picon 1991).

Under the polarising microscope, the bodies show a carbonate-rich clay matrix and abundant inclusions mainly composed of quartz grains. Most of them are small (<0.3 mm across) and angular in shape, but the greater ones (<0.8 mm across) are rounded and characterised by abraded surfaces, well visible under the stereomicroscope or the SEM, which suggests an aeolian origin. Limestone fragments and relics of calcareous microfossils (foraminifera) are present in accessory amounts.

The origin of the raw materials can be related to calcareous marine sediments, whereas the dominant presence of (aeolian) quartz is a typical feature of most of the Tunisian ceramic productions of all ages and sediments (Capelli and Bonifay 2007).

**Table 1.** Chemical analysis by WD-XRF of bodies of lustrewares found in Sabra al-Mansuriya (Tunisia): local production and Egyptian imports. Major and minor elements in oxides wt%, trace elements in ppm; n: number of samples.

| | $Na_2O$ | MgO | $Al_2O_3$ | $SiO_2$ | $P_2O_5$ | $K_2O$ | CaO | $TiO_2$ | MnO | $Fe_2O_3$ | V | Cr | Ni | Zn | Sr | Zr | Ba | Ce |
|---|---|---|---|---|---|---|---|---|---|---|---|---|---|---|---|---|---|---|
| Tunisian lustreware (n=9) | | | | | | | | | | | | | | | | | | |
| m | 2.42 | 2.08 | 9.19 | 64.57 | 0.23 | 1.08 | 15.28 | 0.466 | 0.2964 | 3.76 | 82 | 69 | 26 | 78 | 606 | 134 | 210 | 65 |
| σ | 0.40 | 0.21 | 1.40 | 4.33 | 0.05 | 0.22 | 2.84 | 0.071 | 0.0549 | 0.51 | 9 | 10 | 2 | 10 | 148 | 11 | 29 | 9 |
| Egyptian lustreware (n=22) | | | | | | | | | | | | | | | | | | |
| m | 1.27 | 3.51 | 12.82 | 54.02 | 0.37 | 1.28 | 17.95 | 1.060 | 0.0887 | 7.12 | 134 | 114 | 43 | 92 | 905 | 291 | 335 | 97 |
| σ | 0.28 | 0.30 | 1.10 | 1.71 | 0.07 | 0.29 | 1.77 | 0.147 | 0.0104 | 0.83 | 23 | 14 | 7 | 7 | 101 | 19 | 23 | 7 |

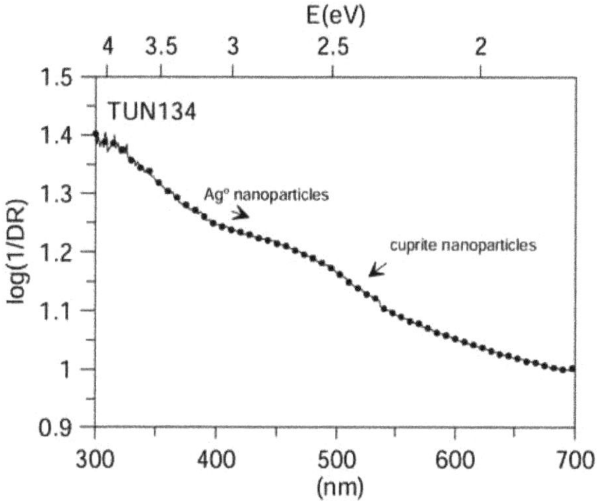


**Figure 2.** UV-Vis spectrum corresponding to sample TUN134. The presence of metal silver nanoparticles, as well as cuprite nanoparticles, is indicated.

The studied lustreware samples are not very homogeneous in their chemical and petrographic features. They are however well integrated in the range of ceramic products attributed to the region of Kairouan (Waksman *et al.* forthcoming), a region within which clays are likely to present similar characteristics (Louhichi and Picon 1983; Ben Amara *et al.* 2005). A more precise attribution of lustrewares to Sabra al-Mansuriya can however be inferred from the glaze analysis (*infra*) and from the discovery at the site of a lustreware waster (TUN118, *supra*).

### Glazes

Glazes have homogeneous thicknesses (0.2–0.3 mm) and include several gas bubbles, while no relic inclusions are present. Cassiterite (the opacifier) forms abundant and relatively large microaggregates. Both glazes are extensively weathered, mainly into Pb-carbonates.

In terms of chemical composition, the lustreware glazes fall in the field of lead-alkali glazes (Tite *et al.* 1998; PbO 23–27%, $SiO_2$ 56–57%, $Na_2O$ 7–8%, $K_2O$ 2–3%, CaO 4–5%, $Al_2O_3$ 1%). They are very similar in microstructure and chemical compositions to glazed wasters from the pottery workshop (Waksman *et al.* forthcoming).

### Lustre layer

Lustre is distinguished from other glaze decorations by its characteristic procedure of production. Lustre technology involves firstly the application of the lustre paint (typically containing sulphur and silver and/or copper-based compounds mixed with clay) on the previously glazed ceramic, and then firing it in a combination of oxidising-then-reducing atmosphere at temperatures between 500 and 600 °C. The temperature is low enough to avoid glaze softening, but high enough to facilitate the ionic diffusion. The formation of molten mixed sulphates in the lustre paint containing silver, copper and alkalis results in the ionic exchange of the $Ag^+$ and $Cu^+$ ions from the paint by $Na^+$ and/or $K^+$ from the glaze (Pradell *et al.* 2005). Silver and copper are incorporated into the glaze, and, following the introduction of a reducing atmosphere, precipitate forming small metal nanoparticles (from 2 to 50 nm in size). The nanoparticles appear concentrated in a thin layer (well below 1 micrometre) close to the glaze surface. This nanostructure, that is the nature and size of the nanoparticles and the thickness and concentration of nanoparticles in the layer, is directly responsible for the characteristic colours and metallic shine shown by lustres. In particular, green, yellow, orange, and brown are the colours shown by silver lustres (containing silver metal nanoparticles), while pinkish, ruby red, and crimson are the colours shown by copper lustres (containing copper metal nanoparticles) (Molera *et al.* 2007). The metallic shine is 'golden' for silver lustres and 'coppery' for copper lustres. The concentration of small metal nanoparticles in a thin layer close to the glaze surface is the chief feature of lustre and is a direct consequence of the lustre production process (Pradell *et al.* 2007; Gutierrez *et al.* 2010).

### UV-Vis spectroscopy

For the determination of the presence of metal nanoparticles, one of the most useful techniques is UV-Vis spectroscopy. Measuring the reflectance by means of an Ulbritch Integrating Sphere, it is possible to evaluate the absorption spectrum. The presence of metal nanoparticles in the layers gives a very characteristic profile (Kreibig and Vollmer 1995). Figure 2 shows the UV-Vis absorption spectra corresponding to the lustre sample TUN134. The profile is related to the presence of silver metal nanoparticles (peak at 430–450Å) and cuprite nanoparticles (shoulder at 500Å), as indicated in the figure.

### Chemical analysis

The presence of metallic nanoparticles in the glaze is certainly an indication of lustre, but it is not sufficient; for instance, red Chinese glazes contain copper nanoparticles, but they are not lustres (Wood 1999, 167–87). However, the exchange of $Ag^+$ and $Cu^+$ from the paint by $Na^+$ and $K^+$ from the glaze, characteristic of the lustre production process, implies that for each atom of silver/copper present in the lustre one atom of Na/K has been removed from the glaze. This inverse unity correlation is seen by plotting the chemical atomic composition of Ag and Cu from the lustre

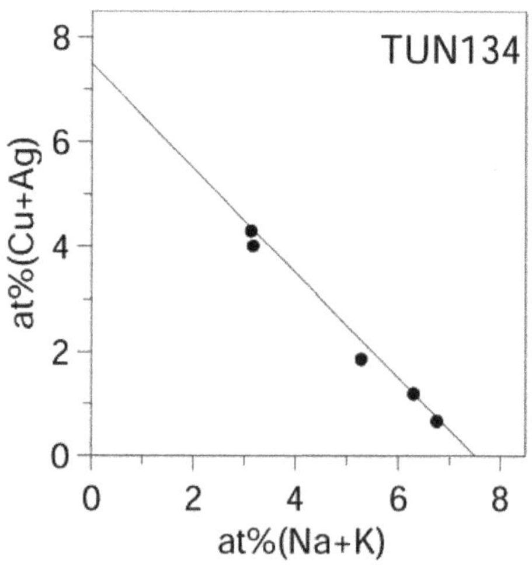


**Figure 3.** Plot of the atomic concentration of copper and silver versus that of potassium and sodium, showing the characteristic inverse unity correlation found in lustres.

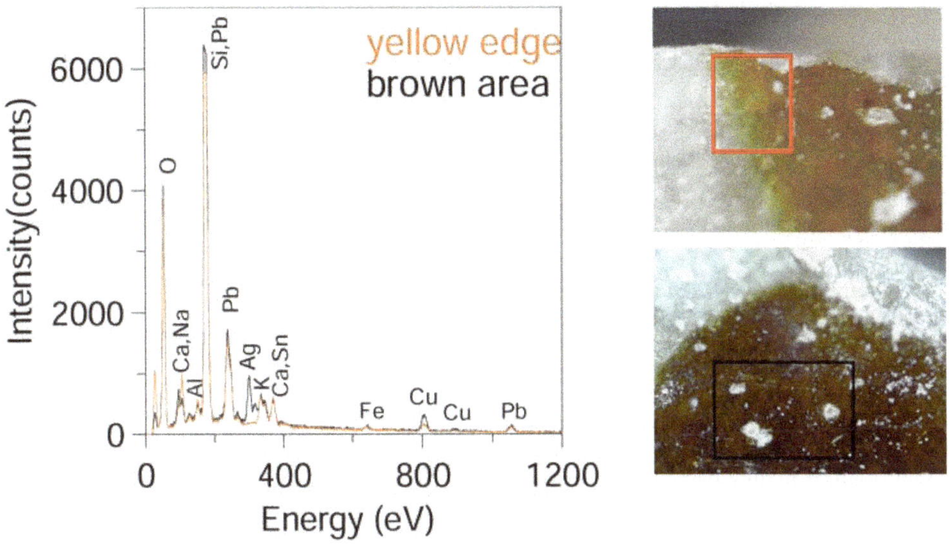


**Figure 4.** Colour and chemical analysis of the brown and yellow areas in sample TUN134.

surfaces versus the alkali atomic content (Molera *et al.* 2007). Figure 3 shows the correlation found for the Tunisian lustre and demonstrates that it was produced following the lustre procedure described above.

The studied Tunisian lustre (TUN134) has a characteristic brownish colour typical of silver lustres. They contain silver nanoparticles and also copper as $Cu^{+}$ and/or $Cu^{2+}$, however without forming metal copper nanoparticles. The lustre also shows a yellow colour at the edges of the lustre decoration, as seen in Figure 4. The edges of lustre decorations often have different colours and chemistry, the most characteristic being red or coppery edges. Red and coppery edges contain a high ratio of $Cu/(Ag + Cu)$, and copper forming $Cu^{\circ}$ and cuprite ($Cu_2O$) nanoparticles. Red and coppery edges are found in some Fatimid (Egypt, 12th century AD), Persian (12th and 13th centuries AD), and Spanish (Paterna, late 13th–14th century AD) lustre productions (Molera *et al.* 2001; Smith *et al.* 2006; Pradell *et al.* 2008a). The reason for the formation of red edges was discussed in detail elsewhere (Smith *et al.* 2006); it is related to the easier access of the furnace atmosphere and heat to the lustre paint/glaze contact area. The initial oxidising conditions facilitate the penetration of copper into the glaze and the further reducing conditions produce the reduction of copper ions into cuprite and metal copper nanoparticles (Molera *et al.* 2007). Something similar happened for TUN134 – a higher relative concentration of copper is found at these yellow edges as compared to the brown central part of the lustre decoration. The chemical analysis of both the brown and the yellow areas are shown in a comparative plot in Figure 4 and in Table 2. We can see that the yellow area is also richer in copper in relative terms. The colour is yellow because in this case the reduction atmosphere has not been strong enough to produce the precipitation of copper metal nanoparticles, but only small cuprite nanoparticles (2–30 nm) that show a characteristic deep yellow colour (Molera *et al.* 2007).

Another feature observed is the reduced amount of lead in some areas of the glaze and lustre surface. This is very typical in samples which have been buried, and may be related to the leakage of lead during burial (the SEM-EDS analysis of the glazes of samples TUN 3 and 8, which are strongly weathered into Pb-carbonates, supports this hypothesis). This has also been found in other Islamic lustres (Pradell *et al.* 2008a; 2008c, Roqué *et al.* 2007). Moreover, they also contain quite high amounts of tin, and because of its heterogeneous distribution in the glaze, some of the analyses may indicate particularly large amounts of tin, which are not the average content of tin of the glazes.

Sample TUN134 shows typical copper contents of about 2–5 wt% and a very heterogeneous distribution of silver, varying

**Table 2.** Chemical analysis by SEM-EDS of the lustre layer in sample TUN134, in wt%. Comparison between the yellow edges and the brown central part of the lustre decoration. Detection limits are below 0.1wt% for both silver and copper.

| | wt% | O | Na | Mg | Al | Si | K | Ca | Fe | Cu | Ag | Sn | Pb | Cu/(Cu + Ag) |
|---|---|---|---|---|---|---|---|---|---|---|---|---|---|---|
| TUN134 | Brown area | 31.9 | 1.7 | 0.4 | 0.8 | 27.7 | 1.4 | 1.0 | 0.7 | 4.6 | 7.0 | 3.8 | 19.2 | 40.0 |
| | | 37.2 | 4.9 | 0.4 | 0.9 | 30.5 | 2.0 | 1.0 | 0.8 | 1.5 | 0.2 | 4.9 | 15.6 | 40.0 |
| | | 32.4 | 1.7 | 0.4 | 0.9 | 28.1 | 1.4 | 1.1 | 0.6 | 5.3 | 7.3 | 3.5 | 17.4 | 42.0 |
| | Yellow edge | 34.0 | 4.1 | 0.4 | 1.2 | 29.9 | 2.2 | 1.6 | 0.9 | 2.7 | 0.3 | 4.5 | 18.3 | 91.5 |
| | | 24.4 | 2.6 | 0.4 | 1.1 | 35.1 | 2.3 | 1.4 | 1.0 | 3.6 | 0.4 | 5.7 | 22.0 | 89.4 |

from 0.2 to 7 wt%. This is very typical in Islamic lustres, with copper being more homogeneously distributed over the entire surface, while silver appears more heterogeneously distributed, forming silver-rich areas. For the studied lustre layer (TUN134), the average copper and silver contents of the brown lustre are 7.1 wt% Ag and 5.0 wt% Cu, respectively. The copper to silver ratio (as defined in Molera *et al.* 2001) is wt% Cu/(Cu + Ag) = 40% (Table 2).

## 4. Comparative data

Comparative data were available both from lustreware imports to Sabra al-Mansuriya and from previous studies on lustreware. The former only include so far body and glaze analyses, and mostly concern ceramics which are likely to be Egyptian productions from Fustat (Waksman *et al.* forthcoming; Capelli *et al.* 2011). The latter involve a larger variety of lustrewares originating from museum collections, whose dating and origin is mainly based on art historical criteria; they are summarised in Table 3 (Borgia *et al.* 2004; Mason 2004; Darque-Ceretti *et al.* 2005; Smith *et al.* 2006; Roqué *et al.* 2007; Chabanne *et al.* 2008; Pradell *et al.* 2008a; 2008b; 2008c; Gutierrez *et al.* 2010; Polvorinos del Rio and Castaing 2010).

Egyptian lustrewares from Sabra al-Mansuriya are dated according to the period of occupation of the site (mid-10th – mid-11th century AD). From a technological viewpoint, their calcareous clay body belongs to the same category of pastes as the Tunisian ones. The two productions can however easily be differentiated in the framework of provenance studies, on the basis of both elemental (Table) and petrographic features of the bodies.

Their glaze features show clear compositional and technological differences (Capelli *et al.* 2011; Waksman *et al.* forthcoming). As for the texture, Egyptian glazes are characterised by many gas bubbles and several angular to rounded relic inclusions of quartz, plagioclase, and K-feldspar, which suggest the use of impure alluvial sand. Cassiterite is less abundant than in Tunisian wares.

SEM-EDS analyses of Egyptian imports found at Sabra al-Mansuriya, including two Egyptian lustrewares, show the presence of homogeneous lead-alkali glazes, but their composition (PbO 40–41%, $SiO_2$ 46–48%, $Na_2O$ 4–5%, $K_2O$ 2–3%, CaO 1%, $Al_2O_3$ 3%) and especially the relative ratios of PbO/$SiO_2$, $Na_2O/K_2O$, and CaO/$Al_2O_3$ are very different from the Tunisian lead-alkali production, confirming the use of different recipes.

These imports at Sabra al-Mansuriya are only one of several lustreware productions attributed to Fustat, which differ in terms of technology and chronology (Mason 2004; Chabanne *et al.* 2006; Pradell *et al.* 2008a; 2008b). A selection of sherds from the Ashmolean Museum, representing four different groups, had been characterised for their body, glaze, and lustre layer (Table 3). In the Fatimid lustreware from Fustat, both stonepastes and calcareous pastes were used, and within the calcareous pastes we encounter three different types: fine calcareous, coarse calcareous, and very high calcareous. It is not clear if this is related to productions of different quality. The bodies corresponding to the end of the Fatimid production tend to be of the very high calcareous type. Stonepastes continued to be used in Persia (Kashan) and also in Syria (Raqqa) in later productions (late 12$^{th}$ and 13$^{th}$ centuries AD) (Mason 2004; Pradell *et al.* 2008a, 2008b). In contrast, calcareous ceramic pastes were used in the later Islamic and Hispano Moresque productions from Spain (Paterna, Seville, and Manises; 14$^{th}$ and 15$^{th}$ centuries AD) (Polvorinos del Rio and Castaing 2010). Tunisian lustreware also has a calcareous body.

The glazes from the Fustat lustreware are generally speaking lead-rich (about 30–40 wt% PbO), although the later very high calcareous pastes have glazes poorer in lead (about 25 wt% PbO). The glazes from later Persian lustreware contain about 25 wt% PbO, while those used in the Syrian lustreware are lead free. In contrast, later Spanish productions have glazes richer in lead (35–45 wt% PbO) and potassium.

Concerning the composition of the lustre layers, the Fatimid lustres from Fustat have a Cu/(Cu + Ag) ratio between 10 and 20%; they are richer in silver than the lustre from Tunisia. This difference may be explained, at least in part, by the use of more oxidising conditions for the Tunisian lustres. However, it may also indicate an intentional decrease in the silver content, either to obtain a cheaper lustreware for the local market, or to modify the colour of the lustre designs. The increase of copper content in the lustre designs is responsible for the change of the colour from green-yellow to orange-brown. The lustreware productions from Persia and Syria (12$^{th}$ and 13$^{th}$ centuries AD) are also richer in copper (Pradell *et al.* 2008b) than the Fatimid lustres from Fustat, and have a characteristic brown colour similar to the Tunisian lustre studied here. In Spain, the lustreware becomes particularly copper-rich in the second half of the 14$^{th}$ century and in the 15$^{th}$ century (Polvorinos del Rio *et al.* 2006; Smith *et al.* 2006; Roqué *et al.* 2007), and in these cases the colour of the lustre designs becomes redder and copper-like.

## 5. Concluding remarks

A preliminary study of Tunisian lustreware manufactured at Sabra al-Mansuriya confirms that it presents the features of lustre, with a mechanism of ionic exchange between the copper and silver atoms of the lustre and the alkalis of the glaze followed by a (partial) reduction to metallic Cu and Ag. Further analyses on an extended assemblage are planned, to confirm these features and characterise the structure and the thickness of the lustre layer.

Another subject of interest is the possible role of the Tunisian lustreware production in the transmission of this technique in the Mediterranean, especially between Egypt and Spain. It actually presents some global resemblance with examples of Egyptian and Spanish lustreware, when compared to others attributed to Iraq, Persia, and Syria: fine calcareous clay body (as opposed to stonepaste), lead-alkali tin-opacified glaze (as opposed to low-lead-alkali or alkali glaze), 'true' lustre with lustre shine. However, they also differ in several respects.

Tunisian and Egyptian lustrewares found at Sabra al-Mansuriya, in mid-10$^{th}$ to mid-11$^{th}$ century AD contexts, show noticeable differences in the chemical and textural features of the glaze. The characterisation of the lustre layer shows that Egyptian lustrewares from the Ashmolean Museum, dated back to the period between the 10$^{th}$ and the 12$^{th}$ century AD, are richer in silver than the Tunisian sample studied. A direct technological connection between

**Table 3.** General characteristics of the Tunisian sample TUN134 and of other Islamic lustreware productions ([1]Pradell *et al.* 2008c; [2]Pradell *et al.* 2008a; [3]Smith *et al.* 2006; Roqué *et al.* 2007 and Chabanne *et al.* 2008; [4]Borgia *et al.* 2004; [5]Darque-Ceretti *et al.* 2005; [6]Mason 2004; [7]Gutierrez *et al.* 2010).

| | Dating | Origin or *supposed origin* | Body | Glaze | Lustre | Lustre colour | Lustre shine | %Cu/ (Cu + Ag) | Oxidation state |
|---|---|---|---|---|---|---|---|---|---|
| Iraq[1,5,6,7] | 9th AD | *Basra* | fine calcareous | low-lead-alkali (5–10% PbO), tin | polychrome | brown-green & amber | only green golden | 40–90 | $Cu^{+}$, $Cu^{2+}$, $Ag^{0}$ |
| | 10th AD | *Basra* | fine calcareous | low-lead alkali (10–15% PbO), tin | monochrome | green-yellow | golden | 0–10 | $Ag^{0}$, $Cu^{+}$ |
| Egypt[2,6,7] | 10th - 11th AD | *Fustat* | fine calcareous | lead-alkali (30–40% PbO), tin | monochrome | green, yellow, amber | golden | 10 | $Ag^{0}$,$Cu^{+}$ |
| | 12th AD | *Fustat* | highly calcareous | lead-alkali (25–40% PbO), tin | monochrome | yellow, amber, brown | golden | 10–30 | |
| | 12th AD | *Fustat* | stonepaste | lead-alkali (25–40% PbO), tin | monochrome | yellow, amber, brown | golden | 10 | |
| Tunisia | mid 10th - mid 11th AD | Sabra al-Mansuriya | fine calcareous | lead-alkali (20–30% PbO), tin | monochrome | brown/yellow edges | golden | 40 | $Ag^{0}$,$Cu^{+}$ |
| Spain[3,5] | late 13th - 1st half 14th AD | Paterna | fine calcareous | lead-alkali (40% PbO), tin | monochrome | brown/red edges | golden | 10–30 | $Ag^{0}$,$Cu^{0}$ |
| | 2nd half of 14th - 15th AD | Paterna | fine calcareous | lead-alkali (30% PbO), tin | monochrome | brown reddish/ red edges | golden | 40–50 | $Ag^{0}$,$Cu^{0}$ |
| Persia[2,4,6] | late 12th - 13th AD | *Kashan* | stonepaste | lead-alkali (20–25% PbO), tin | monochrome | brown/red edges | golden | 40–60 | $Ag^{0}$,$Cu^{0}$,$Cu^{+}$ |
| Syria[2,6] | late 12th - 13th AD | *Ma'arrat al Numan* | stonepaste | alkali | monochrome | red | no | 100 | $Cu^{+}$,$Cu^{0}$ |
| | 13th AD | *Raqqa* | | | monochrome | brown, yellow-green | no | 70 | $Cu^{+}$,$Ag^{0}$ |

the two productions is thus not supported by the presently available data. However, the Egyptian production is complex and diversified in its technological features (Mason 2004), and we cannot rule out any connection whatsoever at the present stage of the study. Further research on ceramics of attested provenance and originating from stratified excavations, such as those at Istabl' Antar, Cairo (Gayraud *et al.* forthcoming), would be particularly helpful.

In addition, the Tunisian lustreware examined in the present study shows more resemblance, in its copper to silver ratio with the later (second half of the 14$^{th}$ – 15$^{th}$ century AD) than with the early lustreware production at Paterna (end of the 13$^{th}$ – first half of the 14$^{th}$ century AD). It does not thus appear as its precursor. However, the chronological gap between the Sabra al-Mansuriya and Paterna productions calls for data on the earliest lustrewares manufactured in Spain.

This preliminary study of a Tunisian lustreware production is important as it brings in new data on ceramics which are well-defined archaeologically – a rare feature in the field of lustreware studies. It however opens more questions than it provides answers at the present stage of research, and makes clear how little we still know about the transmission of the lustre technique between the Eastern and the Western parts of the Mediterranean.

## References

Ben Amara A., Schvoerer M., Thierrin-Michael G. and Rammah M., 2005, Distinction de céramiques glaçurées aghlabides ou fatimides (IX$^{e}$-XI$^{e}$ siècles, Ifriqiya) par la mise en évidence de différences de texture au niveau de l'interface glaçure - terre cuite. *ArcheoSciences, Revue d'Archéométrie*, 29, 35-42.

Bobin O., Schvoerer M., Ney C., Rammah M., Daoulatli A., Pannequin B. and Gayraud R.P., 2003, Where did the lustre tiles of the Sidi Oqba mosque (AD 836–63) in Kairouan come from? *Archaeometry*, 45, 569-577.

Bobin O., Schvoerer M., Chabanne D., Ney C., Lahaye M., Labrugère C. and Guette A., 2005, Localisation et distribution par spectroscopie Auger des constituants colloïdaux des décors de lustre métallique de céramiques glaçurées. In *Décor de lustre métallique et céramique glaçurée*, (eds. M. Schvoerer, C. Ney and P. Peduto), 39-45, Scienze e materiali del patrimonio culturale 7, Edipuglia, Bari.

Borgia I., Brunetti B., Giulivi A., Sgamellotti A., Shokouhi F., Oliaiy P., Rahighi J., Lamehi-Rachti M., Mellini M. and Viti C., 2004, Characterisation of decorations on Iranian (10th–13th century) lustreware. *Applied Physics A*, 79, 257-261.

Capelli C. and Bonifay M., 2007, Archéométrie et archéologie des céramiques africaines: une approche multidisciplinaire. In *LRCW2. Late Roman Coarse Wares, Cooking Wares and Amphorae in the Mediterranean: Archaeology and Archaeometry*, (eds. M. Bonifay and J.-C. Tréglia), 551-568, British Archaeological Reports International Series 1662 (II), Archaeopress, Oxford.

Capelli C., Waksman Y., Cabella R., Gragueb S. and Tréglia J.-C., 2011, Il contributo delle analisi di laboratorio allo studio delle ceramiche nordafricane: l'esempio di Sabra al-Mansûriya (dati preliminari). In *La céramique maghrébine du haut Moyen Âge (VIIIe - Xe siècle). État des recherches, problèmes et perspectives*, (eds. P. Cressier and E. Fentress), 221-232, Roma.

Chabanne D., Aucouturier M., Bouquillon A. and Darque-Ceretti E., 2006, Les céramiques à décor de lustre métallique, un traitement de surface nanométrique traditionnel. Connaissance des productions islamiques depuis le IX$^{ème}$ siècle jusqu'à la Renaissance. *International Conference "Matériaux 2006", Dijon novembre 2006*, Fédération Française des Matériaux, Dijon.

Chabanne D., Bouquillon A., Aucouturier M., Dectot X. and Padeletti G., 2008, Physico-chemical analyses of Hispano-Moresque lustred ceramic: a precursor for Italian majolica? *Applied Physics A*, 92, 11-18.

Cressier P. and Rammah M., 2004, Une autre ville califale: Sabra al-Mansuriya. *Cuadernos de Madinat al-Zahra*, 5, 241-255.

Cressier P. and Rammah M., 2006, Sabra al-Mansuriya: une nouvelle approche archéologique. *Comptes rendus des séances de l'Académie des Inscriptions et Belles Lettres*, 2006, 113-133.

Darque-Ceretti E., Hélary D., Bouquillon A. and Aucouturier M., 2005, Gold-like lustre: a nanometric surface treatment for the decoration of glazed ceramics in ancient Islam, Moresque Spain, and Renaissance Italy. *Surface Engineering*, 21, 352-358.

Gayraud R.-P., Vallauri L., with the collaboration of Guionova G., Tréglia J.-C. and the participation of C. Capelli and S.Y. Waksman, forthcoming, *Fouilles d'Istabl' Antar (Fustat). Céramiques d'ensembles des IXe et Xe siècles*, IFAO, Cairo.

Gragueb S. and Tréglia J.-C., 2005, Observations préliminaires sur la céramique du secteur 6. In *Fouilles de Sabra al-Mansuriya. Kairouan, Tunisie. Programme de coopération archéologique franco-tunisienne, campagne 2005*, (eds. P. Cressier and M. Rammah), 98-116. Unpublished excavations report.

Gutierrez P.C., Pradell T., Molera J., Smith A.D., Climent-Font A. and Tite M.S., 2010, Color and Golden Shine of Silver Islamic Luster. *Journal of the American Ceramic Society*, 93, 2320-2328.

Kreibig U. and Vollmer M., 1995, *Optical properties of metal clusters*, Springer Series in Materials Science 25, Springer, Berlin and New York.

Louhichi A. and Picon M., 1983, Importation de matériel céramique ifriqiyen en Mauritanie. *Revue d'Archéométrie*, 7, 45-58.

Mason R., 2004, *Shine like the sun. Lustre-painted and associated pottery from the medieval Middle East*, Mazda Press, Costa Mesa, California and the Royal Ontario Museum, Toronto.

Molera J., Mesquida M., Perez-Arantegui J., Pradell T. and Vendrell M., 2001, Lustre recipes from a medieval workshop in Paterna. *Archaeometry*, 43, 455-460.

Molera J., Bayés C., Roura P., Crespo D. and Pradell T., 2007, Key parameters in the production of medieval lustre colours and shines. *Journal of the American Ceramic Society*, 90, 2247-2257.

Picon M., 1991, Quelques observations complémentaires sur les altérations de composition des céramiques au cours du temps. Cas de quelques alcalins et alcalino-terreux. *Revue d'Archéométrie*, 15, 117-122.

Picon M., 1993, L'analyse chimique des céramiques: bilan et perspectives. *Archeometria della Ceramica. Problemi di Metodo, Atti 8° Simposio Internazionale della Ceramica (Rimini 1992)*, 3-26, Centro Ceramico, Bologna.

Polvorinos del Rio A. and Castaing J., 2010, Lustre-decorated ceramics from a $15^{th}$ to $16^{th}$ century production in Seville. *Archaeometry*, 52, 83-98.

Polvorinos del Rio A., Castaing J. and Aucouturier M., 2006, Metallic nano-particles in lustre glazed ceramics from the $15^{th}$ century in Seville studied by PIXE and RBS. *Nuclear Instruments and Methods B*, 249, 596-600.

Pradell T., Molera J., Roqué J., Smith A.D., Crespo D., Pantos E. and Vendrell M., 2005, Ionic-exchange mechanism in the formation of medieval lustre decorations. *Journal of the American Ceramic Society*, 88, 1281-1289.

Pradell T., Climent-Font A., Molera J., Zucchiatti A., Ynsa M.D., Roura P. and Crespo D., 2007, Metallic and non-metallic shine in luster: an elastic ion backscattering study. *Journal of Applied Physics*, 101, 1-8.

Pradell T., Molera J., Smith A.D. and Tite M.S., 2008a, Early Islamic lustre from Egypt, Syria and Iran (10th to 13th century AD). *Journal of Archaeological Science*, 35, 2649-2662.

Pradell T., Molera J., Smith A.D., Climent Font A. and Tite M.S., 2008b, Technology of Islamic lustre. *Journal of Cultural Heritage*, 9, e123-e128.

Pradell T., Molera J., Smith A.D. and Tite M.S., 2008c, The invention of lustre: Iraq 9th and 10th centuries AD. *Journal of Archaeological Science*, 35, 1205-1215.

Roqué J., Molera J., Pérez-Arantegui J., Calabuig C., Portillo J. and Vendrell-Saz M., 2007, Lustre colour and shine from the Olleries Xiques workshop in Paterna (Spain), 13th century AD: nanostructure, chemical composition and annealing conditions. *Archaeometry*, 49, 511-528.

Smith A.D., Pradell T., Roqué J., Molera J., Vendrell-Saz M., Dent A.J. and Pantos E., 2006, Colour variations in 13th century Hispanic Lustre - an EXAFS study. *Journal of Non-Crystalline Solids*, 352, 5353-5361.

Thiriot J., forthcoming, L'atelier de potiers du "palais sud-est" de Sabra al-Mansuriya. In *Sabra al-Mansûriya. Capitale fatimide*, (eds. P. Cressier and M. Rammah).

Tite M.S., Freestone I., Mason R., Molera J., Vendrell-Saz M. and Wood N., 1998, Lead glazes in antiquity - Methods of production and reasons for use. *Archaeometry*, 40, 241-260.

Waksman S.Y., 2011, Ceramics of the 'Serçe Limani type' and Fatimid pottery production in Beirut. *Levant*, 43, 201-212.

Waksman S.Y., Capelli C. and Cabella R., forthcoming, Etudes en laboratoire des céramiques de Sabra al-Mansuriya et de la région de Kairouan. In *Sabra al-Mansûriya. Capitale fatimide*, (eds. P. Cressier and M. Rammah).

Walter V. and Besnus Y., 1989, Un exemple de pollution en phosphore et en manganèse de céramiques anciennes. *Revue d'Archéométrie*, 13, 55-64.

Wood N., 1999, *Chinese glazes*, Black Publishers, London.

## ACKNOWLEDGEMENTS

We would like to thank the French-Tunisian mission at Sabra al-Mansuriya, and especially P. Cressier and J.-C. Tréglia, and M. Delavenne for improving Fig.1.

# 5

# Thermal shock resistance of tempered archaeological ceramics

**Noémi S. Müller[a,c], Vassilis Kilikoglou[a], Peter M. Day[b] and George Vekinis[a]**
[a]Institute of Materials Science, N.C.S.R. "Demokritos", Aghia Paraskevi, 15310 Athens, Greece
[b]Department of Archaeology, University of Sheffield, Northgate House, West Street, Sheffield S1 4ET, UK
[c]Fitch laboratory, British School at Athens, Souedias 52, 10676 Athens, Greece

**Abstract** - The principles governing thermal shock resistance are introduced briefly, followed by the presentation of an experimental study conducted to assess the influence of temper and firing temperature on the thermal shock resistance of clay-based ceramics. Focusing on the assessment of a material's capability to resist thermal stresses, difficulties which arise when interpreting experimental results are discussed. We report that thermal shocking may induce a change in fracture mode in the ceramics, a finding which has important implications for the interpretation of experimental results and the assessment of the performance of ceramic vessels when exposed to sudden thermal stresses. Generally, it appears that for coarse cooking ware pottery, thermal shock resistance might not play such an important role as frequently has been assumed. Limited thermal shocking might even be beneficial for a ceramic's subsequent exposure to thermal stresses, as it can result in a change in fracture mode and may increase the toughness of the material.

## 1. Introduction

Cooking vessels are exposed to thermal stresses during daily use. Potters' choices, in particular the selection of temper materials, have been connected to the capability of a cooking pot to resist such stresses. While it has been stated that "entirely adequate cooking pots can be made using a wide range of alternative tempers" (Tite and Kilikoglou 2002), and calls for a more restrained use of the concept when interpreting technological choices in cooking ware have been made by pointing out the production of apparently 'unsuitable' vessels over a long period of time (Woods 1986), thermal shock resistance is drawn upon regularly to explain tempering choices in cooking ware pottery (e.g., Hoard *et al.* 1995; Killebrew 1999; Broekmans *et al.* 2004). Only rarely, however, are such explanations based on experimental studies, and the complexity of the concept of thermal shock resistance is frequently overlooked. Since these shortcomings also have implications for the archaeological interpretation of the findings, the present article aims to outline some of the principles governing the thermal shock resistance of clay-based ceramics and touches upon difficulties which arise when interpreting experimental results.

While the present paper focuses on the response of a ceramic when exposed to thermal stresses, it should be kept in mind that the actual conditions of the thermal shock are also of importance when considering thermal stresses. Thermal stresses that develop during gradual heating (quasi-steady-state conditions) may be studied by analytical techniques or by computational methods. The situation that occurs during thermal shocking, however, is different, as in this case stresses may develop faster than the material is able to dissipate them via mechanisms such as crack propagation, bifurcation, etc.

In order to assess systematically the influence of tempering and firing temperature on the performance of clay ceramics when exposed to sudden thermal stresses, thermal shock resistance was determined on ceramic test bars, bearing a variety of tempering and fired under a range of temperatures. In the following, some of the results of the experimental study are presented briefly, before discussing their implications for theoretical considerations.

## 2. Thermal shock resistance

Thermal shock resistance is a measure of the ability of a material to withstand rapid changes in temperature (Kingery *et al.* 1976, 822). It is important to note that thermal shock resistance, unlike elastic modulus or thermal conductivity for example, is *not* a material property but a complex parameter, which is dependent not only on different physical properties such as thermal expansion coefficient, mechanical strength or toughness, but — crucially — also on the conditions of thermal shock (Hasselman 1969). Due to the complexity of responses of a material that is exposed to thermal shock, and due to the difficulties in assessing heat transfer parameters for specific applications, thermal shock resistance of various ceramic materials is usually measured empirically, by assessing their performance under simulated service conditions (BSI 2004). The thermal shock resistance is assessed by measuring the mechanical strength of a ceramic material after exposure to a series of sudden changes in temperature (e.g., BSI 1998; 2006; ASTM 2003).

In this context, it must be emphasised that no general rating of materials in terms of their thermal shock resistance is possible. Any material's resistance to thermal shock depends critically on the magnitude and conditions of thermal shock (Davidge 1986). Relative order may differ according to the nature of the thermal shock (heat transfer conditions), and, while one material might be more resistant to thermal stresses which arise due to exposure to a certain temperature difference, say 500°C, than another material, when the same materials are exposed to a larger temperature difference, say 1000°C, exactly the opposite might be the case. The reasons for this are complex and are related to microstructural differences, but also to the onset of various fracture energy dissipation mechanisms. For archaeological ceramics, this means that results which have been obtained with an experimental setup designed to assess the performance of cooking ware ceramics cannot directly be employed to evaluate, for example, metallurgical ceramics, which operate under different conditions and are exposed to much larger temperature differences than cooking ware.

As mentioned above, thermal shock resistance is usually assessed by measuring strength reduction after exposure to sudden changes in temperature. The transverse rupture strength is assessed, which can be determined from three-point bending tests (a schematic drawing of a typical set-up is included in Fig. 1). This flexural strength is defined by the maximum stress that can be applied to a specimen without cracks starting to grow. In the case of truly brittle materials which show *unstable fracture*, the application of an external load leads to catastrophic failure once a crack initiates, without the need to apply further stress. Curve a of Figure 1 is a typical load-displacement curve for such a material, where complete failure occurs at maximum load, with the load dropping immediately to zero. Brittle materials show little or no plastic deformation before failure, and fracture absorbs relatively little energy, even in materials of high strength. On the other hand, materials which are able to dissipate energy during crack propagation by mechanisms such as crack deflection require the application of additional force after the initiation of a crack, so as to drive the crack through the material and break the specimen completely. Such materials are said to exhibit *stable fracture*, as, upon failure, the crack is immediately arrested and the load decreases gradually until complete separation (Fig. 1, curve c). In these cases, a considerable amount of energy is dissipated during crack propagation and until final fracture as a result of crack deflection, bifurcation, and arrest, as well as temper fracture or pull out, and the material usually possesses high fracture energies and is therefore said to be tough. (Unlike strength measurements which are performed on un-notched specimens, the quantitative determination of fracture energies requires bending tests on bars with a notch cut, which provides the largest flaw at which the crack is initiated.) The tail of the load-displacement curves of such materials (shaded area under curve c) provides a measure of the energy dissipated during crack propagation. In between the two, for some materials semi-stable fracture (Fig. 1, curve b) is observed. For further information on fracture modes of clay-based ceramics, the reader is referred to Kilikoglou *et al.* (1998) and Tite *et al.* (2001).

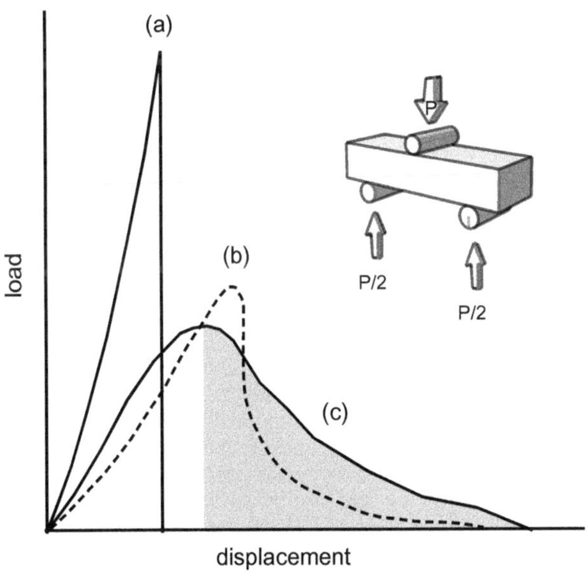


**Figure 1.** Typical load-displacement curves for different types of fracture: (a) unstable, (b) semi-stable, (c) stable, obtained from bending tests on clay-based ceramics.

In the context of clay-based ceramics, unstable fracture is typically associated with high fired porcelain, semi-stable fracture with standard earthenware, and stable fracture with low fired, coarse tempered earthenware. The former are referred to as 'brittle' ceramics, while the latter are frequently said to be 'tough'. However, this designation can be misleading, as it opposes 'brittle' and 'tough'. While the term 'brittle' designates a fracture mode, toughness is used equivalent with the 'fracture energy' of a material. Fracture energy consists of an intrinsic part ($G_{Ic}$), corresponding to the stored elastic energy at the moment of fracture, and — in the case of stable and semi-stable fracture — of an additional part accounting for energy dissipated during crack propagation ($G_t$). Strictly speaking, therefore, 'brittle' and 'tough' are not opposites: brittle materials can also be said to be tough if the stored elastic energy at the moment of fracture is high. It should be noted here that, in engineering applications, frequently the so-called *fracture toughness* $K_{Ic}$, also referred to as *stress intensity factor*, is listed for brittle materials. This property is linked to the stored elastic energy at the moment of fracture ($G_{Ic}$) by $K_{IC} = \sqrt{E}G_{IC}$ with E = modulus of rupture. Although related, this is not the same as the 'toughness' which is used as equivalent to 'fracture energies', and which includes both an intrinsic and a dissipation part. Throughout the present text, the term 'toughness' is used as equivalent to 'fracture energy', i.e., refers to the sum of $G_{Ic}$ and $G_t$.

Strength and toughness are interrelated, as they both depend on a material's microstructure. A material may be strong and tough if it ruptures under high applied loads and simultaneously exhibits high strains. Normally, however, in the case of clay-based ceramics, brittle materials which may be strong usually only show limited deformation, so that they are not tough, while materials which rupture after

**Table 1.** Briquette types examined: five different clay mixtures were fired to three different temperatures each, resulting in a total of 15 different briquettes.

| | | Firing temperature | | |
|---|---|---|---|---|
| | | **550°C** | **850°C** | **1050°C** |
| Tempering | None | untempered | untempered | untempered |
| | Granite | 10 vol% | 10 vol% | 10 vol% |
| | | 40 vol % | 40 vol % | 40 vol % |
| | Phyllite | 10 vol% | 10 vol% | 10 vol% |
| | | 40 vol % | 40 vol % | 40 vol % |

applying relatively small forces may be able to dissipate a substantial amount of energy during crack propagation until final fracture, so that they can be said to be tough but not strong. For traditional ceramics, strategies which result in an increase in strength in most cases reduce the toughness of a material and vice versa (Müller *et al.* 2009).

Generally speaking, strength indicates how much force can be applied to a material until a crack initiates (with crack initiation in the case of brittle ceramic being equivalent to final fracture), while toughness indicates how much energy a material can absorb before fracture, and encompasses both the energy required for crack initiation and for crack propagation.

It is of importance that there are significant differences in the thermal shock behaviour of brittle materials exhibiting unstable fracture and those materials that exhibit stable fracture, since clay-based ceramics can show both fracture modes, as will be detailed below. As illustrated in Figure, unstable brittle ceramics experience an abrupt loss of strength at a critical temperature difference $\Delta T_c$, due to the initiation of cracks to a length close to the critical length (Hasselman 1969). On the other hand, the loss of strength is more gradual in ceramics exhibiting controlled crack propagation, as is shown in the lower curve in Figure. In strong but brittle materials, due to the large amount of energy stored at the point of stress concentration (where fracture initiates, corresponding to the 'peak' in the load-displacement curves, cf. Figure), cracks experience a sudden large expansion when initiation occurs, while in a weaker material which can dissipate energy during crack propa-gation, cracks propagate in a more controlled way. Consequently, the strategy for optimising thermal shock resistance in a ceramic which fractures in a stable way is fundamentally different from the one for a material which fractures unstably, in that it usually involves avoiding crack propagation rather than crack initiation. Clay-based ceramics, depending on a variety of factors, can exhibit different fracture modes. The same clay type may break unstably when it is highly fired, but dissipate fracture energy, resulting in stable crack propagation, when lower fired. Also, the amount of aplastic inclusions can influence fracture modes: for quartz-tempered ceramics fired to intermediate temperatures, a change to stable fracture has been observed at around 20 volume percent of temper material (Kilikoglou *et al.* 1995).

### *2.1. Influence of the thermal expansion factor mismatch between clay matrix and temper material on the thermal shock resistance*

Thermal stresses arise in a material due to changes in temperature. In unrestricted multiphase materials, such as clay-based ceramics, there are two main causes of thermal stress: firstly, the non-uniform expansion of an object, which is caused by temperature gradients, and, secondly, thermal stresses caused by different thermal expansion coefficients between different constituents, such as the matrix and any inclusions (Kingery *et al.* 1976). The latter is cited frequently in archaeological literature to explain the choice of certain temper materials in the production of cooking vessels. It is important to recognise, however, that upon firing in a clay paste which contains temper particles of a different thermal expansion coefficient, micro-damaged zones are produced around these temper particles (Davidge 1986, 86-7). For quartz-tempered ceramics, it has been shown that at volume fractions of temper higher than 10%, the individual damaged zones interact with each other and produce extensive microcrack networks that cover the entire specimen (Fig. 3). During fracture, it is this microcrack network which encourages crack deflection and bifurcation, thus increasing energy dissipation and contributing to the material's toughness (Kilikoglou *et al.* 1995).

Therefore, if a vessel survives firing, as it frequently does, then these damaged zones enhance the toughness of a material. Consequently, thermal shocking of such a fired ceramic material could be expected to be less disastrous than thermal shocking of a material where the thermal expansion coefficients of clay and temper material are very similar, and in which the generation of damaged zones and the related increase in toughness is not observed. Therefore, when we compare a ceramic that contains temper of significantly different thermal expansion coefficient than the matrix (such as quartz) with one whose temper has a very similar thermal expansion coefficient to the clay (such as grog), we would expect that the quartz-tempered ceramic will be better able to accommodate thermal stress than the one containing grog temper. This conclusion contradicts many of the beliefs expressed in archaeological literature to date, especially concerning the frequently cited 'advantage' of calcite temper over others in cooking ware production, which is argued based on the similarity of the thermal expansion coefficients of clay and temper material (e.g., Rye 1976; Hoard *et al.* 1995; Killebrew 1999; Broekmans *et al.* 2004). It should be noted here that while the influence of calcite tempering on the thermal shock resistance of a finished vessel can be debated, its presence can affect the performance of a clay paste in other ways and during different steps in the manufacture sequence. Thus, it is sometimes argued that the addition of calcite is beneficial for the workability of the wet clay paste, since calcium ions released in the clay increase plasticity of the clay water system (Lawrence and West 1982). On the other hand, the presence of calcite temper in a clay paste may potentially cause problems when pottery is fired to intermediate temperatures. When fired above 600-870°C in oxidising atmosphere (temperatures given in the literature for the onset of the process vary substantially, Hoard *et al.*

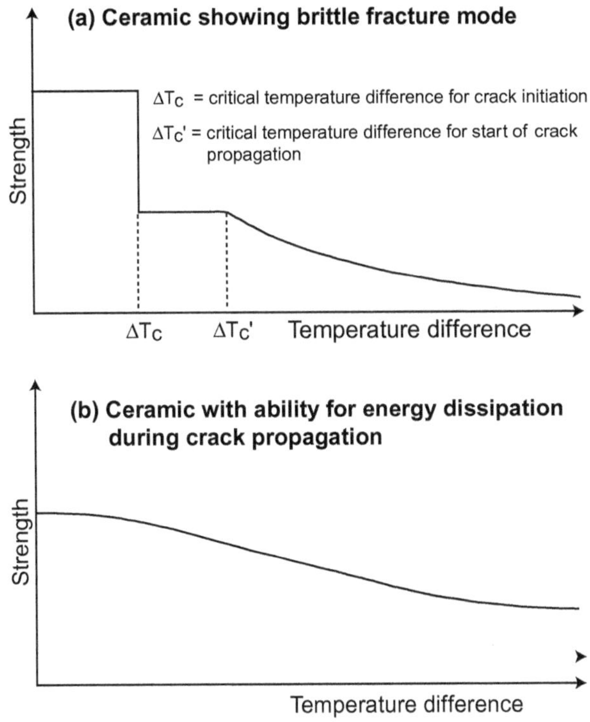


**Figure 2.** Different behaviour of ceramics with different fracture modes upon thermal shocking. While brittle ceramics experience an abrupt strength loss at a critical temperature difference $\Delta T_c$ due to initiation of fracture, the loss in fracture strength in ceramics which break in a more stable manner is gradual, due to more controlled crack propagation.

1995 for an overview), calcite is converted to lime. Upon cooling down and subsequent exposure to moisture, the lime reacts to form calcium hydroxide, a process which is accompanied by a volume expansion, causing cracking and spalling that in extreme cases may destroy a vessel.

As a side note, while fully agreeing with Woods (1986) that thermal shock resistance has been overworked and thermal shock might not be as important for cooking ware pottery as frequently assumed, it seems nevertheless interesting that the above sheds new light on the material she discusses. When questioning the emphasis put on thermal shock resistance of cooking ware, Woods (1986) states (among other issues) that quartz sand was probably more commonly used than calcite temper over a long period of time in the manufacture of cooking ware in Britain. Based on the available literature, she considers this finding as counterintuitive in terms of the thermal performance of the cooking ware. However, keeping in mind the above discussed benefit of adding temper with a different thermal expansion coefficient on thermal shock resistance, her quartz-tempered cooking pots might equally well be argued to perform better than their calcite-tempered counterparts, at least in terms of their thermal shock resistance.

## 3. Experimental

Thermal shock resistance was determined on test briquettes, since, when assessing technological changes in terms of physical properties which are linked to material performance, it is important to carry out measurements under controlled conditions in order to investigate the influence of selected parameters. Additionally, use and subsequent burial of ceramic objects results in microstructural changes which might influence the physical properties measured. Finally, the assessment of material properties normally requires a substantial amount of material for destructive tests, usually not available from archaeological remains. In order to determine the thermal shock resistance of a particular material by measuring its reduction in strength upon exposure to thermal shock, tests on several bars made from that material, both unshocked and thermally shocked, are required.

### *3.1. Materials and processing*

Experimental briquettes were manufactured in order to study the influence of various parameters, such as temper type or firing temperature, on the thermal shock resistance of the ceramic material. Granite and phyllite were chosen as temper materials and added to a non-calcareous base clay from Kalami, Greece. This clay has been used in a previous study of thermal properties, and its chemical and mineralogical composition is described elsewhere (Hein *et al.* 2008). A fraction with a particle size of $<30\,\mu m$ was separated from the raw clay and mixed with granite from

**Table 2.** Changes in fracture mode observed upon thermal shocking. The number of samples tested and examined is indicated. B: brittle, SS: semi-stable, S: stable fracture (cf. Figure 1 for fracture modes).

| | | Before thermal shock | | | After thermal shock | | |
|---|---|---|---|---|---|---|---|
| | fired to: | B | SS | S | B | SS | S |
| untempered | 550°C | 5 | | | | 1 | 2 |
| | 850°C | 5 | | | | 3 | |
| | 1050°C | 5 | | | 2 | 1 | |
| | 550°C | | 5 | | | 2 | 1 |
| 10% granite | 850°C | | 5 | | | 3 | |
| | 1050°C | | 3 | 2 | | 1 | 2 |
| | 550°C | | | 5 | | | 3 |
| 40% granite | 850°C | | | 5 | | | 3 |
| | 1050°C | | | 5 | | | 3 |
| | 550°C | | 5 | | | | 3 |
| 10% phyllite | 850°C | | 5 | | | 3 | |
| | 1050°C | 5 | | | 1 | | 2 |
| | 550°C | | | 3 | | | 3 |
| 40% phyllite | 850°C | | 3 | 2 | | | 3 |
| | 1050°C | 3 | 2 | | | 3 | |

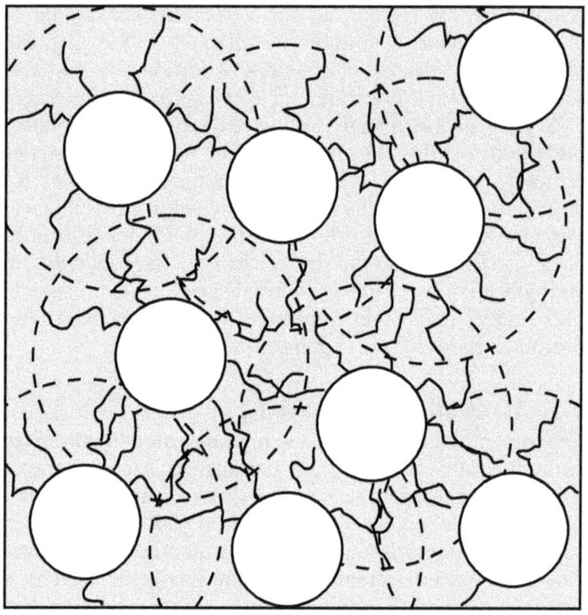

**Figure 3.** Graphical representation of damaged zones (dashed circles) surrounding temper grains (white circles) in a fracture surface. Illustration adapted from Kilikoglou *et al.* 1998.

the island of Naxos (Greece) and phyllite from the Northeast Peloponnese (Greece), respectively. Both temper materials were crushed and screened to obtain temper in the desired size range, i.e., between 1 mm and 0.5 mm. Ceramic briquettes with 10 vol% and 40 vol% temper material were prepared in addition to untempered reference briquettes, using a Perspex mould (*c.* 12 × 7 × 1 cm). For the preparation of the phyllite tempered briquettes, the paste mixture was repeatedly folded and flattened, in order to obtain a preferred orientation of the platy phyllite particles parallel to the largest surface of the briquettes, imitating common archaeological fabrics. The briquettes were dried for over ten days at ambient temperature and humidity, covered by plates made of plaster of Paris to ensure homogenous drying. Briquettes of each composition (i.e., non-tempered, 10% phyllite, 40% phyllite, 10% granite, and 40% granite) were fired to 550, 850 and 1050°C, resulting in a total of 15 different briquette types ((Table 1)). Firing took place at a heating rate of 200°C/h and a soaking time of 1 h in oxidising atmosphere. The large surfaces of the fired briquettes were ground parallel and cut into test bars of an approximate size of 1 × 1 × 7 cm for material testing.

### *3.2. Thermal shock resistance tests*

In the present study, the experimental test bars were placed for 20 minutes in a furnace held at 430°C and then quenched in water (27-30°C), where they were left to cool down for 5 minutes. Before placing them in the furnace again, they were dried for at least 20 minutes at 115°C. Strength reduction was determined by comparing the fracture strength in bending (often referred to as the 'transverse rupture strength' [TRS] or 'modulus of rupture' [MOR]) of test bars exposed to five such cycles with the strength of unshocked specimens. The transverse rupture strength was determined from three-point bending tests on bars with the dimensions 1 × 1 × 7 cm on an INSTRON 1195 universal testing machine, at a constant loading rate of 109 μm/min. This sample size ensures minimal sheer stresses, while still including a statistically adequate number of temper grains in the fracture surface (Kilikoglou *et al.* 1998). The load as a function of displacement was recorded for every specimen. For every ceramic type (corresponding to one set of parameters), TRS was measured on five bars before and three bars after thermal shocking. All measurements on the phyllite-tempered fabrics were performed perpendicular to the alignment axis of the inclusions. After breakage, the fracture area was examined visually, and data from bars with clearly identifiable macroscopic flaws in the fracture surface were not included in the determination of the mean. TRS was calculated according to standard methods (BSI 2002).

## 4. Results and discussion

### *4.1. Strength reduction upon thermal shocking*

When restricting the definition of thermal shock resistance to the *variability* of strength levels upon thermal shocking, the interpretation of the experimental results is rather straightforward. Figure 4 plots the reduction of fracture strength in bending upon thermal shocking for the various materials tested. It can be seen that the parameters that were varied in the manufacture of the briquettes strongly affect the thermal shock resistance. Firstly, high amounts of temper material increase thermal shock resistance. Secondly, firing temperatures below or close to the onset of vitrification result in ceramics that lose much less strength upon thermal shocking than highly fired ones. Finally, bulky granitic temper results in more thermal shock resistant ceramics than the platy phyllitic one, at least at higher firing temperatures.

These results can be explained by the variation of the toughness (stored elastic energy and energy dissipated during crack propagation) between each type of material studied. In other words, by considering the effect of the extent of pre-existing flaws in the materials, as well as the relative bonding strength of the matrix as reflected in its degree of vitrification. High temper contents have been observed to result in increased porosity, due to restricted shrinkage of the matrix. Further, damaged zones are introduced around temper particles upon firing, due to the different thermal expansion factors of clay and temper (Davidge 1986, 86-7), and their extent is dependent on the amount of temper (Kilikoglou *et al.* 1995). Such flaws and microcracks enable the material to better accommodate tensions arising from the thermal expansion mismatch, and thus lead to the measured increased thermal shock resistance. On the other hand, increased vitrification upon firing reduces a material's ability to absorb thermal stresses, because of the increased instability of fracture in the glassy phase. The pronounced differences between platy phyllitic and bulky granitic temper, which are observed at high firing temperatures and low temper content, can be explained with reference to the differing extent of damaged zones. In this case, it has been shown that the damaged zones around granitic grains are far more pronounced than the ones around phyllitic grains with the same lateral expansion, due to different particle geometry. Even when correcting for the higher number of phyllitic particles per unit volume for the same overall amount of temper material, this results in much

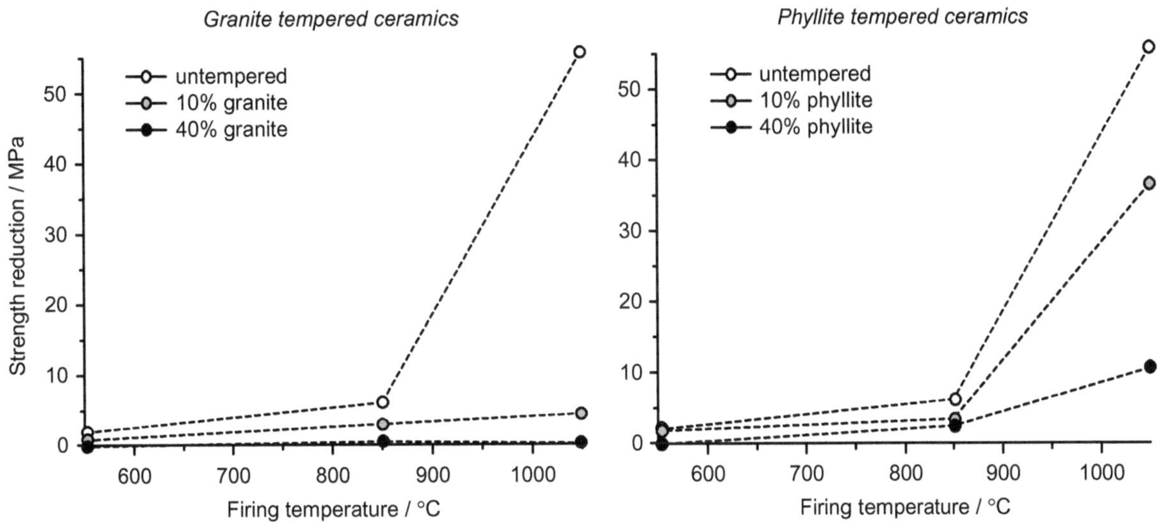


**Figure 4.** Strength reduction due to thermal shocking of non-calcareous ceramic test bars fired at different temperatures.

more extensive damaged zones for granite-tempered ceramics (Müller *et al.* 2010).

## *4.2. Residual strength levels after thermal shocking*

As outlined above, when restricting the definition of thermal shock resistance to the variability of strength levels upon thermal shocking, the interpretation of the experimental results is straightforward. However, the interpretation of the experimental results becomes more complex when taking into account *the residual strength* of the different materials (Fig. 5), since a greater loss in strength does not necessarily imply that the material is weaker after thermal shocking in absolute terms.

This becomes apparent when juxtaposing the high fired ceramics which contain 10 vol% temper material (Fig. 5, inset). Although the phyllite-tempered ceramics lose more strength when exposed to thermal shock and - according to our definition above - have a lower thermal shock resistance, their *residual* strength after thermal shock is still higher than the one of the ceramic bars tempered with granite. In other words, even after losing a significant amount of strength upon thermal shocking, the phyllite-tempered material is still more resistant to crack initiation than its granitic counterpart. It could be argued that the particular example, i.e., a high fired ceramic material with a relatively low amount of aplastic inclusions, is not relevant when assessing the usually relatively low fired and highly tempered archaeological cooking ware. However, the same effect, albeit less pronounced, is also observed with the ceramics fired to 850°C which contain 40% temper material. Finally, the highly fired untempered samples also exhibit a higher strength than most of the tempered sherds fired to similar temperatures, even after thermal shocking. The high initial strength results in relatively high residual strength even though these specimens, due to their

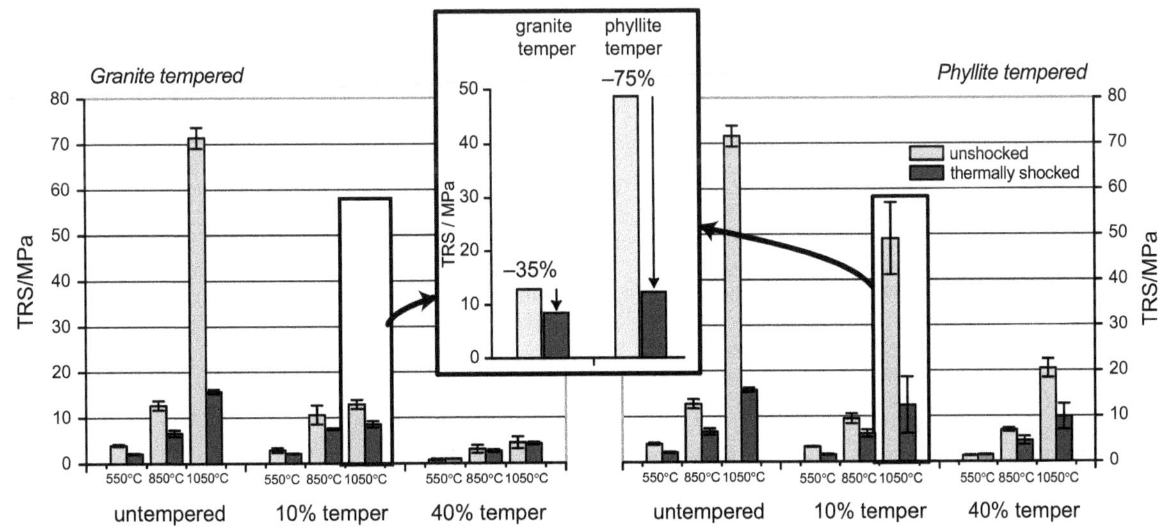


**Figure 5.** Transverse rupture strength of test specimens before and after exposure to thermal shock.

brittleness, experience a substantial loss of strength upon fracture.

### *4.3. Influence of thermal shocking on fracture mode*

In many cases, thermal shocking induced a change in the fracture mode (Table 2), in some cases increasing the toughness of the material. This trend, i.e., a change to a more stable fracture after exposure to thermal shock, is observed for ceramics fired to low as well as to higher temperatures. The change in fracture mode upon thermal shocking is important, since, when exposed to the same temperature differences, the strength reduction through thermal shocking associated with tough ceramics is less severe than with brittle ceramics. This is due to the different response of the material, as illustrated in Figure 3.

## 5. Summary and conclusions

The results of this experimental study have shown that, in terms of *variability* of strength levels upon thermal shocking, high amounts of temper material increase thermal shock resistance. Further, low firing temperatures seem beneficial for a ceramic's performance and, at least at higher firing temperatures, bulky granitic temper results in more thermal shock resistant ceramics than the platy phyllitic one. However, as outlined above, the residual strength level and fracture mode and changes therein need to be taken into account when interpreting the results of experimental tests. While this article has focused on the response of a ceramic when exposed to thermal shock, it should be noted that the magnitude of the thermal stresses that arise in a ceramic vessel to start with is equally important. It is suggested that, in order to produce meaningful conclusions on the performance of cooking ware in terms of their response to thermal shock, it is crucial to consider thermal shock resistance in a more integrated fashion than has been the case to date. This should include an assessment of the material's capability to resist those stresses, taking into account the implications of issues discussed in the present article, but also considerations of the origin of thermal stresses. The experimental outcome of the present study, in particular the observation that thermal shocking may actually induce a change in fracture mode, has important implications for such considerations, since brittle ceramics react in a fundamentally different way to thermal shock than those which fracture in a more stable way.

Finally, based on the experimental results and on the conclusions presented above, it appears that, for coarse cooking ware exposed to stresses that would be expected under cooking conditions, thermal shock resistance does not play such an important role as has frequently been assumed. In fact, we could even speculate that, in some cases, limited thermal shocking might be beneficial for a ceramic's subsequent exposure to thermal stresses, as it increases the energy dissipating micro-damage zone around temper particles. These observations call for more caution when accounting for the occurrence of specific temper materials in ceramic cooking ware or for changes in manufacturing practices over time. Rather than simply explaining away preferred tempering practices with stating their alleged benefits for thermal shock resistance, discussions should be more comprehensive and look beyond this particular performance characteristic to include not only other potentially relevant material properties, but, importantly, also the multitude of non-technological factors influencing technological choice.

## References


ASTM (American Society for Testing and Materials), 2003, Test method for quantitatively measuring the effect of thermal shock and thermal cycling on refractories (C 1171-96). *Annual book of ASTM standards*, 238-240, Volume 15:01, ASTM International, West Conshohocken.

BSI (British Standards Institute), 1998, *Ceramic tiles, part 9: determination of resistance to thermal shock.* British Standard (BS EN ISO 10545-9:1996).

BSI (British Standards Institute), 2002, *Advanced technical ceramics, mechanical properties of ceramics composites at room temperature, part 3: determination of flexural strength.* British Standard (BS EN 658-3:2002).

BSI (British Standards Institute), 2004, *Advanced technical ceramics, methods of testing monolithic ceramics, thermo-mechanical properties, part 3: determination of resistance to thermal shock by water quenching.* British Standard (BS EN 820-3:2004).

BSI (British Standards Institute), 2006, *Dense shaped refractory products, part 11: determination of resistance to thermal shock, draft for development.* (DD CEN/TS 993-11:2003).

Broekmans T., Adriaens A. and Pantos E., 2004, Analytical investigations of cooking pottery from Tell Beydar (NE-Syria). *Nuclear Instruments and Methods in Physics Section B: Beam Interactions with Materials and Atoms*, 226, 92-97.

Davidge R.W., 1986, *Mechanical behaviour of ceramics*, Cambridge University Press, Cambridge.

Hasselman D.P.H., 1969, Unified theory of thermal shock fracture initiation and crack propagation in brittle ceramics. *Journal of the American Ceramic Society*, 52, 600-604.

Hein A., Müller N.S., Kilikoglou V. and Day P.M., 2008, Thermal conductivity of archaeological ceramics: the effect of inclusions, porosity and firing temperature. *Thermochimica Acta*, 480, 35-42.

Hoard R.J., O'Brien M.J., Khorasgany M.G. and Gopalaratnam V.S., 1995, A materials-science approach to understanding limestone-tempered pottery from the Midwestern United States. *Journal of Archaeological Science*, 22, 823-832.

Kilikoglou V., Vekinis G. and Maniatis Y., 1995, Toughening of ceramic earthenwares by quartz inclusions: an ancient art revisited. *Acta Metallurgica et Materialia*, 43, 2959-2965.

Kilikoglou V., Vekinis G., Maniatis Y. and Day P.M., 1998, Mechanical performance of quartz-tempered ceramics: part I, strength and toughness. *Archaeometry*, 40, 261-279.

Killebrew A.E., 1999, Late Bronze and Iron I cooking pots in Canaan: a typological, technological and functional study. In *Archaeology, history and culture in Palestine and the Near East: Essays in memory of Albert E. Glock*, (ed. T. Kapitan), 3-126, ASOR books volume 3, Scholars Press, Atlanta, GA.

Kingery W.D., Bowen H.K. and Uhlmann D.R., 1976, *Introduction to ceramics*. 2nd edition, John Wiley and Sons, Inc., New York.

Lawrence W.G. and West R.R., 1982, *Ceramic science for the potter*. 2nd edition, Chilton, Radnor, PA.

Müller N.S., Kilikoglou V., Day P.M., Hein A. and Vekinis G., 2009, The influence of temper on performance characteristics of cooking ware ceramics. In *Vessels: inside and outside, Proceedings of the conference EMAC'07, 9th European meeting on ancient ceramics*, (eds. K.T. Biró, V. Szilágyi and A. Kreiter), 145-150, Hungarian National Museum, Budapest.

Müller N.S., Kilikogou V., Day P.M. and Vekinis G., 2010, The influence of temper shape on the mechanical properties of archaeological ceramics. *Journal of the European Ceramic Society*, 30, 2457-2465.

Rye O.S., 1976, Keeping your temper under control: materials and the manufacture of Papuan pottery. *Archaeology and Physical Anthropology in Oceania*, 11, 106-137.

Tite M.S. and Kilikoglou V., 2002, Do we understand cooking pots and is there an ideal cooking pot? In *Modern trends in scientific studies on ancient ceramics*, (eds. V. Kilikoglou, A. Hein and Y. Maniatis), 1-8, British Archaeological Reports International Series 1011. Archaeopress, Oxford.

Tite M.S., Kilikoglou V. and Vekinis G., 2001, Strength, toughness and thermal shock resistance of ancient ceramics, and their influence on technological choice. *Archaeometry*, 43, 301-324.

Woods A.J., 1986, Form and function: some observations on the cooking pot in antiquity. In *Ceramics and civilization, Volume 2*, (ed. W.D. Kingery), 157-172, American Ceramic Society, Columbus, OH.

# 6

# The colour and golden shine of early silver Islamic lustre

**P.C. Gutierrez[a], T. Pradell[b], J. Molera[c], A.D. Smith[d], A. Climent-Font[a] and M.S. Tite[e]**

[a]Centro de Micro-Análisis de Materiales and Departamento de Física Aplicada, C-XII, Universidad Autónoma de Madrid, Campus de Cantoblanco, 28049 Madrid, Spain

[b]Departament de Física i Enginyeria Nuclear, Universitat Politècnica de Catalunya, Campus Baix Llobregat, ESAB, Av. Canal Olímpic 15, 08860 Castelldefels, Spain

[c]GR-MECAMAT. Escola Politècnica Superior. Universitat de Vic-Universitat Central de Catalunya . c) de la Laura, 13, 08500 Vic, Spain

[d]STFC, Daresbury Laboratory, Sci-Tech Daresbury, Daresbury, Warrington WA4 4AD UK. Now at : Dalton Cumbrian Facility, University of Manchester, Westlakes Science & Technology Park, Moor Row, Cumbria CA24 3HA, UK

[e]Research Laboratory for Archaeology and the History of Art, Dyson Perrins Building, South Parks Road, Oxford OX13QY, UK

**Abstract** - A selection of lustres including 9th century AD polychrome and 10th century AD monochrome Abbasid lustres from Iraq, and 10th to 12th centuries AD Fatimid lustres from Egypt and Syria is studied in the present paper. The selection is based on previous studies that demonstrated that all of them contain metal silver nanoparticles and copper, which, when present, appears either as $Cu^{+}$ or $Cu^{2+}$ dissolved in the glaze. They show different colours, green, yellow, amber, and brown, and may also show or lack a golden-like reflectivity, which results mainly from average size and concentration in the layer of the silver nanoparticles. In this paper, a depth profile composition of the lustre layers is determined using Rutheford Backscattering Spectroscopy, allowing the determination of the total silver content, concentration of silver, copper to silver ratio, and thickness of the lustre layers. We show that the enhanced golden-like reflectivity occurs only for layers with a high concentration of silver, and that the addition of PbO to the alkaline glaze helps the formation of more concentrated layers. The results obtained provide new hints concerning the lead enrichment of the glazes during this period.

## 1. Introduction

The first lustre decorations (Caiger-Smith 1991) on glazed ceramics can be dated to the 9th century AD in the Abbasid caliphate, and were probably produced in Bashra. The earliest examples were found in the Palace of Samarra. They are polychrome, with the most common combining green, amber, and brown, but also yellow and red. During the 10th century AD, they were replaced by a monochrome green lustre production. At the end of the 10th century AD and contemporary to the Fatimid occupation of Egypt, a new lustre production began in Fustat and in Syria, lasting until the end of the Fatimid rule. These productions are all monochrome, but show colours ranging from green to yellow and from orange to brown. The lustre decorations (Pérez-Arantegui et al., 2001; Padovani *et al.* 2003) are extremely thin layers (well below the micrometre) made of small metal silver/copper nanoparticles inside the glaze matrix (from a few to some tens of nanometres), which appear very close to the glaze surface (some tens of nanometres). The Abbasid and Fatimid lustre layers studied in this paper contain only silver metal nanoparticles, and, although they also contain copper, it appears either as $Cu^{+}$ or $Cu^{2+}$ dissolved in the glassy matrix (Pradell *et al.* 2008a; 2008b).

Lustre was produced following a highly ingenious procedure in which the lustre design (containing copper, silver- and sulphur- containing compounds, and clay) was painted (Molera *et al.* 2001) onto the glaze, which was then subjected to a firing at a relatively low temperature (500–600°C). $Ag^{+}$ and $Cu^{+}$ ions diffused into the glaze, substituting alkalis (either Na or K) (Pradell *et al.* 2005). The introduction of an external reducing atmosphere into the kiln favoured the precipitation of the metal

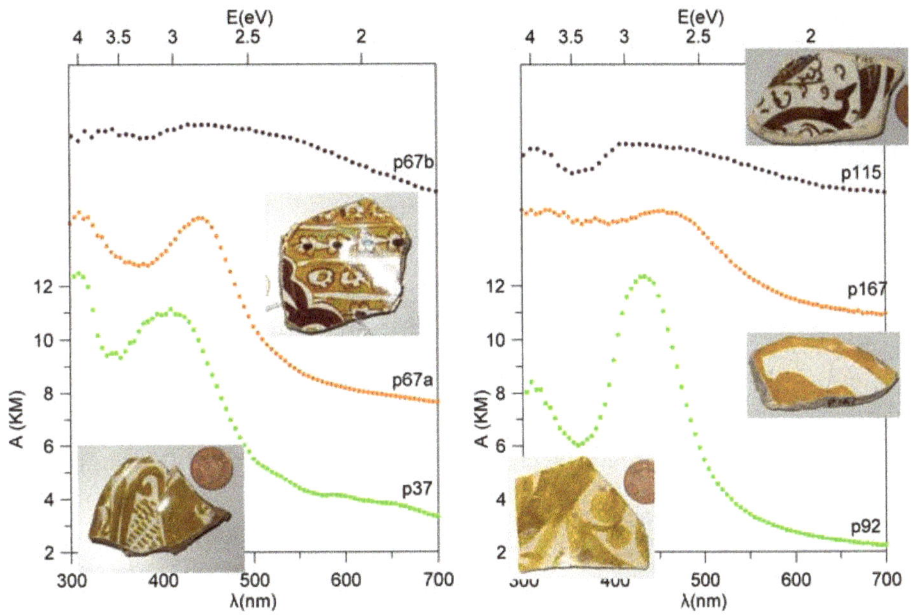


**Figure 1.** Images and UV-Vis spectra corresponding to some of the Islamic lustre layers showing different colours, corresponding to (left) green, amber, and brown Abbasid silver lustres and (right) green, yellow, and brown Fatimid silver lustres.

nanoparticles, which formed in a thin layer close to the glaze surface. The paint was washed off after firing to leave the colourful lustre decoration, which often exhibited a metal-like shine. Previous studies have shown that silver always appears as metal nanoparticles, while copper, being more difficult to reduce to metal than silver, often appears either as $Cu^{+}$ or $Cu^{2+}$ ions dissolved in the glaze. Only when very reducing conditions are applied can the copper ions form nanoparticles of $Cu_2O$ and metal copper.

The colour and shine of these layers are determined by this nanocomposite structure and produce particularly beautiful effects. The metal nanoparticles are essentially transparent to visible light, with the exception of the resonant absorption at a wavelength characteristic of the type, size, and shape of the metal nanoparticles (Berthier *et al.* 2006; Reillon and Berthier 2006; Reillon 2008). They also often show an enhanced reflectivity, responsible for the names 'lustre' given to them in English, and either 'metallic shine' or 'golden' in Spanish. This 'golden' shine may be described as resulting from the increase in reflectivity, in a manner similar to that shown by a metallic surface. Laboratory replicas of lustre (Molera *et al.* 2007) indicated that the size of the nanoparticles, thickness of the layer, and the amount of copper or silver in the layer depend on the composition of the glaze and on the firing protocol, temperature, and atmosphere. Our laboratory replicas of copper lustres showed that the metallic shine was directly related to the presence of a high volume fraction of copper nanoparticles in the layer (Pradell *et al.* 2007), and that the composition of the glaze is the essential parameter that regulates this. The paint composition and firing temperature and atmosphere also play an important role in determining the success rate.

In the present paper, we discuss the reasons for the colour and golden shine shown by Abbasid and Fatimid Islamic silver lustre in terms of the composition and thickness of the lustre layers, which are obtained by using Rutherford Backscattering Spectroscopy (RBS); this method has been applied to the study of other lustre productions (Chabanne *et al.* 2008; Polvorinos del Rio *et al.* 2008;

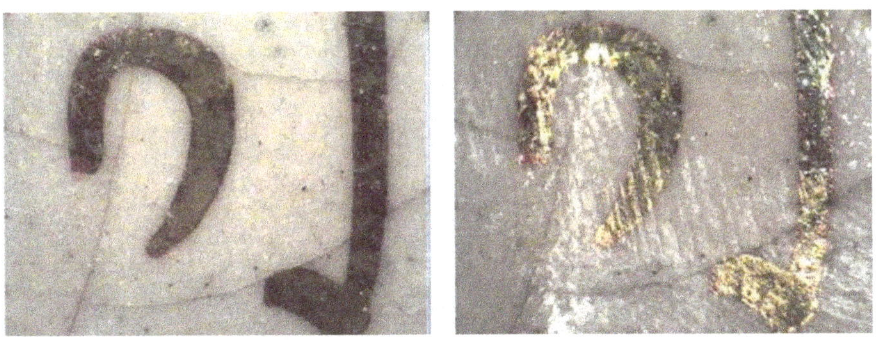

**Figure 2.** Left, photograph of sample p99, Fatimid (11$^{th}$ century AD). Right, photograph of the same area with the surface tilted in order to show the reflected light. This shows the enhanced golden reflectivity of the layer.

Gutierrez *et al.* 2010; Chabanne *et al.* 2012). Optical calculations of the absorption and reflectivity of silver lustre-like layers provide direct clues about the colours and golden shine shown by the lustre layers. The voluntary modification by the Islamic potters of the glaze composition, by increasing the PbO content of the glazes in order to enhance the development of the golden shine, is discussed.

## 2. Materials and methods

We have analysed a set of lustre decorations which are representative of Abbasid lustre from Iraq ($9^{th}$–$10^{th}$ centuries AD) and Fatimid lustre from Egypt (10–$12^{th}$ centuries AD). The selection includes lustre layers showing all the range of colours: green, yellow, orange and brown. Figure 1 shows some of the samples studied, on the left Abbasid and on the right Fatimid lustres, showing green, orange and brown colours. For all the colours, there are samples that show a golden shine and some that do not. Figure 2 shows an image of one of the Fatimid lustres showing an enhanced yellow reflectivity (golden shine).

UV-Vis diffuse reflectance (DR) avoiding the mirror reflected light was measured directly on the surface of the lustre layers by a Spectrophotometer Shimadzu UV-Vis 2401 PC. A small circular spot of 5 mm was used to collect the data. From these data, the colour coordinates and the corresponding dominant wavelength were evaluated using CIE 1931 chromaticity space. The dominant wavelength is given in Table and shows an increase while red-shifting the colour, green to yellow to amber to brown. These measures are related to the total light that is transmitted through the lustre layer (neither scattered nor absorbed), reflected on the ceramic white background and transmitted back to the surface. Figure shows the extinction spectra corresponding to some Abbasid and Fatimid lustres showing different colours.

The thickness and depth composition of the lustre layers are determined by means of Rutherford Backscattering Spectroscopy (RBS). RBS measurements were carried out with the 5 MV tandem accelerators at the CMAM facility (Climent-Font *et al.* 2004). The analyses were performed in vacuum using a 3035 keV He beam in order to take advantage of the $^{16}O(\alpha,\alpha)^{16}O$ elastic resonance occurring at this energy, which increases the sensitivity of oxygen detection by a factor of 23. The RBS beam at the sample was square, of size $0.7\times0.7$ mm$^2$. The backscattered ions were collected by means of two particle surface barrier detectors, at scattering angles of 170° and 165°. A careful quantification of the RBS spectra was achieved using the SIMNRA simulation code (Mayer 1997). Figure 3 shows the RBS fitted spectra from the white glaze surface and from the lustre decoration corresponding to one of the samples (p99). For the glaze, the average composition corresponds well to the Microprobe data presented elsewhere (Pradell et al. 2008a; 2008b), obtained from the glaze surface, but the addition of a thin lead depleted layer (of about $216\times10^{15}$ atm/cm$^2$, equivalent to 22 nm) at the surface is necessary to fit the spectrum. Modelling the RBS spectra of the lustre layer p99 reveals the same starting chemical composition as for the glaze, but with reduced quantities of Na and K, and corresponding increases of Ag and Cu, as determined from the Microprobe data (Pradell *et al.* 2008a; 2008b). As such, it corresponds to the ionic exchange mechanism responsible for the lustre formation.

To determine depth profiles, a sequence of layers has to be modelled. Figure 3 shows the depth concentration profile for the lustre layer, fitted using nine layers of varying thickness and silver content. For this sample, the total thickness of the layer approaches 500 nm, although the silver rich region is thinner, about 205 nm thick. Moreover, the maximum concentration of silver in the layer is 20.0 at%, present at a depth of 72 nm. All the lustre layers have been fitted following the same procedure.

## 3. Results and discussion

The corresponding depth concentration profiles for all the Islamic lustre decorations studied were also obtained, and the silver and copper profiles are shown in Figure 4. Some characteristics of the lustre layers can be determined from the fitted RBS data and are summarised in Table ; that is, the maximum content of silver in the layer (in at%) and depth at which the maximum silver content occurs, i.e., depth at which the lustre layer forms (in nm), the total amount of copper and silver that has entered the glaze by ionic exchange per unit surface (atm/cm$^2$), and, finally, the thickness of the lustre layer (in nm). The silver/copper ratio in the lustre layer, wt% Ag/(Ag + Cu), is also provided.

We can observe that there is great variability in the total amount of copper and silver in the layers. Silver always appears as a well-defined layer, although the characteristics of the profile may vary greatly among samples. The profile shows some asymmetry, being steeper at the surface and showing a longer tail inside the glaze. This profile is characteristic of a diffusion controlled process.

The values corresponding to the total silver and copper in the layer are very different and are mainly related to the original paint composition and firing protocol, but do not influence the reflectivity of the layers (p67b contains a large amount of silver and does not show a metallic golden shine, while p92, containing a very low amount of silver, does show a golden shine). In contrast, the data unambiguously shows that only those layers containing a high concentration of silver show golden shine (p51a, p51b, p92, and p99). What is most interesting of the data obtained is the inverse correlation that clearly exists between the depth at which the lustre layer forms and the lead content of the glaze; this is plotted in Figure 5. This indicates that the diffusivity of silver ions into the glaze is lower for glazes richer in lead, which may explain why, broadly speaking, lustre layers tend to be very thin and contain a higher concentration of silver. It should be noted, however, that the original lustre paint composition and firing protocol followed do also play an important role (differences between p99, p92, and p167). We know from previous replication experiments that copper enters the glaze mainly under oxidising conditions at about 500–600°C, but afterwards it needs a strong reducing atmosphere for the precipitation of metal nanoparticles; conversely, silver can reduce under less reducing conditions. Therefore, even using a glaze richer in lead, the silver in the layer may not reach the necessary concentration to produce a golden shine if a suitable firing protocol is not followed, or if the paint used does not have the appropriate composition. Finally, although the thickness of the lustre layer is a related feature, on its own it does not determine the presence of a golden shine. P51a and p37 have similar thicknesses, and only the first shows a golden shine, while

**Table 1.** The Islamic Abbasid and Fatimid lustre layers studied. Summary of the main characteristics of the lustre layers obtained from the fitted RBS spectra.

| | | Abbasid ($9^{th}$ - $10^{th}$ century AD) | | | | | Fatimid ($10^{th}$ - $12^{th}$ century AD) | | |
|---|---|---|---|---|---|---|---|---|---|
| **Samples** | | **P67b** | **p67a** | **p51b** | **p51a** | **p37** | **p99** | **p92** | **p167** |
| Colour<br>dominant wavelength (nm) | | Brown<br>587.8 | Amber<br>579.0 | Dark green<br>578.1 | Green<br>575.3 | Green<br>577.3 | Yellow<br>580.3 | Green<br>577.0 | Amber<br>581.3 |
| Shine | | No/bluish | No | Golden | Golden | No | Golden | Golden | No |
| PbO glaze (wt%) | | 6.9(0.3) | 4.3(0.3) | 13.9(2.1) | 13.9(2.1) | 7.9(0.4) | 37.0(3.0) | 34.3(1.1) | 40.2(2.3) |
| Lustre layer | Ag/(Cu + Ag) wt% | 57 | 21 | 93 | 100 | 98 | 68 | 79 | 62 |
| | Max. Ag (at%) | 4.7 | 1.1 | 14.5 | 8 | 4.6 | 20 | 8.9 | 4.6 |
| | Position max. (nm) | 353(71.5) | 488(29.7) | 164(12.7) | 156(33.3) | 265(43.3) | 57(6.5) | 34(6.8) | 79(10.8) |
| | Max. volume fraction Ag (%) | 4.4 | 1.2 | 16.3 | 9 | 5.2 | 23.3 | 10.3 | 5.2 |
| | Total Cu × $10^{15}$ (atm/$cm^2$) | 220 | 144 | 11 | - | 3 | 87 | 18 | 31 |
| | Total Ag × $10^{15}$ (atm/$cm^2$) | >154 | 16 | 80 | 56 | 71 | 109 | 38 | 30 |
| | Thickness (nm) | >660 | 402 | 82 | 275 | 287 | 206 | 167 | 120 |

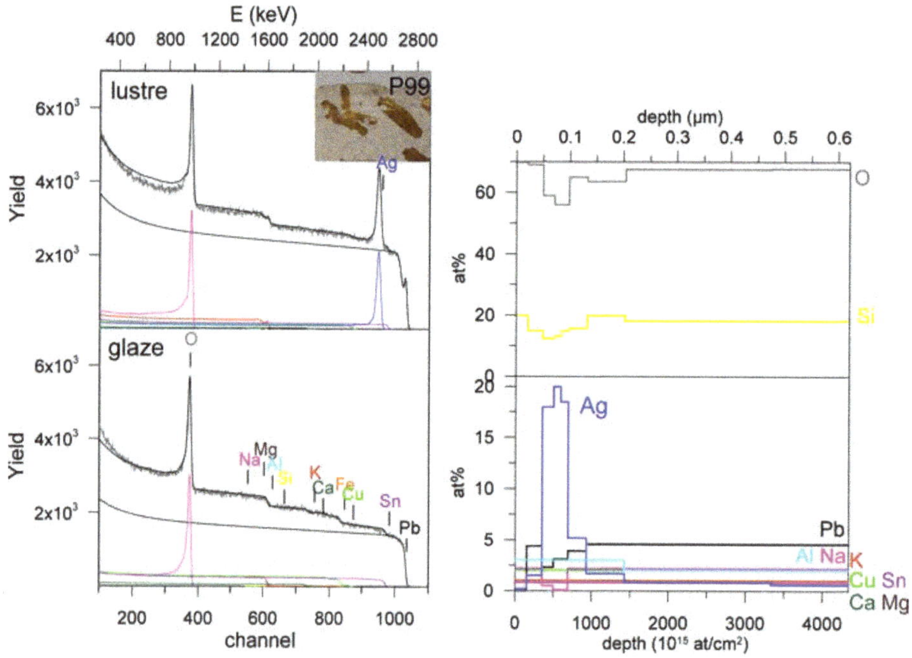


**Figure 3.** Left: Fitted RBS spectra corresponding to the white glaze (bottom) and to the yellow golden lustre decoration (top) for sample p99, Fatimid (11th century AD). Right: The chemical concentration depth profiles obtained by fitting the spectrum (top left) for the yellow lustre from sample p99.

p51a and p51b have different thicknesses and both show golden shine.

In an ionic exchange process, the reduced diffusivity of silver ions in lead-alkali glazes richer in lead has to be related to the reduced ionic mobility of the alkali ions in the glaze. Some studies demonstrated that this ionic mobility is inversely related to the ionic size of the added ion for divalent ions; therefore, the addition of lead in the glaze reduces the alkali diffusivity (Doremus 1994). We can then conclude that the increase of the PbO content of the glazes from earlier Abbasid to later Fatimid lustre

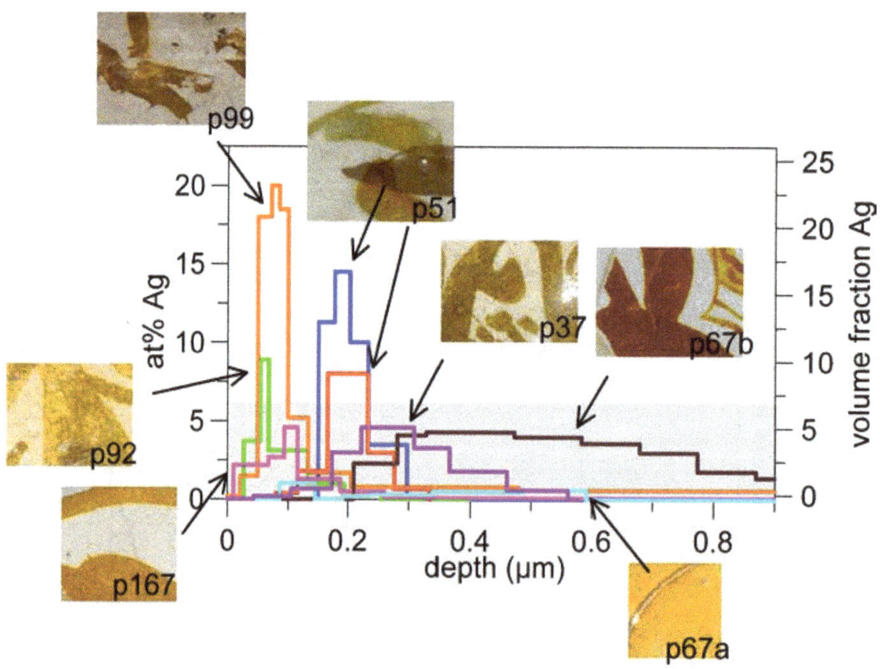


**Figure 4.** Silver depth concentration profiles obtained for all the Islamic lustre decorations studied. The samples showing golden shine (p51, p99, p92) exhibit a high concentration of silver in the lustre layer.

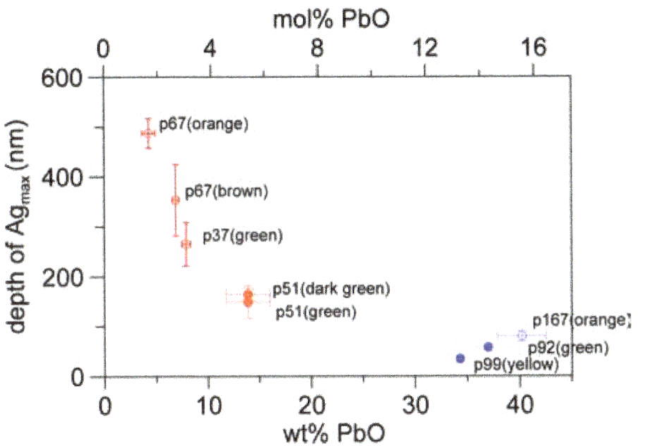


**Figure 5.** Depth at which the silver lustre layer forms inside the glaze as a function of the lead oxide (PbO) content of the glaze. Red symbols correspond to Abbasid lustres and blue symbols to Fatimid lustres. Solid symbols correspond to those lustre layers showing golden shine.

could have been deliberate, carried out in order to improve the success rate in producing golden shining decorations.

Concerning the colour (green, yellow, orange, and brown) shown by the lustre layers, the UV-Vis data (Fig. 1) indicate that there is a shift of the absorption peak related to the silver metal nanoparticles to longer wavelengths, as well as a broadening of the peak itself (Kreibig and Vollmer 1995; Reillon 2008). This shift is responsible for the change in colour from green to yellow and orange. This may be related to the increase in size of the metal silver nanoparticles, from 20 nm up to 40 or 50 nm. However, the peak broadening and the general flattening shown for the brown lustres has to be related to other aspect as well: on the one hand, the presence of a broad particle size distribution of silver nanoparticles, and on the other, to the presence of $Cu^{+}$ and $Cu^{2+}$ dissolved into the glaze, which are known to respectively confer a yellowish or greenish tinge to the glaze. Finally, the presence of copper in solution into the glaze is known to produce the formation of aggregates of silver nanoparticles (Quinten and Kreibig 1993), with the consequent increase in size and a more heterogeneous distribution of sizes. The presence of copper in the lustres is therefore, directly or indirectly, responsible for the development of the yellow, orange, and brown colours.

Finally, concerning the golden shine, it is possible to calculate the reflectivity of the lustre layers using a model developed by Farbman *et al.* (1992), as a function of the concentration of metal silver nanoparticles in the lustre layer. Low concentrations of nanoparticles (up to 5% volume fraction) result in a low bluish reflectivity; with an increase in the concentration of metal silver nanoparticles (between 10 to 20% volume fractions), the reflectivity also increases and shifts to higher wavelengths, to a green-yellow colour. This enhanced yellow reflectivity is therefore responsible for the golden shine shown by the lustre layers.

## 4. Conclusion

We have demonstrated that the use of high lead content lead-alkali glazes in Islamic times helped the formation of more concentrated lustre layers. This seems to indicate an intentional increase in the PbO content of the glazes from earlier Abbasid ($9^{th}$–$10^{th}$ centuries AD) to later Fatimid ($11^{th}$–$12^{th}$ centuries AD) lustre productions. However, the use of an adequate oxidising-reducing protocol and paint compositions is also fundamental.

The studies of Islamic lustre carried out to date have shown the high control of the process attained by its makers, allowing for ionic exchange and the precipitation of metal nanoparticles, with appropriate control of the type, size, and volume fraction of the metal nanoparticles. The Islamic potters produced a *true nanomaterial*, and their success in producing layers shining like gold may have contributed to the ancient alchemical belief in the transmutation of other metals into gold.

## References

Berthier S., Padeletti G., Fermo F., Bouquillon A., Aucouturier M., Charron E. and Reillon V., 2006, Lusters of renaissance pottery: experimental and theoretical optical properties using inhomogeneous theories. *Applied Physics A*, 83, 573-579.

Caiger-Smith A., 1991, *Lustre Pottery*, New Amsterdam Books, New York.

Chabanne D., Bouquillon A., Aucouturier M., Dectot X. and Padeletti G., 2008, Physico-chemical analyses of Hispano-Moresque lustred ceramic: a precursor for Italian majolica? *Applied Physics A*, 92, 11-18.

Chabanne D., Aucouturier M., Bouquillon A., Darque-Ceretti E., Makariou S., Dectot X., Faÿ-Hallé A. and Miroudot D., 2012, Ceramics with metallic lustre decoration. A detailed study of Islamic productions from the 9th century until the Renaissance. *Matériaux & Techniques*, 100, 47-68.

Climent-Font A., Pászti F., García G., Fernández-Jiménez M.T. and Agulló F., 2004, First measurements with the Madrid 5 MV tandem accelerator. *Nuclear Instruments and Methods in Physics Research B*, 219-220, 400-404.

Doremus R.H., 1994, *Glass Science*, John Wiley & Sons, New York.

Farbman I., Levi O. and Efrima S., 1992, Optical response of concentrated colloids of coinage metals in the near-ultraviolet, visible and infrared regions. *Journal of Chemical Physics*, 96, 6477-6485.

Gutierrez P.C., Pradell T., Molera J., Smith A.D. Climent-Font A. and Tite M.S., 2010, Color and golden shine of silver islamic lustre. *Journal of the American Ceramic Society*, 93(8), 2320-2328.

Kreibig U. and Vollmer M., 1995, *Optical Properties of metal clusters*, Springer Series in Materials Science 25, Springer Verlag, Berlin.

Mayer M., 1997, *SIMNRA User's Guide*. at <http://www.rzg.mpg.de>.

Molera J., Mesquida M., Pérez-Arantegui J., Pradell T. and Vendrell M., 2001, Lustre receipts from a medieval workshop of Paterna. *Archaeometry*, 43, 455-460.

Molera J., Bayés C., Roura P., Crespo D. and Pradell T., 2007, Key parameters in the production of medieval luster colors and shines. *Journal of the American Ceramic Society*, 90, 2245-2254.

Padovani S., Sada C., Mazzoldi P., Brunetti B., Borgia I., Sgamellotti A., Giullvi A., D'Acapito F. and Battaglin G., 2003, Copper in glazes of Renaissance luster pottery: nanoparticles, ions, and local environment. *Journal of Applied Physics*, 93, 10058-10063.

Pérez-Arantegui J., Molera J., Larrea A., Pradell T., Vendrell M., Borgia I., Brunetti B.G., Cariati F., Fermo P., Mellini M., Sgamellotti A. and Viti C., 2001, Luster Pottery from the Thirteenth Century to the Sixteenth Century: A Nanostructured Thin Metallic Film. *Journal of the American Ceramic Society*, 84, 442-446.

Polvorinos del Rio A., Castaing J., Roehrs S., Vallejo-Triano A. and Escudero-Aranda J., 2008, Estudio arqueométrico de loza dorada de Madinat al'Zahra, Córdoba. *Cuadernos de Madinat al-Zahra*, 6, 165-179.

Pradell T., Molera J., Roque J., Smith A.D., Crespo D., Pantos E. and Vendrell M., 2005, Ionic-exchange mechanism in the formation of medieval luster decorations. *Journal of the American Ceramic Society*, 88, 1281-1289.

Pradell T., Climent-Font A., Molera J., Zucchiatti A., Ynsa M.D., Roura P. and Crespo D., 2007, Metallic and non-metallic shine in luster: an elastic ion backscattering study. *Journal of Applied Physics*, 101, 1035181-1035188.

Pradell T., Molera J., Smith A.D. and Tite M.S., 2008a, The invention of lustre: Iraq 9$^{th}$ and 10$^{th}$ centuries AD. *Journal of Archaeological Science*, 35, 1201-1215.

Pradell T., Molera J., Smith A.D. and Tite M.S., 2008b, Early Islamic lustre from Egypt, Syria and Iran (10$^{th}$ to 13$^{th}$ century AD). *Journal of Archaeological Science*, 35, 2649-2662.

Quinten M. and Kreibig U., 1993, Absorption and elastic scattering of light by particle aggregates. *Applied Optics*, 32, 6173-6183.

Reillon V., 2008, Caractérisation et modélisation des propriétés optiques des ceramiques lustrées. PhD Thesis, Université Pierre et Marie Curie–Paris 6, Paris, at <http://tel.archives-ouvertes.fr/tel-00374766/fr>.

Reillon V. and Berthier S., 2006, Modelization of the optical and colorimetric properties of lustered ceramics. *Applied Physics A*, 83, 257-265.

## ACKNOWLEDGEMENTS

T. Pradell would like to thank the Ashmolean Museum for providing the Islamic lustre samples. This project is funded by CyCIT grant MAT2010-20129-C02-01 and Generalitat de Catalunya grant 2009SGR01225 and Judit Molera's research by grant 2009SGR1251. P. C. Gutiérrez-Neira, is indebted to the Ministerio de Economía y Competitividad, Spain and to the European Social Fund for the economical support by a JAE-Doc contract.

# 7

# Lessons from the Elephant's Child: Questioning ancient ceramics

**Clive R. Orton**
UCL Institute of Archaeology, 31–34 Gordon Square, London, WC1H 0PY, UK

**Abstract** - This paper presents a brief overview of the present state of the scientific study of ancient ceramics, focussed through six key questions – what, why, when, how, where and who. The relative importance of each question is assessed by means of a brief literature search. The overview is preceded by a summary account of developments over the last 50 years or so, and is followed by a discussion of the role of sampling in such studies, and of the need for good communication between the field and the laboratory.

## 1. Introduction

While looking for a way into the session topic of "Methodological developments" for the tenth anniversary European Meeting on Ancient Ceramics (EMAC'09), I re-encountered Rudyard Kipling's Just-So Story of the Elephant's Child. The Elephant's Child, you may remember, was characterised by his "insatiable curiosity", and this seemed to me to be a good paradigm for our work. However, we need more focus for our curiosity, and this was provided by a rhyme in this story:

> "I keep six honest serving men
> (They taught me all I knew);
> Their names are What and Why and When
> And How and Where and Who."

This rhyme provides a useful alternative typology for our questions and hence our methods – we can group them according to the questions that they seek to answer. We can then go further and examine the progress made in each area, and whether there have been trends in the popularity of the questions. My feelings before I started to look for evidence was that there had been a considerable trend during my working career, from 'when' questions, through 'where' and 'how', leading perhaps to 'why', so I made a very selective literature search to see if this was in fact the case.

It seems useful to follow up this overview with an examination of the applicability of our methods – which methods can be applied to all relevant ceramics, and which (for reasons of cost or time) can only be applied to a sample? A more difficult, but perhaps more important, question, is whether they should be applied universally, or whether there are good reasons for sampling, over and above sheer necessity? Finally, there is the link back from sampling to interpretation, a link which can easily be overlooked or over-simplified, and what this implies for communication between different branches of our discipline.

## 2. Historical perspective

When I started working on archaeological ceramics in the late 1960s and 1970s, pottery tended to be seen primarily as dating evidence, at least by field archaeologists and at least for the historic periods. Paul Tyers once characterised this view as "never mind all that trendy tagspeak stuff, just give me the dates"; the fact it was contemporary with the Theoretical Archaeology Group (founded in 1979) shows the persistence of this view, at least in some quarters. Yes, we had radiocarbon dating, which seemed rather vague for the historic periods, yes, we had coins, which had an annoying habit of being residual (one that I remember was residual by 1000 years!), and we sort of knew about dendrochronology, but no-one in London had done anything useful with it yet (that was to come in the late 1970s and the 1980s). So "when" was the predominant question, and "give me the dates" was the order of the day.

Paradoxically, this position was already being undermined by work that had been done in the 1950s. Anna Shephard (1956) showed (amongst many other things) that pottery, and in particular coarse pottery, had moved around more than people had thought (my simplified paraphrase), thus making "where" a more potent but more difficult question. The methodology was provided at one level by thin-sectioning and low-power microscopy, popularised in the UK by Peacock (1977) and at another level by the introduction in 1957 of Instrumental Neutron Activation Analysis (INAA) to archaeological ceramics (Speakman and Glascock 2007).

Along with this, and again owing much to Anna Shephard, was a new emphasis on ceramic technology and production – the "how" question. Clues to the production processes were sought, often involving experimental archaeology (for example, Mayes 1961) as a way of replicating possible conditions of preparation and firing. It also became clear that sourcing the temper, glaze, and paint was as important as the raw clay itself. The "how" question was broadened out from "how was it made" to "how was production organised", with for example Peacock's (1982) work on modes of production and several studies of craft specialisation (e.g. Costin 1991; Feinman 1999; Hagstrum 1985).

Studies of the function of pottery, which had traditionally been based on the shapes of vessels and the uses that such shapes might imply, received a boost from the introduction of organic residue analysis on the 1950s (Evershed 2008). Here was a more objective way of asking "what (for)" than simply looking at a pot and what might feasibly be done with it.

This was all brought together in a holistic 'cradle to grave' approach by the 'Ceramic Ecology' movement of the 1960s and 1970s (Matson 1965), which sought to track the lifetime (trajectory) of pottery from, literally, dust to dust.

Until the 'post-processual' movement of the 1980s and later, potting had been seen as a very rational activity – pots were made to fulfil certain functions, and were designed to (for example) resist thermal shock, or absorb (or not absorb) liquids (Tite 2008, 223–224). One can almost see the potter setting out with his 'design specification' to create the perfect functional vessel. This was challenged by the realisation that some pots were not actually that good at their supposed functions (for example, badly-attached handles on drinking jugs, Orton 1982, 82) and that other factors, such as their symbolic importance, or craft conservatism, might be at play. With this came the recognition that 'progress' was not inevitable, and that the pattern was sometimes one of regression (*ibid.*). The characteristic question of this phase is "why", including "why not" (e.g. why didn't something happen, such as the establishment of a stoneware industry in London before the late 17th century).

There is in some sense a logical progression to these questions – we cannot ask "why" until we have a fair grasp of the "when" and "where" – but at the same time there are theoretically-based trends in archaeological questioning.

## 3. Where are we now?

To try to answer this question, I have looked at the abstracts of the other papers in this session, and in recent volumes of two journals dedicated to our methodology: *Journal of Archaeological Science* (for 2005 and 2006) and *Archaeometry* (for 2007, 2008 and part of 2009). As usual with typologies, it is not as easy as it seems. Some papers set out to answer more than one type of question, while others answer one question in order to provide a firmer basis from which to answer another. So some element of double-counting is inevitable, and my number of 'questions asked' is more than the number of papers included in the study. The outcome is shown in Table 1.

**Table 1.** Summary of the main questions addressed by papers in the EMAC'09 session on Methodological Developments, the *Journal of Archaeological Science* (for 2005 and 2006) and *Archaeometry* (for 2007–2009).

| | EMAC | J. Arch. Sci | Archaeometry | Total |
|---|---|---|---|---|
| What | 1 | 2 | | 3 |
| What for | | 5 | 4 | 9 |
| Why | 2 | | 6 | 8 |
| When | | 2 | 1 | 3 |
| How | 3 | 3 | 9 | 15 |
| Where | 2 | 15 | 15 | 32 |
| Total | 8 | 27 | 35 | 70 |

### *Where?*

The primary question is clearly "where" (over 40%), in which the 50-year dominance of INAA (Speakman and Glascock 2007) is being challenged by a growing range of analytical techniques (Tite 2008). In my brief review, INAA is still the front-runner, with ICP-MS and ICP-OES and various SEM-based techniques leading the challenge. It is not my place to discuss the relative merits of various techniques, but I will try to draw out some more general points. First, there is an inherent conservatism in the choice of technique, in that large existing databases, particularly of INAA data (Speakman and Glascock 2007) and PIXE data (Grave *et al.* 2005) mean that users of new techniques will be, for a while at least, 'on their own', as their data may not be strictly comparable with earlier data, this is unfortunate as archaeology is in the business of comparison. This mirrors a larger and unresolved problem of archaeological data: changes (e.g. in terminology) have to occur if the discipline is to make progress, but such changes create 'fault lines' between datasets, making the task of comparison that much more difficult.

Another point is the increased emphasis on the portability of equipment (Morgenstein and Redmount 2005). This may be in part a response to political solutions, making the removal of material from its country of origin more difficult. "If the sherd won't go to the equipment, then the equipment must go to the sherd." It is clear that, as scientists, we must see our work in the wider social, economic and political context, to judge what is possible and where methodological advances need to be made.

### *How?*

The second most frequent question is "how" (about 25%). The long-running interest in firing conditions (e.g. temperature, heating and cooling rates, atmosphere) continues, supported by a growing range of analytical techniques (Botti *et al.* 2006; Capel *et al.* 2006; Montana *et al.* 2007). Sometimes the question is more specific, relating to the achievement of a particular effect, for example blue-and white decoration (Wood *et al.* 2007), aspects of glazing (Fortina *et al.* 2008; Pace *et al.* 2008), sgraffito decoration (Brianese *et al.* 2008) and red/black gloss (Walton *et al.* 2009). It's interesting to note that, despite our methodological advances, the production of lustre ware remains something of a mystery (Roqué *et al.* 2007). The links between the selection of raw materials and the technology of production still features seriously, showing that this approach is not yet exhausted.

Broader questions can be seen in the use of the *chaine opératoire* concept to examine production sequences (Hunt 2014) and in the discussion of how potting skills are acquired and transmitted. These questions lead on to issues of how production is organised – is it, for example, centralised or decentralised (Vaughn *et al.* 2006), and, one might add, what do those terms actually mean? The topic of craft specialisation is clearly still an issue (e.g. Orton 2012), even if not represented in the sample. Finally, pottery can be used as a way to examine how units of measurements were perceived and used in various societies (Zapassky *et al.* 2006).

### What (for)?

This question accounts for over 10% of the total questions asked. The use of organic residue analysis dominates, as one might expect from this maturing sub-discipline (Evershed 2008). The majority of the current work seems to focus on dairying, from its origins in the Neolithic (Copley *et al.* 2005a; Spangenberg *et al.* 2006) to its continuation in the Bronze and Iron Ages (Copley *et al.* 2005b; 2005c). Other residues recently identified include fish products (Ribechini *et al.* 2009) and aromatic or medicinal substances (Marangou and Stern 2009). More general studies, often based on form, continue alongside (Skibo 1992; Tite 2008, 228).

### Why?

This question too accounts for just over 10% of the total questions. Like the 'how' question, it divides into detailed questions of technology and broader questions of organisation. The former concentrate on issues of ceramic performance and the fitness of ceramics for certain functions (Sillar and Tite 2000). The choice of particular temper recipes are important (Montana *et al.* 2007; Müller *et al.* 2014), as are choices of surface treatment (Reber and Hart 2008; Stern *et al.* 2008). Changes over time that are observed in the use of raw materials, whether for body or glaze, also raise questions of why that change occurred (Martineau *et al.* 2007).

### When?

Despite popular belief, this question is well down our list, at less than 5% of the total. Pottery is, of course, of immense value for dating deposits and events, but the evidence does not commonly come from scientific analyses. The exception to this rule comprises the various forms of luminescence dating, which have been with us for over 50 years (Wintle 2008). As one might expect in a sub-discipline of this maturity, work is mainly concerned with 'dotting the i's and crossing the t's', for example by comparing the outcomes of TL on ceramics and sediments (Yang *et al.* 2005) or by studying the effects of burial conditions (Zacharias *et al.* 2005). However, this approach remains both expensive and complicated.

A very promising new technique, rehydroxylation, has recently been announced (Wilson *et al.* 2009). Although still in its research phase, it appears to promise dates to within a few years for ceramics (so far, apparently, only brick) of up to 2000 years old, with the possibility of being able to go back 10,000 years. The archaeologist in me says "wonderful", but the statistician in me says "it's too good to be true", and asks questions about inherent variability and error modelling.

### What?

Here, I have rather arbitrarily redefined "what" to cover techniques of automatic capture of pottery forms and decoration by various types of 3D scanning, separating it from the "what for" questions mentioned above. On this basis, this question constitutes less than 5% of the total.

There is considerable interest in the use of 3D scanning to capture the shapes of pots, sherds and other artefacts (such as flints), with a view to replacing and refining hand-drawing (e.g. Saragusti *et al.* 2005 amongst many others). My experience at Novgorod (Orton and Malygin 2001) suggests that hand-drawing can be unreliable, and that even drawings made by skilled ceramicists may not be strictly comparable. One major problem seems to be in assessing the angle that a sherd makes with the horizontal; this is particularly serious because a small angular difference at the rim becomes magnifies the further one proceeds down a rim sherd. There is no doubt that automatic procedures can improve and standardise this process.

Once captured, the data can lead us into very interesting questions of shape analysis, and the discovery of real but subtle differences that would escape detection by the naked eye. One such example is the study of variation in 'torpedo' pots in Iron Age Israel (Gilboa *et al.* 2004).

### Who?

This question did not feature in the survey, but is included here for the sake of completeness. Traditionally regarded as the territory of the historical archaeologist, dealing with makers' marks or similar identifiers, it could nevertheless be opened up to 'anonymous' potters if the shape analysis mentioned above lives up to its promise. We may not be able to give potters names, but we may be able to distinguish their products.

## 4. Where might we go from here?

I have no track record for predicting technological development in archaeology, so I will not try here. It seems likely that ICP-MS will continue to replace NAA for provenance studies, although NAA will not disappear altogether. To make best use of the large and growing bodies at data, ways to integrate them should be sought, though this is likely to be a very difficult task.

Rehydroxylation will need to be watched closely, to see if its use can be extended from brick (for which it seems to be ideally suited) to various types of pottery (for which it may prove to be less well suited). Whether it will be able to cope with a wide range of bodies and surface finishes will be an interesting challenge.

Automatic capture of shape data is likely to expand, and we may see specialist service organisations set up to provide it to the discipline as a whole. Certainly, its use will have to be organised to ensure maximum and continuous throughput for what is likely to remain very expensive equipment. I hope that archaeologists will be able to ask the sorts of questions that will make the generation of data on this scale worthwhile.

## 5. Sampling and selection

If I have hinted that future issues may be organisational rather than technological, that was deliberate. It is clear that we have the ability and the equipment to generate and store vast amounts of data, at relatively low cost. However, "Science is built up of facts, as a house is with stones. But a collection of facts is no more a science than a heap of stones is a house" (Poincaré 1902). This raises the question – do we need all the data that we could (potentially) collect? Are some data now so cheap that we could (if we wish) collect them from *all* archaeological ceramics? If "no", how do we select, and if "yes", do we need to do so? The question splits into two parts: selection (which variables do we measure?) and sampling (on which objects do we measure them?).

I do not have the data, either in terms of cost per technique or number of units (sherds) to answer quantitatively, so I shall have to stick to general principles. My guess is that the answer will almost always be "no", although it has been claimed that total capture of shape data is now possible, at least on a national basis (Smilansky, *pers. comm.*). Two factors that militate against total data capture are (1) the opportunity costs – the true cost of doing something is often the loss of other activities that it precludes us from doing (whether for reasons of time or money) and (2) the law of diminishing returns – in a statistical sense, to get twice as much information requires four times as much data, and so on (it is a 'square law'). Further, collecting a relatively small amount of data, well, may lead to a more satisfactory outcome than collecting more data, badly. For such reasons, I usually advocate the collection of enough data (but no more) to answer the question(s) in hand, while admitting that this may cause severe problems in the formulation of questions in an appropriate way.

This leads on to the widely misunderstood topic of sample design. The first issue is that of sample size. There is a widespread myth that what matters when it comes to sample size is the percentage that it is of the population (know as the 'sampling fraction'). In fact, the crucial factor is the absolute size of the sample, as this defines the standard deviation of the outcome (the 'margin of error'). Second, we need to look beyond simple sampling to (slightly) more complicated sample designs, such as stratified sampling, which can yield serious improvements in performance (Orton 2000). To take full advantage of such schemes, it is likely that one will need to use different sampling fractions in different strata (e.g. types), which may be counter-intuitive to some but which follow logically from the first point.

Perhaps most important is the point that the sample must be representative of its population. Neither a large sample size nor a sophisticated design can compensate for an unrepresentative sample (it can even make matters worse). This is mainly a matter of exactly how the sample is chosen – asking a field archaeologist to choose 'typical' examples may quite innocently open the way to all sorts of bias. In my experience (based on asking students to select 'typical' objects from a collection), a 'typical' sample tends to cluster too closely round the 'average', and not to encompass the full range of variation present in the population. It is surprisingly difficult to achieve a truly representative sample, and a formal random process may be the only practical way.

## 6. Communication and interpretation

It follows from this that there must be good communication between the field and the laboratory if sensible interpretations are to be made. For example, in the Novgorod project I had to select a very small sample of sherds to be sent for thin-sectioning. Some were intended to be 'typical', representing many thousands of sherds, while others were rarer, or even 'oddities', selected simply because they looked different to me, and I wanted to know if they were 'really' (petrographically) different, or had perhaps undergone different firing conditions. Unfortunately, I did not explain this properly when submitting the samples, and the report that came back treated each example as of equal weight and importance. Communication needs to be two-way: the fieldworker needs to explain the process and rationale behind the selection of the sample, and the labworker needs to take this on board and to reflect them fully in their report.

## 7. Conclusion

We have much to learn from the Elephant's Child: we must focus our 'insatiable curiosity' around one or more specific questions, and consider carefully how much data, and of what sorts, we need in order to be able to answer them.

## References

Brianese N., Casellato U., Fenzi F., Mendera M., Messiga B., Nodari L., Riccardi M.P. and Vicato P.A., 2008, Incised slipware production at Castelfiorentino (Florence): the 'Via Bertini' site (15th–17th centuries). *Archaeometry*, 50, 775–796.

Botti A., Ricci M.A., De Rossi G., Kockelmann W. and Sodo A., 2006, Methodological aspects of SANS and TOF neutron diffraction measurements on pottery: the case of Miseno and Cuma. *Journal of Archaeological Science*, 33, 307–319.

Capel J., Huertas F., Pozzuoli A. and Linares J., 2006, Red ochre decoration in Spanish Neolithic ceramics: a mineralogical and technological study. *Journal of Archaeological Science*, 33, 1157–1166.

Copley M.S., Berstan R., Mukherjee A.J., Dudd S.N., Straker V., Payne S. and Evershed R.P., 2005a, Dairying in antiquity. III. Evidence from absorbed lipid residues dating to the British Neolithic. *Journal of Archaeological Science*, 32, 523–546.

Copley M.S., Berstan R., Straker V., Payne S. and Evershed R.P., 2005b, Dairying in antiquity. II. Evidence from absorbed lipid residues dating to the British Bronze Age. *Journal of Archaeological Science*, 32, 505–521.

Copley M.S., Berstan R., Dudd S.N., Straker V., Payne S. and Evershed R.P., 2005c, Dairying in antiquity. I. Evidence from absorbed lipid residues dating to the British Iron Age. *Journal of Archaeological Science*, 32, 485–503.

Costin C.L., 1991, Craft specialization: issues in defining, documenting, and explaining the organisation of production. In *Archaeological Method and Theory*, (ed. M.B. Schiffer), 1–56, vol. 3., Academic Press, New York.

Evershed R.P., 2008, Organic residue analysis in archaeology: the archaeological biomarker revolution. *Archaeometry*, 50, 895–924.

Feinman G.M., 1999, Rethinking our assumptions: economic specialization at the household scale in ancient Ejutla, Oaxaca, Mexico. In *Pottery and People: a dynamic interaction*, (eds. J.M. Skibo and G.M. Feinman), 81–98, University of Utah Press, Salt Lake City.

Fortina C., Memmi Turbanti I. and Grassi F., 2008, Glazed ceramic manufacturing in southern Tuscany (Italy): evidence of technological continuity throughout the medieval period (10th–14th centuries). *Archaeometry*, 50, 30–47.

Gilboa A., Karasik A., Sharon I. and Smilansky U., 2004, Towards computerized typology and classification of ceramics. *Journal of Archaeological Science*, 31, 681–694.

Grave P., Lisle L. and Maccheroni M., 2005, Multivariate comparisons of ICP-OES and PIXE/PIGE analysis of east Asian storage jars. *Journal of Archaeological Science*, 32, 885–896.

Hagstrum M.B., 1985, Measuring prehistoric ceramic craft specialization: a test case in the American Southwest. *Journal of Field Archaeology*, 12, 65–75.

Hunt, A.W., 2014. Assyrian Palace Ware Definition and *Chaîne Opératoire*: Preliminary Results from Nineveh, Nimrud, and Aššur, this volume.

Marangou C. and Stern B., 2009, Neolithic zoomorphic vessels from eastern Macedonia, Greece: issues of function. *Archaeometry*, 51, 397–412.

Martineau R., Walter-Simonnet A.-V., Grobéty B. and Buatier M., 2007, Clay resources and technical choices for neolithic pottery (Chalain, Jura, France): chemical, mineralogical and grain-size analyses. *Archaeometry*, 49, 23–52.

Matson F.R., 1965, Ceramic ecology: an approach to the study of the early cultures of the Near East. In *Ceramics and man*, Viking Fund Publications in Anthropology 41, (ed. F.R. Matson), 202–217, Aldine, Chicago.

Mayes P., 1961, The firing of a pottery kiln of a Romano-British type at Boston, Lincs. *Archaeometry*, 4, 4–30.

Montana G., Fabbri B., Santoro S., Gualtieri S., Iliopoulos I., Guidicci G. and Mini S., 2007, Pantellerian Ware: a comprehensive archaeometric review. *Archaeometry*, 49, 455–481.

Morgenstein M. and Redmount C.A., 2005, Using portable energy dispersive X-ray fluorescence (EDXRF) analysis for on-site study of ceramic sherds at El Hibeh, Egypt. *Journal of Archaeological Science*, 32, 1613–1623.

Müller N.S., Kilikoglou V., Day P.M. and Vekinis G., 2014. Thermal shock resistance of tempered archaeological Ceramics, this volume.

Orton C.R., 1982, The excavation of a late medieval/transitional pottery kiln at Cheam, Surrey. *Surrey Archaeological Collections*, 73, 49–92.

Orton C.R., 2000, *Sampling in Archaeology*, Cambridge University Press, Cambridge.

Orton C., 2012, Indicators of craft specialisation in medieval ceramics from North-West Russia. In *The Archaeology of Medieval Novgorod in Context: Studies in Centre/Periphery Relations*, (eds. M. Brisbane, N. Makarov and E. Nosov), 435–447, Oxbow, Oxford.

Orton C.R. and Malygin P., 2001, Approaches to a large, relatively uniform assemblage of ceramics. In *Novgorod: The Archaeology of the Medieval Town and its Hinterland*, (eds. M. Brisbane and D. Gaimster), 59–66, British Museum Occasional Paper 141, London.

Pace M., Bianco Prevot A. and Mirti P., 2008, The technology of production of Sasanian glazed pottery from Veh Ardašõ r (central Iraq). *Archaeometry*, 50, 591–605.

Peacock D.P.S., 1977, Ceramics in Roman and medieval archaeology. In *Pottery in early commerce*, (ed. D.P.S. Peacock), 21–34, Academic Press, London.

Peacock D.P.S., 1982, *Pottery in the Roman World: an Ethnoarchaeological Approach*, Longmans, London.

Poincaré J.H., 1902, *La Science et l'hypothèse*.

Reber E.A. and Hart J.P., 2008, Pine resins and pottery sealing: analysis of absorbed and visible pottery residues from central New York State. *Archaeometry*, 50, 999–1017.

Ribechini E., Colombini M.P., Giachi G., Midugno F. and Pallecchi P., 2009, A multi-analytical approach for the characterization of commodities in a ceramic jar from Antinoe (Egypt). *Archaeometry*, 51, 480–494.

Roqué J., Molera J., Perez-Arantegui J., Calabuig C., Portillo J. and Vendrell-Saz M., 2007, Lustre colour and shine from the olleries xiques workshop in Paterna (Spain), 13th century AD: nanostructure, chemical composition and annealing conditions. *Archaeometry*, 49, 511–528.

Saragusti I., Karasik A., Sharon I. and Smilansky U., 2005, Quantitative analysis of shape attributes based on contours and section profiles in artifact analysis. *Journal of Archaeological Science*, 32, 841–853.

Shephard A.O., 1956, *Ceramics for the archaeologist*, Carnegie Institute of Washington, Washington D.C.

Sillar B. and Tite M.S., 2000, The challenge of 'technological choices' for materials science approaches in archaeology. *Archaeometry*, 42, 2–20.

Skibo J.M., 1992, *Pottery function: a use-alteration perspective*, Plenum Press, New York.

Spangenberg J.E., Jacomet S. and Schibler J., 2006, Chemical analyses of organic residues in archaeological pottery from Arbon Bleiche 3, Switzerland – evidence for dairying in the late Neolithic. *Journal of Archaeological Science*, 33, 1–13.

Speakman R.J. and Glascock M.D., 2007, Acknowledging fifty years of neutron activation analysis in archaeology. *Archaeometry*, 49, 179–183.

Stern B., Connan J., Blakelock E., Jackman R., Coningham R.A.E. and Heron C., 2008, From Susa to Anuradhapura: reconstructing aspects of trade and exchange in bitumen-coated ceramic vessels between Iran and Sri Lanka from the third to the ninth centuries AD. *Archaeometry*, 50, 409–428.

Tite M.S., 2008, Ceramic production, provenance and use – a review. *Archaeometry*, 50, 216–231.

Vaughn K.J., Conlee C.A., Neff H. and Schreiber K., 2006, Ceramic production in ancient Nasca: provenance analysis of pottery from the Early Nasca and Tiza cultures through INAA. *Journal of Archaeological Science*, 33, 681–689.

Walton M.S., Doehne E., Trentelman K., Chiari G., Maish J. and Buxbaum A., 2009, Characterization of coral red slips on Greek Attic pottery. *Archaeometry*, 51, 383–396.

Wilson M.A., Carter M.A., Hall C., Hoff W.D., Ince C., Savage S.D., Mckay B. and Betts I.M., 2009, Dating fired-clay ceramics using long-term power law rehydroxylation kinetics. *Proceedings of the Royal Society Series A*, 465, 2407–2415.

Wintle A.J., 2008. Fifty years of luminescence dating. *Archaeometry*, 50, 276–312.

Wood N., Tite M.S., Doherty C. and Gilmore B., 2007, A technological examination of ninth-tenth century AD Abbasid blue-and-white ware from Iraq, and its comparison with eighth century AD Chinese blue-and-white Sancai ware. *Archaeometry*, 49, 665–684.

Yang X.-Y., Kadereit A., Wagner G.A., Wagner I. and Zhang J.-Z., 2005, TL and IRSL dating of Jiahu relics and sediments: clue of $7^{th}$ millennium BC civilization in central China. *Journal of Archaeological Science*, 32, 1045–1051.

Zacharias N., Buxedai Garrigos J., Mommsen H., Schwedt A. and Kilikoglou V., 2005, Implications of burial alterations on luminescence dating of archaeological ceramics. *Journal of Archaeological Science*, 32, 49–57.

Zapassky E., Finkelstein I. and Benenson I., 2006, Ancient standards of volume: negevite Iron Age pottery (Israel) as a case study in 3D modelling. *Journal of Archaeological Science*, 33, 1734–1743.

8

# Pots as signals: Explaining the enigma of long-distance ceramic exchange

**Hector Neff**
Department of Anthropology and IIRMES, California State University Long Beach, 1250 Bellflower Blvd.
Long Beach, CA90815, hector.neff@csulb.edu

**Abstract** - Long-distance transport of ceramic vessels entails high costs, especially in pre-modern situations where motorised or animal-borne transport are unavailable. Thus, in many parts of the world, including Mesoamerica, exchange of ceramics beyond more than a few tens of kilometres is difficult to understand. In this paper, I first discuss ceramic provenance results from Mesoamerica that demonstrate unequivocally that, contrary to expectations, long-distance ceramic exchange was, in fact, common.

One way to approach the enigma of long-distance ceramic transport is to build on the evolutionary theory of costly signalling. I propose that pots sometimes serve as signals that convey information about hidden qualities of those who make the pots, export them, or import them. If pots serve as signals, design considerations differ from considerations involved in producing pots that serve primarily as tools. In particular, there must be a means for ensuring signal reliability, because without this, potential receivers of the signal have no incentive to pay attention.

One type of information that can be encoded reliably in ceramics is information about control of resources. Signallers benefit from dissemination of this information because of the deference shown by others to rich and powerful individuals. Receivers benefit because they gain accurate information that can help them rank individuals with whom they interact. The reliability of a signal of high resource access can be guaranteed by building high cost into a pot, and one way to do this is to transport the pot a long distance from its origin. As noted, costs of transport rise dramatically with distance, especially in the absence of water-borne, animal, or motorised transport, so if a pot is recognisably foreign, it is guaranteed to be a high-cost vessel and thus a reliable signal of an individual's control of or access to resources.

Ceramic technology, however, is often fairly simple, and convincing fakes are often easy to produce. Thus, the reliability of foreign pots as signals of access to resources is easily undermined. This leads to a hypothesis that is testable in the archaeological record: long-distance exchange of ceramics should in general be short-lived, but technologies that are more difficult to duplicate should be associated with longer-lived exchange systems. Comparison of long-distance ceramic exchange patterns in Early Formative, Classic, and Early Postclassic Mesoamerica appear to accord with this hypothesis, and the hypothesis also serves as a guide for future research.

## 1. Introduction

In a seminal 1983 article entitled "Pots as Tools", David Braun argued that if pots are tools for cooking, carrying, storage, and serving, pottery-making practices should be shaped primarily by consideration of *how to ensure adequate performance with minimum cost*. Other things being equal, water storage jars will be designed to keep water cool and clean, cooking vessels will be resistant to thermal stress, and so on. Braun (1983, 111–2) argues that adaptive engineering design emerges through a selective process in which practices that yield favourable results increase in frequency within a population of potters. This has been a very productive model, as Braun (1983), Feathers (1989), Schiffer et al. (1994, 2001), and others (Steponaitis 1984; Bronitsky and Hamer 1986; Bronitsky 1989; Skibo *et al.* 1989; O'Brien *et al.* 1994; Pierce 1999) have shown in a variety of studies of how the design of tool-pots is shaped by conditions of use.

Braun (1983) did not discuss ceramic exchange within his 'pots-as-tools' framework, but a few additional considerations lead to some clear predictions. To begin with, weathering processes that produce ceramic raw materials are ongoing everywhere on the planet. As a result, primary clays are available on the surface within a short distance of almost any place on earth. Erosion and re-deposition create secondary clay deposits, which may be excellent raw material sources because of the natural size-separation process that takes place during transport. In short, while the suitability of clays and tempers for ceramic production obviously varies, raw material availability is a very small if not completely insignificant limitation on where ceramic production can take place.

Not only are clays ubiquitous, but ceramic technology, while perhaps not something everyone can master, is the oldest additive technology invented by humans. People have been using clay and firing it for more than 20,000 years, and ceramic container technology was invented and refined independently in numerous places all over the world within the past 8000 years or so. Thus, from a technological perspective, there is no reason to think that ceramic production should be localised either, or that people anywhere could not make pots to fulfil all of their needs.

Ceramics are also heavy and bulky and can be very difficult to transport depending on specifics of their size and shape. Our expression 'bull in a china shop' alludes to the fragility of ceramics. And the historical record of ceramic breakage is evident on archaeological sites all over the world, where the dominant artefact class is often broken pottery. Weight, bulk, and fragility dramatically increase the transport cost of ceramics.

The fact that the cost of tool-pots increases rapidly as distance from the manufacturing spot increases places severe constraints on the geographic scope of ceramic circulation. This does not preclude circulation of ceramics in regional exchange systems altogether, of course. Below some threshold distance, given some level of environmental stability, patterns of differential comparative advantage can be expected to arise, such that people in different locations would tend to specialise in different productive activities (Neff and Larson 1997). In the Postclassic Basin of Mexico, for instance, consistently higher rainfall in the southern Basin coupled with the development of highly productive *chinampa* agriculture gave the southern Basin an edge in agricultural production, so that ceramic production ultimately declined in this region, as production of Aztec Black-on-Orange serving vessels became concentrated in other centres, less favourably situated with respect to agricultural production (Nichols *et al.* 2002; Neff and Hodge 2008). Abbott (2009) makes a similar argument about ceramic specialisation and regional exchange among the Classic-period Hohokam of the Southwest United States. Ethnographic observations (e.g., Arnold 1985) also show that ceramics often circulate within regional exchange systems.[1]

Above some threshold distance, however, the cost of importing a pot will exceed the opportunity cost of making one locally. The specific threshold distance beyond which ceramic transport would be unexpected would depend on local geographic and economic conditions, such as presence or absence of animal transport and possibility of water-borne transport. Additionally, superior performance of imports can extend the threshold distance. At some point, however, any performance advantage will be outweighed by the high cost of transportation. Thus, much more enigmatic than circulation of ceramics within regional exchange systems are cases of ceramics transported across great distances by human muscle power.[2]

In Mesoamerica (Fig. 1), where there is no animal transport and canoe transport is largely confined to the coasts, threshold distances for ceramic transport should be comparatively small. From this perspective, the numerous cases of interregional ceramic exchange in Mesoamerica, some of which are summarised in the following section, stand out as particularly puzzling and cry out for an explanatory framework other than 'pots as tools'. Later sections of this paper develop and apply a 'pots as signals' approach to long-distance ceramic circulation that relies on the evolutionary theory of costly signalling.

## 2. Evidence of long-distance ceramic exchange in prehistoric Mesoamerica

Some of the strongest evidence for long-distance ceramic exchange in Mesoamerica has accumulated through provenance studies based on instrumental neutron activation analysis (INAA) undertaken since the 1960s. The basic technique of standard-comparator INAA was developed by Perlman and Asaro (1969, 1971) at Lawrence-Berkeley Laboratory and by Harbottle and Sayre (Sayre *et al.* 1971; Harbottle 1976) at Brookhaven National Laboratory (BNL). When the national laboratory programs began to wind down in the late 1970s, university research reactor-based programs, especially the one at University of Missouri (MURR; Glascock 1992), and a collaborative program between the Smithsonian and National Institute of Standards and Technology (NIST) stepped in to fill the vacuum. The examples discussed below are based mainly on data generated at BNL, MURR, and NIST.

### *Early Formative Ceramic Exchange*

One of the most surprising cases of Mesoamerican long-distance ceramic exchange involves some of the earliest pottery made within the region. Pottery appears soon after 1800 BC in the lowland region stretching from the Gulf Coast down along the Pacific coast of southern Mexico and Guatemala and into El Salvador (e.g., Arroyo 1995; Blake *et al.* 1995). By 1400 BC, distinctive carved-incised grey (Fig. 2) and fine white vessels were being produced on the Gulf Coast, in the vicinity of the growing Olmec centre of San Lorenzo, epicentre of the famed Olmec sculptural style (Blomster *et al.* 2005). Around the same time, carved-incised grey and fine white pottery also appeared in other regions, from the Pacific coast of southern Mexico to highland Oaxaca and the Basin of Mexico. The obvious question that arises is whether the wide distribution

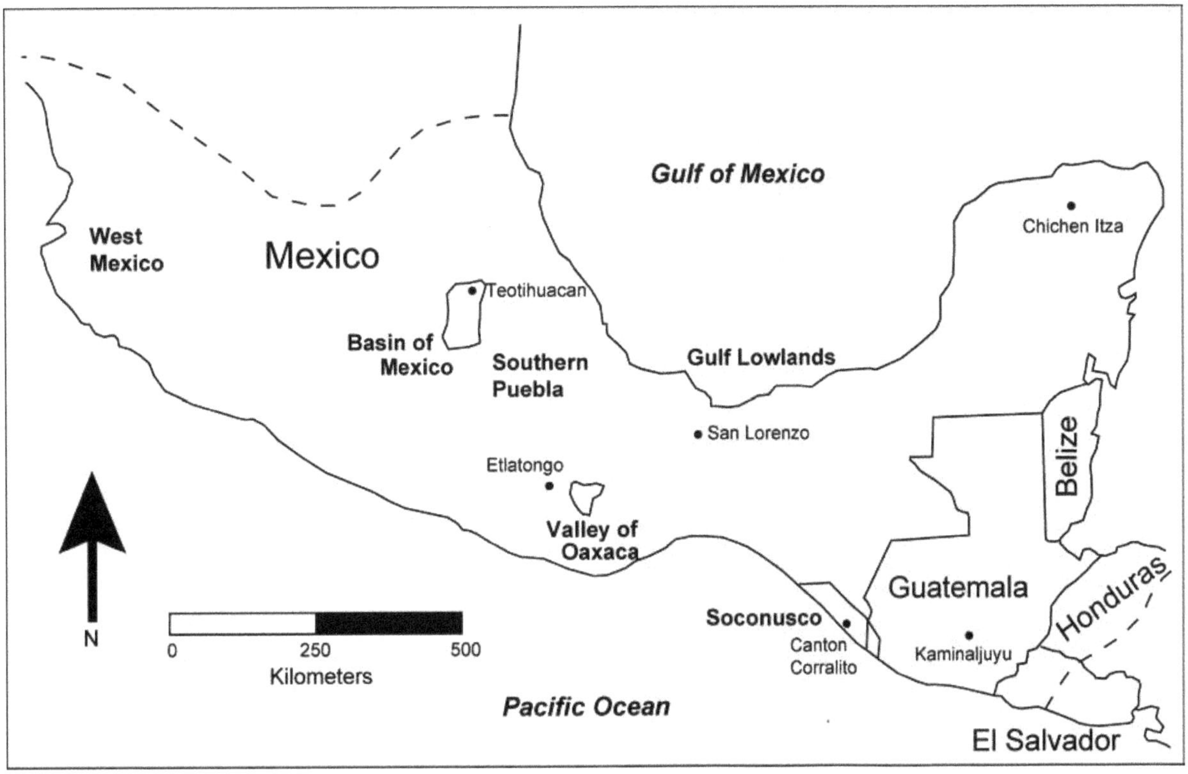


**Figure 1.** Map of Mesoamerica showing locations and regions mentioned in the text.

of similar ceramic vessels originates in wide dissemination of tastes and production practices, or whether there was some component of ceramic exchange involved.

A project initiated by Jeffrey Blomster (Blomster *et al.* 2005) sought to disentangle the movement of goods from the dissemination of tastes and decorative conventions. Samples of pottery from widespread Early Formative centres were analysed in an effort to identify long-distance ceramic exchange. The sampling strategy emphasised vessels exhibiting widely-distributed stylistic elements, but locally distinctive vessels were analysed as well, in order to provide a basis for discriminating local from imported artefacts at the various sites. Raw materials from the Gulf Coast, the eastern Soconusco region, Oaxaca Valley, and the Basin of Mexico were also analysed, in order to provide direct links to geographic locations for these key regions.

Results of the Olmec ceramic provenance project were surprising, even shocking. In the sample of over 700 Early Formative ceramics analysed by INAA, over 80 were determined to have been moved long distances across the Mesoamerican landscape (Blomster *et al.* 2005). Even more surprising, all instances of long-distance ceramic exchange were of vessels made in the vicinity of the Gulf Olmec centre of San Lorenzo and consumed elsewhere. Gulf Coast imports were identified at Canton Corralito, at Etlatongo, in the Valley of Oaxaca, in the Basin of Mexico, and at other Early Formative sites distant from the Olmec heartland.

The results of the Early Formative ceramic provenance project were disputed by anti-Gulf Coast primacy partisans (Flannery *et al.* 2005; Stoltman *et al.* 2005), but the critiques are marred by statistical fallacies and tiny sample sizes that degrade their credibility (Neff *et al.* 2006a, 2006b). Thus, the pattern identified in the INAA study – high volume export of symbolically charged carved grey pottery together with fine white pottery from the Gulf Coast plus undetectable exports from anywhere else – remains intact. That is, not too long after the invention of pottery in Mesoamerica, people were moving this bulky, fragile product over substantial distances and in substantial quantities, almost exclusively in a one-way flow out of the Gulf lowland Olmec heartland.

The 'pots-as-tools' model offers little help toward understanding Early Formative ceramic exchange. As discussed earlier, the expense of long-distance transport of fragile ceramics was considerable in Mesoamerica: Olmec pots exported from San Lorenzo to Canton Corralito might have travelled part of the way by canoe, but pots exported to highland regions such as the Valley of Oaxaca and the Basin of Mexico must have travelled on the backs of humans. Such an expenditure of human labour to procure pots from a long distance cannot be explained by techno-functional superiority: tool-pots for cooking, carrying, storage, and serving were being produced locally in all of the regions sampled, and in several regions local versions of the Olmec carved and incised grey vessels were produced, probably in imitation of the Gulf Coast imports.

### *Teotihuacan-Horizon Tradewares*

The emergence of the Classic-period (200–800 AD) empire centred at Teotihuacan entailed interregional interaction on a geographic scale comparable to that associated with

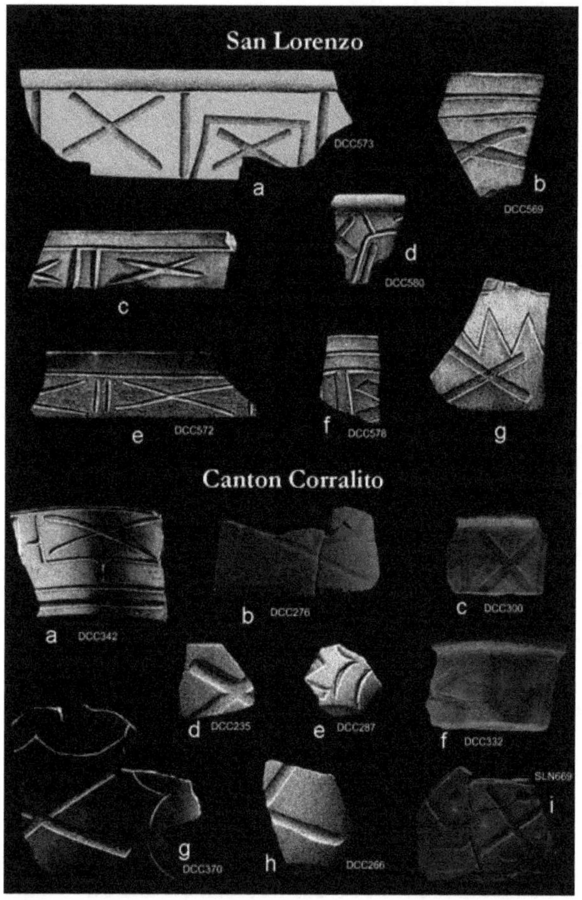


**Figure 2.** Calzadas Carved pottery from San Lorenzo (the Gulf Olmec heartland) and Canton Corralito (Pacific coast, approximately 400 km from the heartland). These illustrations show parallel examples of a motif referred to as St. Andrew's Cross. Courtesy of David Cheetham (2010, Fig. 7.14), with permission.

the Early and Middle Formative Olmec Horizons. The quintessential Teotihuacan tradeware was Thin Orange, a hard-fired ceramic produced in southern Puebla (Rattray and Harbottle 1992). Several less well-known tradewares found at Teotihuacan are most likely derived from the Gulf Coast (Cowgill and Neff 2004). Of particular interest is a fine, light-firing ware found in Early and Middle Classic contexts at Teotihuacan and attributed to a source zone in the southern Gulf Coast region.

Thin Orange and southern-Gulf Coast fine paste ceramics were exported south at least 600 kilometres, to the Pacific coastal plain and southern highlands of Guatemala. Both tradewares stand out within the Early Classic assemblages of this region, first because of their rarity and second because of their technological and decorative contrasts with indigenous Terminal Formative – Early Classic ceramics (Neff 2006). INAA confirms their foreign origin (Bove and Medrano 2003; Cowgill and Neff 2004; Neff 2006). Bove and Medrano (2003) argue that both wares, together with green (central Mexican) obsidian reflect an early stage of Teotihuacan's political interest in the region. As in the case of Olmec ceramic exports, export of Thin Orange and Gulf Coast fine paste ceramics from Mexico to the Guatemalan Pacific coast makes little sense as a response to demand for superior tool-pots made in specialised centres. The expenditure of human labour in transport certainly would have outweighed any performance advantage, especially considering that both wares are dominated by serving vessels (open bowls or cylindrical vessels), which would not be subjected to severe thermal or mechanical stresses.

Also in parallel with the Olmec case, importation of Thin Orange and Gulf coast fine paste ware to coastal Guatemala stimulated efforts to develop local copies. Imitation Thin Orange became Esmeralda Flesh ware, a regionally distributed pottery type of the Early Classic period (Neff et al. 1994) that is similar enough to Thin Orange in form and technology that the two are sometimes mistaken for one another (Kidder et al. 1946, 177; Bove and Medrano 2003; Neff 2006). Copies of Gulf Coast fine paste pottery (originally designated 'proto-Tiquisate' ware by Bove [Bove and Medrano 2003]), meanwhile, became Tiquisate ware, a key diagnostic of the Middle through Terminal Classic period (Chinchilla Mazariegos *et al.* 2009). These Classic-period Guatemalan ceramic industries contrast with earlier ceramic industries on numerous dimensions, including resource choice, regional distribution patterns, style, and technology (Neff *et al.* 1994, 1999; Neff 1995, 2006). As viable local copies became available, importation of Thin Orange and Gulf coast fine paste ceramics ceased, neither appearing in Pacific coastal assemblages post-dating 400 AD (Bove and Medrano 2003).

### *Plumbate*

Another ceramic class that provides an interesting comparison to the Olmec and Teotihuacan cases is Plumbate, a tradeware studied long ago by Anna Shepard (1948). One main variety of Plumbate, called Tohil, is dominated by human, animal (Fig. 3), and deity effigies, some in miniature, and most with a highly lustrous, vitrified surface. Shepard noted the technological uniqueness of the highly lustrous, glazed surface and advanced

**Figure 3.** Tohil Plumbate zoomorphic effigy in the collection of the Museo Nacional de Antropología e Historia, Guatemala. Photograph by the author.

a hypothesis to explain it. She also demonstrated that Plumbate from tombs and offerings from all over Mesoamerica, from Panama to Chichen Itzá and West Mexico, had the same suite of non-plastic tempering materials. The implication was clear: Plumbate was made in a single, restricted source zone and then transported all over lowland and highland Mesoamerica, from the Maya area to the Toltec empire and beyond.

Elemental characterisation via INAA has reinforced Shepard's petrographic study, showing that there are two main Plumbate chemical groups that correspond to Shepard's petrographic groups (Neff 1984; Neff and Bishop 1988). One chemical group corresponding to Shepard's San Juan Plumbate is dominated by plain serving vessels and occurs in assemblages on the coast of southern Mesoamerica and in the adjacent highlands. The other chemical group includes miniatures, effigies, and abstract-incised vessels and appears as whole-vessel offerings across virtually the whole of Mesoamerica.

Archaeological survey from the 1940s through the 1960s (Lowe and Mason 1965; Shook 1965; Coe and Flannery 1967) showed that sherd collections from the Pacific coast of Guatemala and adjacent southern Mexico have the highest frequencies of Plumbate. In accord with the suspicion that this was the Plumbate production zone, initial INAA results (Neff 1984; Neff and Bishop 1988) demonstrated that both elemental groups (San Juan and Tohil) are represented among the plain Plumbate sherds in this region. Later, a raw material survey within this region recovered clays that match both the clays and slips of the two Plumbate chemical groups (Neff 2002, 2003). Specifically, the San Juan group, which is generally earlier in time, matches the compositional profile of clays sampled along the lower Rio Naranjo, in far-southwestern Guatemala, while the Tohil group, which includes the fancy effigies and miniatures, closely matches clay samples from along the Rio Cahuacan in southern Mexico.

Based on the evidence just summarised, the trajectory of Plumbate long-distance exchange unfolded as follows. During the Terminal Classic Period, San Juan Plumbate was the dominant ceramic locally in the eastern Soconusco source region, and it was also exported to areas lying further east, especially the central Pacific coast and the highlands of Guatemala and western El Salvador. Later, during the Early Postclassic Period, Tohil, a technologically improved and aesthetically enhanced Plumbate, was exported to many of the same locations in southern Mesoamerica, but it also reached much more far-flung locations, such as Chichen Itzá, central Mexico, and West Mexico. Probably around 1100 AD, Plumbate ceased to circulate in interregional distribution networks.

Plumbate history parallels the Olmec and Teotihuacan cases in that substantial numbers of vessels were transported over distances long enough to elevate their procurement costs dramatically. As in the previous cases, superior performance as tools for cooking, carrying, storage, or serving is an implausible explanation for the large expenditures on procurement of these vessels. Plumbate differs from the Olmec and Teotihuacan cases, however, in that convincing copies of Plumbate were never produced in the regions to which it was exported. In the case of Plumbate, instead, efforts to produce copies yielded obvious fakes, such as the polished black effigy from Offering V in the Casa de Las Aguilas in Tenochtitlan (Lopez Lujan *et al.* 2000, 225).

## 3. Pots as signals

Many additional examples of long-distance ceramic exchange in Mesoamerica could be listed. The above examples were selected because they highlight inadequacies in the 'pots-as-tools' framework. The examples suffice to demonstrate that long-distance ceramic exchange is a pervasive feature of the Mesoamerican archaeological record, and it was pervasive right from the time when people first started making and using ceramics (i.e., from the Early Formative on). Yet, as previously argued, under a 'pots-as-tools' model, the investment in transporting ceramics is difficult to understand, since high transport costs coupled with the ubiquity of ceramic raw materials and simplicity of ceramic technology are expected to limit the geographic scope of ceramic circulation. How then, can we account for the ancient Mesoamericans' frequent practice of carrying fragile ceramic pots to distant corners of the Mesoamerican world?

An alternative to 'pots-as-tools' is to view pots as signals. This subsumes long-distance ceramic exchange under the evolutionary theory of costly signalling (e.g., Bliege Bird and Smith 2005; Zahavi and Zahavi 1997). Bliege Bird and Smith (2005) note the potential applicability of costly signalling theory to ceramic production, and Plourde (2008) has developed a model based on costly signalling that explains the origin of prestige goods, which may include ceramics. Elsewhere (Neff 2011), I present some of the ideas elaborated below in a discussion of Early Formative Olmec cultural evolution. My focus here is on costly signalling as an explanation of long-distance ceramic exchange and how archaeologically testable hypotheses can be derived from the theoretical framework.

As signals, pots may convey information about such things as group affiliation, network connections, special access to resources, or prestige and status. These are hidden qualities of the signaller that, if perceived accurately by audience members, would predispose them to attitudes or actions that benefit the signaller. The design considerations for signalling systems are different from the design considerations involved in producing a pot that is a tool. For signal-pots, the main consideration is how to ensure that the signal is a reliable indicator of the underlying trait that is being communicated. The reason why a signal has to be reliable is that, otherwise, receivers of the signal have no incentive to pay attention, and if receivers do not pay attention, signallers have no incentive to signal. That is, if reliability cannot be ensured, the signalling system will disappear.

Under pots as signals, long-distance transport of pots, which adds cost, can be seen as a mechanism for ensuring signal reliability – making sure that the signal is honest. The principle that added cost can ensure signal reliability has been discovered independently in biology (Zahavi and Zahavi 1997) and the social sciences (Spence 1973; Veblen 1979 [1899]). It was introduced to archaeology by Neiman (1997), who argued that Maya elite activity (investment in monumental art and architecture, especially) could be understood as 'wasteful advertising' of the ability of individuals and lineages to come out on top in political competition. The high cost of acquiring literacy and controlling labour and other resources ensured that only individuals with *actual* power and ability could erect monuments, build temples, and so on.

Based on costly signalling theory, long-distance transport of pots takes place *precisely because* it adds cost. Most ceramics start out as low-cost items, since, as discussed, raw materials are widely available and the basic technology is simple. Possession of such low-cost items says little about prestige or other qualities of individuals. A pot can become a reliable signal of access to resources, however, if some extra investment is made that elevates a vessel's cost. One obvious and open-ended means to add cost is to invest human labour in long-distance transport.

The above argument applies whether importers or exporters shoulder the cost of long-distance transport. If it is the exporters, the ceramic exports constitute gifts to one's competitors, and gifts incur obligations. The more resources one has to invest in gifts, the larger and more numerous the obligations he/she will be owed. Expensive gifts and attendant obligations are thus a *reliable* signal of an individual's power, wealth, and influence. They are given in order to accumulate 'symbolic capital' (Bourdieu 1977). The reliability of the information conveyed, which is ensured by the *actual* high cost of lavish gifts, is the basic incentive for receivers to participate in the signalling system. Long-distance export of distinctive pots can thus be seen as a strategy for interregional advertising of prestige.

If importers rather than exporters are assumed to pay the costs of long-distance transport, the foreign vessel is a signal of the importer's ability to bear costs that are beyond the reach of most local people. The imported pot also signals extra-local connections, thus further augmenting the information conveyed about prestige-related attributes of the possessor. Again, the *actual* high cost of imports makes them reliable signals of resource control and/or network connections, so audience members gain information about local political and status relations if they pay attention to the signal.

A further consideration is that signal-pots must be *recognisably* foreign outside of their production region. Otherwise, audience members will not be able to appreciate their added cost, which in turn would obscure the information content of the signal. If receivers misinterpret the signal and fail to pay attention, the incentive to signal disappears. This design constraint favours distinctive decorative practices and/or technology that impart distinctive aesthetic qualities to the finished vessels and make them recognisable as foreign, and thus costly, to an audience familiar with locally produced pots.

Finally, duplication of a foreign technology or decorative style will cause the foreign features to lose their value as indicators of high cost. Once this happens, the signalling system, in which reliability is ensured by added cost due to long-distance transport, is likely to disappear. If, at the other extreme, the combination of raw materials, decoration, and technology cannot be duplicated, the signalling system should persist indefinitely. Chinese porcelain is an example of difficult-to-imitate technology and decorative style (Kerr and Wood 2004). Exported to Europe during the $16^{th}$ through $18^{th}$ centuries, it stimulated many efforts to reproduce the technology. Not until the early $18^{th}$ century, however, did European ceramists come up with viable European versions. Fine Chinese porcelains thus retained their value, providing a clear and reliable statement about prestige and taste, over several hundred years. As European porcelains developed, demand for Chinese imports declined (Munger and Frelinghuysen 2003).

## 4. Pots as signals in Mesoamerican prehistory

Whereas the cost involved in interregional transport of Olmec, Teotihuacan, and Plumbate pottery rules out accounts based on any hypothetical superiority as tool-pots, that same trait – cost added by the labour invested in long-distance transport – marks them as components of signalling systems, in which signal reliability is the overriding design consideration. Additionally, accurate interpretation of the signal by foreign audiences depends on aesthetic distinctiveness, which may be achieved by unique decorative elements, unique raw materials, unique technology, or some combination thereof. These principles shed light on the cases of long-distance ceramic exchange discussed earlier and lead to some additional hypotheses that can be tested in the future.

### *Olmec*

Consider first the carved grey Olmec pottery transported long distances across Early Formative Mesoamerica. These vessels were reduction fired, they were decorated with a distinctive set of symbols that originated on the Gulf Coast, and the designs were rendered with an unusual carved, sometimes excised technique. When they first appeared in regions outside the Gulf Coast, they would have contrasted with pottery familiar to the local populace in colour (grey rather than orange or red, the dominant colours achieved with oxidation firing) and in features of the carved and incised designs. These contrasts would have made them unfamiliar and thus recognisable as imports from some unknown foreign place. Alternatively, they might have been brought in as gifts by Olmec traders, in which case they would have reinforced messages about power and control over resources brought by these same traders from the Gulf Olmec heartland. Either way, being easily recognisable as imports, they were also immediately recognisable as high-cost items, because of the expense of transporting pots discussed above. Thus, while we cannot be certain who shouldered the costs of long-distance transport, whoever did sent a reliable signal about power and access to resources.

But since ceramic technology is fairly simple, creating grey surfaces and carving flame eyebrows could be easily duplicated, and the high status associated with the high-cost imports would have created an incentive to produce lower-cost imitations that were indistinguishable from the originals. Imitation Olmec carved grey ware then should begin to appear in places that received imports. This prediction seems to be borne out, since local-paste variants of carved grey pottery were identified alongside the San Lorenzo imports in the Valley of Oaxaca, Basin of Mexico, and elsewhere.

A further implication here is that the imitations would undercut the reliability of imports as signals of access to resources. Once carved grey vessels no longer reliably signalled access to resources, local audience members no longer had an incentive to pay attention, which would have eliminated the incentive for signallers to pay the high costs of acquiring true imports. This, then, leads to a prediction about the historical trajectory of Olmec grey-ware exchange, namely that the flow of exports from the Gulf Coast would have been relatively short-lived. Available evidence is consistent with this prediction, but needs to be tested further, with improved chronological control in the assemblages containing imports together with local copies.

The fine-white pottery (Xochiltepec White) exported from the Gulf Olmec heartland may have been more difficulty to copy, since fairly pure, high-kaolinite clays were required for its production. Thus, a different evolutionary trajectory (longer-lived interregional exchange) would be predicted for the white wares, since their reliability as signals might not have been as easily undermined as that of the carved grey ware. This prediction too can be tested once chronological control is improved for Early Formative assemblages.

### *Teotihuacan Trade Wares in Guatemala*

The histories of Thin Orange ware and Gulf Coast fine paste ware ('Proto-Tiquisate') in Guatemala provide interesting parallels to the Olmec case just described. Both Teotihuacan-related imports qualified as reliable signals of investment in long-distance transport when they were first transported to Pacific coastal and highland Guatemala during the Early Classic period. As required for accurate interpretation of the signal, both wares contrasted markedly with locally made pottery. The unique qualities of Thin Orange are due in part to technology, whereas the Gulf Coast fine paste ware was distinguished by the use of fine, light-firing clay. Signal reliability was short-lived however, for, despite the fact that their aesthetic qualities depended on special technology and resources, viable local versions (Esmeralda Flesh ware and Tiquisate ware) eventually appeared.

Esmeralda Flesh, the Guatemalan imitation of Thin Orange, seems to have appeared very shortly after the first Thin Orange imports arrived in Guatemala, and the imports probably disappeared within a short period of time, judging from their rarity. The largest Thin Orange sample is from high-status tombs in Mounds A and B at Kaminaljuyu, which were apparently constructed over a fairly short period of time during the 4$^{th}$ and 5$^{th}$ centuries AD. A total of 15 whole Thin Orange vessels were found in seven tombs, and 12 Esmeralda Flesh vessels were found in four tombs in these mounds (Kidder *et al.* 1946, Table 1). Tomb A-I contained one Thin Orange vessel together with three Esmeralda Flesh vessels. On the central Pacific coast, in the heartland of Teotihuacan influence, Thin Orange occurs as a rare import in Early Classic sherd assemblages that also contain Esmeralda Flesh (Bove and Medrano 2003, 51–2). As predicted by the costly signalling model, the flow of Thin Orange imports was short-lived because the reliability of the signal declined rapidly as a similar local ware became available. The local variant, Esmeralda Flesh, thereafter evolved in situ, eventually becoming the Amatle ware of the Late Classic Period (Neff *et al.* 1994).

Gulf Coast fine paste pottery is also very rare, a total of only six examples having been found on the Pacific coast, where they occur in association with green (Pachuca) obsidian and Thin Orange ware in Early Classic assemblages (Bove and Medrano 2003). Later, during the Middle Classic (400–600 AD), the earliest Tiquisate ware, a red-painted variant called Caulote by Bove (Bove and Medrano 2003, 71) appeared on the Pacific coast in the region where Gulf Coast fine paste had been imported earlier (some of the Gulf fine paste specimens, both from Pacific Guatemala and Teotihuacan, are decorated with red paint [Cowgill and Neff 2004]). Duplication of the Gulf fine paste ware may have been delayed by the need to locate fine, light-firing clays in coastal Guatemala; discovery of a residual clay developing on rhyolitic volcanic ash along the Rio Nahualate supplied the appropriate raw materials (Neff 1995; Neff and Bove 1999). This source was then used from Middle through Terminal Classic times to produce several Tiquisate-ware variants (Neff and Bove 1999; Chinchilla Mazariegos *et al.* 2009).

In sum, as in the Olmec case, long-distance transport of Teotihuacan-related tradewares seems to have been a fairly brief historical phenomenon, as predicted for signalling media that can be copied. In both cases, local variants of the style and technology of the imports were developed, and this undermined the reliability of the imports as signals of high cost, which led quickly to the disappearance of the imports. What is clearer in the Teotihuacan case is that production of the local variants outlasted the influx of imports. Further, as hypothesised for the fine white Olmec tradeware (Xochiltepec), it appears that development of the local variant of fine, light-firing ware (Tiquisate) stalled until suitable raw materials were discovered. Once the Early Classic imports disappeared, the new resource-procurement and technological practices adopted to create copies were no longer constrained by the need for fidelity to the imports, and this seems to have led to the evolution of regionally distinctive ceramic industries of the Middle through Late Classic periods (Neff 2006). While current understanding of the Teotihuacan tradewares and their copies is consistent with the 'pots as signals' framework, more precise chronological control over both the imports and the local ceramic industries would provide a more detailed picture of the longevity of the imports, of how quickly the local copies were developed, and how rapidly the new Guatemalan industries diverged from the introduced Mexican canons once the fidelity constraint disappeared.

### *Plumbate*

Like Olmec and Teotihuacan pottery, Plumbate vessels were aesthetically distinctive and readily identifiable as foreign, and therefore costly, when they appeared in locations far from their source. The large component of cost added by long-distance transport could not be borne by everyone, so possession of one of these shiny grey effigies was a reliable signal of an ability to invest large amounts of human labour.

But Plumbate differed from lower-tech Olmec and Teotihuacan pottery in that special raw materials combined with special technology complicated efforts to duplicate its aesthetic qualities. Plumbate potters had apparently discovered the fluxing action of alkalis when added to slip material (Neff 2010). Depth profiling by LA-ICP-MS shows enrichment of sodium and potassium at the surfaces.

X-ray fluorescence analysis of surfaces of Tohil and San Juan vessels accentuate this point, showing potassium concentrations in the two to three percent range, which is uniformly higher than corresponding paste concentrations. Also, since the paste and surface potassium concentrations are uncorrelated, the surface would seem to be an artefact of technology rather than a reflection of natural variability in source raw materials.

In light of these findings and the fact that Plumbate was produced within the lower-coastal estuarine zone, Plumbate technology may have been discovered as a by-product of salt production, the other main activity documented for estuary dwellers of the Classic period (Neff 2010). Multiple non-exclusive routes toward such a discovery can be suggested: accidental over-firing during salt production in ceramic containers (the *sal cocida* method) is one possibility; alternatively, if mangrove wood ash generated either in salt or ceramic production was added purposely (e.g., to promote deflocculation) or accidentally to slip clay, levels of soda and potash could have been elevated to the point that vitrification occurred within the normal firing range. Either way, since unusual raw materials had to be combined with a special pyrotechnology, re-discovery of Plumbate technology and duplication of a Plumbate pot outside its production region would have been difficult if not impossible.

In fact, no other glazed ware was ever invented in Mesoamerica, and Plumbate surfaces were never duplicated. Since Plumbate pots were impossible to fake, the reliability of the signal would not have been undermined in the same way as in the Olmec and Teotihuacan cases. Because of this added stability of the signalling system, it would be expected to persist.

In light of the evidence that reliability of Plumbate as a prestige symbol was never undermined, what might account for its disappearance by around 1100 AD? Costly signalling systems are maintained by selection, because signallers gain from the attention paid by the audience (possessors of Plumbate gain prestige in this case), while receivers gain because they learn about important qualities of the senders (Neiman 1997; Zahavi and Zahavi 1997; Bliege Bird and Smith 2005). For such a system to disappear (as Plumbate did during the Early Postclassic), the selective pressures that created and maintained the costly signalling system would have to be disrupted in some way. One possible disruption, viable local imitations, has already been ruled out. Other possibilities include innovations in transport technology (which reduce the cost added by long-distance exchange) or disruption of the interaction of signallers and receivers, for example by population loss and migration (as Neiman [1997] has argued for Classic Maya monument dedication). However, no transport innovations are suspected for Postclassic Mesoamerica, and disruption of the interaction between senders and receivers of signals seems unlikely to have occurred simultaneously in all of the various regions where fancy Tohil Plumbate arrived as a trade good.

Still another alternative explanation for the disappearance of Plumbate interregional exchange is that the ability to make the signal (the supply of Plumbate pots) disappeared for reasons unrelated to the selective pressures that maintained the signalling system. That is, demographic shifts or political upheaval within the eastern Soconusco source region eventually eliminated the population of Plumbate potters. This hypothesis is, of course, testable through archaeological investigation within the Plumbate source region. The basic prediction is that the archaeological and paleoenvironmental records should show a demographic collapse within the region sometime around 1100 AD.

The pots-as-signals framework thus makes different predictions about the historical trajectory of Plumbate production than it does about the Olmec and Teotihuacan tradewares. The differences arise from differences in the technological sophistication of the wares. Olmec and Teotihuacan tradewares could be duplicated with a little effort invested in finding adequate raw materials (e.g., the clay source used for Tiquisate ware). Plumbate, the pinnacle of the potter's craft in Precolumbian Mesoamerica, required raw material properties and technological processes discovered only once and never duplicated. As a result, Plumbate's reliability as a signal of cost added by long-distance transport could not be undermined. Plumbate production and interregional exchange is thus predicted to show greater stability and longevity than the lower-tech Olmec and Teotihuacan cases.

The pots-as-signals framework also points the way toward future research on Plumbate. As noted, the prediction that Plumbate trade disappeared because of demographic disruption in the source zone is subject to straightforward archaeological testing. Further tests of the alkaline glaze hypothesis are also possible; for instance, experimentation with raw materials from the production region combined with further analytical work could be used to reproduce the evolutionary process through which Plumbate technology was discovered and refined and to show what made it so difficult to duplicate. Finally, improved chronological control both within the source zone and in the assemblages in which the tradewares appear would permit more convincing tests of the prediction that Plumbate trade was more stable and long-lasting than the more easily copied Olmec and Teotihuacan tradewares.

## 5. Conclusion

I draw the following conclusions from reviewing and thinking about the results of chemistry-based ceramic provenance research in Mesoamerica. First, a number of studies over the past 40 years demonstrate conclusively that ceramics did, in fact, circulate widely within and even beyond Mesoamerica (a Plumbate vessel has been reported from Panama [Dutton 1943]). Countering expectations based on the ease of ceramic manufacture coupled with the difficulty of transporting ceramics, many widespread ceramic traits were the result at least partly of movement of materials rather than movement of ideas. As I have argued, interregional transport is difficult to explain under a 'pots-as-tools' understanding of ceramics and ceramic production.

However, if we conceive of pots as signals, long-distance circulation is predictable because the cost added by investing in transport ensures reliability of the pots as signals of hidden qualities of the producers, exporters, or consumers. The high cost of long-distance transport could only be borne by individuals or groups with superior access to resources.

The 'pots as signals' perspective does not *just* explain long-distance circulation of ceramics; it also serves as a fruitful source of hypotheses about the design and distribution of pots and about the evolutionary history of pots designed as signals. As any good theory should do, it makes *different predictions for different historical-ecological contexts*, the example here being the different predictions for Olmec and Teotihuacan tradewares versus the technologically more sophisticated Plumbate.

Finally, an objection might be raised here that archaeologists routinely discriminate between 'luxury' or 'elite' wares (signal-pots) and utilitarian wares (tool-pots), so there is really nothing new in the 'pots as signals' framework. The problem is that these categories are normally defined by common sense and empirical generalisation, with no theoretical justification given for why such categories *should* exist. The principle that costly features are engineered by evolution to ensure signal reliability provides the theoretical basis for defining these categories. This is a common feature of organisms shaped by natural selection (Zahavi and Zahavi 1997). My claim here (and Neiman 1997) is that selection operating on culturally inherited information produces convergent adaptations that are observable in the archaeological record.

### Notes

1. The ubiquity of clay, the relative simplicity of ceramic technology, and the high transport cost of pottery, have often led archaeologists to assume that ceramics would rarely have been moved much distance through space and that specialisation in ceramic production should be infrequent. This assumption is especially clear in the US Southwest, where Kidder (Kidder and Shepard 1936, xxiii) once suggested that "[it] has always been assumed that potting was one of the regular household tasks of every Pueblo woman; that each town was in this regard self sufficient..." and Abbott (2009, 531) more recently asserted that, up until just two decades ago, "most analysts believed that ancient Southwestern households were largely self sufficient for pottery manufacture..."

   We know from ethnographic cases, of course, that pots do circulate in regional markets (e.g., Arnold 1985), and archaeological evidence affirms the existence of ceramic exchange in prehistory as well. In Kidder's (Kidder and Shepard 1936) Pecos study, for instance, the initial hypothesis of local self-sufficiency was rejected on the basis of petrographic study by Anna O. Shepard, which demonstrated the presence of fabrics that originated 25 or more km from the Pueblo.
2. To be clear, 'cost' here refers not to any monetary cost but to investment of energy and time. In biological terms, any investment of energy and time should always promote reproductive success, i.e., fitness. Making ceramics always has a cost, but these costs can be outweighed by the fitness benefits of cooking foods to enhance their nutritional value or the benefits of local storage of grain, clean water, etc. The enigma of long-distance exchange arises when the costs of transport exceed any conceivable direct fitness benefit of this kind. The problem addressed in this paper, then, is what *fitness benefit* could outweigh the very high cost in terms of time and energy of long-distance ceramic exchange.

## References


Abbott D., 2009, Extensive and long-term specialization: Hohokam ceramic production in the Phoenix Basin, Arizona. *American Antiquity*, 74, 531–557.

Arnold D.E., 1985, *Ceramic Theory and Cultural Process*, Cambridge University Press, Cambridge.

Arroyo B., 1995, Early ceramics from El Salvador: the El Carmen site. In *The Emergence of Pottery*, (eds. W.K. Barnett and J.W. Hoopes), 199–208, Smithsonian Institution Press, Washington DC.

Blake M., Clark J.E., Voorhies B., Michaels G., Love M.W., Pye M.E., Demarest A.A. and Arroyo B., 1995, Radiocarbon Chronology for the Late Archaic and Formative Periods on the Pacific Coast of Southeastern Mesoamerica. *Ancient Mesoamerica*, 6, 161–183.

Bliege Bird R. and Smith E.A., 2005, Signaling Theory, Strategic Interaction, and Symbolic Capital. *Current Anthropology*, 46, 221–248.

Blomster J.P., Neff H. and Glascock M.D., 2005, Olmec pottery production and export in ancient Mexico determined through elemental analysis. *Science*, 307, 1068–1072.

Bourdieu P., 1977, *Outline of a Theory of Practice*, Cambridge University Press, Cambridge.

Bove F.J. and Medrano S., 2003, Teotihuacan, militarism, and Pacific Guatemala. In *The Maya and Teotihuacan: Reinterpreting Early Classic Interaction*, (ed. G.E. Braswell), 45–79, University of Texas Press, Austin, TX.

Braun D.P., 1983, Pots as Tools. In *Archaeological Hammers and Theories*, (eds. J. Moore and A. Keene), 107–134, Academic Press, New York.

Bronitsky G., 1989, Ceramics and Temper: Response to Feathers. *American Antiquity*, 54, 589–593.

Bronitsky G. and Hamer R., 1986, Experiments in Ceramic Technology: The Effects of Various Tempering Materials on Impact and Thermal-Shock Resistance. *American Antiquity*, 51, 89–101.

Cheetham D., 2010, *America's First Colony: Olmec Materiality and Ethnicity at Canton Corralito, Chiapas*, Unpublished PhD Thesis, Department of Anthropology, Arizona State University.

Chinchilla Mazariegos O., Bove F.J. and Genovez J.V., 2009, La cronología del período Clásico en la Costa Sur de Guatemala y el fechamiento del estilo excultórico de Cotzumalguapa. In *Cronología y periodización en Mesoamérica y el norte de México*, (ed. A. Daneels), 435–471, Universidad Nacional Autónoma de México, Instituto de Investigaciones Antropológicos, México, DF.

Coe M.D. and Flannery K.V., 1967, *Early Cultures and Human Ecology in South Coastal Guatemala*, Smithsonian Contributions to Anthropology 3, Smithsonian Institution, Washington DC.

Cowgill G.L. and Neff H., 2004, Algunos resultados del análisis por activación neutrónica de la cerámica foránea de Teotihuacan. In *La costa del Golfo en tiempos teotihuacanos: propuestas y perspectives*, (eds. M.E. Ruiz Gallut and A. Pascual Soto), 63–75, Memoria de la

Segunda Mesa Redonda de Teotihuacan, Instituto Nacional de Antropología e Historia, México, DF.

Dutton B.P., 1943, *A History of Plumbate Ware*, Papers of the School of American Research 31, School of American Research, Santa Fe.

Feathers J.K., 1989, Effects of Temper on Strength of Ceramics: Response to Bronitsky and Hamer. *American Antiquity*, 54, 579–588.

Flannery K.V., Balkansky A.K., Feinman G.M., Grove D.C., Marcus J., Redmond E.M., Reynolds R.G., Sharer R.J., Spencer C.S. and Yaeger J., 2005, Implications of New Petrographic Analysis for the Olmec "Mother Culture" Model. *Proceedings of the National Academy of Sciences*, 102, 11219–11223.

Glascock M.D., 1992, Characterization of archaeological ceramics at MURR by neutron activation analysis and multivariate statistics. In *Chemical Characterization of Ceramic Pastes in Archaeology*, (ed. H. Neff), 11–26, Prehistory Press, Madison, WI.

Harbottle G., 1976, Activation analysis in archaeology. In *Radiochemistry: A Specialist Periodical Report*, (ed. G.W.A. Newton), 33–72, Vol. 3., The Chemical Society, London.

Kerr R. and Wood N., 2004, *Science and Civilisation in China, Volume 5, Part XII: Ceramic Technology*, Cambridge University Press, Cambridge.

Kidder A.V. and Shepard A.O., 1936, *The Pottery of Pecos, Vol. II*, Papers of the Southwestern Expedition, Department of Archaeology, Phillips Academy, Andover (MA), Yale University Press, New Haven.

Kidder A.V., Jennings J.D. and Shook E.M., 1946, *Excavations at Kaminaljuyu, Guatemala*, Publication 561, Carnegie Institution of Washington, Washington DC.

Lopez Lujan L., Neff H. and Sugiyama S., 2000, The 9-Xi vase: a Classic Thin Orange vessel found at Tenochtitlan. In *Mesoamerica's Classic Heritage: From Teotihuacan to the Aztecs*, (eds. D. Carrasco, L. Jones and S. Sessions), 219–249, The University Press of Colorado, Boulder, CO.

Lowe G.W. and Mason J.A., 1965, Archaeological survey of the Chiapas coast, highlands, and upper Grijalva Basin. In *Handbook of Middle American Indians*, (ed. G.R. Willey), 195–236, Vol. II., University of Texas Press, Austin.

Munger J. and Frelinghuysen A.C., 2003, East and West: Chinese Export Porcelain. *Heilbrunn Timeline of Art History*, The Metropolitan Museum of Art, New York. http://www.metmuseum.org/toah/hd/ewpor/hd_ewpor.htm

Neff H., 1984, *Developmental History of the Plumbate Pottery Industry in the Eastern Soconusco Region, A.D. 600A.D. 1250*, Unpublished PhD Thesis, Department of Anthropology. University of California, Santa Barbara.

Neff H., 1995, A role for "sourcing" in evolutionary archaeology. In *Evolutionary Archaeology: Methodological Issues*, (ed. P. Teltser), 69–112, University of Arizona Press, Tucson.

Neff H., 2002, Sources of raw material used in Plumbate pottery. In *Incidents of Archaeology in Central America and Yucatan*, (eds. M. Love, M.P. Hatch and H. Escobedo), 217–231, University Press of America, Lanham, MD.

Neff H., 2003, Analysis of Plumbate pottery surfaces by laser ablation-inductively coupled plasma-mass spectrometry (LA-ICP-MS). *Journal of Archaeological Science*, 30, 21–35.

Neff H., 2006, Orígenes y Evolución de las Tradiciones Cerámicas del Periodo Clásico en la Costa del Pacífico de Guatemala. In *Iconografía y Escritura Teotihuacana en la Costa Sur de Guatemala y Chiapas*, U Tzíb, Serie Reportes, Volumen 1, Número 5, (eds. O. Chinchilla Mazariegos and B. Arroyo), 17–34, Asociación Tikal, Guatemala.

Neff H., 2010, Plumbate technology revisited. *Proceedings of the 2nd Latin-American Symposium on Physical and Chemical Methods in Archaeology, Art and Cultural Heritage Conservation and Archaeological and Arts Issues in Materials Science*, IMRC 2009, Mexico.

Neff H., 2011, Evolution of the Mesoamerican Mother Culture. *Ancient Mesoamerica*, 22, 107–122.

Neff H. and Bishop R.L., 1988, Plumbate origins and development. *American Antiquity*, 53, 505–522.

Neff H. and Bove F.J., 1999, Mapping ceramic compositional variation and prehistoric interaction in Pacific coastal Guatemala, in 'Proceedings of the International Symposium on Archaeometry, University of Illinois at Urbana-Champaign (UIUC), Urbana, Illinois, 20–24 May, 1996. Special Issue'. *Journal of Archaeological Science*, 26, 1037–1051.

Neff H. and Hodge M.G., 2008, Serving vessel production at Chalco: evidence from neutron activation analysis. In *Place of Jade: Society and Economy in Ancient Chalco*, (ed. M.G. Hodge), 187–226, Serie Arqueología de Mexico, Instituto Nacional de Antropología e Historia, Mexico, D.F. and University of Pittsburgh, Pittsburgh.

Neff H. and Larson D.O., 1997, Methodology of comparison in evolutionary archaeology. In *Rediscovering Darwin: Evolutionary Theory in Archaeological Explanation*, Archaeological Publications of the AAA 7, (eds. C.M. Barton and G.A. Clark), 75–94, American Anthropological Association, Arlington, VA.

Neff H., Blomster J., Glascock M.D., Bishop R.L., Blackman M.J., Coe M.D., Cowgill G.L., Cyphers A., Diehl R.A., Houston S., Joyce A.A., Lipo C.P. and Winter M., 2006a, Smokescreens in the Provenance Investigation of Early Formative Mesoamerican Ceramics. *Latin American Antiquity*, 17, 104–118.

Neff H., Blomster J., Glascock M.D., Bishop R.L., Blackman M.J., Coe M.D., Cowgill G.L., Diehl R.A., Houston S., Joyce A.A., Lipo C.P., Stark B.L. and Winter M., 2006b, Methodological issues in the provenance investigation of Early Formative Mesoamerican ceramics. *Latin American Antiquity*, 17, 54–76.

Neff H., Bove F.J., Robinson E. and Arroyo B., 1994, A ceramic compositional perspective on the Formative to Classic transition in southern Mesoamerica. *Latin American Antiquity*, 5, 333–358.

Neff H., Cogswell J.W., Kosakowsky L.J., Estrada Belli F. and Bove F.J., 1999, A new perspective on the relationships among cream paste ceramic traditions of southeastern Mesoamerica. *Latin American Antiquity*, 10, 281–299.

Neiman F.D., 1997, Conspicuous Consumption as Wasteful Advertising: A Darwinian Perspective on Spatial Patterns in Classic Maya Terminal Monument Dates. In *Rediscovering Darwin: Evolutionary Theory in Archaeological Explanation*, Archaeological Publications of the AAA 7, (eds. C.M. Barton and G.A. Clark), 267–290, American Anthropological Association, Arlington, VA.

Nichols D.L., Brumfiel E.M., Neff H., Charlton T.H. and Glascock M.D., 2002, Neutrons, markets, cities, and empires: a millennial perspective on ceramic production and distribution in the Postclassic Basin of Mexico at Cerro Portezuelo, Chalco, and Xaltocan. *Journal of Anthropological Archaeology*, 21, 25–82.

O'Brien M.J., Holland T.D., Hoard R.J. and Fox G.L., 1994, Evolutionary Implications of Design and Performance Characteristics of Prehistoric Pottery. *Journal of Archaeological Method and Theory*, 1, 259–304.

Perlman I. and Asaro F., 1969, Pottery analysis by neutron activation. *Archaeometry*, 11, 21–52.

Perlman I. and Asaro F., 1971, Pottery analysis by neutron activation. In *Science and Archaeology*, (ed. R.H. Brill), 182–195, MIT Press, Cambridge, MA.

Pierce C., 1999, *Explaining Corrugated Pottery in the American Southwest: An Evolutionary Approach*, Unpublished PhD Thesis, University of Washington, Seattle.

Plourde A.M., 2008, The origins of prestige goods as honest signals of skill and knowledge. *Human Nature*, 19, 374–388.

Rattray E.C. and Harbottle G., 1992, Neutron activation analysis and numerical taxonomy of thin orange ceramics from the manufacturing sites of Rio Carnero, Puebla, Mexico. In *Chemical characterization of ceramic pastes in archaeology*, (ed. H. Neff), 221–232, Prehistory Press, Madison, WI.

Sayre E.V., Chan L.-H. and Sabloff J.A., 1971, High-resolution gamma ray spectroscopic analysis of Mayan Fine Orange pottery. In *Science and Archaeology*, (ed. R.H. Brill), 165–181, MIT Press, Cambridge, MA.

Schiffer M.B., Skibo J.M., Boelke T.C., Neupert M.A. and Aronson M., 1994, New Perspectives on Experimental Archaeology: Surface Treatments and Thermal Response of the Clay Cooking Pot. *American Antiquity*, 59, 192–217.

Schiffer M.B., Skibo J.M., Griffitts J.L., Hollenback K.L. and Longacre W.L., 2001, Behavioral Archaeology and the Study of Technology. *American Antiquity*, 66, 729–737.

Shepard A.O., 1948, *Plumbate: A Mesoamerican Tradeware*, Carnegie Institution of Washington Publication 573, Washington DC.

Shook E.M., 1965, Archaeological Survey of the Pacific Coast of Guatemala. In *Handbook of Middle American Indians*, (ed. G.R. Willey), 180–194, Vol. II., University of Texas Press, Austin.

Skibo J.M., Schiffer M.B. and Reid K.C., 1989, Organic-Tempered Pottery: An Experimental Study. *American Antiquity*, 54, 122–146.

Spence M., 1973, Job Market Signaling. *Quarterly Journal of Economics*, 87, 355–374.

Steponaitis V.P., 1984, Technological studies of prehistoric pottery from Alabama: Physical properties and vessel function. In *The Many Dimensions of Pottery: Ceramics in Archaeology and Anthropology*, Cingvla 7, (eds. S.E. van der Leeuw and A. Pritchard), 79–122, Universiteit van Amsterdam, Albert Egges van Giffen Instituut voor Prae- et Protohistorie, Amsterdam.

Stoltman J.B., Marcus J., Flannery K.V., Burton J.H. and Moyle R.G., 2005, Petrographic Evidence Shows that Pottery Exchange Between the Olmec and Their Neighbors Was Two Way. *Proceedings of the National Academy of Sciences*, 102, 11213–11218.

Veblen T., 1979 [1899], *Theory of the Leisure Class*, Penguin Books, New York.

Zahavi A. and Zahavi A., 1997, *The Handicap Principle: A Missing Piece of Darwin's Puzzle*, Oxford University Press, New York.

## ACKNOWLEDGEMENTS

Ron Bishop, Fred Bove, and Mike Glascock contributed in various ways to my interest in long-distance ceramic exchange in Mesoamerica. I hasten to add the customary disclaimer that I alone am responsible for errors of fact, interpretation, logic, etc. in this paper. Thanks also to Marcos Martinón-Torres for the invitation to participate in the 2009 EMAC Conference and to two anonymous reviewers who made extremely astute and valuable suggestions about how to improve the paper. A number of grants from the US National Science Foundation (including BNS-8911580, SBR-9503035, and SBR-9802366) provided funding for ceramic provenance investigations that generated the empirical basis for this paper.

9

# Capodimonte porcelain: A unique manufacture

**Bruno Fabbri, Sabrina Gualtieri and Francesca Amato**
CNR-ISTEC, National Research Council, Institute of Science and Technology for Ceramic Materials, Faenza, Italy

**Abstract** - Established in 1743 in Naples following a commission of Charles of Bourbon, the factory of Capodimonte was dedicated to the production of porcelain. The history of the manufacture and its products are well known, but not the production techniques and the raw materials used. This work presents the results of the archaeometric characterisation of Capodimonte porcelain. Chemical analyses by means of SEM-EDS, observation by optical microscopy in thin section, and crystalline phase composition by X-ray diffraction were carried out in order to investigate the characteristics of the raw materials used for the ceramic body and to establish the firing conditions. The pastes of Capodimonte porcelain are siliceous and show a homogeneous chemical composition, permitting us to infer that the more likely formulation of the starting batch was: 70-75% of quartz-rich sand, 15-20% of clay, and 5-10% of soda and maybe tartar. In contrast, the pastes show different proportions of crystalline phases (quartz, cristobalite, and tridymite), which allow us to evaluate a firing temperature range of 1050-1300°C. Capodimonte porcelain is significantly different from all other contemporary European porcelain. It would thus probably be appropriate to design a new category for it, which could be called 'siliceous porcelain'.

## 1. Introduction

Capodimonte porcelain was born following a commission of Charles of Bourbon, who became King of Naples and Sicily in 1734. The passion for porcelain, which was widespread at the time in all European Courts, also affected Charles. After the marriage with Maria Amalia of Walpurgis, granddaughter of Frederick Augustus the Strong, Elector of Saxony and founder of the Meissen Porcelain Factory (1710), Charles received as a dowry beautiful collections of porcelain produced in Germany and decided to start his own production. Consequently, the Capodimonte porcelain factory was established in 1743 and installed in the Royal Factory in Naples, operating until 1759, when Charles became King of Spain and moved the manufacture to Madrid, in the Buen Retiro Park.

Capodimonte porcelain has been the subject of numerous studies, but almost all of them focused on art-historical matters. Investigations of the chemical-physical and technological aspects have been performed only by the chemist Orazio Rebuffat (Rebuffat 1905) and, in more recent years, by the staff of the Professional Institute "G. Caselli" in Capodimonte, with the collaboration of the University Federico II of Naples and the Superintendence of Artistic and Historical Heritage in Naples (Mascolo 1993; Dell'Agli *et al.* 1995; Pepe 1995).

The scarcity of information concerning the manufacture of Capodimonte, due to the destruction of the documentary archives of the factory, has certainly made the investigations more difficult than those performed on other European porcelain manufactures. The available documentation is mainly constituted by the reports presented to the Pontaniana Academy by Camillo Minieri-Riccio (Minieri Riccio 1878a, 1878b) and Giuseppe Novi (Novi 1878) after studying the eleven original collections of correspondence of the Royal Factory of Capodimonte, subsequently destroyed during World War II.

According to the historians, the Capodimonte manufacture was characterised by a very intense production (30,000-40,000 items per year) and by a strong effort to elaborate new models and to experiment innovative technological procedures in order to industrialise the production (Musella Guida 1983). In the beginning, the products mainly consisted of items for domestic use, such as plates, cups, vases, coffee-pots, etc. (Fig. 1a), with a minority of ornamental objects. The artefacts were shaped by hand, on a lathe, or by molding, depending on the complexity of the shape. The first firing was carried out at the maturation temperature of the paste, and the second one likely after the application of lead-rich glazes.

However, the results were not entirely satisfactory, mainly due to a non-favourable production context. Although a lively intellectual environment and an ancient tradition of majolica production were present in Naples, the skill level of the workers was not suitable for the production of porcelain, and various attempts were made to 'steal' personnel from other manufactures. It is not a coincidence that the key roles of the entire manufacturing organisation of the newly

**Figure 1.** Objects typical of the Royal Factory of Capodimonte (Naples, Duca di Martina Museum): a) sugar bowl, 1745 - 1750; b) figures of Goffredo on the tomb of Dudone, 1758 - 1759.

established porcelain factory were performed by non-Neapolitan workers, arrived together with the new king.

Despite the use of several technical devices, the production of Capodimonte was characterised by a very high percentage of waste, around 30%. Over the years, the high amounts of waste and the low technical characteristics of the product, especially the poor thermal shock resistance, led to orienting the production towards ornamental objects, like the large groups of figures of which an example is shown in Figure 1b.

The objective of this study is to compensate for the previous scarcity of information, mainly by supplying technological information pertaining to this porcelain. Therefore, the archaeometric investigation presented here was aimed at defining the chemical and microstructural peculiarities of this material, in order to obtain information on the raw materials used and the manufacturing processes involved in its production. This study is an expansion of previous archaeometric studies (Mascolo 1993; Dell'Agli *et al.* 1995), as it significantly increases the total number of samples analysed. In addition, our selection criteria for the samples, as illustrated below, provide a more reliable and evident connection between the analytical results and the specific characteristics of the production context.

## 2. Materials

The analysed samples were selected from the many fragments found in 1950 in ancient discharge pits of the factory, located in the park of Capodimonte in Naples (Liverani 1959). The present study considers 22 fragments whose dimensions allow recognising, in most cases, the shape and the use of the original objects, mainly cups and plates (Fig. 2). The authenticity of the finds is confirmed by the presence of the brand of the Bourbons, represented by the under-glaze blue fleur-de-lis, located on the bottom of most of the objects.

According to naked-eye observations, the sherds to be analysed were selected based on the presence or absence of glaze and the colour of the glaze and paste:

A) 7 sherds without glaze:
- white paste: samples 104, 161, 180
- very light blue paste: samples 42, 144
- pale brown paste: sample 140
- sherd made of two joined parts, coloured in white (183w) and violet (183v), respectively

B) 7 sherds with colourless glaze:
- white paste: samples 19, 27, 177, 182, 185, 186
- very light blue paste: sample 9

C) 8 sherds with blue glaze of different shades:
- white paste: samples 172, 173, 176, 188, 191, 192
- very light blue paste: samples 170, 190

## 3. Analytical techniques

The pastes were analysed in order to determine their chemical and crystalline phase compositions. The quantitative chemical compositions were obtained by using scanning electron microscopy equipped with an energy dispersive microanalysis system (SEM/EDS). Polished sections of the samples were prepared and analysed with a Cambridge - INCAx - Sight Oxford Instrument (mod. 7060). The protocols used were as follows: standard element = Cobalt, energy range 0 - 10 KeV, working distance (WD) = 25 mm, process time = 4. The results are expressed as oxides and the total amount of the oxides was normalised to 100%. In order to obtain representative data, at least five analyses were carried out for each sample, with an analysis area of approximately 1 $mm^2$.

The crystalline phase composition was obtained by X-ray diffraction, using a Rigaku-Geigerflex diffractometer with an angular range 2θ from 4 to 64° and a scanning rate of 2° per minute.

Finally, observations in thin section at the optical microscope were carried out with the purpose of confirming the presence or absence of the glaze and the characteristics of the macroporosity already observed with the scanning electronic microscope. The observations were performed on a selection of samples, for which the available material was sufficient for a thin section. A Leitz microscope, model Laborlux 11 POL S, was used, providing magnification up to 400x.

## 4. Results

### 4.1. Chemical composition

Table 1 shows the chemical composition of the paste of the analysed samples. It indicates that the samples have very

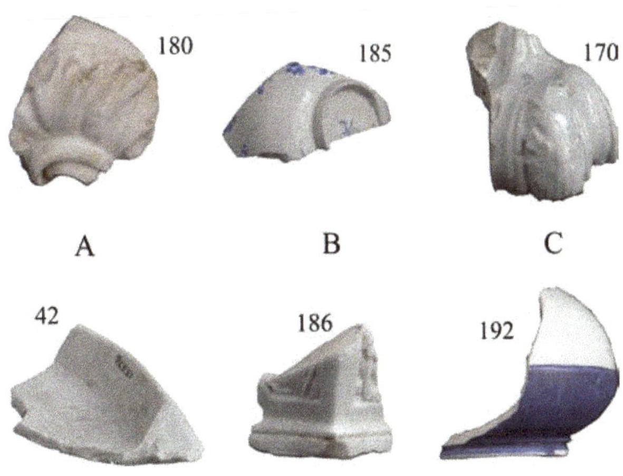


**Figure 2.** Some examples representative of the analysed samples.

**Table 1.** Chemical composition of the paste.

| | $SiO_2$ | $Al_2O_3$ | $TiO_2$ | $Fe_2O_3$ | MgO | CaO | $Na_2O$ | $K_2O$ | $P_2O_5$ | $SO_3$ | Cl | PbO | $SnO_2$ | MnO | $As_2O_3$ | SrO |
|---|---|---|---|---|---|---|---|---|---|---|---|---|---|---|---|---|
| 9 | 84.83 | 5.65 | 0.42 | | 0.31 | 2.20 | 1.83 | 2.93 | 0.32 | 0.24 | 0.16 | | 0.78 | | | 0.34 |
| 19 | 86.29 | 5.58 | 0.12 | 0.24 | 0.42 | 1.24 | 2.30 | 2.93 | 0.29 | 0.15 | 0.07 | | 0.37 | | | |
| 27 | 84.27 | 6.48 | 0.11 | 0.34 | 0.35 | 1.76 | 2.80 | 2.96 | 0.31 | 0.17 | 0.07 | 0.12 | 0.27 | | | |
| 42[1] | 85.54 | 6.43 | 0.11 | 0.49 | 0.39 | 1.30 | 2.26 | 2.70 | 0.13 | 0.14 | 0.07 | | 0.23 | 0.04 | 0.09 | |
| 104 | 85.86 | 6.42 | 0.09 | 0.41 | 0.39 | 1.21 | 2.21 | 2.34 | 0.26 | 0.13 | 0.12 | 0.06 | 0.49 | | | |
| 140 | 82.18 | 8.61 | 0.11 | 0.53 | 0.50 | 0.55 | 1.46 | 2.83 | 0.33 | 0.19 | 0.12 | 0.82 | 1.35 | | 0.14 | 0.30 |
| 144 | 82.56 | 7.60 | 0.12 | 0.37 | 0.35 | 1.19 | 4.17 | 2.56 | 0.35 | 0.18 | 0.08 | | 0.19 | | | 0.27 |
| 161[2] | 84.68 | 7.48 | 0.23 | 0.39 | 0.51 | 1.19 | 1.97 | 2.92 | 0.25 | | 0.07 | 0.03 | 0.29 | | | |
| 170 | 83.75 | 7.27 | 0.13 | 0.44 | 0.43 | 1.09 | 2.40 | 2.66 | 0.26 | 0.16 | 0.09 | 0.82 | 0.51 | | | |
| 172 | 84.84 | 5.94 | 0.03 | 0.49 | 0.33 | 1.81 | 2.74 | 3.22 | 0.15 | 0.17 | 0.06 | | 0.20 | | | |
| 173 | 84.70 | 5.55 | 0.21 | 0.32 | 0.41 | 1.46 | 3.46 | 2.45 | 0.39 | 0.19 | 0.21 | | 0.66 | | | |
| 176 | 83.88 | 6.10 | 0.11 | 0.38 | 0.40 | 1.60 | 2.61 | 2.81 | 0.34 | 0.10 | 0.10 | 0.19 | 1.37 | | | |
| 177 | 84.27 | 5.45 | 0.16 | 0.52 | 0.50 | 1.61 | 3.15 | 2.71 | 0.24 | 0.21 | 0.14 | | 1.04 | | | |
| 180 | 84.34 | 6.05 | 0.20 | 0.21 | 0.41 | 1.20 | 3.76 | 2.35 | 0.40 | 0.21 | 0.15 | | 0.72 | | | |
| 182 | 83.95 | 6.48 | 0.13 | 0.42 | 0.44 | 1.69 | 2.50 | 2.74 | 0.31 | 0.31 | 0.18 | | 0.84 | | | |
| 183v | 78.51 | 4.52 | 0.04 | 0.45 | 0.40 | 4.49 | 1.69 | 6.07 | 0.34 | 0.16 | 0.26 | 2.03 | 0.35 | 0.18 | 0.28 | 0.26 |
| 183w[3] | 82.53 | 5.26 | 0.08 | 0.32 | 0.42 | 1.54 | 1.31 | 5.18 | 0.37 | 0.26 | 0.08 | 1.13 | 0.46 | | 0.44 | 0.32 |
| 185 | 79.72 | 3.55 | 0.18 | 0.27 | 0.49 | 1.81 | 3.16 | 3.41 | 0.22 | 0.40 | 0.46 | 2.82 | 3.50 | | | |
| 186 | 86.09 | 6.93 | 0.01 | 0.03 | 0.25 | 1.42 | 2.62 | 2.45 | 0.03 | | | | 0.17 | | | |
| 188 | 84.72 | 7.17 | 0.15 | 0.34 | 0.35 | 1.04 | 2.97 | 2.57 | 0.41 | 0.12 | 0.06 | | 0.10 | | | |
| 190 | 82.89 | 7.63 | 0.15 | 0.34 | 0.35 | 1.19 | 3.76 | 2.75 | 0.28 | 0.25 | 0.06 | | 0.35 | | | |
| 191 | 85.35 | 6.37 | 0.03 | 0.39 | 0.34 | 1.48 | 2.13 | 3.00 | 0.36 | 0.09 | 0.07 | | | | 0.12 | 0.19 |
| 192 | 84.63 | 6.19 | 0.09 | 0.41 | 0.36 | 1.76 | 2.61 | 2.86 | 0.27 | 0.21 | 0.06 | | 0.55 | | | |

[1] $Cr_2O_3$ = 0.08%; [2] BaO = 0.03%; [3] ZnO = 0.32%.

homogeneous compositions, characterised by very abundant silica, between 82 and 86% $SiO_2$ approximately, except for only two samples (183v and 185), with an amount of silica of less than 80%. Another abundant component is alumina (mostly between 5 and 8% $Al_2O_3$), followed by potassium (2-3% $K_2O$), sodium (1-4% $Na_2O$), and calcium (1-2% CaO). The white and the violet parts of sample 183 show a very high content of potassium, between 5 and 6% $K_2O$; in addition, the violet part contains an unusually high amount of calcium oxide (more than 4% CaO).

The constant presence of tin dioxide is very curious, although in rather low concentrations, usually below 1% $SnO_2$, with the exception of sample 185 (3.50% $SnO_2$). This presence should be indicative of an intentional introduction of tin oxide in order to make the paste whiter. The low concentration of tin can raise some doubts about its real effectiveness, but the fact that tin is distributed only in the glassy matrix should be taken into account, and thus its percentage in this phase should be at least doubled. Lead oxide was only found in 9 samples: four with trace levels (<0.2% PbO), three around 1% PbO, and two (samples 183v and 185) which exceed 2% PbO. The accidental presence of lead, its low concentration, and the lack of correlation with the tin would exclude the hypothesis of an intentional introduction into the paste. As a consequence, it can be hypothesised that the presence of lead is due to pollution during the preparation of the paste or during firing. Considering that the body becomes impermeable during the first firing, so that it cannot absorb any vapour during the second firing of the artefacts, the first hypothesis would seem the most likely. Alternatively, it could be assumed that unglazed items were fired together with items on which a lead-rich glaze had been applied. This solution is actually known from other contexts, where it was carried out in order to make the most of the uneven distribution of the heat inside the kilns (Kingery 1986).

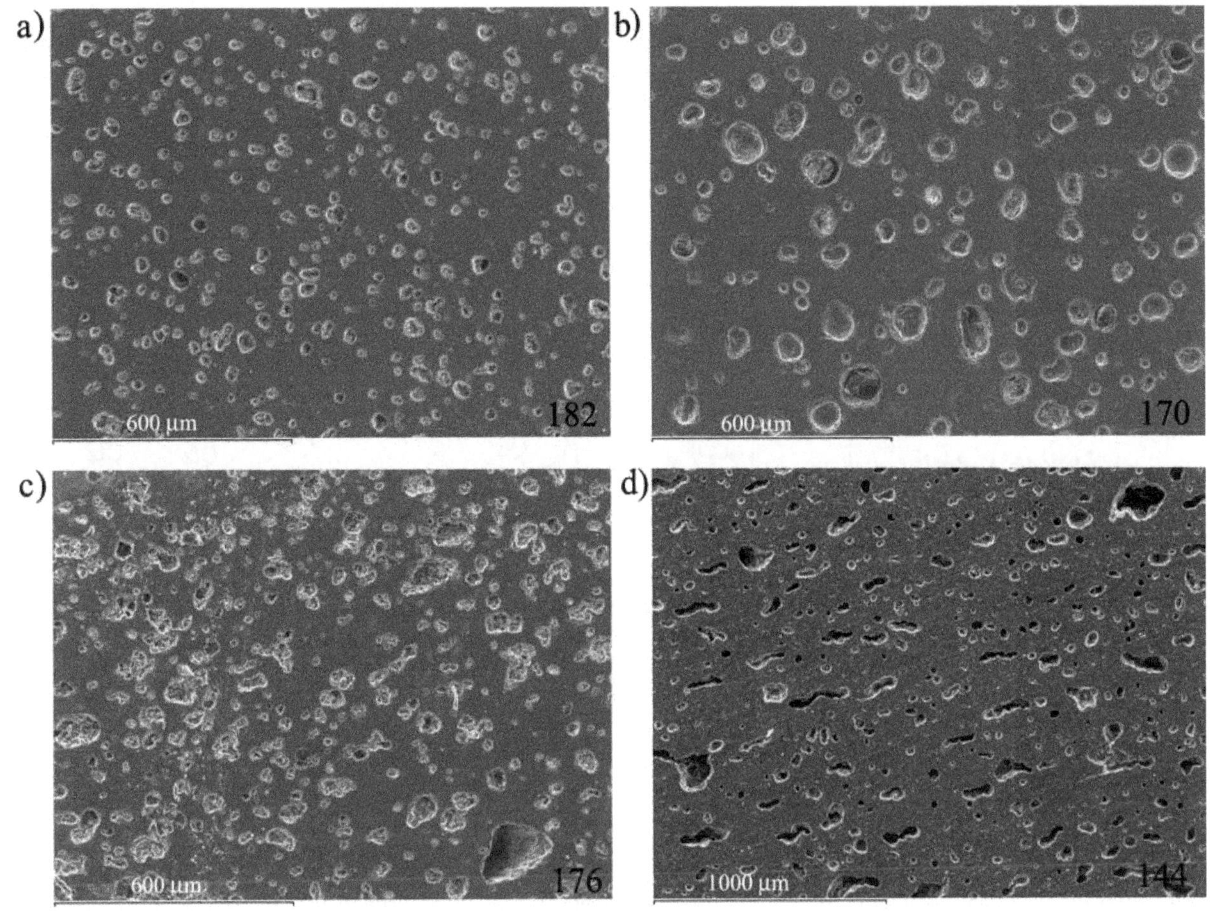


**Figure 3.** SEM micrographs of the paste: a) Closed porosity only made up of small and rounded pores; b) Closed porosity also including larger pores; c) Closed porosity made up of pores with rough borders; d) Closed porosity including elongated pores.

For this purpose, the unglazed items were placed in the warmest areas of the kiln.

All the other detected elements are very scarce: MgO, $Fe_2O_3$, and $P_2O_5$ values are all inferior to 0.5%; $SO_3$ and Cl < 0.3% each. However, sample 185 shows relatively higher values of sulphur and chlorine (0.40 and 0.46%, respectively). Some other elements (arsenic, barium, chromium, manganese, strontium and zinc) were sporadically detected in low quantities. The manganese content (0.18% MnO) identified in the violet portion of sample 183 justifies its colour, thus excluding other possible hypotheses. In contrast, no element was found that can justify the very light blue colour of some pastes (144, 170, and 190), but it is possible that it is related to very low concentrations of some elements that are not clearly detectable by chemical analysis using SEM-EDS. Finally, taking into account the absence of other specific colouring elements, the pale brown colour of sample 140 can be attributed to iron, as a result of firing in an oxidising atmosphere.

### *4.2. Microstructure and mineralogical composition*

The twenty-two artefacts observed using electron microscopy have been subdivided into three groups based on the dimension and shape of the pores, also supported by observations in thin section. The first group includes eight samples (27, 140, 161, 170, 182, 185, 186, and 188) which are mainly characterised by roundish pores, of the order of 50 µm in diameter, with rounded borders (Fig. 3a); sample 170 shows slightly larger pores, around 80-100 µm in diameter (Fig. 3b). The second group includes 10 finds (9, 42, 104, 172, 173, 176, 177, 180, 191, and 192), in which the pores are larger on average and more elongated, with rough borders. The pore sizes vary from a few tens of microns to over a hundred microns (Fig. 3c). The third group consists of only four samples (19, 144, 183, and 190); the characteristics of the pores are similar to those of the second group, but their sizes are significantly larger, up to a maximum of 300-400 µm (Fig. 3d). It can be hypothesised that the different shapes and the size distribution of the pores are to be mainly associated to different firing conditions of the artefacts.

Since the pastes contain more than 80% of silica, their diffraction patterns show only the presence of polymorphous phases of this oxide, i.e., quartz, cristobalite and tridymite (Fig. 4), of course accompanied by a certain amount of glass. Quartz and cristobalite are present in all the samples, albeit in different proportions, while tridymite has only been found in 6 samples (104, 172, 180, 182, 183v, and 190). Because cristobalite is formed through the phase transformation of primary quartz, it can be assumed

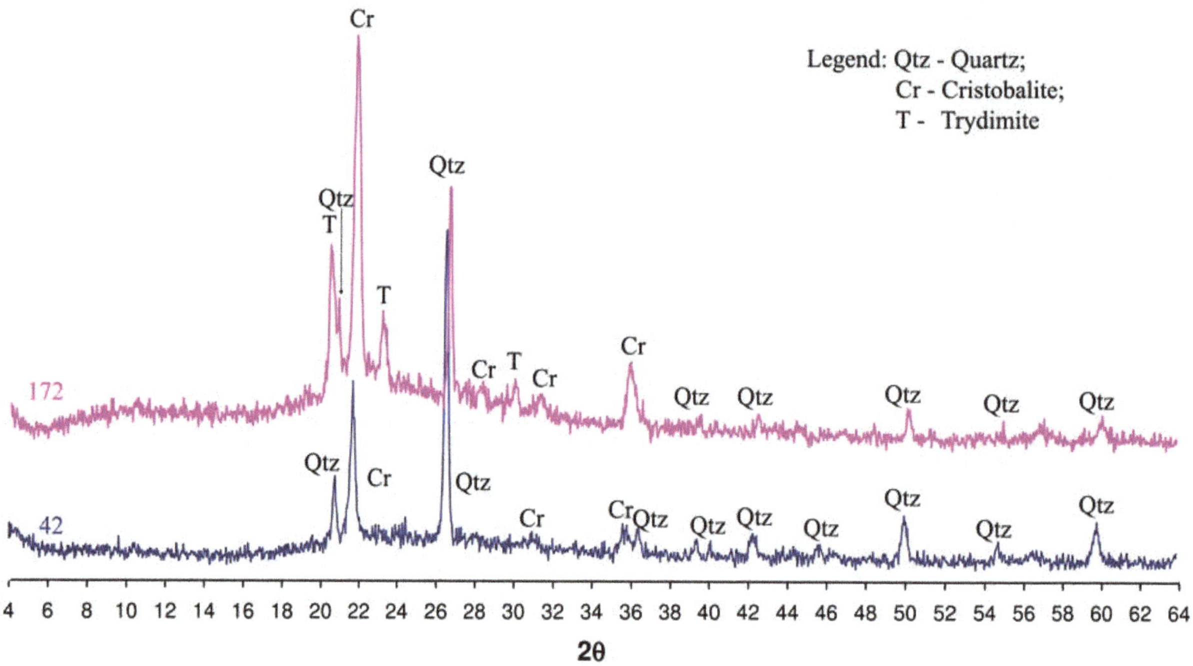


**Figure 4.** Typical XRD patterns of low fired (42) and high fired (172) bodies.

that the samples containing less cristobalite were fired at the lowest temperatures, around 1050°C (Wu *et al.* 2002). In contrast, the samples with the lowest contents of quartz were fired at the highest temperatures, around 1250°C. Taking into account the peak intensities in XRD patterns, most of the samples have a cristobalite/quartz ratio slightly lower or slightly higher than 1, so that a firing temperature of 1100-1200°C can be attributed to them. As concerns tridymite, it is hypothesised that it formed through crystallisation from the melt during the cooling phase of the firing cycle. Therefore, the samples richest in tridymite should be those which developed the greatest amounts of melted material, as a result of a higher temperature or a greater fusibility of the paste. In reality, the crystallisation of tridymite from a melt is also strongly influenced by the cooling rate. Therefore, the samples containing tridymite are the ones which underwent a slower cooling phase after having attained the maximum temperature. It is possible that the firing temperature deduced for the coated products is slightly in excess, as they underwent a double firing, for paste and glaze respectively.

## 5. Discussion

The results obtained from the observations with the scanning electron microscope allowed highlighting micro-structural differences related to the characteristics of the pores, which can be attributed to several causes. The grain size distribution and milling degree of the raw materials, for example, mainly affect the size and size distribution of the pores. On the other hand, the viscosity of the melted phase and the firing conditions exert their main influence on pore morphology, i.e. their sphericity and roundness.

As a general rule, the samples characterised by well-rounded pores with high sphericity could have originated from a less viscous melted phase at high temperatures. However, the effects of the above-mentioned causes are often in opposition to each other, so that it is practically impossible to determine which are the most important causes in each case. In addition, during the firing of the glaze, at a lower temperature than the previous one, the melted phase may retain low viscosity to permit a restructuring of the pores, probably with the aggregation of several small bubbles into a few larger bubbles. Our results confirm these indications; in fact, no correlation emerges between the morphology of the pores and either the estimated firing temperature or the composition of the chemical paste.

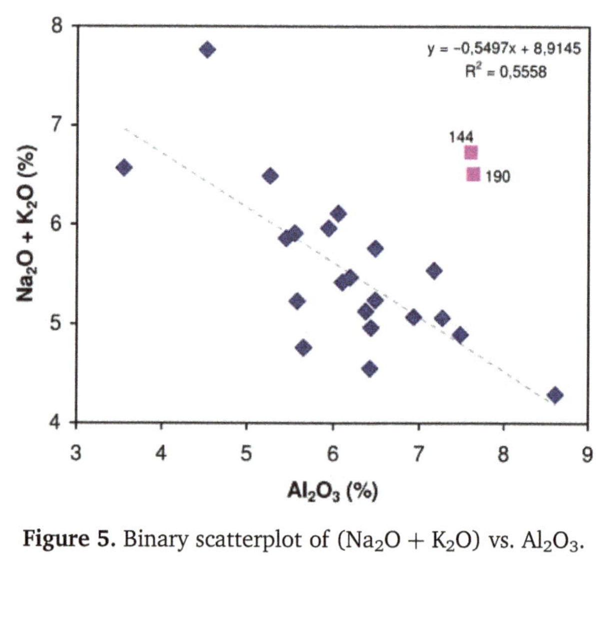


**Figure 5.** Binary scatterplot of ($Na_2O + K_2O$) vs. $Al_2O_3$.

**Table 2.** Comparison between Capodimonte porcelain and other European manufactures.

| Manufacture | $SiO_2$ | $Al_2O_3$ | $K_2O$ | $Na_2O$ | PbO | CaO | MgO | $P_2O_5$ | $SnO_2$ |
|---|---|---|---|---|---|---|---|---|---|
| *Capodimonte 1750-1760* | 84.4 | 6.5 | 2.8 | 0.3 | - | 1.4 | 0.4 | 0.3 | 0.5 |
| *Medici 1575-1587* [(a)] | 77.6 | 9.4 | 5.4 | 3.4 | 0.3 | 2.4 | 0.7 | 0.1 | - |
| *Bottger 1712* [(b)] | 61.5 | 30.1 | 0.3 | 0.1 | - | 6.3 | 1.1 | - | - |
| *Meissen 1731* [(b)] | 59.0 | 35.2 | 3.9 | 0.8 | - | 0.3 | 0.2 | - | - |
| *St. Cloud 1722-1750* [(c)] | 72.9 | 2.1 | 4.5 | 2.1 | <0,1 | 14.5 | 1.6 | - | <0.1 |
| *Vincennes circa 1750* [(c)] | 72.8 | 2.1 | 4.3 | 2.7 | 0.2 | 15.7 | 0.8 | - | <0.1 |
| *Chantilly circa 1750* [(c)] | 75.0 | 2.2 | 2.6 | 3.1 | 0.1 | 13.6 | 1.3 | - | <0.1 |
| *Limehouse 1745-48* [(d)] | 72.4 | 10.7 | 3.0 | 2.8 | 1.0 | 7.1 | 0.7 | <0.2 | - |
| *Chelsea 1745-49* [(d)] | 62.8 | 4.9 | 5.3 | 0.8 | 0.4 | 20.1 | 0.3 | 0.3 | - |
| *Bow 1755-60* [(d)] | 51.2 | 5.6 | 0.6 | 0.6 | 0.4 | 23.2 | 0.6 | 15.3 | - |
| *Worcester 1760* [(d)] | 72.3 | 3.4 | 3.3 | 1.4 | 5.7 | 1.9 | 11.0 | 0.3 | - |

[(a)] Kingery and Vandiver, 1984. [(b)] Kingery, 1986. [(c)] Kingery and Smith,1985. [(d)] Freestone, 1999.

It would seem that the thermal treatment aimed at the firing of the glaze also influenced the crystalline phases, in particular their proportions. The second firing lengthens the total time of the firing process and this could facilitate the formation of a larger amount of silica phases corresponding to the highest temperature ranges (cristobalite and tridymite). In any case, the maximum firing temperature has been evaluated in the range 1050-1250°C. Such a wide temperature range is not only due to differences between a batch and the other, but perhaps especially to differences between the various areas of the kiln, where variations up to 150°C can be considered normal (Kingery and Smith 1985; Kingery 1986).

Starting from the chemical data shown in Table, some considerations related to the composition of the raw paste can be developed. The paste should contain at least three components which perform the three fundamental roles of the ceramic process: a quartz-rich raw material for the structure, a clay for plasticity, and an alkaline raw material for the melting. As for the definition of the type of alkaline material, some help is provided by the correlation diagram ($Na_2O + K_2O$) vs. $Al_2O_3$ (Fig. 5). Excluding two apparently anomalous samples (144 and 190), this correlation is negative, although the correlation coefficient (0.556) is not very significant from a statistical point of view. This tendency allows hypothesising that alumina and alkalis are contained in different raw materials. As a consequence, it would be possible to exclude the use of feldspar and to propose the use of alumina-free minerals, like salt and soda for sodium and barrel tartar for potassium, as reported by Novi (1878) and Pepe (1995). On the other hand, soda was also used during the same period in the Tuscan porcelain manufacture of Doccia, near Florence (Biancalana 2006). Finally, soda and barrel tartar were usually employed in the preparation of glazes for majolica (Piccolpasso 1879), whose production was widespread at that time in Naples and the surrounding area.

Taking into account the concentrations of silica, alumina, sodium, and potassium, it can be deduced that the formula of the raw mixture would be not very different from the following: siliceous sand 70-75%, clay 15-20%, soda plus tartar 5-10%. This is a composition with very little plasticity, which probably entailed substantial processing difficulties, especially in the shaping phase. The high amount of quartz, with its phase transitions (especially β-quartz to α-quartz during cooling), could be one of the principal causes of the large percentage of wastes that characterised the Capodimonte manufacture throughout its period of activity.

Finds 183 and 185 appear slightly anomalous, perhaps due to the difficulty of keeping constant the characteristics and the initial proportions of raw materials. We exclude the possibility that the two finds are products deliberately obtained from mixtures with different composition. In particular, the two parts of sample 183 are significantly different from each other with respect to some of the major elements, such as calcium (4.5 and 1.5% CaO, respectively), silica (78 and 82%, respectively), and to a lesser extent lead

**Table 3.** Average chemical composition of Capodimonte pastes.

| | $SiO_2$ | $Al_2O_3$ | $TiO_2$ | $Fe_2O_3$ | MgO | CaO | $Na_2O$ | $K_2O$ | $P_2O_5$ | $SnO_2$ | Others |
|---|---|---|---|---|---|---|---|---|---|---|---|
| M | 84.39 | 6.51 | 0.13 | 0.35 | 0.39 | 1.41 | 2.62 | 2.85 | 0.29 | 0.52 | 0.55 |
| s | 1.15 | 0.87 | 0.08 | 0.14 | 0.06 | 0.35 | 0.74 | 0.58 | 0.09 | 0.38 | — |

and alkalis, as well as with regards to the presence of manganese in the violet portion. This situation could be explained by the combination of two different pastes, prepared at different times, without accurate mixing.

Neglecting the anomalous samples 183v and 185, it is possible to calculate an average chemical composition (M) and the corresponding standard deviation (s), very representative for the paste of Capodimonte porcelain (Table 3).

These chemical data are in agreement with those reported in the literature (Dell'Agli *et al.* 1995). However, there is some incongruence as regards the crystalline phase composition. In fact, we recognise a considerable amount of quartz and cristobalite, and little or no tridymite, whereas previous studies always identified tridymite in addition to quartz, but never cristobalite. Notwithstanding this, Dell'Agli and co-authors (1995) have suggested a firing temperature of 1200°C, in agreement with the range 1050 - 1250°C that we hypothesise in this study.

The chemical composition of Capodimonte porcelain is very typical and renders it distinct from all other productions, permitting the formulation of a solution to questionable attributions. By comparing the Capodimonte factory with other European manufactures (Table), it clearly becomes visible that the various productions are very well characterised by their chemical composition. The results for Medici porcelain are very similar to those obtained for the Capodimonte production (Kingery and Vandiver 1984), even though it was produced more than 150 years before. Both are very silica-rich, but Medici porcelain does not reach 80% $SiO_2$, while Capodimonte porcelain is close to 85%; in addition, the latter contains a smaller amount of alkaline and earth-alkaline elements ($Na_2O + K_2O +$ - $CaO + MgO = 7.2\%$, compared to 11.9%).

Bottger and Meissen hard porcelain is characterised by a high quantity of alumina, due to the use of kaolin in the raw mixture. In contrast with the first Bottger production, the subsequent production of the Meissen manufacture does not indicate the use of calcium compounds, which are substituted by potassium minerals.

The soft French porcelain produced in St. Cloud, Vincennes, and Chantilly is homogeneous and characterised by the presence of calcium, in a proportion of around 15% CaO. In contrast, the composition of the soft English porcelain is more variable and different typologies can be identified (Freestone 1999). In Chelsea, it is characterised by a high amount of calcium, like in France, and by the presence of lead (more than 4% PbO). In Limehouse, on the contrary, the quantity of calcium is very low, while the silica and alumina contents are significantly higher.

The use of bone ash is attested for the first time in Bow, so that the product (Bone China) has a very high amount of phosphorous (15.3% $P_2O_5$) and calcium. In Worcester, instead, porcelain is produced by employing a mixture of raw materials which includes soapstone, as indicated by the high magnesium content; in addition, the paste contains a significant amount of lead.

The above observations clearly underline that the so-called 'soft porcelain' includes a series of products completely different between one another from the chemical point of view. They probably also differ in terms of their production requirements and the characteristics of the finished product. Therefore, the simplistic division of porcelain into 'hard' and 'soft' should be abandoned, and replaced with a more complete classification based on the chemical composition.

## 6. Conclusions

The paste of Capodimonte porcelain is siliceous and shows a very homogeneous composition, while the different proportions of the silica phases (quartz, cristobalite, and tridymite) and the amount of glassy phase indicate different firing temperatures. Therefore, it appears evident that in the manufacture, the control of the raw materials was easier to accomplish than the control of the firing conditions. The more likely formula of the starting batch appears to be: 70 - 75% of quartz-rich sand, 15 - 20% of clay, and 5 - 10% of soda plus tartar.

The chemical composition of Capodimonte porcelain is very different from that of the other contemporary European porcelains. This makes it possible to propose a new category of porcelain, which could be called 'siliceous porcelain'. Furthermore, we also propose that devising a new classification of the so-called 'soft porcelain' would be useful, because the term includes completely different formulas, which of course give rise to final products with different specifications.

The firing temperatures were mostly estimated to be in the range of 1100 to 1200°C according to the relative proportions of quartz and cristobalite. The presence of tridymite in some samples seems to be also related to the cooling rate, in addition to the effects of the high firing temperature.

The ceramic body is characterised by an abundant closed porosity formed due to the high viscosity of the melt, so that the gases originated during firing could not escape from the paste. The size, shape, and morphology of the pores vary from one sample to another according to the variations in firing conditions and the different viscosity of the melt.

The violet colour of one body is due to the occasional presence of manganese, while the light brown colour of another sample is attributable to iron, as a result of firing in an oxidising atmosphere.

The investigation of Capodimonte porcelain is ongoing, with the aim of characterising the glazes and obtaining new information with regards to the manufacturing process of this unique soft porcelain typology.

## References

Biancalana A., 2006, Terre, massi, vernici e colori della Manifattura Ginori dalla sua nascita agli albori del XIX secolo. *Faenza*, 92, 48-92.

Dell'Agli G., Mascolo G. and Marino O., 1995, The Capodimonte porcelain: a myth turns into a reality. In *Proceedings of the 3rd EMAC Conference*, (ed. B. Fabbri), 283-292, Gruppo Editoriale Faenza Editrice, Faenza.

Freestone I.C., 1999, The science of early British porcelain. *British Ceramic Proceedings*, 60, 11-17.

Kingery W.D., 1986, The development of European porcelain. In *High Technology Ceramics Past, Present and Future: The Nature of Innovation and Change in Ceramic Technology*, (ed. W.D. Kingery), 153-180, Ceramics and Civilization 3, The American Ceramic Society, Columbus, OH.

Kingery W.D. and Smith D., 1985, The development of European soft-paste (frit) porcelain. In *Ancient Technology to Modern Science*, (ed. W.D. Kingery), 273-292, Ceramics and Civilization 1, The American Ceramic Society, Columbus, OH.

Kingery W.D. and Vandiver P.B., 1984, Medici porcelain. *Faenza*, 70, 441-453.

Liverani G., 1959, Porcellane di scavo da Capodimonte. *Faenza*, 45, 120-127.

Mascolo G., 1993, *La porcellana di Capodimonte: tra mito e realtà: Atti Giornate di Studio sulla Porcellana Europea, Napoli, 9/10 Dicembre 1993*, Istituto Professionale di Stato per l'Industria e l'Artigianato della Porcellana e della Ceramica "Giovanni Caselli", Napoli.

Minieri Riccio C., 1878a, *La fabbrica della porcellana in Napoli e sue vicende*, Manuscript read at the Pontaniana Academy in Naples on 27 January 1878.

Minieri Riccio C., 1878b, *Notizie intorno alle ricerche fatte dalla R. Fabbrica della Porcellana di Napoli per rinvenire materiale ...* , Manuscript read at the Pontaniana Academy in Naples on 10 February 1878.

Musella Guida S., 1983, La Real Fabbrica della Porcellana di Capodimonte: la sperimentazione, la struttura produttiva, la commercializzazione del prodotto. *Manifatture in Campania, dalla produzione artigiana alla grande industria*, 68-115, Guida, Napoli.

Novi G., 1878, *La fabbricazione della porcellana in Napoli e dei prodotti ceramici affini*, Manuscript read at the Pontaniana Academy in Naples on 3 November 1878.

Pepe I., 1995, A historical and bibliographical study on Capodimonte porcelain production. In *Proceedings of the 3rd EMAC Conference*, (ed. B. Fabbri), 305-312, Gruppo Editoriale Faenza Editrice, Faenza.

Piccolpasso C., 1879, *I tre Libri dell'arte del vasaio*, (16th century manuscript), Nobili, Pesaro.

Rebuffat O., 1905, *Studi chimici sulla porcellana di Napoli*, Società Cooperativa Tipografica, Napoli.

Wu S.W., Wong D.S.H. and Lu S.Y., 2002, Size effect on silica polymorphism. *Journal of the American Ceramic Society*, 85, 2590-2592.

# 10

# Late Neolithic pottery productions in Syria evidence from Tell Halula (Euphrates valley): A technological approach

**Anna Gómez, Walter Cruells and Miquel Molist**
Department of Prehistory, Universitat Autònoma de Barcelona, Seminar on Prehistoric Archaeology of the Near East (SAAPO), anna.gomez@uab.cat, walter.crvells@uab.cat, miquel.molist@uab.cat

**Abstract** - The pottery production from Tell Halula, dating from the late Neolithic period, reveals the economic and social complexity of this community in the middle Euphrates valley. Throughout the study of pottery assemblages in the context of field work projects carried out from 1996 to 2007, we combined traditional morpho-stylistic analysis with technological ones. In this paper, we present the current situation of the research on the final stage of the Halaf culture, commonly known as Late Halaf. The results are focussed on manufacturing techniques, with the aim to identify several pottery productions and variability patterns.

## 1. Introduction

Like many other concepts, the Late Halaf culture, a term originating from an artificial terminology used to define the Neolithic communities who lived in northern Syria, southern Anatolia, and northern Iraq during the 6$^{th}$ millennium cal BC, arose from an emerging local substrate (Cruells and Nieuwenhuyse 2004; Nieuwenhuyse 2007). The environmental framework reveals, for the entire span of the 7$^{th}$ and 6$^{th}$ millennia cal BC, a climatic optimum with minor fluctuations around 5000 cal BC (calibrated with Oxcal V.3.5 and calibration curve INTCAL98 1, sigma ranges in Gérard and Thiessen 2002). Recent multi-disciplinary studies have revealed more information about these agro-pastoralist communities who lived between 5500-5300 cal BC in semi-sedentary settlements, and who developed a particular material culture, as well as several economic and social competences (Campbell 2007). The symbiosis of these strategies forged the future complex communities of the Chalcolithic period in the Near East (Akkermans and Schwartz 2003; Frangipane 2007).

These groups were characterised by an opportunistic settlement pattern, sealing practices, specific funerary rituals, and painted pottery showing geometric and animal designs. Archaeological remains such as their pottery productions give us a way to approach the economic and social practices of these farming communities and a way to reconstruct, to a certain extent, their production system and social organisation (Davidson 1977; Campbell 1992; Cruells 2001, 2006, 2009; Wilkinson 2003).

## 2. Archaeological context: Tell Halula

Tell Halula is located on the right bank of the middle Euphrates river valley, in the region of Djerabus, Northern Syria. The site is situated between two small seasonal *wadis*, which are becoming a single *wadi* flowing into the Thisrin Dam. The site, which is 360 m by 300 m and 11 m high, includes two elevations with anthropic deposits and has been excavated from 1989 until the present by a team from the Autonomous University of Barcelona led by Dr. Miquel Molist (Molist 1996, 2001; Molist *et al.* in press) (Fig. 1).

The archaeological sequence includes strata from a Pre-Ceramic Neolithic B until the complete Ceramic Neolithic levels (Pre-Halaf and Halaf), covering a timeline from 7900 to 5300 cal BC. Late Halaf (HLVII) layers are located in sectors 30, 32, 36, 38, 39, and 49, dating to 5550-5300 cal BC. The pottery assemblages ascribed to this period are found in several archaeological contexts, principally ash levels, external areas, domestic areas, and an important group of pits situated at the top of the tell (Cruells 2005).

Remains dating back to the Late Halaf period were recovered from various test pits at the site. A first trench opened on the south western slopes in 1996 and 1999 (sectors 30 and 32) yielded a series of stratified open areas occupied during the Late Halaf periods. The recent stratigraphic sequence and related information comes from the new trench opened in 2006 and 2007 in what is also termed sector 30, with an extension known as sector 49. This trench confirms an uninterrupted sequence from

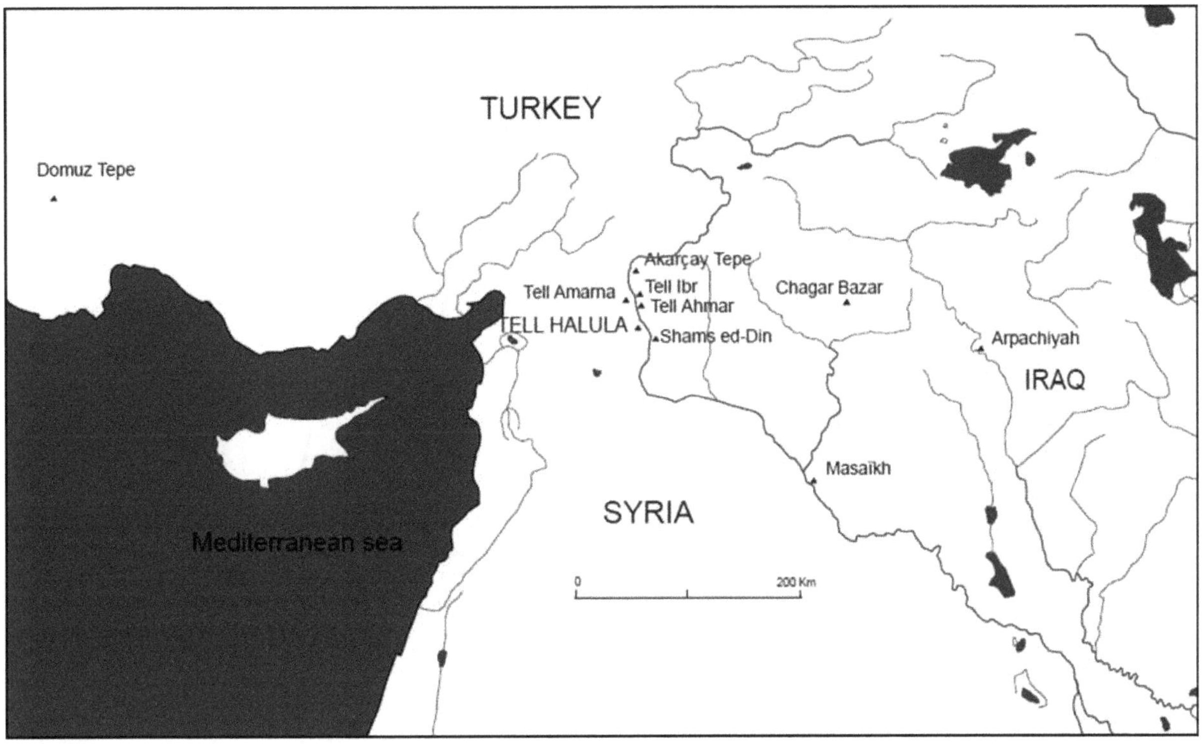


**Figure 1.** Map showing the location of Tell Halula.

Late Pre-Halaf to Late Halaf, but not always at the same place (Gómez 2011). At this point, we must emphasise the mobility pattern of these groups, which did not always settle on formerly occupied layers, but around them (Akkermans and Schwartz 2003). This enables a differentiated settlement pattern to be documented for each site.

The sample presented in this paper constitutes a ceramic assemblage unequally dispersed around the site, as both secondary material, mainly in pits and holes, and external floors related to open areas which constitute a succession of architectural levels formed by rectangular buildings with stone foundations. Halula Late Halaf ceramic assemblages include a large variety of wares. There are fine wares, plain or decorated, as well as simple and burnished coarse wares, either with mineral or plant inclusions (although only a few sherds of the latter). This group also shows incised or impressed decoration.

Using macroscopic observations, this assemblage was initially divided into four main categories: fine painted wares, fine unpainted wares, coarse mineral wares, and coarse vegetal tempered wares. The assemblage also shows modifications in technology, vessel shape, and decorative style. As in earlier periods (Cruells and Nieuwenhuyse 2004; Nieuwenhuyse 2007), fine wares were made of well-levigated clay, fired mostly in oxidising conditions, and commonly having a light or white slip on the exterior surface.

## 3. From intuitive typology to a pottery classification

Generally, there are no unitary criteria that can be employed to arrange a comprehensive typology of published Halaf pottery from the middle Euphrates valley. It is also very difficult, on the basis of the published material, to differentiate the individual uses of cooking and storage wares. Materials from stratigraphic contexts discovered during recent excavations do not allow a detailed typology with varieties and variants to be drawn up, because they are extremely scarce and discontinuous throughout the region. However, prioritising the technological approach, a sufficient number of samples were found to provide an idea of the most typical shapes in different local sequences.

The assemblage presented in this paper consists of 21,164 sherds, minimum number of individuals (mni) 16,688, which were first classified according to their macropetrographic characteristics, manufacture techniques, and main surface treatments. Secondly, a morphological and stylistic typology was established for each technological group using the descriptions of the main inflected points. Although the number of reconstructed vessels is not large, 8 basic typologies have been identified (Fig. 2).

A morphological classification was conducted for each type of ware. The rim was the first morphological feature used to sort the sherds. This first selection separated closed vessels from open ones. In some cases we could calculate the maximum diameter, and, where possible, the volume. Variabilities in the lip, rim slope, and neck height were used to identify different morphological types and enabled us to clarify the typological corpus. Dishes and lids (group VI), straight sided bowls (group II), concave sided bowls (group III), sinuous sided bowls (group I), jars with short or high neck (group IX), simple open and closed bowls (groups III and VII), and closed bowls with short neck (group VIII) are

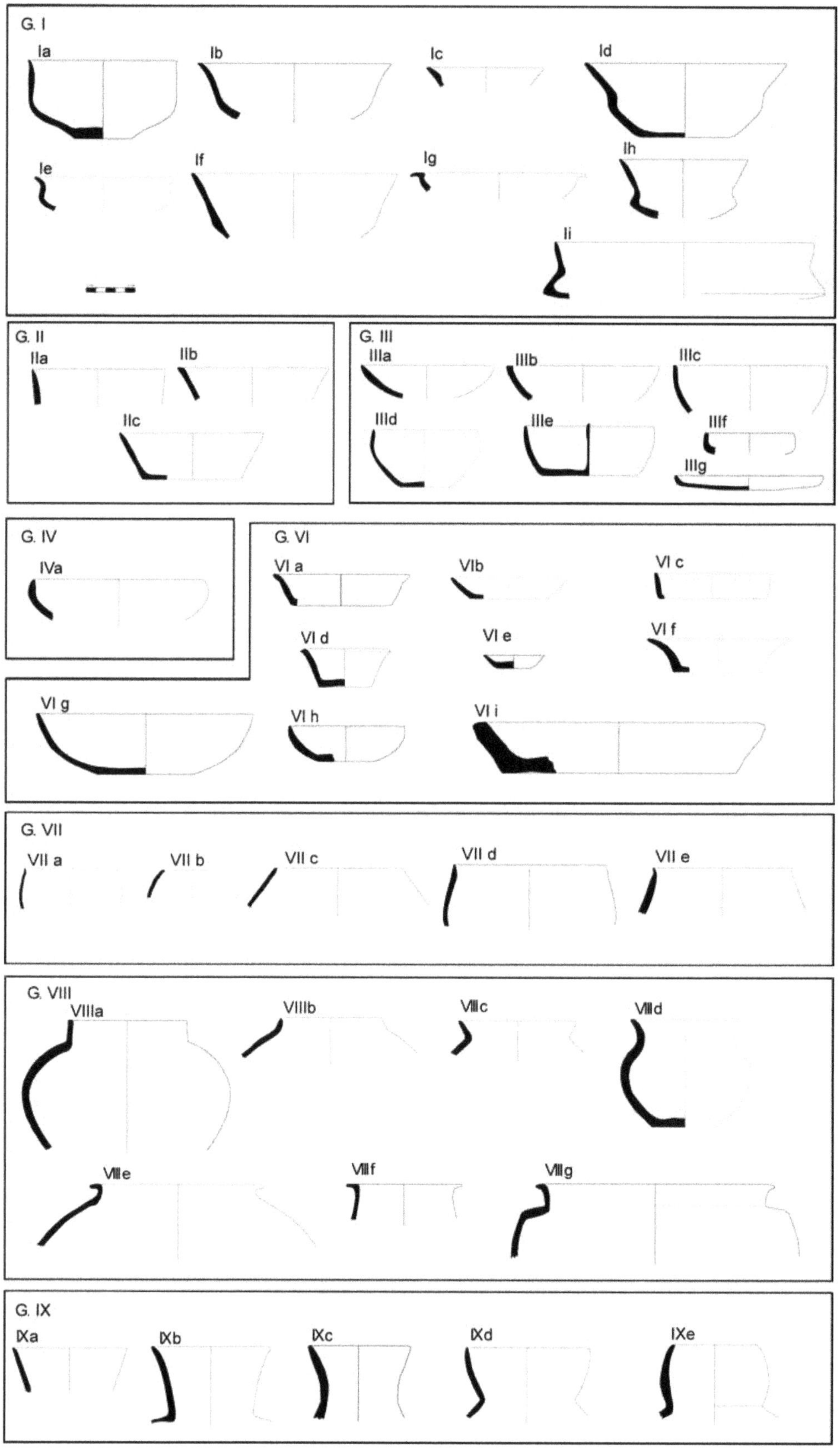


**Figure 2.** Typological chart of Late Halaf ware at the Tell Halula site. Note that Group V, corresponding to *Champagne Vase*, is a form not identified at Halula.

the predominant shapes. Moreover, metric variants led to the distinction of several subgroups according to the different inflection points of the sherds. On the basis of the analysed material, it was quite easy to distinguish a common base typology for these types of vessels, but the technological parameters are almost all different.

This work reveals that 76% of the sherds that are part of the assemblage are open types, and 24% are closed shapes. Diagnostically the most common sherds are lips, dominating the assemblage, with diameters measuring 150-200 mm and bases measuring 100-150 mm in diameter, and with a wall thickness ranging from 5 to 10 mm. The most documented shapes are small containers, such as bowls; 22% are IVa bowls with a concave profile; 18% belong to the Ia and IIb groups, and to closed forms, especially VIIa. We have also identified two types of dishes, VIb (38%) and VI (13%).

The large and medium sized containers comprise pots and jars. The VIIIc is the most common pot shape, although other pots also appear. The IXd jar shape is the most frequently documented, making up 50% of the sample, followed by the IXe shape or bow rim jar, and shape IXb.

## 4. Technological variability

When we studied this assemblage from the technological point of view, we combined data pertaining to manufacture traces and the results of physical and chemical analyses (part of them still in progress) with the aim of identifying not only the main pottery traditions, but also several fashioning and finishing techniques related to different potters' abilities.

Among the basic set of operations which transformed the raw material into a finished product, we have registered several main stages: manufacture and paste preparation, fashioning, finishing (including decoration), and firing (Fig. 3). In this operation chain, Halaf pottery productions generally appear as fine walled vessels with relatively complex shapes, for which a great level of skill must have been required. These pots were constructed using several methods, sometimes in combination: coiling, paddling, pinching, and shaping. In some sherds we found traces of rotation kinetic energy (RKE). It is also possible that baskets were used as a form of *tournette*, or as a stable base upon which pots could be made and dried before they could be used over a flame. This technique is suggestive of identifiable cultural traditions (Roux 2003a; Roux and Courty 2005).

## 5. Petrographic approach

A preliminary mineralogical and petrographic study of the materials at the site has been carried out. Samples were collected in an effort to determine the geological variability of clay sources in the Euphrates region and to redefine the compositional matching of clays to paste groups from previous provenance analyses (X. Clop, work in progress). Geological clays were collected and processed. The preliminary results show that clays in the region were compositionally heterogeneous, and that the clay sources for most of the sampled plain pottery can be traced to locations near the site.

Petrography shows vacuole texture and a low temper content (<5%) in cream coloured core samples, but a clastic texture and higher temper content (10-30%) in brown and red sherds. From a mineralogical point of view, three distinct groups have been observed: the main group made of calcite, quartz, feldspar (phyllosilicates - hematites - gehlenite - hornblende); a second one made of quartz - feldspar (phyllosilicates - hematites - dolomite) and a third one made of quartz, feldspar, calcite, and gehlenite (augite - phyllosilicates - hematites - dolomite), with added biotite.

The dominant wares were made of fine clay that, on the whole, shows non-plastic, small, mineral inclusions only. In addition to the clay fabric, it is important to consider the way in which the clays were fired. The general appearance of the sherds indicates that the potters had probably used kiln types where oxygen could be highly controlled. To date, such kilns have not been well-documented in reliable

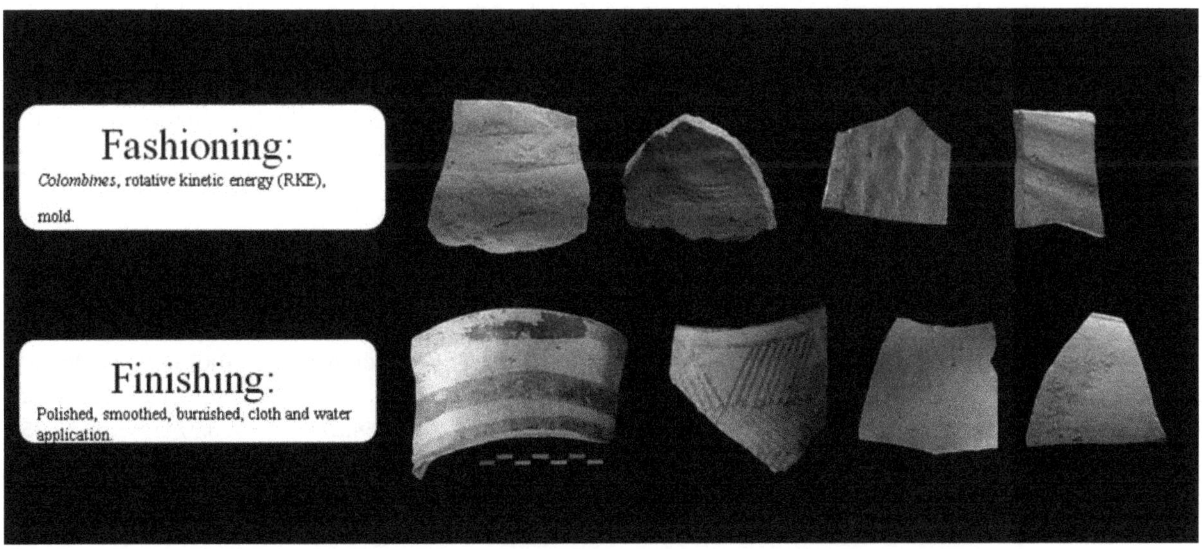


**Figure 3.** Table of the manufacturing processes related to pottery shapes.

contexts in Syria, and the few examples come from Iraq, at the Yarim Tepe site (Munchaev and Merpert, 1973; Merpert and Munchaev 1993a; 1993b). In general, there is a predominance of incomplete oxidation, in fine pastes with mineral inclusions, and an incomplete reduction in coarse pastes with mineral and vegetable temper. Most of the pieces (65%) have their surfaces finished with a slip, followed by smoothing, brushing and burnishing treatments.

The results have implications for the understanding of the nature of plain and coarse ceramic vessels (present on a large scale at the site), as well as decorated ones, and they also help explain why plain ware is well-represented in these assemblages. Since petrographic analyses confirmed the use of several local clays, the differences between them suggest a variety of factors affecting the selection and manipulation of the clays used in vessel manufacture. Further research on these data from Tell Halula vessels is ongoing and will compare the fabrics from Tell Halula with those published from other sites. Chemical analysis will also be conducted on the sherd samples in order to add this information to a growing comparative database.

## 6. Manufacturing techniques

The identification of manufacturing techniques has been carried out on the surface fragments that presented sufficiently large dimensions to preserve clear traces. An initial framework of analysis is seeking technological features that are noticeable in the typological study, not only in the treatment of external or internal parts, or in terms of finishing and decorating, but also with respect to the inflection points that will shape the pot (Bril and Roux 2002). This is the case of the modelling of the edge, an important feature revealing differences between edges in vessels with the same overal shapes. As for the the moulding of the base, it has only rarely been possible to document the complete technical sequence. These have been developed primarily by the technique known as *colombine*, a technique for modelling a flat or slightly concave disk from which the wall starts; this is made by mould or from a clay lump (Table 1).

The ceramics recovered have some physical characteristics in common; mainly, walls have a thickness of about 0.5-0.8 cm, and this feature is generally associated to fine mixed or partially reducing firing treatment. Exterior and interior surfaces are smoothed in all cases, and numerous sherds show polishing/smoothing and burnishing combinations.

With regards to the finishing methods and their location on the vessels, these make it possible to identify a variety of finishes for each type, but also to infer the relationship between sherd and type of finishing strategy (Roux 2007). The orientation of the traces may also indicate the presence of specific patterns of production, especially if these are repeated in several sequences.

The establishment of manufacturing sequences was mostly based on the analysis and evaluation of complete vessels. This procedure has its limitations, since the majority of pieces recovered complete correspond to small containers, mainly bowls and plates; we managed to obtain the complete shape only in 17 cases.

The manufacturing sequence shows that all pots (except Id) rose from the base. In the case of globular shapes, it is possible that the difficulty of modelling a spherical body resulted in a more streamlined sub-spherical profile when the level of skill involved was limited. The bowls and vessels such as dishes also seem to have been moulded from the base. The lip is often enhanced by superimposing a strip at the edge as a decorative solution. The bowls and plates seem to be suitable containers for food service and consumption, although in the case of the former, their use as cooking pots cannot be ruled out.

Most pieces appear modelled with more or less large *colombines* (in small containers these have 0.2 cm frequency tapes, while for mid-sized containers this proportion can vary from 1 to 2.5 cm, with interspaces created during the homogenisation process). Small and medium-sized cointainers show differentiated edges with a deliberate aperture to the exterior made by applying a final strip.

In medium-sized pots, the join between the upper and lower part that indicates that the body and the neck were modelled in different stages is more easily recognisable. In general, this area shows a high rate of breakage, showing that it is one of the weaker areas of the vessel, which therefore was easier to break by indirect impact. Thus, it is possible to infer a multi-stage manufacturing procedure for these.

For closed shapes, we documented different treatments on the inner and outer surfaces. Inside closed shapes, we found an intentional reduction in wall thickness and a regularisation of the inner face. The outside of closed shapes usually shows a simple smooth finish, and this application seems not to be corresponding to any specific technical or morphological criteria other than the intention to make pots with a small investment of time. Also, this element could be indicative of manufacture by an inexperienced potter (Bernbeck 1999).

**Table 1.** Main manufacturing techniques at Tell Halula by category (fine ware FW, fine painted ware FPW, coarse organic-tempered ware COW, and coarse mineral ware CMW). RKE stands for 'rotative kinetic energy'.

| | Balls of clay % | Flat slabs % | Colombine/coils % | Mould % | RKE % | Mixed % |
|---|---|---|---|---|---|---|
| FW | 2.44 | 5.77 | 40.65 | 54.27 | 3.77 | 42.75 |
| FPW | 1.25 | 2.48 | 42.41 | 39.66 | 0.48 | 25.44 |
| COW | 3.01 | 0.76 | 26.55 | 28.19 | 0.21 | 24.17 |
| CMW | 8.4 | 3.03 | 30.33 | 21.56 | 0.09 | 40.25 |

There are different techniques to complete the surface treatment of walls that reduce stress and increase thermal conductivity, improving the performance of vessels for culinary uses (Hein *et al.* 2008). The burnishing treatment, the compaction of the surface, or friction and pressure produce a glossy finish and also reduce porosity, thus increasing the resistance and the waterproof qualities of the pot; we documented these features in the coarse ware assemblages.

Except for large closed vessels, it seems that the decision to employ one type of finishing technique rather than another did not depend on the modelling option, but it rather shows a clear relationship with the typological groups. This suggests that for medium-sized vessels, more regular patterns of execution were needed in the manufacturing process, which could be associated with a higher level of skill in the craft production.

The great diversity in finishing treatments makes us think that the products are the work of several artisans. While modelling and smoothing show a pattern of homogeneous conduct, bases and exterior surface treatments show great diversity of skills, particularly notable in small containers. At this point, we should consider the extent to which all of these productions derive from a common pool of knowledge, and to what degree we can speak of variable technical traditions.

## 7. Volumetric approach

The study was conducted using MicroStation validated by Potutility (a program from the Arcane project), which, unlike other available programs, produced a reasonably reliable table of values concerning the volume of each complete profile recovered.

Prior to the volumetric study, we took into account various parameters associated with archaeological evidence. As shown in the graph (Fig. 2), we have recovered more profiles associated with small vessels than with large ones. These larger containers appear on site, but are very fragmented, and it is very difficult to infer the entire typology. Thus, this sample is mainly made up of bowls, plates, and drinking cups associated to domestic use (Table 2).

The technical-morphological features of the small containers indicate that they were intended for individual consumption of liquids or semi-solid foods. The medium/large pieces are ideal containers for the storage and transportation of liquid provisions (small-rimmed jars) or solid provisions (wide-rimmed jars). Dish-covers or collar-rimmed jars have a depression on the inside of their necks which acts as a closing device. Large bowls, the majority of which are open pots, could have been used to transfer provisions.

## 8. Decorative motifs

It has been demonstrated that Halaf pottery decoration was highly structured (Steinberg and Kamilli 1984). The choices governing the motifs used appear to be based around decisions such as the location on the vessel's interior/exterior, or the type of pottery being decorated, in a harmonic conceptualisation and always related to surface treatment. Pictorial decoration is present in 47% of the productions, which are monochrome, applied over a finished surface (slip application in 65%, followed by smoothing, brushing, and burnishing treatments). We found black, red and brown as the predominant colours, with the additional presence of polychrome vessels (8%) that combine black and red occasionally with white.

The decorative style emphasises the use of bounded design motifs (Wobst 2004). With few exceptions, the motifs were attached to the horizontal, structural lines that divided the empty vessel surface. Figurative motifs are painted along the interior rim in open shapes and on the middle wall in closed ones. The decorative motifs on the top of the vessels, on the outer and inner lips, are mainly geometric motifs (89%), floral (4%), and zoomorphic motifs, as well as horizontal representations of bucrania (7%). The geometric motifs include various stepped patterns, meanders, and variations on the basic themes of crosshatching and zigzag. In general, we did not find a sharp break in decorative patterns between periods. This seems to indicate a lasting tradition of repeating the same pattern designs within the same context (Fig. 4).

## 9. Pigment analyses

The pigment analysis was carried out using Particle Induced X-ray Emission (PIXE), which is a powerful yet non-destructive elemental analysis technique used to identify raw materials and their provenance. This instrument allows rapid non-destructive or minimally destructive identification of most of the pigments used in ancient times, and it has greatly increased the amount of pigment identification work carried out by the Archaeometry group at the University of Liège, under the direction of Helena Calvo del Castillo (Gómez *et al.* 2012) (Fig. 5).

So far, 17 pottery samples have been analysed with the aim of identifying the composition of each pigment, and of making several observations regarding their origin and manipulation process. The results show that in 13 out of the 17 sherds, the black pigment is composed of manganese oxide or manganese phosphate. Manganese oxide is found as a free element in nature (often mixed with iron) and as a component in many minerals. Manganese ions have various colours depending on their oxidation state, and they are usually used as pigments. Black and red pigments are all composed of iron oxides and in few cases charcoal. These results are important because the published pigment analyses for one of the nearby sites, Tell Amarna (Gilbert 2004), situated nearly 40 km north of Halula, also in the Euphrates valley, show a composition consisting primarily of hematite ($Fe_2O_3$), hematite (alfa), and magnetite ($Fe_3O_4$), the rare presence of $CaCO_3$ and charcoal. The results from Tell Amarna emphasised that the main component of pigments used in the dark zones is magnetite, while in Halula manganese is the main mineral employed for this purpose. Other observations made in Tell Amarna are that the paler zones are made with mixtures of magnetite and hematite, and the bright character of some samples is not the consequence of the use of different pigments, but of polishing. Thus, continuing this study and increasing the number of samples appears necessary in order to formulate further questions.

**Table 2.** Volumetric results of complete vessels from Tell Halula. Note that values indicate raw densities.

| Id Number | Typology | Volume (litres) | Water | Oil | Wine | Wheat | Barley |
|---|---|---|---|---|---|---|---|
| THL1182 | VIc | 0.51 | 0.47 | 0.47 | 0.49 | 0.38 | 0.32 |
| THL1221 | VIc | 0.55 | 0.55 | 0.51 | 0.53 | 0.41 | 0.34 |
| THL1234 | Ib | 0.42 | 0.42 | 0.38 | 0.40 | 0.31 | 0.26 |
| THL1243 | VIb | 0.33 | 0.33 | 0.31 | 0.32 | 0.25 | 0.21 |
| THL1245 | Ic | 1.04 | 1.04 | 0.96 | 1.00 | 0.77 | 0.65 |
| THL1431 | VIe | 0.15 | 0.15 | 0.14 | 0.15 | 0.11 | 0.10 |
| THL1464 | Ic | 0.54 | 0.54 | 0.50 | 0.52 | 0.40 | 0.34 |
| THL1476 | VIb | 0.14 | 0.14 | 0.13 | 0.14 | 0.11 | 0.09 |
| THL1901 | VIb | 0.24 | 0.24 | 0.22 | 0.23 | 0.17 | 0.15 |
| THL1902 | IIIb | 0.17 | 0.17 | 0.16 | 0.16 | 0.12 | 0.10 |
| THL1907 | VIc | 0.40 | 0.40 | 0.37 | 0.38 | 0.30 | 0.25 |
| THL1927 | VIIId | 0.25 | 0.25 | 0.23 | 0.24 | 0.18 | 0.15 |
| THL1928 | VIc | 0.50 | 0.50 | 0.46 | 0.48 | 0.37 | 0.30 |
| THL1940 | VIc | 0.21 | 0.21 | 0.19 | 0.20 | 0.15 | 0.13 |
| THL1942 | VIIa | 1.52 | 1.52 | 1.40 | 1.46 | 1.13 | 0.94 |
| THL1957 | Ib | 2.49 | 2.49 | 2.29 | 2.39 | 1.84 | 1.54 |
| THL1960 | Id | 0.38 | 0.38 | 0.35 | 0.36 | 0.28 | 0.23 |
| THL1967 | Ib | 0.33 | 0.33 | 0.31 | 0.32 | 0.25 | 0.21 |
| THL1998 | IIIa | 0.57 | 0.57 | 0.52 | 0.55 | 0.42 | 0.35 |
| THL2007.1 | Id | 1.03 | 1.03 | 0.95 | 0.99 | 0.76 | 0.64 |
| THL2026 | VIc | 0.10 | 0.10 | 0.09 | 0.09 | 0.07 | 0.06 |
| THL2101 | VIIIc | 0.54 | 0.54 | 0.49 | 0.52 | 0.40 | 0.33 |
| THL2103 | VIb | 0.54 | 0.54 | 0.50 | 0.52 | 0.40 | 0.34 |
| THL2134 | VIc | 0.36 | 0.36 | 0.33 | 0.34 | 0.27 | 0.22 |
| THL2141 | Ic | 1.40 | 1.40 | 1.29 | 1.34 | 1.03 | 0.87 |
| THL2145 | Id | 1.99 | 1.99 | 1.83 | 1.91 | 1.47 | 1.23 |
| THL2146 | IIIc | 0.49 | 0.49 | 0.45 | 0.47 | 0.36 | 0.31 |
| THL2165 | IIId | 0.30 | 0.30 | 0.28 | 0.29 | 0.22 | 0.19 |
| THL2166 | VIIa | 1.79 | 1.79 | 1.65 | 1.72 | 1.33 | 1.11 |
| THL2168 | Id | 4.88 | 4.88 | 4.49 | 4.68 | 3.61 | 3.03 |
| THL2170 | IIId | 0.09 | 0.09 | 0.08 | 0.08 | 0.06 | 0.05 |
| THL2175 | IIIa | 0.70 | 0.70 | 0.65 | 0.68 | 0.52 | 0.44 |
| THL3024 | VIc | 0.23 | 0.23 | 0.21 | 0.22 | 0.17 | 0.14 |
| THL3055 | IIb | 1.02 | 1.02 | 0.93 | 0.97 | 0.75 | 0.63 |

**Figure 4.** Some painted sherds from Tell Halula.

## 10. Preliminary considerations

This study has enabled us to identify several pottery traditions at the Tell Halula site. These are interpreted as the result of different production and distribution systems, involving production at local and regional levels, well documented in nearby sites, such as Tell Amarna (Cruells 1998; 2004), Shams ed-Din (Al-Radi and Seeden 1980; Seeden 1982) and Tell Masaikh (Robert *et al.* 2008). Thus, in further works, attesting the complex relations that the local populations of the middle Euphrates valley may have

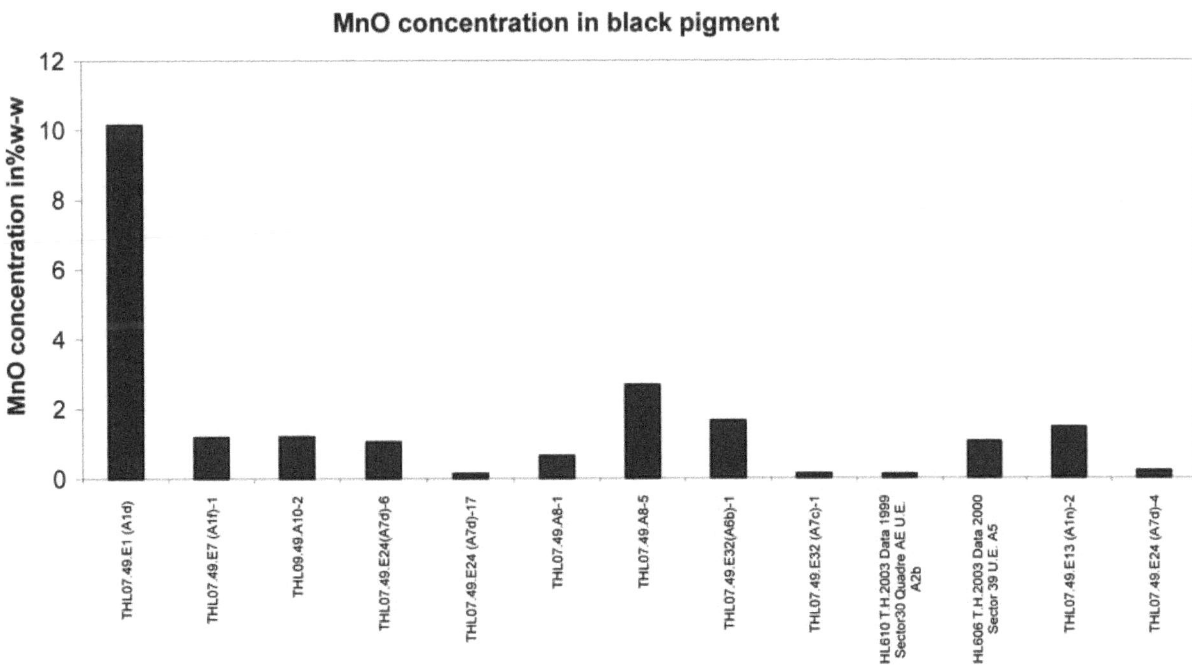


**Figure 5.** Composition of black pigment.

maintained with groups living in neighbouring regions appears as a very important endeavour.

The Late Halaf pottery from the Euphrates region shows an 'apparent' homogeneity, although this disappears when we examine the technological aspects, such as the manufacturing and production features of the pieces. In this context, the pottery in the Euphrates area is considered to be the result of local developments, with obvious contact with surrounding communities.

Such variability could be attributed to manufacturing techniques, individual knowledge, functional purposes, and site formation processes. It was concluded that several ceramic production practices existed within the population of Halula, which transmitted technological knowledge through the manipulation of pottery. The presence and abundance of these pottery types, especially the painted productions, in different archaeological contexts indicate that pottery production was increasing, and some shapes appear to be the work of specialist manufacturers. The use of special tools and standardised techniques supports this idea (Roux and Matarasso 1999; Roux 2003b).

At the same time, during the Late Halaf period, we report an increase in complex shapes, such as the bow rim jar, or jars with arched necks and a flat base. Also, some middle Halaf shapes increase in number, such as shallow open bowls and bowls with a globular carination, and others, known as cream bowls, appear with less frequency and become careened shapes that show us a clear continuity with regards to previous pieces (Cruells 1996). These new shapes are not related to new fashioning trends, as we could see in our work, and the answers seem to be related to dietary practices or storage needs.

By using binocular microscope techniques, we found that some pottery sherds from the latest stages seem to have been made with rotative kinetic energy movement. It has been suggested, however, that pottery was decorated using a slow *tournette* leading to the increased use of strongly linear, larger geometric designs. Oval mouths and rounded bases on other pottery shapes seem to have been finished with a type of rotative kinetic energy. These sherds belong to small wares, bowls, and open shapes. Although Tell Halula offers good contextualised contexts, it is important to underline that similar processes can be observed at sites such as Chagar Bazar (A. Gómez, work in progress), Domuztepe (Irving 2001), etc.

It seems reasonable to suggest that these different manufacturing techniques were associated to different pathways of learning. In addition, evidence of combination techniques has now been confirmed, as has the use of wheel coiling. Wheel coiling and wheel-throwing co-existed. The use of such methods required that the potters had specialised skills, denoting a particular transmission of knowledge that affected several dimensions: object function, technological innovation, and social identity.

Also, the large number of painted decorative motifs, and their high quality, as well as the presence of polychrome examples, seem to indicate the consolidation of increasingly complex societies (Davidson 1977), as a consequence of the social inequalities initially documented in the formative Halaf. The way in which these communities managed their resources and consolidated and socially assimilated Ubaid influence is another major pending issue.

## References


Akkermans P.M.M.G. and Schwartz G.M., 2003, *The Archaeology of Syria. From Complex Hunter-Gatherers to Early Urban Societies (cal. 16,000-300 BC)*, Cambridge University Press, Cambridge.

Al-Radi S. and Seeden H., 1980, The AUB Rescue Excavations at Shams ed-Din Tannira. *Berytus*, 28, 88-126.

Bernbeck R., 1999, Structure strikes back: intuitive meaning of ceramics from Qale Rostam, Iran. In *Material symbols: Culture and economy in prehistory*, (ed. J. Robb), 90-111, Southern Illinois University, Carbonadale.

Bril B., and Roux V. (eds.), 2002, *Le geste technique. Réflexions méthodologiques et anthropologiques*, Editions Erès, Ramonville Saint-Agne.

Campbell S., 1992, *Culture, Chronology and Change in the Later Neolithic of North Mesopotamia*, PhD Thesis, University of Edinburgh.

Campbell S., 2007, Rethinking Halaf Chronologies. *Páleorient*, 33, 103-136.

Cruells W., 1996, La cerámica pintada Halaf. In *Tell Halula (Siria). Un yacimiento Neolítico del Valle Medio del Éufrates. Campañas de 1991 y 1992*, (ed. M. Molist), 99-114, Ministerio de Educación y Cultura, Madrid.

Cruells W., 1998, The Halaf levels of Tell Amarna (Syria). First preliminary report. *Akkadica*, 106, 1-21.

Cruells W., 2001, Nuevas aportaciones a la cultura Halaf en Siria. In *Actas del I Congreso de Arqueología e Historia Antigua del Oriente Próximo*, (eds. J.L. Montero, J. Vidal and F. Maso), 135-157, Monografies Eridu 1, Barcelona.

Cruells W., 2004, The pottery. In *Tell Amarna (Syrie) I. La période de Halaf*, (eds. Ö. Tunca and M. Molist), 41-200, Publications de la Mission archéologique de l'Université de Liège en Syrie, Peeters, Louvain, Paris, and Dudley, MA.

Cruells W., 2005, *Orígens, emergència i desenvolupament de la ceràmica Halaf a Síria*, PhD Thesis, Universitat Autònoma de Barcelona.

Cruells W., 2006, The pottery. In *Tunca, in Chagar Bazar (Syrie) I. Les sondages préhistoriques (1999-2001)*, (eds. Ö. Tunca and A-M. Bagdho), 25-79, Publications de la Mission archéologique de l'Université de Liège en Syrie, Peeters, Louvain, Paris, and Dudley, MA.

Cruells W., 2009, Nouvelles données sur les origines et le développement de la céramique Halaf en Syrie. In *Méthodes d'approche des premirères productions céramiques: étude de cas dans les Balkans et au Levant. Table-ronde de la Maison de l'Archéologie et de l'Ethnologie (Nanterre-France), 28 février 2006, Vol. 12*, (eds. L. Astruc, A. Gaulon and L. Salanova), 93-103, Verlag Marie Leidorf, Rahden/Westfalen.

Cruells W. and Nieuwenhuyse O., 2004, The Proto-Halaf Period in Syria, New Sites, New Data. *Paléorient*, 30, 47-68.

Davidson T.E., 1977, *Regional Variation within the Halaf Ceramic Tradition*, PhD Thesis, University of Edinburgh.

Frangipane M., 2007, Different types of egalitarian societies and the development of inequality in early Mesopotamia. *World Archaeology*, 39, 151-176.

Gérard F., and Thiessen L. (eds.), 2002, *The Neolithic of Central Anatolia. Internal Developments and External Relations during the 9th-6th Millennia cal BC*, Proceedings of the International CANeW Table Ronde, Istanbul, 23-24 November 2001, Ege Yayinlari, Istanbul.

Gilbert B., 2004, Analysis of the painting on six Halaf period pottery samples from Tell Amarna (Syria). In *Tell Amarna (Syrie) I. La période de Halaf*, (eds. Ö. Tunca and M. Molist), 213-219, Publications de la Mission archéologique de l'Université de Liège en Syrie, Peeters, Louvain, Paris, and Dudley, MA.

Gómez Bach A., 2011, Caracterización del producto cerámico en las comunidades neolíticas de mediados del VI milenio cal BC: el valle del Éufrates y el valle del Khabur en el Halaf Final, PhD Thesis, Universitat Autònoma de Barcelona.

Gómez Bach A., Calvo del Castillo H., Cruells W., Strivay D. and Molist M., 2012, Black, red and white: characterizing Neolithic ceramic productions in middle Euphrates Valley. In *7th International Congress on the Archaeology of the Ancient Near East. British Museum, London 12-16 April 2010*, (eds. R. Matthews and J. Curtis), Vol. III, 527-536, Harrassowitz, Wiesbaden.

Hein A., Müller N.S., Day P.M. and Kilikoglou V., 2008, Thermal conductivity of archaeological ceramics: The effect of inclusions, porosity and firing temperature. *Thermochimica Acta*, 480, 35-42.

Irving A., 2001, *A Contextual Study of Ceramic Evidences for Social Relations and Change during the Halaf-Ubaid Transition*, PhD Thesis, University of Manchester.

Merpet N.Y. and Munchaev R.M., 1993a, Yarim Tepe II: the Halaf levels. In *Early Stages in the Evolution of Mesopotamian Civilization*, (eds. N. Yoffee and J.J. Clark), 128-162, Soviet Excavations in Northern Iraq. University of Arizona Press, Tucson, AZ.

Merpet N.Y. and Munchaev R.M., 1993b, Yarim Tepe III: the Halaf levels. In *Early Stages in the Evolution of Mesopotamian Civilization*, (eds. N. Yoffee and J.J. Clark), 163-206, Excavations in Northern Iraq. University of Arizona Press, Tucson, AZ.

Molist M. (ed.), 1996, *Tell Halula (Siria). Un yacimiento neolítico del valle medio del Éufrates. Campañas de 1991 y 1992*, Ediciones del Ministerio de Educación y Ciencia, Madrid.

Molist M., 2001, Halula, village néolithique en Syrie du Nord. In *Communautés villegeoises du Proche Orient à l'Atlantique 8000-2000 avant notre ère*, (ed. J. Guilaine), 35-52, Editions Errance, Paris.

Molist M., Anfruns J., Borrell F., Buxo R., Clop X., Cruells W., Faura J.M., Ferrer A., Gómez A., Guerrero E., Saña M., Tornero C. and Vicente O., in press, Tell Halula: new data on VII and VI millennia cal BC occupation in Upper Mesopotamia. *Interpreting the Late Neolithic of Upper Mesopotamia, Leiden 26-28 mars 2009*.

Munchaev R.M. and Merpert N., 1973, Excavations at Yarim Tepe 1972: fourth preliminary report. *Sumer*, 29, 3-16.

Nieuwenhuyse O., 2007, *Plain and painted pottery. The rise of late Neolithic ceramic styles on the Syrian and northern Mesopotamian plains*, Paper on archaeology of the Leiden Museum of Antiquities, Brepols, Turnhout.

Robert B., Blanc C. and Massetti-Rouault M-G., 2008, Characterizing the Halaf-Ubaid Transitional Period by studying ceramic from Tell Masaikh, Syria. Archaeological data and archaeometry investigations. In *Proceedings of 4th International Congress of the Archaeology of the Ancient Near East, 29 March - 3 April 2004, Freie Universität Berlin, Vol 2*, (eds. H. Kühne, R.M. Czichon and F.J. Kreppner), 225-234, Harrassowitz, Wiesbaden.

Roux V., 2003a, A dynamic systems approach to technological change: application to the emergence of the potter's wheel in South Levant. *Journal of Archaeological Method and Theory*, 10, 1-30.

Roux V., 2003b, Ceramic standardization and intensity of production: quantifying degrees of specialization. *American Antiquity*, 68, 768-782.

Roux V., 2007, Ethnoarchaeology: a non-historical science of reference necessary for interpreting the past. *Journal of Archaeological Method and Theory*, 14, 153-178.

Roux V. and Courty M.-A., 2005, Identifying social entities at a macro-regional level: Chalcolithic ceramics of South Levant as a case study. In *Pottery Manufacturing Processes: Reconstruction and Interpretation*, (eds. A. Livingstone-Smith, D. Bosquet and R. Martineau), 201-215, British Archaeological Reports International Series 1349. Archaeopress, Oxford.

Roux V. and Matarasso P., 1999, Crafts and the evolution of complex societies: new methodologies for modelling the organization of production, a Harappan example. In *The social dynamics of technology, practice, politics and world*, (eds. M.-A. Dobres and C. Hoffman), 46-70, Smithsonian Institution Press, Washington and London.

Seeden H., 1982, Ethnoarchaeological Reconstruction of Halafian occupational units at Shams ed-Din Tannira. *Berytus*, 30, 55-96.

Steinberg A. and Kamilli D.C., 1984, Paint and Paste Studies of Selected Halaf Sherds from Mesopotamia. In *Pots and Potters. Current Approaches in Ceramic Archaeology*, (ed. P.M. Rice), 187-208, Institute of Archaeology, University of California, Monograph 24, Los Angeles.

Wilkinson T.J., 2003, *Archaeological Landscapes of the Near East*, University of Arizona Press, Tucson, AZ.

Wobst H.M., 2004, Stylistic behaviour and information exchange. In *Material culture. Critical concepts in the social sciences, Vol I*, (ed. V. Buchli), 327-349, Routledge, London.

## ACKNOWLEDGEMENTS

We are grateful to the General Direction of Antiquities and Museums (DGAM) of Syria, the Spanish Ministerio de Ciencia y Tecnologia (HUM2007-66237), and the Generalitat de Catalunya (SGR-2009-00607). This paper was reviewed at UAB Language Service.

# 11

# The use of andesite temper in Inca and pre-Inca pottery from the region of Cuzco, Peru


**Rob Ixer, Sara Lunt and Bill Sillar**

UCL Institute of Archaeology, 31-34 Gordon Square, London WC1H 0PY, United Kingdom, r.ixer@btinternet.com, saralunt13@gmail.com, b.sillar@ucl.ac.uk



**Abstract** - Using ceramic petrography, a fabric of Cuzco Inca pottery is compared with those of two pre-Inca wares, Killke and Lucre. Andesite temper is identified in the Lucre and Cuzco Inca fabrics. This is compared with andesite from the Rumicolca Formation and a match is found for some of the material. Lucre is identified as a technological precursor to Cuzco Inca. The possibility that the temper and Inca building stone came from the same source is mooted but cannot be confirmed based on the present evidence.


## 1. Introduction

Amongst the pre-Conquest pots from the Cuzco region of Peru (Fig. 1), there is a group made from a distinctive clay body. In thin section, the predominant inclusions are identified as andesite, and the evidence suggests that this is temper, introduced by the potters to the clay in order to create an artificial body, or fabric, from which to make their pots (Ixer and Lunt 1991).

This andesite-tempered fabric (hereinafter 'ATF') was used to make Cuzco Inca pottery, as defined stylistically by John Rowe (1944, 47-9). Although other fabrics were also used for these wares (Ixer and Lunt 1991, 152, 162), the conjunction of the ATF and Cuzco Inca pottery is noteworthy: the standardisation of the fabric and the very high quality of the wares mark a step-change in the ceramic production of the region.

Although in these respects the pottery is highly innovative, some of its decorative, formal, and technical features can be identified in the pre-Inca wares of the south-central Andes. Killke pottery, in particular, has long been considered a precursor to the pottery of the Incas. However, there are important elements of Cuzco Inca wares that are not found in Killke, including the very large size of some of the vessels (Chatfield 2007, 75), and, most significantly here, the exclusive use of one specific rock type, andesite, as temper.

To examine this issue, this paper presents two groups of evidence. These indicate, first, that another pre-Inca ware, Lucre, also contributed to the development of Cuzco Inca pottery. Secondly, one of the sources of andesite temper in Inca and Lucre pottery can be broadly identified. Andesite from an outcrop in the same geological formation was used for Inca buildings in Cuzco. Although the evidence presented here cannot confirm that the sources of temper and building stone were the same, it raises the possibility of an interesting relationship between two state-controlled activities, which should be pursued in future work.

## 2. Pre-Inca and Inca pottery in the Cuzco region

The production of Killke pottery began in the Late Intermediate Period (c. 1000-1400 AD). Stratigraphically, it occurs in deposits underlying Inca contexts, sometimes continuing through them, and there is some evidence that production continued into the colonial period (Chatfield 2007, 334 ff). It is the dominant ceramic of the pre-Inca period in the valley of Cuzco, and both Killke itself and an entire family of 'Killke-related' wares, are distributed throughout the Cuzco region.

The relationship between Killke and Cuzco Inca pottery was discussed by John Rowe in an influential paper (Rowe 1944), in which he identified the 'Killke Series' as 'Early Inca' on the basis of stylistic similarities, distribution and relative date (Rowe 1944, 61). Although this view can be challenged based on good evidence (Chatfield 2007, 70ff, 334ff), it still has currency, underpinning recent work on the significance of Killke for studying the emergence of the Incas (summarised in Bauer 2004, 74-5) and providing the starting-point for the technological comparison between Killke and Cuzco Inca wares in Ixer and Lunt (1991).

The Killke fabric technology, as described by Ixer and Lunt in 1991, was based on the use of clean clay, tempered with a limited range of medium-grained, feldspathic rocks, similar in appearance to one another; there was no andesite. The fabrics form a fairly coherent group (Ixer and Lunt 1991, 156), although there are variations, suggesting

**Figure 1.** Classic Inca pottery displayed in the Museum of the Universidad de San Antonio Abad del Cuzco. The largest vessels are over one metre high.

that Killke was produced in a number of workshops, albeit all working within a common tradition. The Killke fabric preparation technique - the use of a clean clay and the addition of plentiful temper from a limited range of rocks - is comparable to that of the ATF. It is notable that an early Cuzco Inca sub-style, previously identified as Ware 29, was made from a fabric identical to the Killke fabric (Ixer and Lunt, 152, 162-3). Killke wares were coil-built. The forms include open bowls and jars, some of constricted form. They were painted in red, black and, very occasionally, white, in freely-executed rectilinear motifs built into patterns arranged vertically and diagonally on the body of the pot, and then polished. They were sometimes modelled, the 'face-neck' type. They were fired in oxidising conditions (Ixer and Lunt, 140-1).

Cuzco Inca pottery (also called 'Imperial Inca', 'Classical Inca', or 'Inca Fine Wares') was made during the Late Horizon Period (c. 1400-1532 AD). It was probably made in or around Cuzco, the capital of the Inca Empire, to judge by the quantity found there and the occurrence of shapes and designs unknown elsewhere. The pottery is highly recognisable, very standardised in technology, form and design, and of excellent quality. This is all in marked contrast to earlier wares and to other Late Horizon wares of the region, and suggests a unique level of production control. The sudden appearance of these wares, fully-formed, in the archaeological record of the Late Horizon, and the nature of many contexts in which they occur, give credence to the widely-held view that Cuzco Inca pottery was a political artefact, created by the Inca state for the benefit of the Inca elite and their favoured clients, and that its production was state-controlled. The Cuzco Inca style was influential in the development of pottery-making across the Inca Empire, where 'provincial Inca' wares, based on a selection of Cuzco prototypes, were produced in state-controlled potteries.

Rowe's descriptions of Cuzco Inca pottery (1944, 47-9) are the standard guide. He identified seven stylistic sub-groups, which together he called the Cuzco Series (Rowe 1944, 60). The fabric ('paste') was also described (Rowe 1944, 47). This fabric, the ATF, has been subsequently studied in detail (Ixer and Lunt 1991, Ware 13), and its salient features identified. It is highly standardised, based on very clean clay with few accessory minerals, and tempered with andesite in uniform sizes, shapes and quantities throughout the samples. Although other fabrics were also used to make Cuzco Inca pottery, the ATF predominates. The pottery was coil-built. The forms are standardised in a range of modular sizes and include the pointed-based, long-necked aryballus, wide-mouthed storage vessels and small open bowls. It was painted, predominantly with small rectilinear motifs, in the same colours and shapes that occur in Killke, built into fine,

**Figure 2.** A large Lucre style face neck jar with conical base, 870 × 600 mm diameter. (Photograph courtesy of Melissa Chatfield and the Chokepukio Project)

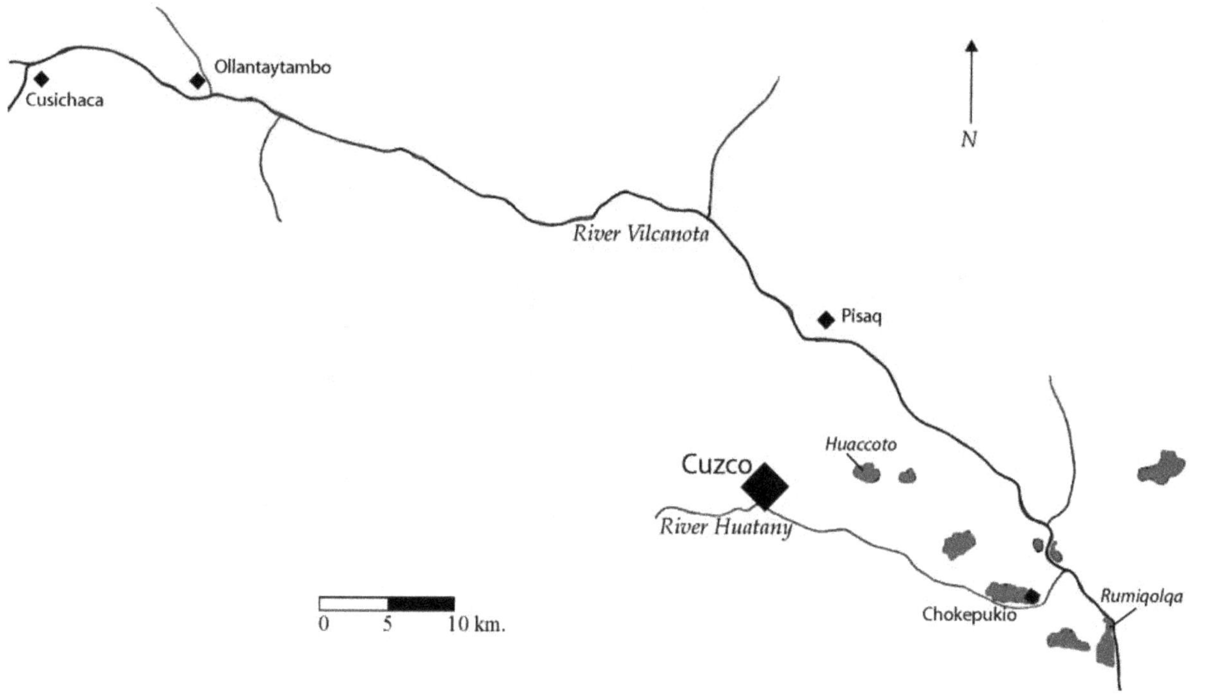


**Figure 3.** Map of the Cuzco region, showing the location of the Rumicolca Formation andesite sources and sites mentioned in the text.

precise and richly-textured patterns across the body of the pot; the surfaces were highly burnished to a smooth, sometimes mirror-like, finish. In cross-section, sherds have a uniform grey core, revealing a pyrotechnology consistent with a final phase of oxidation, to display the colours and decorative detail.

Lucre pottery is by far the least well-known of the three wares described here (Bauer 2004, 75 and references). It has been found in high concentrations at the site of Chokepukio, 25 km south-east of Cuzco and close to Rumiqolqa ; sherds also appear elsewhere, as at Cusichaca, 80 km north-west of Cuzco, but they are very rare and their identification is hampered by the lack of a comprehensive published account. The pottery was produced during the Late Intermediate Period (and possibly later): at Cusichaca, it occurs in the same pre-Inca deposits as Killke.

There is no comprehensive published account to date of the fabric of Lucre pottery. However, the few sherds found at Cusichaca, examined in the field in hand-specimen, contained a temper unique in the pre-Inca collections. This closely resembled the andesite of the ATF. The overall appearance of the ATF and the Lucre fabric was extremely similar (Wares 13 and 15, Lunt 1987, 163): in unstratified and undecorated material, they were difficult to distinguish from one another. The Lucre designs have their roots in the earlier Middle Horizon Period, and, although they share the pallet and rectilinearity, they differ in execution and arrangement from Killke designs, being boldly set four-square on the pot (McEwan 1987, 80,102-4; Chatfield 2007, 260; Fig. 3). The shapes include very large, pointed-based, face-neck jars and Chatfield likens their paste technology to the one used for the Cuzco Inca aryballus (Chatfield 2007, 75), making the valuable point that large vessels benefit from certain paste (fabric) characteristics (Chatfield 2007, 81; Ixer and Lunt 1991, 160); (Fig).

Cuzco Inca wares are generally perceived not as technical products or functional objects, but as vital chronological and cultural markers in provincial stratigraphies, their value a function of their identity as the Inca state ceramic. For our purposes here, however, it is important to identify the various ceramic traditions that contributed to the creation of these wares and thereby to place them in a developmental sequence. Killke pottery offers several significant comparanda in its technology and style, but we should note that Rowe and Bauer, inter alia, use this evidence, as well as distribution and stratigraphy, to identify the makers of Killke as 'early Incas', suggesting that there may be more at stake here than tracing ceramic traditions. One of the most notable aspects of the Cuzco Inca potting technology, the exclusive use of andesite temper, is absent in Killke but is characteristic of Lucre. We now turn to the sources of andesite in the geology of the Cuzco region.

## 3. Andesite in the Cuzco region

The volcanic andesites of the Rumicolca Formation are dated to the late Pleistocene. The nine principal outcrops lie to the south-east of Cuzco (INGEMMET 1999), and there is also a small number of outcrops to the north. The south-eastern outcrops are all within 40 km of Cuzco and are the nearest and most accessible sources of andesite to the Inca capital.

Two of the Rumicolca Formation outcrops supported ancient quarries. At Rumiqolqa (we follow Ogburn's spelling to distinguish the quarry from the geological formation: Ogburn 2004, 439, n.3), there is plentiful and dramatic evidence of Inca quarrying; some scantier evidence survives at Huaccoto. Other andesite outcrops were also quarried in the Inca period, at Ollantaytambo and Pisaq (Hunt 1990, 30-1), although these are far more difficult to access from Cuzco (Fig. 3).

The Incas' knowledge and use of the Rumiqolqa and Huaccoto quarries has focused much attention on these sites. Amongst the more recent papers, Protzen's study of Inca stone-masonry was based on Rumiqolqa (Protzen 1985). He described the different qualities of andesite at various locales across the outcrop and then concentrated his attention on one quarry area, the Llama Pit, where extensive evidence of Inca working survives, commenting that one of the three qualities of andesite found there 'seems to correspond to....(one)...used in Cuzco' (Protzen 1985, 168). Ogburn (2004), in his extraordinary work identifying the Rumicolca Formation as the source of the carved andesite found at Saraguro in Ecuador, characterised the Rumiqolqa and Huaccoto andesites geochemically. His results confirmed that the formation as a whole is chemically distinctive and, as could be expected, that the andesites from the two sites could not be distinguished geochemically (Ogburn 2004, 432).

Hunt (1990) provides the most recent published petrographical descriptions. He describes the Rumiqolqa samples as hornblende andesite, with primary phenocrysts of plagioclase feldspar, basaltic hornblende, and biotite; there is a little pyroxene and occasional xenoliths of apatite and ilmenite. The Huaccoto samples are described as biotite andesite, with phenocrysts of plagioclase feldspar and biotite; hornblende is present but infrequent (Hunt 1990, 26-7). Hunt matches these petrographical differences with differences in colour and physical properties, which serve to distinguish between the products of the two quarries in the buildings of Cuzco and to confirm the Rumiqolqa origin of the imperial building stone. These are influential descriptions, which have been quoted subsequently (e.g. Bauer 2004, 157).

## 4. Methods and aims of the study

The study of pottery fabrics is one of the classical archaeological approaches to understanding and classifying ancient pottery, its value lying in the unique relationship that exists between a suite of techniques used to make pottery and a potting community. The use of ceramic petrography to identify the non-clay components and the technological features of a pottery fabric is a long-established and effective technique. Here, petrography is used to characterise and compare the three principal fabrics used to make Killke, Lucre, and Cuzco Inca pottery. The petrography of the tempering materials is then compared with that of geological samples from the Rumicolca Formation, with the aim of identifying the source of the andesite temper in the Cuzco Inca and Lucre fabrics. Although largely outside the scope of this paper, this is the essential evidence on which to base ideas about how the temper was obtained and why it was chosen; and for discussing the transmission of a ceramic technology through time and between different production systems.

### *The samples*

The origins of this study lie in an adventitious comparison made between the petrography of a sample of andesite from an imperial building in Cuzco and a sample of the ATF. This gave weight to an earlier idea, i.e., that the andesites of the ATF originated in the basalt-andesites of the Cuzco area (Ixer and Lunt 1991, 160); and it also provided a context for studying the similarities between the Lucre fabric and the ATF. The samples of Lucre pottery and Rumicolca Formation andesite are far fewer compared with those of Killke and Cuzco Inca, and they were not collected primarily with this study in view; but they are sufficient to construct a hypothesis, which can be tested more fully in the future.

### *The pottery samples*

The pottery samples discussed here are from two groups, those that were studied as part of an earlier work to characterise the petrography of Killke and Inca wares (Ixer and Lunt 1991); and new material, added to provide data from other contexts/areas and to provide the data on Lucre, which was not included in the earlier study. As part of this work, all of the 1991 samples have been re-assessed and some lithological designations altered. The numbering of samples from the earlier work has been retained, and the numbering of the new samples follows the same procedure, being the original excavation or collection designations.

### *Killke samples*

27 thin sections are considered here. From the 1991 study, there are 19 samples: F7AG; S162; S162A; Ware 45 (F14-S99; S99; S209); Ware 46 (S312); Ware 47 (47-1; 47-2; 47-3; 47-4); Ware 48 (S48-1; 48-1-1; 48-2; 48-2-1; 48-2-2; 48-3); and Ware 49 (S49, 49-1). F4 is now missing. Each group represents a decorative sub-style of Killke. All are from the pre-Inca deposits at Cusichaca.

To these have been added 8 further samples: Ware 47 (47-1-3; 47-3-1; 47-3-2); Ware 48 (48-2-3; 48-2-4); Sample 312-2 (all from Cusichaca); and Samples CO129B and CO149 (from Brian Bauer's surface collections in the Cuzco area).

### *Cuzco Inca samples*

34 thin sections of the ATF have been studied or re-studied for this paper. These include eight samples from the earlier study, all Ware 13: F12; W13; W13-1; W13-2; S127; S129A; S229; and S306. Excluded are S205 (now missing); and 14 other samples (Ware 26, Ware 27 and Ware 29), none of which are the ATF. The groups sampled here represent all of the form/style variations within the ATF - Cuzco Inca collection from the Inca levels at Cusichaca.

To these have been added 26 further samples: S109; S123; XRF 1 to 6, XRF 11; XRF 12; CUZ 99-2 to CUZ 99-10 (all from Cusichaca); CUZ 99-11; CUZ 99-12 (collected by one of the authors, SL, from surface material at Limatambo); AN 337; CO 270; CO 31-1008; CO 241B; and CO 6B (from the Bauer surface collections in the Cuzco area).

### *Lucre samples*

11 thin sections of Lucre pottery have been studied for this paper. These are: AN 36B; AN 235B; CO 97; CO 129A; CO 242; CHOQ 1; CHOQ 2; (from the Bauer surface collections); 95; 100; 108; and 120 (from Choquipukio,

donated by Melissa Chatfield). The Choquipukio samples are from layers below those dominated by Inca sherds and above the Middle Horizon layer. These samples represent the dominant fabric of Lucre, used for the whole range of wares.

### *Andesite samples*

15 thin sections of andesite from two outcrops of the Rumicolca Formation were studied for this paper. From Rumiqolqa, there are nine samples: INKA 11; INKA 12 (collected by one of the authors, BS); AND 10A; AND 10B (collected during the Bauer surface collection project); RQ-7; RQ-10; RQ-11; RQ-18 and R21 (donated by Dennis Ogburn from the samples used in his own work: Ogburn 2004).

From Huaccoto, there are six samples: AND 12A; AND 12B; AND 12C (Bauer); H2; H3; and H4 (collected for Ogburn 2004 by Jose Gonzales, Universidad Nacional San Antonio Abad del Cuzco).

## 5. Results

The details of the results, including the individual thin-section data, the macro-descriptions and the provenance data, are held in archive.

### *The andesites*

All of the samples from Rumiqolqa and Huaccoto carry abundant biotite and plagioclase phenocrysts. They carry hypersthene in their groundmass, together with distinctive quartz xenocrysts and pleonaste/hercynite-corundum-bearing xenoliths. No amphibole (namely hornblende/ basaltic hornblende/oxyhornblende) was identified, and apatite phenocrysts were also absent (Fig. 4a).

There is no significant petrographical difference between the samples from the two sites: this is consistent with Ogburn's (2004) geochemical results. Most significantly here, no amphibole or apatite has been recognised. This is in contrast with Hunt's results, where hornblende phenocrysts were identified as being more abundant than biotite in the Rumiqolqa samples, pyroxene was uncommon, and apatite phenocrysts were present. He contrasted this with the samples from Huaccoto, where he stated that hornblende was rare.

### *The Killke fabric*

(In re-assessing the 1991 samples, 'syenite' is now re-defined as 'altered diorite' and 'arkose' as 'arkose-litharenite'.)

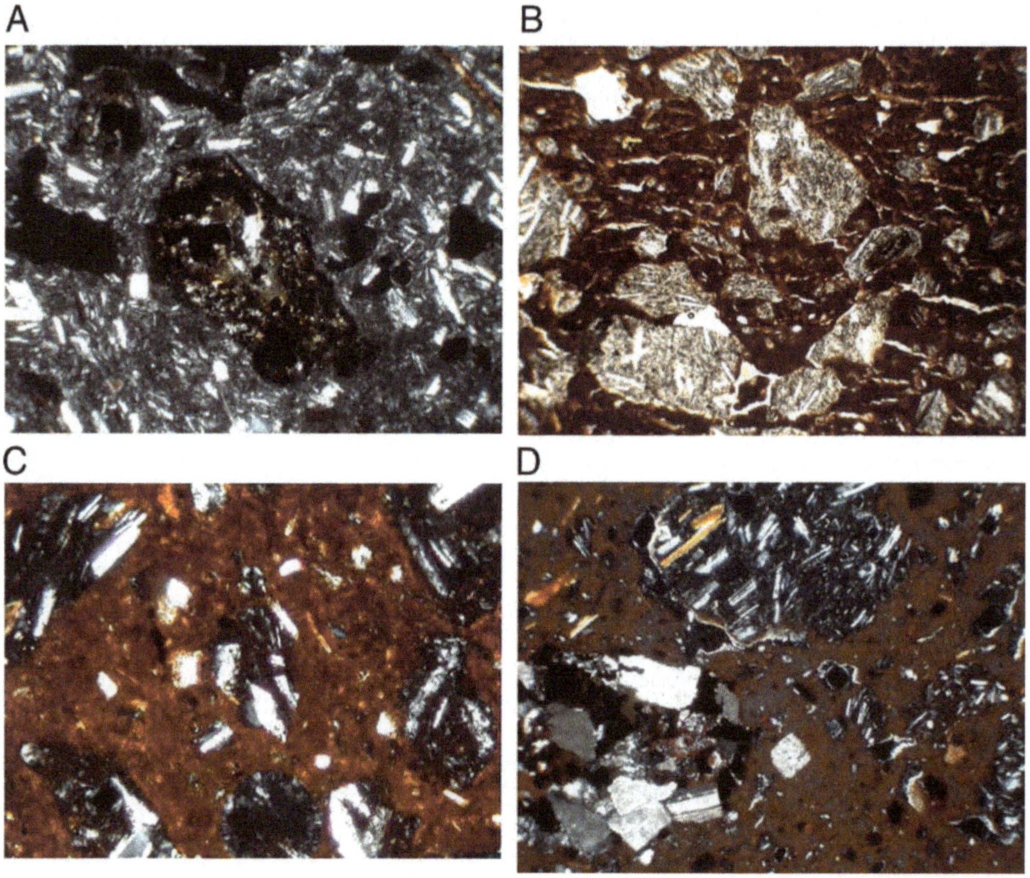


**Figure 4.** (a) Rumiqolqa Formation Andesite, sample R21: biotite- and biotite-pyroxene-bearing andesite, with distinctive pleonaste-bearing xenolith. (b) Inka Andesite Tempered Fabric, sample CUZ 99-4: restricted size-range of sub-rounded to sub-angular biotite andesite within a clean clay. (c) Inka Andesite Tempered Fabric, sample CUZ 99-3: showing hornblende. (d) Lucre Andesite Tempered Fabric, sample CO97: tempered with angular to sub-angular biotite andesite, andesite with amphibole (hornblende), and a large sandstone clast.

The re-assessment has confirmed the lithologies but reveals more variation in the tempering practices. The principal lithologies are altered diorite and arkose/litharenite, both highly feldspathic; the largest group (14 samples) is tempered with either or both of these materials, angular to sub-rounded in form, with irregular surface topographies; the mean size of the temper and its quantity are consistent. Much of the fine-grained material in the groundmass is from the same rocks as the temper. Microporphyry (one sample) and sand (two samples) were also used as temper. Four samples may be untempered. Grog occurs as temper, with other material in four samples and alone in two samples.

The character of the grog (fired clay fragments) is variable. Some pellets resemble the parent sherd in their non-plastics, but others are very different and there can be wide variation within a single sample. Grog is identified as temper on the basis of its prevalence in six samples; it also occurs in small amounts in a further six samples, where most of it matches the main paste and is probably adventitious.

Although not the focus of this paper, it is noteworthy that the variations in the tempering practice show some co-variance with decorative sub-style. Thus, F7 and all the samples of Wares 45, 46 and 47, which together represent the principal decorative sub-style of Killke at Cusichaca, are tempered with the same rock types, although they display a varied use of grog. The samples of Ware 48, a distinctive and unusual sub-style, vary markedly from one another, being tempered with arkose (1 sample), microporphyry (1), and grog (1); three samples are polylithic and may be untempered. Ware 49, another unusual sub-style, is grog-tempered. The evidence is generally consistent with the idea that Killke was made in a number of workshops, all employing a similar technology but using a variety of (local?) materials.

Andesite occurs in two samples of Ware 47. However, this is not temper but part of the groundmass. It is a distinctive spherulitic clinopyroxene-hornblende-biotite andesite, which was not recognised in the Rumicolca Formation samples from Rumiqolqa or Huaccoto.

### *The Cuzco Inca fabric*

The ATF samples are tempered with sub-rounded to sub-angular/angular andesite in very clean clay. Most samples carry only trace amounts of chert/rhyolite, single quartz grains, fine-grained sandstone/arkose, and mud clasts within the groundmass of the clay. All other rock types are extremely rare (Fig. 4, b-c).

Within a single sample, only one type of andesite occurs. The whole group is dominated by the use of hornblende-bearing andesites (27 samples; Fig. 4c), in which there is petrographical variation. The remaining 7 samples are tempered with biotite and biotite-pyroxene andesites (Fig. 4b), again with some petrographical differences. There is no co-variance with size or shape of the parent vessel, context or provenance.

The hornblende-bearing andesites have not been recognised in the samples from Rumiqolqa or Huaccoto studied here. The biotite and biotite-pyroxene andesites can all be matched to samples from both of the sampled andesite sources.

### *The Lucre fabric*

The fabric is tempered with fresh, angular andesite. The clay groundmass varies quite considerably, most samples containing weathered/altered feldspars, some clasts being as large as the temper. Microporphyritic felsite and mudstone/siltstone also occur (Fig. 4d).

Only one type of andesite occurs in any one sample. Across the group, two andesites occur: hornblende (5 samples) and biotite (6 samples), both displaying some petrographical variations. The hornblende-bearing andesites have no equivalent in the Rumiqolqa or Huaccoto quarry samples; the biotite-bearing andesites can all be matched to samples from both sites.

The sample is too small to identify any meaningful co-variance with decorative sub-style, size or form of pot, or context.

## 6. Discussion

Considering the Cuzco Inca fabric, the ATF, the fabric is highly standardised in its preparation, regardless of the type of andesite (hornblende-bearing, biotite, or biotite-pyroxene) or the other petrographical variations. These petrographical differences are not detectable in hand-specimen. No differences can be recognised in provenance, context, form, size or decorative sub-style of the pots containing the dominant, hornblende-bearing andesites compared with those containing biotite or biotite-pyroxene andesites. This suggests that the properties of all the andesites used to temper the ATF were identical from a potting point of view and therefore, for our present purposes, we consider the ATF to be a single entity.

The ATF was used for the entire range of Cuzco Inca forms, from the smallest bowls to the largest serving and storage vessels. There is some indication that, through time, some of the finest wares were made from a fabric with a more delicate texture, although andesite remained the tempering material (Lunt 2000); and it is important to note that other fabrics were also used for these wares, although they are far less common (Ixer and Lunt 1991; Lunt 2000). The ATF has not been firmly identified in pottery made outside the Cuzco region and, within the region, it was used exclusively for Cuzco Inca pottery. For the major production of these wares, the potters followed an identical technological blue-print for their fabric.

Using a standardised fabric in a workshop guarantees consistency in working and finishing properties and in firing behaviour, all of which would be important if, as the visual appearance of Cuzco Inca wares suggests, uniformity was one of the goals of production. This could have been achieved with other tempering materials, provided that they were available in sufficient quantity, and with a different texture: but it is clear that the specific characteristics of the ATF were of value.

The texture of the ATF is relatively coarse, although not uniquely so: it shares this trait with other fabrics from the Cuzco region. It has been suggested that firing pottery at high altitudes, in open fires, may have necessitated an open-textured fabric for successful oxidation (Lunt 1987, 49). Andesite itself has particularly valuable properties that contribute to the serviceability of large vessels: the sharp-edged, angular grains and their irregular surface topographies make strong bonds with the clay (Ixer and Lunt 1991, 160).

ATF, then, could be described as a 'dream fabric', one that promoted manufacturing and firing consistency; could be used for a range of forms and sizes; was appropriate for the traditional firing techniques and local conditions; and was especially suitable for the largest, and therefore most demanding, pots in the repertoire.

The fabric preparation technology of Cuzco Inca pottery existed in the region before the Late Horizon Period. It can be identified in Killke; and although the rocks used in the ATF and in the Killke fabric are entirely different, the use of clean clay and the introduction of plentiful rock fragments from a limited repertoire were common to both. Ware 29, an early non-ATF Cuzco Inca ware discussed above, was tempered with the same rock types as Killke, suggesting that it was transitional. However, for the major production of Cuzco Inca wares, the use of andesite temper seems to have been essential; and therefore the case for identifying Killke as a direct precursor to Cuzco Inca wares rests far more persuasively on its decorative style, relative date, and distribution than on its choice of temper. In identifying a specific precursor for this, we must turn to another pre-Inca ware.

Considering Lucre, the sample is extremely small, which limits our inferences. It is not possible to distinguish in hand-specimen between the hornblende-bearing and the biotite andesite temper; both types were used in the same way, the grains being extremely similar in size, shape, and quantity. There is no evidence of discrimination in the use of the two andesites for particular kinds of pots or through time. Consistent with the approach to the ATF, therefore, we will treat the Lucre fabric as a single entity. The outcrop of Rumiqolqa lies very close to Choquipukio, the principal finds-place of Lucre pottery. All of the biotite andesite in the Lucre fabric samples can be matched to samples from this local source; the provenance of the hornblende-bearing andesite has not been identified.

The comparison between the Lucre fabric and the ATF shows that the Lucre clays are less clean than the ATF, the degree of cleanness varies, and much of the accessory material is absent from the ATF. However, the similarities in temper are extremely striking. The Lucre fabric contains either hornblende-bearing or biotite andesites, identical to those in the ATF: this is persuasive evidence that the sources of andesite were the same (albeit an unknown one for the hornblende-bearing andesite). In both fabrics, the temper is used in exactly the same way: the grains are similar in size, shape, and prevalence, and there is only one type of andesite per sample. It seems clear, based on this evidence, that the Lucre technique of sourcing andesite and using it in the pottery fabric was a significant element in the creation of Cuzco Inca pottery.

The viability of an andesite-tempered fabric was proven by its successful use for a wide range of Lucre wares, especially in the manufacture and serviceability of the large jars. From the view-point of technological pragmatism, it is suggested therefore, following Chatfield (2007), that it was the desire to produce similar vessels that encouraged the transfer in fabric technology from Lucre to Cuzco Inca pottery.

In seeking a source for the andesite used in the ATF and in Lucre, the quarry of Rumiqolqa was seen as a likely candidate. This site has much to recommend it. The quarry is adjacent to Choquipukio, the provenance of Lucre pottery ; and it was known and used by the Incas for building stone. Another quarry, Huaccoto, was also used by the Incas. An earlier work (Hunt 1990) had distinguished petrographically between samples from the Rumiqolqa and Huaccoto quarries, which suggested the possibility of a fine discrimination in sourcing the andesite tempers.

The analysis of our samples shows that the biotite and biotite-pyroxene andesites in the Lucre and Inca ceramics can all be matched to the rock samples from Rumiqolqa and Huaccoto. However, the samples from these two andesite sources are petrographically identical; therefore, contrary to Hunt's claim, it is not possible to distinguish between Rumiqolqa or Huaccoto as the source of the temper based on this evidence.

It has not been possible to characterise either outcrop comprehensively, and we have no samples from the rest of the Rumicolca Formation. Therefore, although we can be confident that the biotite and biotite-pyroxene andesite temper used in the Lucre fabric and the ATF both originate in the Rumicolca Formation, we cannot attribute all of it to the same locality, or be certain that Rumiqolqa and Huaccoto were the only possible sources. Most significantly, we cannot identify a geographical provenance, even in the broadest terms, for the hornblende-bearing andesites in the ceramic samples.

For both the ATF and the Lucre fabric, the exclusive use of one type of andesite per sample suggests a deliberate selection, most marked in the overwhelming use of hornblende-bearing andesite in the ATF. As discussed above, this selection has no corollary, which makes it hard to explain on the evidence of these samples alone. The different andesites may be widely separated geographically, their locales offering different ease of access, and they may look different in their natural state: this would be consistent with Protzen's identification of differences in colour and working properties in the andesites of Rumiqolqa (Protzen 1985). Clarifying these issues, and perhaps locating the elusive hornblende-bearing andesites, requires a comprehensive sampling programme of the Rumicolca Formation, particularly of the andesite outcrop most adjacent to the site of Choquipukio.

## 7. Conclusion

Cuzco Inca pottery is a hybrid. Two distinct elements can now be identified, each contributed by wares that are dissimilar to one another in form and decorative style, although they share a pre-Inca date, some distributional overlap, and a Cuzco-centric location. There are also other features of Cuzco Inca wares that are not part of the Lucre or Killke repertoires and that may derive from other pottery-making traditions. Identifying a precursor to Cuzco Inca pottery is hardly new - Killke has long held that position - but recognising the complexity of its origins is an advance, and, whilst not refuting the Killke contribution, it does make the idea of a single strand of evolution, from Killke pottery to Inca pottery, untenable. There is now a clearer idea of what 'Cuzco Inca' comprises, which will illuminate how, where and when the creative process took place.

As to why it should have occurred: the identification of Cuzco Inca wares as the Inca state ceramic is well-supported by the visual appearance of the pots and by the chronological evidence. If Cuzco Inca wares were political artefacts, then perhaps the selections of elements employed to create them were also political statements, the ceramic representations of the subjugations and alliances that formed the Inca state. As to how the mechanism of state control of the production actually worked - this has been well-studied elsewhere but not here, at the centre of the empire. Who, for instance, were the potters?

Turning to the andesites of the Rumicolca Formation: it is clear that a comprehensive sampling programme is needed, to identify the specific sources of the Lucre and Cuzco Inca temper. Depending on the results, there may be more to consider than ceramic pragmatism: one possible scenario is presented below.

The Rumiqolqa outcrop lay in the territory of the Pinahua, a significant polity whose principal settlement was Choquipukio. These were the makers and users of Lucre pottery. They were subjugated by the Incas in c. 1430 AD ( Bauer 2004, 84-6), and thereafter, the andesites of Rumiqolqa came under Inca state control. Judging by the extensive quarrying evidence at the site, the use of this rock in the most prestigious imperial buildings of Cuzco, and its exclusive use for state projects, the Incas valued their prize highly. The idea of 'value' may have resided in the physical properties of the andesite, its accessibility and, perhaps, in its political significance following the subjugation of the Pinahua: an idea that has a familiar ring here.

If a Rumiqolqa source for all of the ATF temper were established, then this would provide, first, a date (albeit an estimate) *ante quem* for the introduction of the ATF into the manufacture of Cuzco Inca pottery. Secondly, more significantly, it would provide a cultural context for this introduction: the subjugation of the Pinahua and the building of imperial Cuzco.

Finally, we would be presented with a unique scenario: a liaison between two state-controlled crafts. We might view this as pragmatism - the availability of abundant, effective temper with a proven track record - and also, maybe, something more nuanced, reflecting an Inca perception of 'good stone', perhaps. Or was the use of andesite in imperial buildings and imperial pottery a political flaunting of exclusivity?

## References

Bauer B.S., 2004, *Ancient Cuzco*, University of Texas Press, Austin.

Chatfield M., 2007, *From Inca to Spanish Colonial: Transitions in Ceramic Technology*, Unpublished PhD thesis, University of California, Santa Barbara.

Hunt P.N., 1990, Inca volcanic stone provenance in the Cuzco Province, Peru. *Papers from the Institute of Archaeology, University College London*, 1, 24-36.

Instituto Geologico Minero y Metalurgigo (INGEMMET), 1999, *Mapa Geologico del cuadrangulo de Cusco*, INGEMMET, Lima.

Ixer R.A. and Lunt S., 1991, The petrography of certain pre-Spanish pottery from Peru. In *Recent Developments in Ceramic Petrology*, (eds. A. Middleton and I. Freestone), 137-164, British Museum Occasional Paper 1, British Museum, London.

Lunt S., 1987, *Inca and pre-Inca pottery: Pottery from Cusichaca, Department of Cuzco, Peru*, Unpublished PhD thesis, Institute of Archaeology, University of London.

Lunt S., 2000, *A study of the Inca pottery in the Museo Arqueologico, Cuzco*, Unpublished archive of descriptions and photographs of the Inca pottery in the Museum collections and a summary of the Museum Catalogue entries.

McEwan G.F., 1987, *The Middle Horizon in the Valley of Cuzco, Peru: The Impact of the Wari Occupation of the Lucre Basin*, British Archaeological Reports, Oxford.

Ogburn D.E., 2004, Evidence for long-distance transportation of building stones in the Inka Empire, from Cuzco, Peru, to Saraguro, Ecuador. *Latin American Antiquity*, 15, 419-439.

Protzen J.-P., 1985, Inca quarrying and stone-cutting. *Journal of the Society of Architectural Historians*, 44, 161-182.

Rowe J.H., 1944, *An introduction to the archaeology of Cuzco*, Expeditions to Southern Peru Report No. 2, Papers of the Peabody Museum of American Archaeology and Ethnology, University of Harvard 27, The Museum, Cambridge (MA).

## ACKNOWLEDGEMENTS

We extend our thanks to the following: Museo Inka, Universidad de San Antonio Abad del Cuzco, for allowing Sara Lunt and Bill Sillar to research and photograph their collections; Brian Bauer, for providing pottery samples from his survey work in the Cuzco region and for including the collection of rock samples in his work programme; Melissa Chatfield and the Choquipukio Project, for supplying four samples of Lucre pottery, for providing the context data, and for permission to use the illustration of a Lucre jar; Ann Kendall, Director of the Cusichaca Archaeological Project, under whose auspices most of the pottery discussed here was collected; Dennis Ogburn, for providing the Huaccoto and Rumiqolqa andesite samples; and Andy Tindle, for his mineral identification of the exotica within the xenoliths. The uses we make of their contributions do not necessarily represent the donors' own views. For their generosity and support, we are very grateful.

12

# Parallel developments in Chinese porcelain technology in the 13th-14th centuries AD

**Nigel Wood[a] and Sabrina Rastelli[b]**

[a]Research Laboratory for Archaeology and the History of Art, University of Oxford, UK, OX1 3QY

[b]Ca' Foscari University, Department of East Asian Studies, Dorsoduro 3462, 30123 Venezia, Italy

**Abstract** - By the 14th century AD, Jingdezhen in Jiangxi province and the Longquan region of Zhejiang province were China's main centres for the manufacture of porcellaneous wares. Jingdezhen specialised in blue and white porcelain, while the Longquan kilns typically used opaque porcellanous bodies, and smooth and thick, greenish celadon glazes. A pattern of parallel technological development is evident at Jingdezhen and Longquan in the 13th-14th centuries, but probably driven by rather different causes. Jingdezhen is representative of a simpler and more local technological history that seems entirely southern. Longquan however adopted a glaze-type that had already seen many prestigious applications in north China. Unlike the white-bodied Jingdezhen wares, the Longquan bodies were sometimes deliberately discoloured with red clays, perhaps to harmonise with their jade-like greenish glazes.

## 1. Introduction

In the history of Chinese ceramics, two giant kiln complexes have dominated the mass production and export of high-fired wares - namely Jingdezhen in Jiangxi province and the Longquan region of Zhejiang province. Both sites are located in south China and in adjoining provinces. Jiangxi is land-locked, but Zhejiang has a long rocky coastline that borders the East China Sea. For centuries, Jingdezhen's main production has been blue and white porcelain, while the Longquan kilns specialised in grey-bodied porcellaneous wares with thick greenish glazes (Li Dejin 1994; Carswell 2000).

The huge productions of both Jingdezhen and Longquan wares were managed by what were essentially craft techniques — that is mainly by throwing on the potter's wheel and by hand-moulding, using plastic clay pressed into porous clay moulds. Firing was achieved in large high-temperature wood-burning kilns of simple but subtle design (Li Dejin 1994; Xiang Hai Tang 1995). Any equipment used tended to be straightforward, such as water-powered trip-hammers for pulverising rocky raw materials, and settling tanks for separating the powdered stones. However, within these craft-based workshops, extensive division of labour was often practised, and a single piece of blue and white porcelain, for example, may have passed through some seventy pairs of hands in the early 18th century (Du Halde 1736).

Despite the important visual differences between Jingdezhen blue-and-white and Longquan celadon wares, both kiln sites used similar raw materials for their clay bodies — namely low-iron rocks consisting largely of quartz and hydromica, known as *cishi* (porcelain stone) in China (Vogt 1900; Stannard 1986). As hydromica has a composition and crystal structure close to the clay mineral kaolinite, the rock could be rendered plastic by crushing, followed by levigation. At Jingdezhen, the *cishi* material was often mixed with lesser amounts of a white-firing primary clay known as *gaoling* (kaolin), while at Longquan 10 to 20% of primary red clay often rendered the near-white porcelain stone grey and non-translucent, or less pure rocks were used unmodified (Vogt 1900; Chou Jen *et al.* 1973). During firing, the more ferruginous Longquan material could re-oxidise superficially to a rust-red colour when unglazed — an effect that contrasted admirably with the thick, jade-like glazes applied to the wares.

## 2. Southern greenware: a common background to Jingdezhen and Longquan wares

In terms of the ceramics that they replaced, both Jingdezhen white porcelains and porcellaneous Longquan celadon wares shared a common ancestry — namely a stoneware known generically as 'southern greenware' (Guo Yain-yi *et al.* 1980). This was an opaque, grey-bodied material with a thin grey-green glaze. Like the porcellaneous materials that came to replace them, south Chinese greenwares often used rocky materials as the bases for their bodies. These tended to be pulverised and 'washed' microgranites that were already in a state of decomposition from chemical weathering. Rocks of this type are abundant throughout China's southern provinces (Wood *et al.* 2005).

The same (or similar) processed rocks were mixed with about 40% of calcareous wood ash to make thin grey-green stoneware glazes (Wood *et al.* 2005, 194-5). The main kiln type used for the southern greenware industries dated from the Bronze Age and consisted essentially of long sloping tunnels built up hillsides (Xiang Hai Tang 1995). These wood-burning kilns were fired by large fire-boxes at the kilns' feet, followed by successive side-stoking through 'ports' along the kilns' sides. By the 12$^{th}$ century AD, these *long* ('dragon') kilns were approaching 140 metres in length, although kilns of 20-40 metres were perhaps more typical (Zeng Fang 1997, 30-2).

Just such a 'southern greenware' technology was operating at Jingdezhen in the early 10$^{th}$ century, when a white-firing porcelain stone was discovered there in the form of a compacted volcanic ash — a natural mixture of quartz, hydromica, and a small amount of kaolinite (Wood 1978; 1986). After processing, this gave a complete white-firing translucent porcelain body. As the usefulness of this material was established, it gradually replaced the old green-grey stonewares of the district. Eventually, limestone displaced calcareous wood-ash as the main glaze-flux at Jingdezhen, and plain bluish-white porcelains of great quality were developed from these two materials — using a plastic porcelain stone for the body and the same rock mixed with limestone for the glaze. This material is known as *qingbai* ware.

In the late 12$^{th}$ century, the clay-rich material *gaoling* started to be added to the Jingdezhen body, allowing more fusible and less plastic porcelain stones richer in soda feldspar (albite) to be exploited (Tite *et al.* 1984; Wood 2000b). Sometime in the 1320s, underglaze-blue painting with an imported iron-cobalt ore was introduced, to create the first Jingdezhen blue and white wares (Liu Xinyuan 1993). At much the same time, Jingdezhen potters turned from using their rather fluid lime-glazes and semi-opaque alkali-lime 'Shufu' glazes to those of the lime-alkali type. These latter compositions had sufficient transparency to show the underglaze painting clearly, but also sufficient viscosity to prevent the blue cobalt pigments' spreading into the molten glazes at high temperatures.

These were the main milestones in Jingdezhen's technological history — from the traditional southern greenwares of the early 10$^{th}$ century AD, through *qingbai* and Shufu wares, to the blue and white porcelains of the 14$^{th}$ century.

## 3. The origins of Longquan celadon wares

In the west, the technological history of Jingdezhen has attracted a good deal of attention, as its wares are more familiar, and their influences on European porcelain history so profound. In contrast, the origins of Longquan porcellaneous celadon ware are less clear-cut. In particular, a material previously known as *di* ware seems to represent an early use of porcellaneous materials at Longquan (Chou Jen *et al.* 1973; Vandiver and Kingery 1986; Ren Shilong 1994), and the understanding of *di* ware appears to be the key to Longquan's later porcellaneous productions.

## 4. *Di* wares and Longquan celadons compared

In Chinese, *di* means 'younger brother', while *ge* means 'older brother'. According to a story written in the first half of the 16$^{th}$ century by Lu Shen in his *Chunfengtang suibi* and repeated many times in Ming and Qing literature (Zhu Bojian 1998, 21-3, 41-3), in the Song dynasty there were two brothers, Zhang Sheng the first and Zhang Sheng the second, native of Chuzhou, who both worked at the Liutian (today's Dayao) kilns in Longquan. The greenware produced by the younger brother had the jade-like quality typical of Guan ware, while that fired by the older sibling was lighter in colour and was called *ge* ware. Because the story first appeared three hundred years after the manufacturing of this type of ware, it is understandably treated with scepticism by modern scholars. For this reason, the term *di* has rather fallen into desuetude. However, as it is associated with a very specific type of Southern Song Longquan celadon of exceptional quality (a key subject in this paper), it seems suitable for this discussion.

The *di* ware body-material was based on a near-white porcelain stone — a significant departure from the traditional grey-bodied southern greenware that had been the standard material in the Longquan area for most of the 12$^{th}$ century (Chou Jen *et al.* 1973). The bluish *di* ware glazes were also unusually thick, and, like later Longquan wares, were applied in a series of thin coats to maximise both depth and overall smoothness (Sundius and Steger 1963, 430). A further parallel between *di* wares and the later Longquan celadons lay in the use of lime-alkali glazes. This was a more sophisticated glaze-type than the old ash-rich compositions of the region, with their predominantly calcareous fluxes. These earlier, ash-fluxed glazes are known generally as 'lime glazes' (Zhang Fukang 1986). In lime-alkali glazes, the calcium oxide content is almost halved, and the alkalis (potassium and sodium oxides) increased somewhat in compensation. These changes resulted in stable and more viscous glazes with a smooth jade-like nature.

Thus, with its porcellaneous body and its smooth and thick lime-alkali glaze applied in many layers, *di* ware embodied three essential principles that were so successfully exploited in the later Longquan celadon production. All of this raises the following key questions: How did *di* ware come to be made at Longquan? And what were the origins of this sophisticated material that appeared fully-formed in a rather remote ceramics-producing region of Zhejiang province?

Perhaps the best approach to these problems is to divide the discussion into two parts. The first concerns the origin of the bluish lime-alkali glaze used on *di* ware, together with its multi-layer application. The second considers the introduction of white porcelain stone as a body material to the Longquan region.

## 5. The *di* ware glaze

Research by the present authors and colleagues has traced the background of the *di* ware glaze to the 10$^{th}$ century AD Huangbao kiln site in north China, hundreds of miles to the north-east of the Longquan region (Rastelli *et al.* 2002;

**Figure 1.** Early 10th century AD cupstand from the Huangbao kilns (Yaozhou kiln complex). Yaozhou Kiln Museum, Shaanxi.

Wood *et al.* 2004a; 2004b). Huangbao was part of the great Yaozhou kiln complex in Shaanxi province that was established during the Tang dynasty (618-907).

In the early 10th century AD, Huangbao's productions included very high quality grey-bodied stonewares bearing thick bluish celadon glazes of both the lime and the lime-alkali types (Fig. 1). From the Huangbao kiln complex (which ran for some three miles along the Qishui river), this technology seems to have spread westwards within north China to three important kiln sites in the Henan province, known as Qingliangsi, Wenmiao, and Zhanggongxiang (Zhu Wenli and Zhu Yufeng 2002; Henansheng 2008). In the late 11th and early 12th centuries, all three kiln centres produced a type of fine bluish celadon ware that is known as Ru ware.

Many examples of Ru ware were true imperial ceramics, and commissioned for use within the palaces of the Northern Song capital of Kaifeng. Once again, both lime and lime-alkali compositions were adopted as compositional bases for the blue-green Ru ware glazes (Henansheng 2008, Table, 447-8).

In 1127 AD, north China fell to the Jurchen Tartars. The surviving imperial court fled to south China and eventually established a new 'provisional capital' at Hangzhou in the northern Zhejiang province in 1138.

In the later 1130s and early 1140s, the displaced imperial household commissioned two nearby southern greenware kilns (Silongkou and Dilingtou) to make ritual wares for the court - no doubt stipulating the use of the 'imperial quality' bluish celadon glazes with which they were already familiar (Jin Zhiwei and Chen Huqing 2005). Unfortunately these 'official' *yue* wares seem to have lacked the sophistication and exemplary craftsmanship of the true Ru wares of north China (Fig. 2) (Wood *et al.* 2004a, 222-3).

Nonetheless, and with the proper establishment of the new Southern Song capital in Hangzhou, the situation improved. Within Hangzhou itself, and actually within the grounds of the new imperial palace, an official kiln was established specifically to recreate the finest quality Ru wares of north China. Technologically and aesthetically, the venture was a triumph — despite the potters' having to work within an entirely different geological domain (Wood 2000a; Li Jiazhi *et al.* 2001). Given the superlative quality of these new Hangzhou 'official' wares, it seems safe to assume that refugee craftsmen from the north were involved in their creation. This new ware of Hangzhou is now known as 'Guan ware' (i.e., 'official' ware) or, more accurately, 'Southern Guan ware', as 'Northern Guan ware' (an exceptionally fine style of Ru ware) had already been developed in the north a few years before its production was curtailed by the Tartar invasion (Krahl 1993). The establishment of the Guan ware kiln in Hangzhou is now dated to about 1145 AD, and its site, located in a steep wooded valley at Laohudong (Old Tiger Cave) was finally discovered and excavated in 1996 (Du Zhengxian 2002; Qin Dashu and Du Zhengxian 2004), after some decades of fruitless exploration (Fig. 3).

What makes Hangzhou Guan ware so special is the extraordinary combination of glaze and form achieved by its makers. The thick, bluish lime and lime-alkali glazes applied to the wares are of outstanding quality and the forms tend to follow Ru ware designs quite closely.

**Table 1.** Blue celadon glazes of the lime-alkali type from 10th century Yaozhou to 13th century Longquan (Analyses from Wood *et al.* 2004a, 225 (SEM - WDA); Zhu Wenli and Zhu Yufeng 2002, 448 (PIXE); Li iazhi *et al.* 2001, 35 (EPMA); Vandiver and Kingery 1986, 188 (SEM - EDA)).

| | $Na_2O$ | MgO | $Al_2O_3$ | $SiO_2$ | $P_2O_5$ | $K_2O$ | CaO | $TiO_2$ | MnO | FeO |
|---|---|---|---|---|---|---|---|---|---|---|
| Early 10th C Yaozhou ware | 0.3 | 1.4 | 15.4 | 69.9 | 0.5 | 2.9 | 8.1 | 0.1 | 0.1 | 1.5 |
| 11th C Wenmiao Ru ware | n.d | n.d | 14.2 | 67.7 | 0.3 | 3.9 | 10 | 0.2 | 0.07 | 2.1 |
| 12th C Zhangongxiang Ru ware | n.d | n.d | 14.7 | 66.1 | 0.7 | 4.1 | 10.8 | 0.2 | 0.06 | 1.7 |
| 12th C Laohudong Guan ware | 0.4 | 0.7 | 15.2 | 67.4 | 0.6 | 4.2 | 9.6 | 0.1 | 0.3 | 1.1 |
| Early 13th C Longquan Guan ware | 0.2 | 0.8 | 15.9 | 66.0 | n.d. | 4.2 | 12.1 | 0.2 | 0.03 | 1.0 |
| Early 13th C Longquan di ware | 0.1 | 0.6 | 14.3 | 67.2 | 0.2 | 4.2 | 10 | 0.1 | n.d. | 1.3 |

**Table 2.** Outline chronology for the transfer of blue celadon technology to Longquan (Sources used: Wood *et al.* 2004a; 2004b; Jin Zhewei *et al.* 2005, 398; Jin Zhiwei and Chen Huqing 2005, 411; Zhu Wenli 2005, 253).

| Date (AD) | North China (Site, province) | Ware type | Lime glazes | Lime-alkali glazes |
|---|---|---|---|---|
| $10^{th}$ C | Yaozhou, Shaanxi | Yaozhou celadon | Yes | Yes |
| Late $11^{th}$ early $12^{th}$ C | Qingliangsi, Henan | Ru ware | Yes | |
| Late $11^{th}$ C | Wenmiao, Henan | Ru ware | Yes | Yes |
| Early $12^{th}$ C | Zhanggongxiang, Henan | Ru ware (Guan Ru?) | | Yes |
| **Movement of imperial court to south China after Jurchen invasion of north China** | | | | |
| 1131-1138 | Silongkou, n. Zhejiang | 'official' Yue ware | ? | ? |
| 1136-1145 | Dilingtou, n. Zhejiang | 'official' Yue ware | ? | ? |
| c. 1145 - $13^{th}$ C | Laohudong, Hangzhou, n. Zhejiang | Guan ware | yes (most) | Yes (a few) |
| Later $12^{th}$ - $13^{th}$ C | Jiaotanxia, Hangzhou. n. Zhejiang | Guan ware | Yes | Yes |
| 1180s to mid $13^{th}$ C | Guan ware | Dayao, and Xikou, s.w. Zhejiang | Yes | Yes |
| 1180s to mid $13^{th}$ C | *di* ware | Dayao, and Xikou, s.w. Zhejiang | No | Yes |

These forms are an eclectic mixture of the most valuable vessel shapes of the time, and include copies of earlier Chinese bronzes, of imported Islamic glass, of fine Northern Song gold and silver vessels, and of Chinese lacquer wares of supremely simple design.

However, Hangzhou Guan ware soon came to differ from northern Ru wares in two important regards — the thickness of its glaze and the colour of its body material. With regard to its glazes, Southern Guan wares exploited to an unprecedented extent the potentials of multi-layer application. Indeed, with some examples the bodies are wafer-thin, but the glazes can be thicker even than the bodies beneath, with some wares appearing to consist almost entirely of glaze (Sundius and Steger 1963, 438; Li Jiazhi *et al.* 2001, 32). As for the body material, this was initially pale grey stoneware, much like Ru ware, but it soon became much darker, through the use of more ferruginous clays in the recipes. In some examples of Hangzhou Guan ware, the bodies are almost black.

So successful was the new Southern Guan ware that further *long* ('dragon') kilns were established a few years later at Hangzhou, on the other side of the hill from Laohudong, at a site called Jiaotanxia. The first of these was uncovered in 1956 and was 23.5 m long, while a larger structure at Jiaotanxia was excavated from October 1985 to January 1986 and was 40.6 m long (Yao Guifang 1989, 407).

**Figure 2.** Shard of imperial-quality Yue ware. Silongkou kilns, Zhejiang province, ca. 1130 – 1150 AD.

**Figure 3.** Imperial quality Southern Guan wares, excavated from the Laohudong kiln site. Hangzhou, Zhejiang province (mid-12th century AD). Zhejiang Provincial Museum, Hangzhou.

This larger kiln is now roofed-over and serves as the focus for the Guan Ware Kiln Museum in Hangzhou.

## 6. The Longquan connection

Soon after the establishment of the Hangzhou Jiaotanxia kiln, manufacture of a Guan-type ware was also initiated some 200 miles to the south of Hangzhou, in the Longquan region of the south-western Zhejiang province. Two kiln sites in particular are associated with the production of this provincial Guan ware - Dayao and Xikou, both near the market town of Longquan (Chou Jen *et al.* 1973, 153-5). Official planning rather than local enterprise should perhaps be suspected in the creation of the Longquan Guan ware, and in the sudden appearance of this advanced ceramic in this region. Indeed, compositional analysis can hardly distinguish the Longquan copies from the Hangzhou originals (Chen Xianqiu *et al.* 1986, 163-4, Tables).

**Table 3.** Compositions of *di* ware glazes, compared with average Yue ware and *qingbai* types (Sources used: Vandiver and Kingery 1986, 188 (SEM - EDA); Sundius and Steger 1963, 492 (analytical technique not reported); Wood *et al.* 2005, 194 (SEM - EDS)).

| | $Na_2O$ | MgO | $Al_2O_3$ | $SiO_2$ | $P_2O_5$ | $K_2O$ | CaO | $TiO_2$ | FeO |
|---|---|---|---|---|---|---|---|---|---|
| *Di* ware glaze Dayao | 0.1 | 0.6 | 14.3 | 67.2 | 0.2 | 4.2 | 10.0 | 0.07 | 1.3 |
| *Di* Ware glaze Dayao | 0.1 | 0.7 | 13.9 | 68.9 | 0.2 | 4.4 | 8.7 | 0.02 | 0.9 |
| *Di* ware glaze Dayao | 0.1 | 0.4 | 14.3 | 68.6 | 0.1 | 5.0 | 10.4 | 0.02 | 0.7 |
| *Di* ware glaze Dayao | 0.3 | 0.5 | 15.5 | 68.1 | 0.2 | 5.1 | 8.1 | 0.02 | 1.5 |
| *Di* ware glaze Dayao | 0.1 | 0.7 | 15.7 | 67.6 | 0.3 | 3.9 | 10.4 | 0.1 | 0.9 |
| Yue glaze av. Shanglinhu | 0.9 | 2.4 | 13.2 | 61.3 | 1.2 | 1.7 | 15.2 | 0.6 | 2.1 |
| *Qingbai* glaze Jingdezhen | 1.0 | 0.6 | 14.0 | 65.4 | 0.07 | 2.0 | 15.4 | 0.05 | 1.1 |

**Table 4.** Compositional changes in Dayao celadon bodies: 12th-14th century AD (Analyses RLAHA Oxford (AAS)).

| Place & Date of manufacture | $Na_2O$ | MgO | $Al_2O_3$ | $SiO_2$ | $K_2O$ | CaO | $TiO_2$ | MnO | FeO |
|---|---|---|---|---|---|---|---|---|---|
| Dayao late 12th C | 0.2 | 0.24 | 22.7 | 69.0 | 4.7 | 0.04 | 0.2 | 0.04 | 2.8 |
| Dayao late 12th/13th | 0.2 | 0.26 | 22.3 | 69.8 | 4.2 | 0.06 | 0.2 | 0.03 | 3.0 |
| Dayao 13th C | 0.2 | 0.22 | 20.9 | 71.4 | 4.8 | 0.16 | 0.2 | 0.02 | 2.0 |
| Dayao late 13th/early 14th C | 0.3 | 0.15 | 22.1 | 70.1 | 5.2 | 0.04 | 0.2 | 0.05 | 1.7 |
| Dayao early 14th | 0.4 | 0.18 | 21.2 | 70.6 | 5.4 | 0.05 | 0.2 | 0.02 | 1.8 |
| Dayao 14th | 0.4 | 0.22 | 18.9 | 73.3 | 5.1 | 0.05 | 0.1 | 0.04 | 1.8 |
| Dayao 14th | 0.4 | 0.28 | 20.1 | 71.1 | 5.5 | 0.14 | 0.25 | 0.03 | 2.2 |

But why should a new Guan ware manufactory have been established so far from the capital? The answer may be that the region was chosen because wood-fuel was so abundant at Longquan and running short at Hangzhou, which was fast becoming one of the larger cities of the world, and also an important port and shipbuilding centre. There was also an established tradition for making fine Yue-type wares in the south-western Zhejiang province, so good ceramic skills were already in place (Vainker 1993, 37). Suitable raw materials for Guan ware manufacture must have been discovered there too — materials quite different from those then used in the local greenware tradition.

## 7. The appearance of *di* ware

Almost simultaneously with the appearance of Longquan Guan ware at the Dayao and Xikou kilns, another new ceramic type started to be produced in the Longquan district — one that may have been pioneered at the Longquan kilns themselves. This was *di* ware. This typology shared many of the forms and the thick, even, multi-layered bluish glazes of Hangzhou Guan ware, but used a pale grey porcellaneous body instead. Heavy crazing (an admired feature of true Guan ware) tended to be absent from *di* wares, and the glazes were invariably of the lime-alkali rather than the lime glaze type. This combination of a fine pale porcellaneous body, refined Guan-type forms, and thick, jade-like, un-crazed glazes, resulted in a ceramic of quite exceptional quality.

With regard to the time of its first appearance, Ren Shilong writes: "The picture is clear: ceramics with thin paste and thick glaze [i.e., *di* ware and 'Longquan Guan ware'] first occurred in the Chunxi reign of the middle Southern Song period, but not earlier' (Chunxi reign was 1174-1189) (Ren Shilong 1994, 37). Likewise, Jin Zhiwei and Bao Kanjie wrote: " ... it can be concluded that a few delicate Longquan celadon porcelains have likely appeared in the royal palace of Southern Song dynasty during the late reign of the Emperor Song Xiazhong", again during the Chunxi reign (Jin Zhiwei and Bao Kanjie 2005, 405).

On the basis of the above, we can trace the *di* ware glaze technology back to the Huangbao kilns at Yaozhou, in Shaanxi province, of the early 10th century — and then through the various Ru ware kilns that operated in the Henan province in the late 11th and early 12th centuries AD. Following the fall of the Northern Song dynasty in 1127, these important ceramic principles were transferred to south China, and to the Laohudong and Jiaotanxia Guan ware kilns at Hangzhou. From Hangzhou, the blue celadon technology was then transmitted to Longquan — both for glazing 'Longquan Guan wares' and also for the new *di* ware material (Fig. 4). Analysis shows the remarkable consistency that was achieved with this prestigious blue celadon glaze over some three hundred years of production (Table).

With the use of data from previous publications, it is now also possible to propose a provisional outline chronology for the history of Chinese blue celadon glazes of exceptional quality — a process that led eventually to the *di*-type celadon wares of the Longquan kilns (Table).

## 8. Titanium dioxide and the *di* ware glaze

The colour of the *di* ware glaze derives from dissolved ions of iron (mainly $Fe^{2+}$ with some $Fe^{3+}$) in the glaze matrix. However, such fine bluish tones are easily changed to green

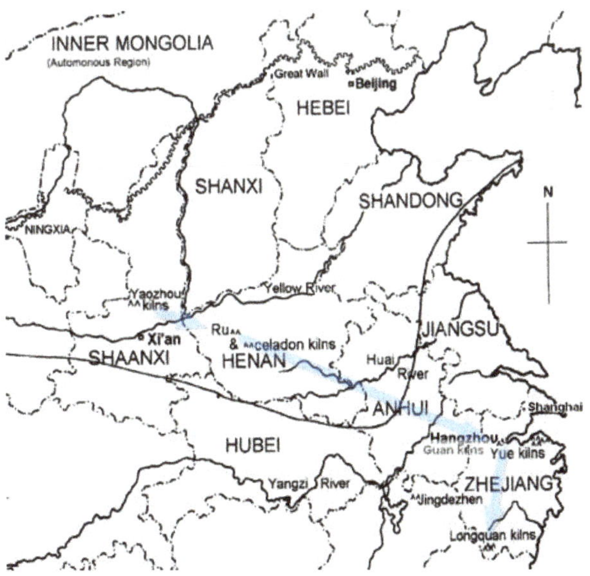


**Figure 4.** Transfer south-east of blue celadon technology in China, 10th to 12th century AD. (Line marks the tectonic division between north and south China; Wood 2000a).

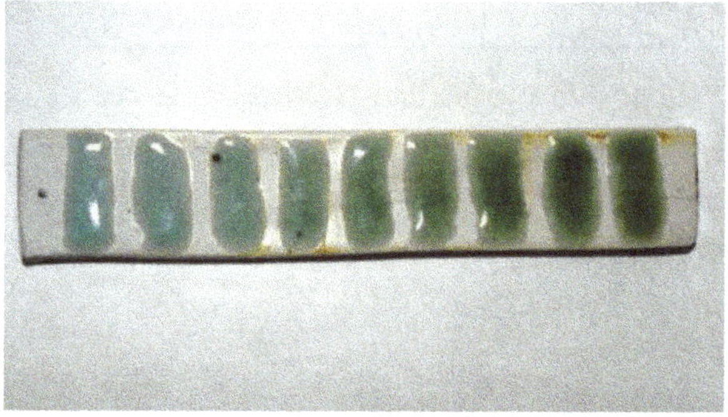

**Figure 5.** Effect of adding titanium dioxide to a lime-alkali celadon glaze containing 1.5% iron oxide (far left), up to 0.8% $TiO_2$ (far right), in 0.1 % increments. Reduction firing to 1250°C.

by a titanium dioxide content higher than 0.2 % in the glaze's overall composition (Fig. 5; Ishii Tsuneshi 1930, 357; Table for typical titania levels in the glazes discussed).

Titania is a common impurity in most glaze raw materials, and one that is difficult to remove by refining. The low levels found in these glazes suggest that low titania raw materials were used to make them, and analyses of rocks local to the corresponding kilns tend to support this view. Reduction firing was also essential for maximising the blue tones of the glazes. As all the key wares in this story have now been analysed, it is possible to show that *di* wares have some of the lowest titania levels in China's blue celadon tradition - an effect further enhanced by the light-coloured porcellaneous bodies used beneath the glazes (Fig. 6).

## 9. Multi-layer glaze application

The colour of blue celadon glazes can be maximised by thickness. However, thick plastic glazes in their raw states tend to crack as they dry, and such cracks can lead to 'crawling' in firing — that is, bare patches appearing in the glaze surfaces. Applying the glaze as a series of thin layers can overcome this problem, particularly with higher alumina, and therefore more plastic, clay- or mica-rich glazes, such as those used on Guan wares and *di* ware. As many as four separate layers may be seen in a single thickness of a Laohudong Guan ware glaze, and the same technique was applied to both Longquan Guan ware and *di* wares (Sundius and Steger 1963, 430, Fig. 39; 438, Fig. 46).

Just how long this multi-layer glazing technique had been practised in China is difficult to say, but one can suspect its use on luxury celadon wares in north China before the technique was transferred south. In particular, a very fine, but broken and repaired, example of Ru ware in The British Museum was dismantled recently for restoration. This showed a wafer-thin body, and a very smooth, thick and even, blue celadon glaze. This bowl has been identified as a likely candidate for the description 'Northern Guan ware', and its thick glaze and thin body appear very close indeed to the Guan wares that were made a few years later in south China (Krahl 1993; Wood 2007). Analysis indicated a northern composition for this British Museum bowl (Ian Freestone personal communication), and recent research has proposed Zhanggongxiang (one of the last Ru ware kiln sites in north China) as a likely site for its manufacture (Sun Xinmin 2006, 16-7).

## 10. Was the *di* ware glaze fluxed mainly by wood ash or by limestone?

As to the main glaze-flux used for *di* ware, the best indication for whether wood ash or limestone was the dominant flux is found in the $P_2O_5$ content of the glaze (Zhang Fukang 1986, 44). Southern greenware glazes tend to average 1.0% $P_2O_5$, which is high, and suggests wood ash. In contrast, in most *qingbai* glazes, $P_2O_5$ levels average about 0.1%, which is low, and suggests limestone. Southern Song Dayao *di* glazes show phosphorous oxide levels of about 0.2%. These are low, but not entirely negligible, and thus a predominance of limestone, plus a minor amount of wood ash, may explain these $P_2O_5$ levels. The major glaze ingredient would have been a porcelain stone, much like the one used for the *di* ware body.

## 11. The *di* ware body

This proposed use of limestone in the *di* ware glaze has interesting parallels with *qingbai* technology, which, by the time that *di* ware first appeared, was already some two centuries old in southern China. *Qingbai* technology may therefore relate in turn to the use of porcelain stone as the main body material in *di* ware. However, the use of a rather ferruginous porcelain stone, or the mixture of porcelain stone with red clay, as proposed by Chou Jen *et al.* (1973), was unprecedented in the *qingbai* tradition, where purity was everything. There are some rare instances of white porcelain stone usage in the Longquan region during the Northern Song dynasty, but these are few and far between. Likewise, bluish celadon glazes appear occasionally on some rare Northern Song Longquan celadons with grey bodies - but the '*di* ware' combination of the thick bluish lime-alkali glaze on a thin pale porcellaneous body seems to have been a major innovation of the late 12$^{th}$ century AD.

Perhaps a more likely source for porcelain stone usage in *di* ware might be found in Longquan Guan ware. As discussed above, *di* ware is contemporary with Longquan Guan ware,

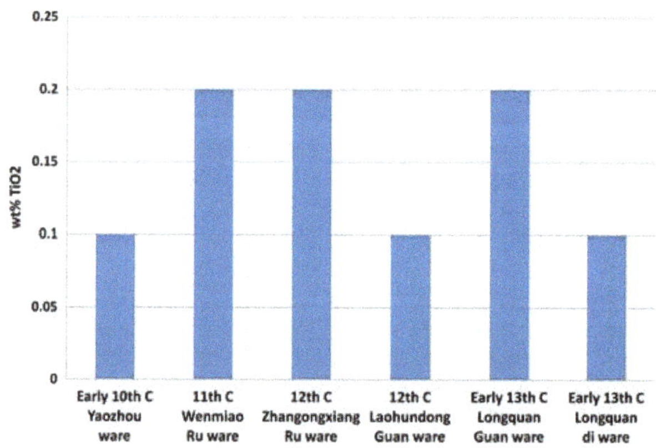


**Figure 6.** Titanium dioxide levels in Chinese blue celadons, 10$^{th}$-13$^{th}$ centuries AD (Analyses from Wood *et al.* 2004a, 225 (SEM-WDA); Zhu Wenli and Zhu Yufeng 2002, 448 (PIXE); Li Jiazhi *et al.* 2001, 35 (EPMA); Vandiver and Kingery 1986, 188 (SEM-EDA)).

which used a body made from a refined primary red clay, and a glaze that was probably a mixture of white porcelain stone with limestone and/or wood ash (Kerr and Wood 2004, 583-4). With Longquan *di* ware, the body is probably a naturally slightly ferruginous pale porcelain stone, or a white porcelain stone mixed with a small amount of refined primary red clay. Thus, two materials used for making Longquan Guan ware could have become the basis for the Longquan *di* ware body. This may be a coincidence, or it may be that *di* ware was developed through experience at Longquan with the new Longquan Guan ware raw materials.

## 12. After *di* ware

*Di* ware was produced in relatively small quantities, for a discerning market. A version with a greener glaze, but sharing the same superior craftsmanship, was devised in the 13$^{th}$ century. This in turn led to the mass-produced, thick-bodied, green-glazed material that became a huge commercial success with the establishment of the Mongol Yuan dynasty in the late 13$^{th}$ century (Ren Shilong 1994, 40). Hundreds of new 'Longquan ware' kilns sprang up in this region, with Dayao itself remaining an important production site well into the Ming dynasty period (Krahl 2004, 70-1).

## 13. Changes occurring in the Dayao body material from the late 12$^{th}$ to the 14$^{th}$ century AD

While recognisably the same material, the Dayao body gradually became richer in potassium and lower in iron and titanium oxides over the centuries — that is, whiter and slightly more vitreous. This could be explained by the use of deeper and less weathered porcelain stones and/or by the use of less red clay in the body recipes. This later Longquan material was made in colossal quantities and exported on an equally massive scale.

## 14. Summary and conclusion

Despite the obvious visual differences between the two wares, Jingdezhen porcelains and Longquan celadon wares are closely related technologically 'beneath the skin'. Both used bodies based on micaceous porcelain stones, often with primary clay additions, and both used the advanced and stable lime-alkali style of high-temperature glaze. For much of the 14$^{th}$ century AD, the two kiln sites were practically matched ware-for-ware, although the huge dishes and vases produced at Longquan in the 14$^{th}$ century, measuring as much as a metre in diameter or height, did not appear at Jingdezhen until the 15$^{th}$ century AD.

However, when the origins of these body types and glaze types are considered, the reasons for their adoption seem rather different. For example, the use of kaolin at Jingdezhen was probably related to resource-depletion, and to the need to exploit reserves of non-kaolinised porcelain stone (Tite *et al.* 1984, 153). At Longquan, in contrast, the occasional additions of iron-rich primary clays to white porcelain stones was more likely to have occurred in the search for a visually harmonious body to use with the jade-like celadon glazes that were adopted there in the late 12$^{th}$ century, inspired by Guan ware. With regard to the glazes themselves, the development of underglaze blue painting at Jingdezhen meant that the fluid lime glazes that had been staples for *qingbai* wares for some three hundred years had to be replaced by the more viscous style of lime-alkali glaze, to prevent the blue painting from running at full heat. At Longquan, the use of the lime-alkali glaze can be traced back to the 10$^{th}$ century blue celadons of Yaozhou. This glaze then saw a complex train of succession on imperial-quality stonewares, before its eventual adoption at Longquan. In this case, lime-alkali glazes were adopted for the fine colours they gave with small amounts of iron oxide in solution, and for their firing stability when applied in successive layers to substantial thicknesses.

Thus, we see a pattern of parallel technological development at Jingdezhen and Longquan — but probably driven by rather different causes. In some ways, Jingdezhen can be seen as representing a simpler and more local technological history that was essentially southern, and unconnected with the history of imperial-quality north Chinese stoneware. In contrast, *di* ware can be regarded as the pinnacle of one major ceramic tradition in China and the beginning of another - namely the culmination of high-quality bluish celadon-making, initiated and largely developed in north China, and the start of large-scale Longquan celadon production, using local porcellaneous materials.

In the end though, the success of Jingdezhen blue and white ceramics led to the demise of the Longquan celadon industry. As the Jingdezhen kiln complex expanded in the 15$^{th}$ and 16$^{th}$ centuries, the manufacture of Longquan-type celadon ware declined through the effects of competition, and also following the withdrawal of official patronage (Krahl 1986, 55; Kerr and Wood 2004, 581). By the 17$^{th}$ century, the Longquan material was practically extinct — marking the decline and eventual collapse of one of the world's most successful and productive ceramic traditions.

## References

Carswell J., 2000, *Blue & White. Chinese Porcelain Around the World*, British Museum Press, London.

Chen Xianqiu, Chen Shiping, Zhou Xuelin, Li Jiazhi, Zhu Boqian, Mou Yunkang, Wang Jiying, 1986, A fundamental research on ceramic of Southern Song Altar Guan ware and Longquan Ge ware. *Scientific and Technological Insights on Ancient Chinese Pottery and Porcelain*, (ed. Shanghai Institute of Ceramics 161-169, Science Press, Beijing.

Chou Jen, Chang Fu-K'ang and Che ng Yung-Fu, 1973, Technical Studies on the Lungchuan Celadons of Successive Dynasties. *Khao-ku Hsu eh-pao*, 1, 131-156.

Du Halde, J-B., ed., 1736, *The General History of China containing A Geographical, Historical, Chronological, Political and Physical Description of the Empire of China, Chinese Tartary, Corea and Thibet*, Printed for and by John Watts, London.

Du Zhengxian, 2002, *Hangzhou Laohudong yaozhi ciqi jingxuan*, Wenwu chubanshe, Beijing.

Guo Yain-yi, Wang Shon-ying and Chen Yao-cheng, 1980, A study on the northern and southern celadons of ancient Chinese dynasties. *Journal of the Chinese Silicate Society*, 8, 232-243, (In Chinese).

Henansheng, 2008, Henansheng wenwu kaogu yanjiusuo. *Baofeng Qingliangsi Ruyao*, Daxiang publishing, Zhengzhou.

Ishii Tsuneshi, 1930, Experiments on the Tenriuji Yellow Celadon Glaze. *Transactions of the British Ceramic Society, Wedgwood Bi-centenary Memorial Number*, 352-359.

Jin Zhiwei and Bao Kunjie, 2005, Li-Styled Stove of Longquan Kiln Used in Royal palace of Southern Song Dynasty (A.D. 1127-1279). *Gu Taoci Kexue Jishu 6: ISAC '05*, 405-410, Shanghai Research Society of Science and Technology of Ancient Ceramics, Shanghai.

Jin Zhiwei and Chen Huqing, 2005, A Study on the "Products of the Guan Kiln at Dilingtou". *Gu Taoci Kexue Jishu 6: ISAC '05, 411-5*, Shanghai Research Society of Science and Technology of Ancient Ceramics, Shanghai.

Jin Zhiwei, Chen Huqing, and Li Qian, 2005, A Brief Discussion on Yue Porcelain for Royal Ritual of the Southern Song Dynasty. *Gu Taoci Kexue Jishu 6: ISAC '05*, 397-404, Shanghai Research Society of Science and Technology of Ancient Ceramics, Shanghai.

Kerr R. and Wood N., 2004, In *Joseph Needham: Science and Civilisation in China, Volume 5: Chemistry and Chemical Technology: Part 12 - Ceramic Technology*, Ed. (R. Kerr, with contributions from Ts'ai Mei-Fen and Zhang Fukang), Cambridge University Press, Cambridge.

Krahl R., 1986, Longquan celadon of the Yuan and Ming dynasties. *Transactions of the Oriental Ceramic Society*, 49, 40-57.

Krahl R., 1993, The 'Alexander Bowl' and the question of Northern Guan ware. *Orientations*, 24, 72-75.

Krahl R., 2004, Famous Brands and Counterfeits: Problems of Terminology and Classification in Song Ceramics. *Percival David Foundation Colloquy on Art & Archaeology in Asia No. 22, Song Ceramics: Art, History, Archaeology and Technology*, 61-75, University of London, School of Oriental and African Studies, London.

Li Dejin, 1994, Technology of Longquan Ware Manufacturing. In *New Light on Chinese Yue and Longquan Wares. Archaeological ceramics Found in Eastern, Southern Asia, A.D. 800-1400*, (ed. Ho Chuimei), 84-99, Centre of Asian Studies, The University of Hong Kong, Hong Kong.

Li Jiazhi, Deng Zequn, and Xu Jiming, 2001, Technical studies and replication of Guan ware, an ancient Chinese ceramic. *Materials Research Society Bulletin*, 26, 31-37.

Liu Xinyuan, 1993, Yuan dynasty Official wares from Jingdezhen. In *The Porcelains of Jingdezhen*, (ed. R.E. Scott), 33-46, Colloquies on Art and Archaeology in Asia 16, University of London. School of Oriental and African Studies, London.

Qin Dashu and Du Zhengxian ed., 2004, *Nan Song Guanyao yu Geyao*, Zhejiang Daxue chubanshe, Hangzhou.

Rastelli S., Wood N. and Doherty C., 2002, Technological development at the Huangbao kiln site, Yaozhou, in the 9th to 11th centuries AD - Some analytical and microstructural examination. In *Gu taoci kexue jishu 5. Guoji taolunhui lunwenji (ISAC '02), Shanghai*, (ed. Guo Jingkun), 179-193, Shanghai kexue jishu wenxian chubanshe, Shanghai.

Ren Shilong, 1994, The Dual Nature of Longquan Wares: Further Discussion. In *New Light on Chinese Yue and Longquan Wares*, (ed. Chuemei Ho), 30-45, Centre of Asian Studies, Hong Kong.

Stannard D., 1986, A Study of Chinese and Western Porcelain-Stones. In *Scientific and Technological Insights on Ancient Chinese Pottery and Porcelain*, (ed. Shanghai Institute of Ceramics), 249-254, Science Press, Beijing.

Sun Xinmin, 2006, The Excavation and Preliminary Study of the Zhanggongxiang Kilnsite in Ruzhou. *Transactions of the Oriental Ceramic Society*, 69, 11-17.

Sundius N. and Steger W., 1963, The constitution and manufacture of Chinese ceramics from Sung and earlier times. In *Sung Sherds*, (ed. Nils Palmgren, Nils Sundius and Walter Steger), 375-505, Almqvist and Wiksell, Stockholm.

Tite M.S., Freestone I. and Bimson M., 1984, A technological study of Chinese porcelain of the Yuan dynasty. *Archaeometry*, 26, 139-154.

Vainker S., 1993, New Light on Zhejiang Greenwares. A note on Greenwares in the Ashmolean Museum. *Oriental Art*, 39, 34-37.

Vandiver P. and Kingery W.D., 1986, Song Dynasty Celadon Glazes from Dayao Near Longquan. In *Scientific and Technological Insights on Ancient Chinese Pottery and Porcelain*, (ed. Shanghai Institute of Ceramics), 187-191, Science Press, Beijing.

Vogt G., 1900, Recherches sur les porcelaines Chinoises. *Bulletin de la Societe d'encouragement pour l'industrie national*, 99, 560-612.

Wood N., 1978, Chinese Porcelain. *Pottery Quarterly*, 47, 101-128.

Wood N., 1986, Some implications of recent analyses of Song yingqing wares from Jingdezhen. In *Scientific and Technological Insights into Ancient Chinese Pottery and Porcelain*, (ed. Shanghai Institute of Ceramics), 261-264, Science Press, Beijing.

Wood N., 2000a, Plate tectonics and Chinese ceramics - new insights into the origins of China's ceramic raw materials. In *Taoci n$^o$ 1. Revue Annuelle de la Socie te française d'E tude de la Ce ramique orientale, Actes du colloque "Le 'Bleu et Blanc' du Proche-Orient a la Chine."*, (ed. M. Crick), 15-24, Muse e Cernuschi, Paris.

Wood N., 2000b, The kaolin question - compositional changes in Jingdezhen porcelain bodies in the Yuan dynasty. In *Taoci n$^o$ 1. Revue Annuelle de la Socie te française d'E tude de la Ce ramique orientale, Actes du colloque "Le 'Bleu et Blanc' du Proche-Orient a la Chine."*, (ed. M. Crick), 25-32, Muse e Cernuschi, Paris.

Wood N., 2007, *Chinese Glazes - Their Chemistry, Origins and Re-Creation*. 2nd edition, A & C Black, London.

Wood N., Rastelli S. and Doherty C., 2004a, Five Dynasties Yaozhou Celadon - a True Ancestor to Laohudong Guan Ware? *Percival David Foundation Colloquy on Art & Archaeology in Asia No. 22, Song Ceramics: Art, History, Archaeology and Technology*, 221-232, University of London, School of Oriental and African Studies, London.

Wood N., Rastelli S. and Doherty C., 2004b, Zaoqi Yaozhou qingci he Nansong Laohudong yao de jishu guangxi (Technological relationships between Yaozhou celadons and Southern Song Laohudong ware). *Proceedings of the International Conference of Southern Song Dynasty Official Ware (Guan Ware) of Laohudong Kiln, Hangzhou*, 190-195, University of Zhejiang Press, Hangzhou.

Wood N., Doherty C. and Rastelli S., 2005, Some aspects of Yue ware production at Shanglinhu in the late Tang dynasty. *Gu Taoci Kexue Jishu 6: ISAC '05*, 185-198, Shanghai Research Society of Science and Technology of Ancient Ceramics, Shanghai.

Xiang Hai Tang, 1995, *Study about the Techniques, Development and Distribution and History of the East Asian Kiln*, Nanjing University, Nanjing.

Yao Guifang, 1989, Brief account of Southern Song Guan kiln's firing date and development. In *Proceeding of 1989 International Symposium on Ancient Ceramics*, (ed. Li Jiazhi and Chen Xianqiu), 389-393, Shanghai Research Society of Science and Technology of Ancient Ceramics, Shanghai.

Zeng Fang, 1997, New Archaeological Discoveries at Jian Kiln Sites. *Transactions of the Oriental Ceramic Society*, 29-33.

Zhang Fukang, 1986, The Origin of High-Fired Glazes in China. In *Scientific and Technological Insights on Ancient Chinese Pottery and Porcelain*, (ed. Shanghai Institute of Ceramics), 40-45, Science Press, Beijing.

Zhu Bojian, 1998, *Celadons from Longquan kilns*, Chu ban, Tao ci wen wu cong shu.

Zhu Wenli and Zhu Yufeng, 2002, A Discussion on Wenmiao and Zhanggongxiang Kiln Sites. In *Gu Taoci Kexue Jishi, Proceeding of ISAC '02 2002 International Symposium on Ancient Ceramics*, (ed. Li Jiazhi and Chen Xianqiu), 445-449, Shanghai Research Society of Science and Technology of Ancient Ceramics, Shanghai.

Zhu Wenli, 2005, Re-discussion on Zhanggongxiang and Wenmiao Kiln Sites. *Gu Taoci Kexue Jishu 6: ISAC '05, Proceeding of 2002 International Symposium on Ancient Ceramics*, 253-257, Shanghai Research Society of Science and Technology of Ancient Ceramics, Shanghai.

# 13

# The second life of ceramics: A new home in a lime environment

**Ioannis Karatasios[a], Konstantinos Alexiou[b], Noémi S. Müller[a,c], Peter M. Day[d] and Vassilis Kilikoglou[a]**

[a]Institute of Materials Science, NCSR "Demokritos", Greece, ikarat@ims.demokritos.gr, kilikog@ims.demokritos.gr, noemi.s.mueller@gmail.com

[b]21[st] Archaeological Ephorate of Prehistorical and Classical Antiquities, Ministry of Culture, Greece, alexiou.konstantin@gmail.com

[c]Science and Technology in Archaeology Research Center (STARC), The Cyprus Institute, Cyprus

[d]Department of Archaeology, University of Sheffield, UK, p.m.day@sheffield.ac.uk

**Abstract** - This paper presents the results of a systematic investigation of the pozzolanic activity of different types of ceramic admixtures, as a function of their chemical composition (calcareous, non-calcareous) and vitrification state. The study was carried out on experimental ceramic-lime mixtures. The evaluation of the reactivity of the different ceramic admixtures includes conductivity measurements of the saturated calcium hydroxide solutions to which they were added, assessment of the amount of hydrated calcium alumino-silicate phases formed, and determination of the microstructure of the resulting solid material. The experimental results of this study provide a baseline against which the performance of different archaeological mortars can be discussed. In conclusion, low-fired calcareous ceramics with no or limited vitrification have been found to exhibit good reactivity with slaked lime, while high-fired non-calcareous ceramics present no reactivity.

## Introduction

Lime mortars are one of the most commonly used materials for structural, protective, and decorative purposes (Zacharopoulou 1998). Generally, mortars are constituted of a binding agent mixed with different types of aggregates and water. After mixing, they harden and gradually develop their adhesive and mechanical properties. The quality and final performance characteristics of archaeological and historic lime-based mortars depend on many parameters. These include the type and quality of binder, the binder to aggregate ratio, the water to binder ratio, the granular distribution of aggregates, the shape and type of aggregates, as well as any admixtures or additives used. With regards to pozzolanic materials, it appears that their ability to enhance the long-term strength and durability of typical lime-based mortars has been known since antiquity (Baronio *et al.* 1997).

Pozzolanas are natural or artificial alumino-silicate materials that are not hydraulic by themselves, but form hydraulic compounds when they are mixed with hydrated lime (Charola and Henriques 1999). The most common pozzolanic materials used historically in mortars include volcanic ash deposits from Pozzuoli in Italy, from the islands of Santorini (Thera) and Milos in Greece, as well as different types of finely ground ceramic powder. The addition of pozzolanas to lime mixtures promotes a chemical reaction between lime and pozzolanic admixtures, which enables them to harden and develop strength, especially in humid conditions (Charola and Henriques 1999). In such cases, the setting of the mortars is affected mainly by the dissolution of the siliceous phases in the aqueous solution (Moropoulou *et al.* 2004a). At the same time, the weak calcium hydroxide crystals are gradually converted to the strong calcium silicate hydrate and calcium aluminate hydrate fibrous gel (Reda Taha and Shrive 2001; Moropoulou *et al.* 2004a). These phases have a similar microstructure to the products developed by the hydration of modern cements and they attribute to pozzolanic mortars their hydraulic properties.

During the late Roman period, ceramic powder and fine fragments of crushed bricks were added to lime mortars to act as artificial pozzolanas and produce a new mixture with hydraulic properties, the so called '*cocciopesto*' (Biscontin *et al.* 2002). These mortars were mainly used for rendering and for the upper layers of floors, while mortars that contain larger fragments of ceramic material were used for masonry (Moropoulou *et al.* 2005). The Romans have been credited with the spread of such mortars throughout their empire, including Europe, northern Africa, and northern Asia (Malinowski 1981). During the early Byzantine period, the use of crushed bricks and/or ceramic roof tiles in the joints of load-bearing walls became more frequent. The amount and the size of ceramic fragments (to act as aggregates) was increased, while consequently the thickness of the horizontal joints gradually increased from 10-15 mm to 40-60 mm, depending on the size of the fragments used (Papayianni and Karaveziroglou 1993; Baronio *et al.* 1997). Representative masonries of Byzantine buildings in Italy, Turkey and Greece are built with very thick mortar joints, using a mortar/brick thickness ratio equal to 1/1. This aspect gives the Byzantine walls a peculiar appearance, with evident horizontal strips. However, the reason for using thick joints is not yet clear (Binda *et al.* 2000).

In all those cases, it is generally assumed that ceramic wasters, sherds from broken ceramic vessels, or construction material (bricks) were used, effectively recycling material that no longer fulfilled its original purpose. However, not all types of ceramics react in the same way when in contact with lime, but their reaction depends on their mineralogical and chemical composition, as well as the degree of vitrification. Consequently, the selection of different types of ceramic material to be used as admixtures is expected to affect the service life and durability of the mortar constructions.

Although there is much work devoted to the study of the reactivity of natural and artificial pozzolanas, only little systematic data are available on the exact effect of their mineralogical and chemical composition on the reactivity of ceramic admixtures with lime. Usually, the pozzolanicity of individual clay minerals (mainly kaolinite and metakaolinite) and zeolites is studied (Sabir *et al.* 2001; Fernandez *et al.* 2011), as is the effect of firing or calcination temperature on the reactivity of the above phases (He *et al.* 1995). It has been suggested that reactivity is affected by the dehydroxylation of the clay minerals. When exposed to temperatures in the range of 600-800°C, this results in a collapsed and disarranged clay structure (Sabir *et al.* 2001), which is believed to enhance pozzolanicity. Similarly, studies which looked at the pozzolanicity of ready-made commercial bricks and ceramic products (Baronio and Binda 1997) and fired clays in lime-based mortars (Moropoulou *et al.* 2004b; Zendri *et al.* 2004) suggested that the firing of clay in the range of 500-900°C results in enhanced reactivity. However, it is clear (Baronio and Binda 1997) that neither all types of clays nor all ancient bricks exhibit the same pozzolanic behaviour. Finally, a study of ceramic materials used in historic lime mortars concluded that low-calcareous, low-fired ceramic admixtures exhibit high pozzolanicity (Böke *et al.* 2006).

Aiming to further the existing knowledge on this subject, this paper presents the results of a systematic monitoring of the pozzolanic activity of different ceramic types on laboratory prepared ceramic-lime mixtures. It aims to identify the degree to which ceramic manufacturing parameters affect the performance of ceramic fragments in lime mortars. In particular, the influence of the chemical composition and the firing temperature of ceramic fragments on their pozzolanic activity was examined. On the basis of the experimental results, the performance of some archaeological mortars is briefly discussed. The archaeological samples examined include two groups of lime-ceramic mortars from Greek monuments, dated to the Roman-Hellenistic and the Byzantine periods, respectively.

## Methodology

### *Preparation and firing of ceramic pozzolanas*

The two types of clays used in the experimental work originate from the natural deposits of Thrapsano (TH) and Agios Syllas (AS), both from the island of Crete, Greece. According to previous work (Hein *et al.* 2004a, 2004b), the above clays contain 3 and 12% CaO respectively, and are therefore expected to develop different microstructures upon firing (Maniatis and Tite 1981). The initial clay batches were crushed gently with a wooden mortar and pestle and subsequently sieved in order to remove any inclusions and impurities, discarding the material with particle diameters larger than 100 μm. The remaining material was mixed with excess water, left to mature, and formed into small briquettes, which were dried in ambient conditions. All briquettes were fired at three different temperatures, aiming to achieve a non-vitrified, an extensively vitrified, and a totally vitrified microstructure, respectively. The briquettes made of non-calcareous clay (TH) were fired at 700°C, 950°C, and 1050°C, while the ones made of calcareous clay (AS) were fired at 700°C, 950°C, and 1100°C. A different highest firing temperature was chosen for the two types of clays, since non- or low calcareous clays are fully vitrified when fired to 1050°C, while above this temperature they start melting. In contrast, calcareous clays require a slightly higher temperature (1100°C) to achieve full vitrification (Maniatis and Tite 1981). Finally, the fired briquettes were ground to powder and separated according to grain size. For the pozzolanicity test, the fraction with grain size below 63 μm was used.

### *Pozzolanic activity*

The reactivity of different ceramic pozzolanas with lime was determined by a pozzolanicity test (McCarter and Tran 1996), measuring the conductivity of a saturated calcium hydroxide ($Ca(OH)_2$) solution that was mixed with different types of ceramic powders. In each case, the amount of ceramic powder added to the solution was 2.5% w/w. All mixtures were thoroughly stirred and stored under laboratory conditions. Altogether, 24 mixtures of approximately 100 ml each were prepared. More specifically, for each of the six ceramic pozzolanas manufactured, four mixtures were prepared in order to enable us to stop the pozzolanic reactions at four preset intervals (3, 7, 14, and 28 days). Conductivity measurements were carried out daily, for a period of 28 days. At the time intervals mentioned above, the solid material was filtered from the solution, and washed out twice with de-ionised water to remove any excess of calcium hydroxide. The solid material was then rinsed with acetone and di-ethyl-ether to stop the hydration reactions and dried at 90°C for 22 hours. In addition to the conductivity measurements, the solid fraction was qualitatively and semi-quantitatively characterised by differential thermal analysis (DTA/TG), scanning electron microscopy (SEM/EDX), and X-ray diffraction (XRD).

**Table 1.** Elemental and mineralogical composition of clays used for the preparation of different ceramic pozzolans (tr = traces, n.d. = not determined, **** = major phase, * = secondary/minor phase). The concentration of elements is expressed as weight percent (wt%) of the relevant oxides, where the oxygen is combined by stoichiometry.

| Chemical composition (wt%) as determined by SEM/EDX | | | Mineralogical composition | | |
|---|---|---|---|---|---|
| **Oxides** | **Agios Syllas** | **Thrapsano** | **Mineral phases** | **Agios Syllas** | **Thrapsano** |
| $Na_2O$ | 0.4 | 1.3 | quartz | **** | **** |
| MgO | 9.7 | 1.6 | calcite | **** | * |
| $Al_2O_3$ | 14.2 | 25.1 | dolomite | *** | * |
| $SiO_2$ | 47.7 | 54.8 | illite/muscovite | **** | **** |
| $P_2O_5$ | 0.0 | 0.3 | plagioclase | ** | *** |
| $SO_3$ | 0.3 | 0.4 | chlorite | *** | *** |
| $Cl_2O$ | 0.1 | 0.3 | kaolinite | * | * |
| $K_2O$ | 3 | 2.7 | paragonite | n.d. | **** |
| CaO | 13.2 | 3.1 | K-feldspars | ** | *** |
| $TiO_2$ | 1.0 | 0.9 | amphiboles | * | * |
| MnO | 0.6 | 0.2 | hematite | tr. | tr. |
| $Fe_2O_3$ | 8.8 | 8.5 | zeolites | tr. | tr. |
| CuO | 0.6 | 0.4 | serpentine | * | n.d. |
| ZnO | 0.4 | 0.3 | smectite | * | n.d. |
| Total | 100 | 100 | | | |

Elemental analysis was carried out with a FEI-Quanta Inspect electron microscope, coupled with an energy dispersive X-ray analyser. X-ray spectra were collected by scanning at least three areas of 30 × 30 μm for 3 minutes, using a 25 kV beam. Thermal analysis was carried out with a Perkin-Elmer Diamond TG/DTA unit, using 20-25 μg of the solid fraction. Measurements were performed in air with a heating rate of 10°C/min in the range 30-1000°C. Diffraction patterns of the samples were collected on a Siemens D-500 Diffractometer, using Cu-Kα radiation (λ = 1.5406 Å, 40 kV, 30 mA) in the range of 2-60° 2θ, with a step of 0.03°/sec. Analyses of all samples were carried out at 1, 3, 7, 14, and 28 days.

## *Archaeological samples*

For the analyses of the archaeological lime-ceramic mortars, ceramic fragments of approximately 1-2 mm diameter were removed from the samples and examined by SEM/EDX, in order to determine their composition and firing temperature (vitrification state). Any lime remains on the surface of the ceramic fragments were manually removed before the fragments were broken, in order to study a freshly fractured surface. Thin sections of the above mortars were examined with a petrographic microscope. Samples of at least 10 $cm^3$ bulk volume were used for the approximate estimation of their compressive strength. A 2 mm layer of high strength epoxy patty was applied on the two sample surfaces to be compressed, in order to achieve two parallel square planes of specific area. The samples were measured with an Instron testing machine, using a displacement rate of 109 μm/min, until failure. The archaeological mortars analysed were sampled from a range of Roman and Byzantine monuments in Greece.

Roman mortars were collected from two archaeological sites on Corfu island, Sammi (S) and Fiskardo (F). S-10 is a mortar used for filling the space between the walls of the two leaf masonry at the Acropolis of Sami. The mortar is actually the binding material of the large agglomerate fragments used as masonry filler. It is lime rich and contains a small amount of fine ceramic fragments. F-06 is a masonry join mortar used in the Roman baths at Fiskardo. It is a lime-ceramic mortar, rich in binder, which consists of both ceramic and natural stone aggregates. F-07 is a masonry join mortar used in a Roman burial complex at Fiskardo. It is a very compact lime-ceramic mortar, rich in binder, which consists of both ceramic and natural stone aggregates.

The mosaic mortars were sampled from three mosaic pavements excavated in the archaeological site of Nea Anghialos (NA), Greece. Samples were collected from the Basilicas complex of Prelate Peter, dated to the 6$^{th}$ century AD. The complex is built on the ruins of two previous basilicas which date to the 4$^{th}$ and 5$^{th}$ century AD

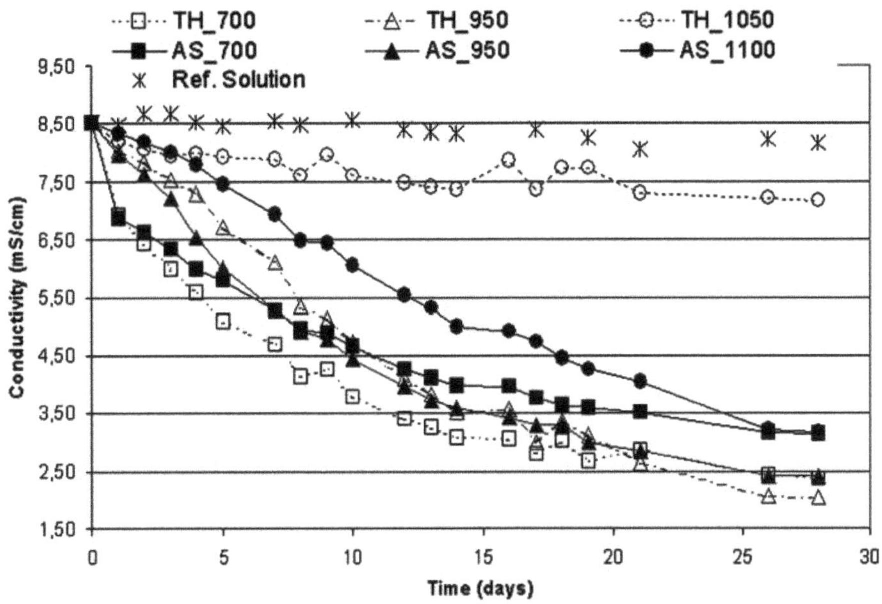


**Figure 1.** Conductivity values of the ceramic-lime mixtures measured during pozzolanicity test.

respectively (Ntina 1994). Since the mosaics were not removed from the archaeological site upon excavation, sampling was carried out from lacunae, i.e., areas where tesserae have been lost. The samples (NA-4, NA-5, and NA-6) were named according to the dating of the archaeological layer they derive of, namely $4^{th}$, $5^{th}$, and $6^{th}$ century AD, respectively. In all samples, the binding material consists of lime mixed with fine- to medium-grained ceramic fragments.

## Results and discussion

### *Mineralogy of ceramic pozzolanas*

The mineralogical and chemical compositions of the initial clays used for the preparation and firing of briquettes at different temperatures are presented in Table 1. It can be seen that the main difference between the two clays is the content of calcite and dolomite, which are significant in Agios Syllas and very low in Thrapsano. These two mineral phases are responsible for the different behaviour of the clays upon firing, with the first developing typical calcareous porous microstructures and the second a dense and less porous matrix (Maniatis and Tite 1981).

### *Pozzolanicity test*

Conductivity values of the ceramic-lime mixtures measured during the pozzolanicity test for different types of ceramic pozzolanas were plotted against time (Fig. 1), in order to compare and contrast the reaction rate of different ceramics with calcium hydroxide. These values were plotted together with those of a reference solution (saturated calcium hydroxide solution) in order to verify that, apart from the lime-ceramic reaction, there is no other mechanism that affects conductivity values (e.g., the carbonation of lime solution).

From the results presented in Figure 1, it seems that the reaction rate of ceramic pozzolanas follows an exponential trend, with the only exceptions being those of Agios Syllas (AS_1100) fired at 1100°C, which follows a first order trend, and Thrapsano fired at 1050°C (TH_1050), which presents no conceivable reactivity with lime. Considering that low conductivity values indicate increased lime consumption, the pozzolanas that show the greatest reactivity according to the conductivity tests are TH_700 and AS_700. However, apart from the TH_1050 that exhibited no reactivity, the ceramic pozzolanas present a very similar reactivity trend, since the differences in lime consumption after 28 days do not seem significant. The slight decrease of conductivity values observed over time in the reference solution may be attributed to the partial carbonation of the saturated solution.

In conclusion, the conductivity monitoring of the solution during the pozzolanicity test was not able to reveal any significant differences in the reactivity of the different types of ceramic materials. Therefore, it was decided to study the reactivity further, through differential thermal analysis and scanning electron microscopy of the ceramic fraction.

### *Thermal analysis and microscopy*

In differential thermal analysis, the temperature difference that develops between the sample and an inert reference material is measured, when both are subjected to an identical heat-treatment. Physical or chemical changes of the sample which lead to the absorption or evolution of heat can be detected relative to the inert reference. In the resulting graph, energy and mass transformations are plotted versus temperature and allow the subsequent characterisation of the samples. The evaluation of the DTA/TG plots was based on the interpretation of the first derivative plots for both DTA and TG. Based on the DTG curves, the formation of the hydrated phases can be monitored. The main area of interest for the evaluation of the reactivity of different ceramic powders with lime was in the range of 100-200°C, where dehydration of poorly crystallised calcium-silicon and calcium-aluminum hydrates (C-S-H, C-A-H) is taking place. Peaks in the region of 100-120°C are usually taken to indicate the presence of C-S-H, while those around 140-180°C are indicative for C-A-H and mixed hydrates (Ramachandran 1988; Moropoulou *et al.* 2004b). Those phases are produced in the saturated calcium hydroxide solution through the reaction of lime ($Ca(OH)_2$) with the ceramic particles. Other areas of interest are in the range of 450-550°C and 700-800°C, where dehydration of calcium hydroxide and decarbonation of calcium carbonate are respectively taking place (Ramachandran 1988). Thermal transformations up to 120°C are also indicative of the evaporation of physically absorbed water, while in the range of 200-600°C chemically bound water of hydraulic compounds (and of other hydroxides) is lost, and in the range of 550 to 830°C that of clay minerals is also lost (He *et al.* 1995; Bakolas *et al.* 1998). Aiming to distinguish the dehydration of C-S-H and C-A-H phases from the evaporation of free water and the dehydration and dehydroxylation phenomena of clays, the samples were dried at 60°C to constant weight prior to measurement. Furthermore, the data obtained from DTA/TG measurements were normalised based on the weight loss of the ceramic material in the same temperature range, as obtained from comparative measurements on untreated samples.

From the weight loss (% w/w) recorded in the range of 100-200°C, the total amount of hydrates was calculated, and two different graphs were plotted, for calcareous and non-calcareous ceramics (Fig. 2). The interpretation of the relevant diffraction patterns indicated that the main hydrate phases formed in the lime solution are tobermorite (low crystallinity C-S-H) and calcium monocarboaluminate hydrate (C-A-H).

In Figure 2, it is apparent that hydrated phases are present in a sufficient amount from the third day of the pozzolanicity test, while their amount increases continuously up to 28 days, where, based on conductivity measurements, the maximum amount of lime consumption occurs. In contrast to the conductivity measurements, thermal analysis reveals that there are significant differences between calcareous and non-calcareous clays, while firing temperature also plays an important role in the reactivity of ceramics. The most active material proved to be the low-fired calcareous ceramic (AS_700), which shows only very limited vitrification. In contrast, high-fired non-calcareous ceramics (TH_1050) seem to be inert and present no reactivity, which is in agreement with the pozzolanicity test.

The crystalline phases of the highest fired samples of each type of clay have been determined by XRD and compared to the initial compositions of the clays (Table 1). The diffraction patterns of AS_1100 indicate the formation of pyroxenes (diopside), gehlenite and leucite (potassium-rich feldspathoid) phases in the calcareous clay. The peak of plagioclase (anorthite) is more intense, while illite/muscovite peaks have disappeared. In the case of non-calcareous clays, the diffraction patterns indicated the decomposition of illite/muscovite and plagioclase phases, whereas the K-feldspar peak was not affected. Hematite becomes more crystalline, while mullite ($3Al_2O_3 \cdot 2SiO_2$) is the new phase formed. Mullite is characterised by a very high chemical stability and no reactivity with calcium hydrate (Baronio and Binda 1997), and its presence might

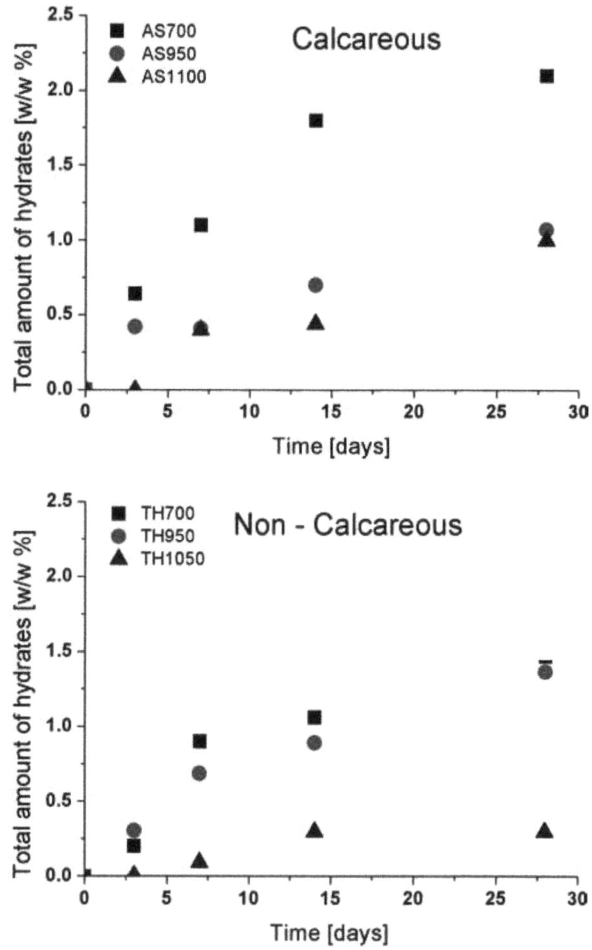


**Figure 2.** The total amount of hydrates as determined by thermal analysis for calcareous (upper graph) and non-calcareous (lower graph) ceramics.

contribute to the very low pozzolanic activity observed in the high-fired non-calcareous ceramic (TH_1050).

The results presented above were confirmed on the SEM, by examination of the solid fraction, filtered from the saturated calcium hydroxide solution. In Figure 3, the dense, angular, gel-like microstructure of the hydrates formed in sample AS_700 can be seen. These new phases fill the empty space of meso- and macro-pores in mortars, creating a denser microstructure with decreased porosity, and therefore increase the strength properties of lime-ceramic mortars. In the case of high-fired non-calcareous ceramics, the extensive vitrification of the ceramic body creates a dense, very stable matrix that deters the reaction with lime (Fig. 4). Concerning the rest of the samples, it seems that low- and medium-fired, non-calcareous ceramics present better reactivity than medium- and high-fired, calcareous ceramics (Figs. 3, 4).

### *Characterisation of archaeological mortars*

The examination of thin sections revealed that the majority of mortars contained both calcareous and non-calcareous ceramic fragments. The grain size and the total amount of

**Figure 3.** Microstructure of the hydrates formed in AS_700 sample (calcareous clay, fired at 700°C) after 28 days in the lime solution.

**Figure 4.** Microstructure of the TH_1050 sample (non-calcareous clay, fired at 1050°C). No hydrates are formed after 28 days in the lime solution.

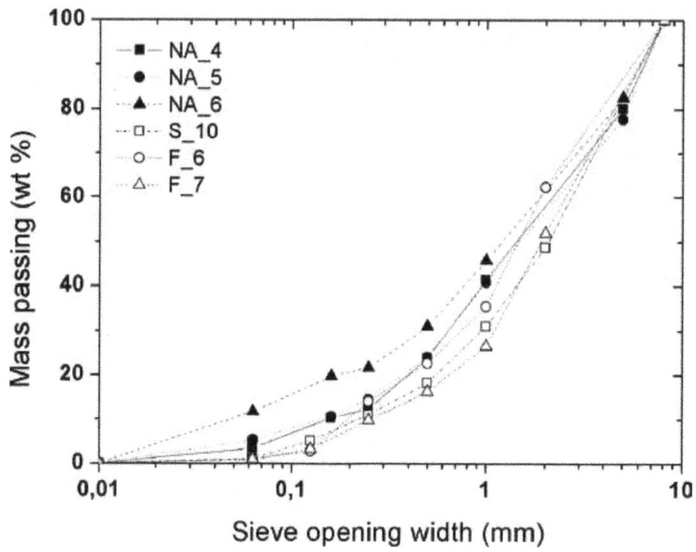


**Figure 5.** Grain size distributions of aggregates incorporated in mortars sampled from archaeological monuments.

the ceramic admixtures in the examined samples appear to depend mainly on the use of mortar. Thus, in the mosaic mortars examined, ceramic fragments range from 1 to 3 mm, while in masonry mortars ceramics exceed 5-10 mm. Both petrographic examination and analyses by SEM indicate that ceramic fragments were generally angular, of high sphericity, and moderately to well fired. In many cases, ceramic fragments contained secondary calcite in their pores. The amount and size of aplastic inclusions in the ceramic fragments varied, and identifiable minerals consisted mainly of quartz, quartzites, and mica. The grain size distribution of the aggregate fraction participated in the mortars studied is presented in Figure 5. Finally, the estimate of the approximate compressive strength for the six mortar samples examined (Fig. 6) indicated a clear trend for superior strength characteristics as compared to typical lime mortars (Hughes and Valek 2003). Considering the geometric parameters that may influence the estimated strength values (Drdácký *et al.* 2008), as well as the differences in mix proportions and grain size distribution of the different mortars (Fig. 5), these values are considered indicative for their strength. However, the presence of considerable amounts of ceramic powder in the binder fraction, along with the experimental results outlined by this experimental study, can further support the notion of the beneficial role of ceramic fragments as admixtures to lime mortars. Considering their ability to harden and set in extremely wet conditions, the addition of ceramic material also increases the types of applications of such mortars.

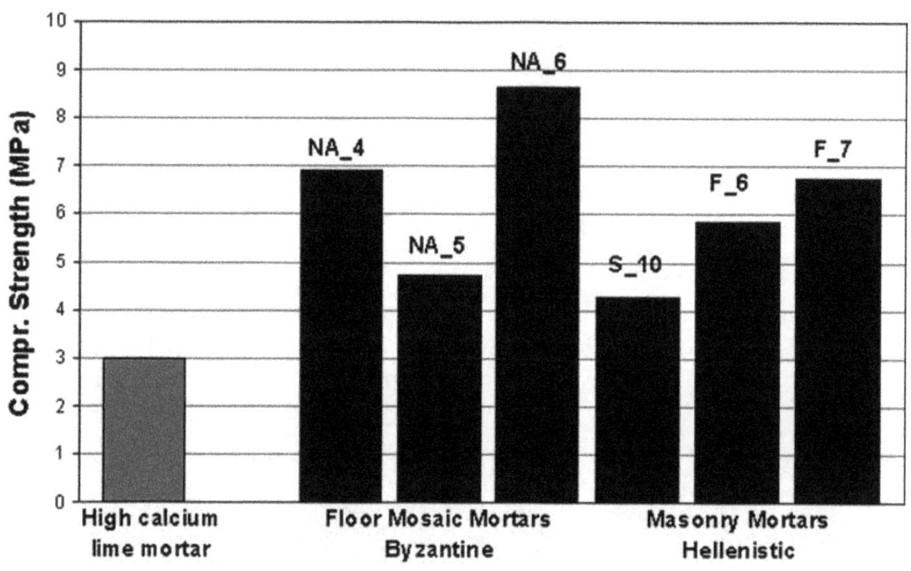


**Figure 6.** Comparison of the approximate compressive strength of the mortar samples studied with those of a high calcium (non-hydraulic) lime mortar.

## Conclusions

The main factors affecting the setting reactions and pozzolanicity of lime-ceramic mortars, besides the grain size and quantity of the ceramic admixtures, are their chemical and mineralogical composition, as well as firing temperature. The conductivity monitoring of the ceramic powder - lime mixture during the pozzolanicity test was not able to reveal any significant differences in the reactivity of the different types of ceramic materials, in contrast to differential thermal analysis and scanning electron microscopy.

Generally, it could be stated that the reactivity of ceramic admixtures in lime mortars is higher for lower fired (non-vitrified) ceramics; while high-fired (totally vitrified) non-calcareous ceramics present limited or no reactivity. The reactivity of the ceramic material was found to be affected by the chemical stability of the phases formed during heating, and, consequently, their cations exchange capacity with lime. The experimental results of the laboratory experiments provide a baseline against which the performance of different archaeological mortars can be discussed. In the present case, a preliminary examination of lime-ceramic mortars from different time periods (Roman, Hellenistic, and Byzantine) indicated that both calcareous and non-calcareous ceramics were used, and the measurement of the compressive strength of those samples confirmed their superior performance as compared to conventional lime mortars. Based on the results of the experimental study and on the examination of the archaeological samples, it appears that the use of ceramic admixtures ensured in all cases enhanced performance and durability. Therefore, it is evident that in the cases where ceramic powder was added in the lime, this was done intentionally, in order to increase the strength and attribute a hydraulic character to the mortars. However, archaeological results and references mentioned in the text suggest that crushed wasters were not used as pozzolanas. In contrast, it seems that partially vitrified domestic pottery and architectural ceramics were recycled in mortars, ensuring the required reactivity and enhancing the performance of lime-ceramic mortars.

## References

Bakolas A., Biscontin G., Moropoulou A. and Zendri E., 1998, Characterization of structural byzantine mortars by thermogravimetric analysis. *Thermochimica Acta*, 321, 151-160.

Baronio G. and Binda L., 1997, Study of the pozzolanicity of some bricks and clays. *Construction and Building Materials*, 11, 41-46.

Baronio G., Binda L. and Lombardini N., 1997, The role of brick pebbles and dust in conglomerates based on hydrated lime and crushed bricks. *Construction and Building Materials*, 11, 33-40.

Binda L, Baronio G and Tedeschi C, 2000, Experimental study on the mechanical role of thick mortar joints in reproduced byzantine masonry. In *Historic Mortars: Characteristics and Tests*, (eds. P Bartos, C Groot and JJ Hughes), 227-247, RILEM Publications, Cachan.

Biscontin G., Birelli M. and Zendri E., 2002, Characterization of binders employed in the manufacture of Venetian historical mortars. *Journal of Cultural Heritage*, 3, 31-37.

Böke H., Akkurt S., Ipeko lu B. and U urlu E., 2006, Characteristics of brick used as aggregate in historic brick-lime mortars and plasters. *Cement and Concrete Research*, 36, 1115-1122.

Charola A.E. and Henriques F.M.A., 1999, Hydraulicity in lime mortars revisited. In *Historic Mortars: Characteristics and Tests*, (ed. P. Bartos), 95-104, RILEM Publications, Cachan.

Drdácký M., Mašín D., Mekonone M.D. and Slížková Z., 2008, Compression tests on non-standard historic mortar specimens. *Historical Mortars Conference HMC08*, CD proceedings, Lisboa, Portugal.

Fernandez R., Martirena F. and Scrivener K.L., 2011, The origin of the pozzolanic activity of calcined clay minerals: A comparison between kaolinite, illite and montmorillonite. *Cement and Concrete Research*, 41, 113-122.

He C., Osbaeck B. and Makovicky E., 1995, Pozzolanic reactions of six principal clay minerals: Activation, Reactivity assessments and technological effects. *Cement and Concrete Research*, 25, 1691-1702.

Hein A., Day P.M., Cau Ontiveros M.A. and Kilikoglou V., 2004a, Red clays from Central and Eastern Crete: Geochemical and mineralogical properties in view of provenance studies on ancient ceramics. *Applied Clay Science*, 24, 245-255.

Hein A., Day P.M., Quinn P.S. and Kilikoglou V., 2004b, The geochemical diversity of neogene clay deposits in Crete and its implications for provenance studies of Minoan pottery. *Archaeometry*, 46, 357-384.

Hughes J. and Valek J., 2003, *Mortars in Historic Buildings. A Review of the Conservation, Technical and Scientific Literature*, Historic Scotland, Edinburgh.

Malinowski R., 1981, Ancient mortars and concretes-Durability aspects. *Mortars, Cements and Grouts Used in the Conservation of Historic Buildings, Proceedings of the ICCROM Symposium, 3-6 November 1981*, 341-349, International Centre for the Study of the Preservation and the Restoration of Cultural Property, Rome.

Maniatis Y. and Tite M.S., 1981, Technological examination of Neolithic-Bronze Age pottery from central and southeast Europe and from the Near East. *Journal of Archaeological Science*, 8, 59-76.

McCarter W.J. and Tran D., 1996, Monitoring pozzolanic activity by direct activation with calcium hydroxide. *Construction and Building Materials*, 10, 179-184.

Moropoulou A., Cakmak A., Labropoulos K.C., Van Grieken R. and Torfs K., 2004a, Accelerated microstructural evolution of a calcium-silicate-hydrate (C-S-H) phase in pozzolanic pastes using fine siliceous sources: Comparison with historic pozzolanic mortars. *Cement and Concrete Research*, 34, 1-6.

Moropoulou A., Bakolas A. and Aggelakopoulou E., 2004b, Evaluation of pozzolanic activity of natural and artificial pozzolans by thermal analysis. *Thermochimica Acta*, 420, 135-140.

Moropoulou A., Bakolas A. and Anagnostopoulou S., 2005, Composite materials in ancient structures. *Cement and Concrete Composites*, 27, 295-300.

Ntina A., 1994, Recherches modernes a la paliochristanique ville de Fthiotides Thebes. *Quinze années de Recherches Archéologiques, 1975 - 1990. Bilans et Perspectives. Colloque international*, 357-370, Editions Kapon, Lyon.

Papayianni I. and Karaveziroglou M., 1993, Aggregate gradation of ancient mortars. Relationship to strength and porosity. In *Conservation of Stone and Other Materials. Proceedings of the RILEM/UNESCO Congress, Paris, 29 June - 1 July 1993*, (ed. M.J. Thiel), 493-499, Spon, London.

Ramachandran V.S., 1988, Thermal analyses of cement components hydrated in the presence of calcium carbonate. *Thermochimica Acta*, 127, 385-394.

Reda Taha M.M. and Shrive N.G., 2001, The use of pozzolans to improve bond and bond strength. In *9th Canadian Masonry Symposium - Spanning the Centuries with Masonry*, (eds. P.H. Bischoff, J.L. Dawe, A.B. Schriver and A.J. Valsangkar), Department of Civil Engineering. University of New Brunswick, Fredericton. URL: http://www.nrc.ca/irc/fulltext/mortar/ (accessed September 2003).

Sabir B.B., Wild S. and Bai J., 2001, Metakaolin and calcined clays as pozzolans for concrete: a review. *Cement and Concrete Composites*, 23, 441-454.

Zacharopoulou G., 1998, The renascence of lime based mortar technology. An appraisal of a bibliographic study. In *Compatible Materials for the Protection of European Cultural Heritage*, (eds. A. Moropoulou, G. Biscontin, J.D. Rodrigues and M. Erdik), 89-114, PACT 55., Technical Chamber of Commerce, Athens.

Zendri E., Lucchini V., Biscontin G. and Morabito Z.M., 2004, Interaction between clay and lime in "cocciopesto" mortars: a study by $^{29}$Si MAS spectroscopy. *Applied Clay Science*, 25, 1-7.

14

# The ceramic technology of the architectural glazed tiles of Huangwa Kiln, Liaoning Province, China

**Baoqiang Kang[a], Simon Groom[b], Hongying Duan[a], Yinzhong Ding[a], He Li[a], Jianmin Miao[a] and Guanglie Lu[c]**
[a]Palace Museum, Beijing, 100009, PRC, kangbaoqiang@hotmail.com
[b]UCL Institute of Archaeology, 31-34 Gordon Square, London WC1H 0PY, United Kingdom, s.d.groom@gmail.com
[c]Central Laboratory, Zhejiang University, Hangzhou, 310028, PRC

**Abstract** - An assemblage of architectural glazed tiles from Huangwa Kiln, Haicheng in Liaoning Province, was analysed using XRF, XRD, and petrography. Microstructure and chemistry indicated the mixing of multiple components. Two potential raw materials for the tile body, a red argillaceous component and a white metamorphic component, were identified and analysed. Experimental replicas were prepared with varying recipes, investigating firing, refinement, and proportions of material, and compared with archaeological compositions. One of the proposed recipes offers a technological justification for the use of two raw materials. This work has clear implications for restoration work on two World Cultural Heritage sites in Liaoning Province.

## 1. Introduction

Glazed tiles have been used for the decoration of imperial palaces in China since the times of the Northern Wei Dynasty, 386-534 AD (Pang 1998/1999). Until the Song Dynasty (960-1279 AD), this was a relatively minor ornamental form, with small amounts of glazed tiles used architecturally on important buildings, such as palaces and temples. Their use then grew to real prominence during the Ming (1368-1644 AD) and Qing (1644-1911 AD) Dynasties, when application, quality, and scale of production increased significantly (Liu 1984). Historical records of official production indicate that kilns were established at Jubaoshan in Nanjing, Jiangsu Province, and Liulichang and Liliqv in Beijing, along with Huangwa Kiln in Liaoning Province (Wang 2004).

Huangwa Kiln, located in North East China (Fig. 1), is a site showing evidence of glazed tile production for the palaces, tombs, and temples of Liaoning Province from the late Ming Dynasty to the beginning of the Republic of China, 1912 AD. Excavations at the site have uncovered the presence of a domestic production kiln in use prior to the official production of architectural glazed tiles (Fu and Lu 2007).

The site has been subject to multiple campaigns of excavation, with the initial field surveys conducted by a team from the Shenyang Palace Museum in the 1970s resulting in unpublished reports (Fu and Lu 2007). Following this, field projects carried out from 2002 to 2006 by archaeologists from the Anshan Museum led to the discovery of a biscuit firing kiln (Fu and Lu 2007). In the course of these excavations, significant quantities of architectural glazed tiles were collected, notably including a subsample inscribed with buildings' names, indicating the intended sites of tile consumption. Inscriptions found include "太庙", translated as *Taimiao*, indicating the Tai Temple - a temple in the Shenyang Imperial Palace - or "永陵", translated as *Yongling*, indicating the Yongling Tomb, and so on. This type of glazed tiles can be seen throughout sites in Liaoning Province: Beizhen Temple in Beizhen, Yongling Tomb in Xinbin, Shenyang Imperial Palace, Fuling Tomb and Zhaoling Tomb in Shenyang. Among the materials in the earlier strata of the site, pottery has been identified, including vats, bowls, and dishes dated to a period from the Jin Dynasty (1115-1234 AD) through to the Ming Dynasty. These tiles and domestic wares are currently interpreted as local productions.

In the conservation and understanding of the cultural monuments of the region, including World Cultural Heritage sites at Shenyang Imperial Palace and at three Imperial Tombs, these glazed tiles present major aesthetic and functional considerations. A technological study into these tiles was initiated by the Palace Museum. They identified two deposits in the region of the kiln site as potential raw materials for the body fabric of the architectural glazed tiles. Preliminary research into these deposits and the composition of the bodies and glazes of the tiles was undertaken (Kang *et al.* 2009), leading to the

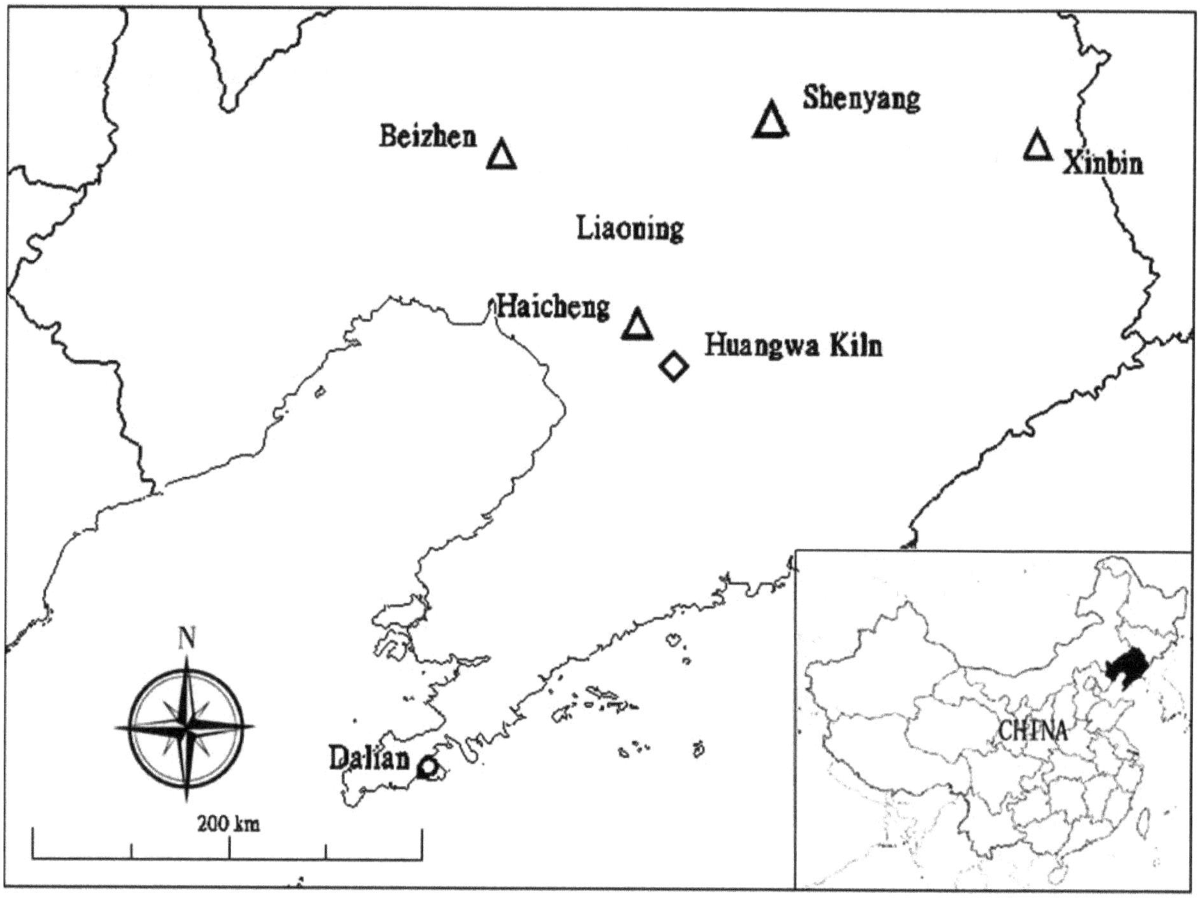


**Figure 1.** The location of Huangwa Kiln, Haicheng, Liaoning Province (diamond). The triangles represent consumption sites.

proposed identification of two separate material types within the tile fabric.

Our aims in conducting this further study were to use a variety of analytical techniques to determine the composition and ceramic technology of the archaeological assemblage, and to produce experimental ceramics for compositional comparison to local raw materials. It is hoped that by providing insights into the technological choices of the imperial potters, conservation procedures at the Imperial Palaces and related buildings may be improved.

## 2. Materials

A total of 34 tiles from the Huangwa Kiln site and a variety of regional tile consumption sites were provided by Anshan Museum; they are labelled as WLLM-0008 to WLLQ-0062. This assemblage was intended to be representative of material covering four production periods (Fu personal communication 2010):

Period I, from Nurhaci's reign to Huang Taiji's reign (1616-1643 AD);
Period II, from Shunzhi's reign, through Kangxi's and Yongzheng's reign, to Qianlong's reign (1644-1795 AD);
Period III, from Jiaqing's reign to Daoguang's reign (1796-1850 AD);
Period IV, from Xianfeng's to Xuantong's reign (1851-1911 AD)

In addition, three items belonging to the pre-imperial domestic ceramic production were analysed: a vat, a bowl, and a dish.

Within the local region, outcrops of several possible raw materials were identified and sampled. The most significant were a red, clay-rich raw material (*hongtu*) and a white crystalline rock (*baitu*), collected from exposures to the northwest and northeast of the kiln site, respectively. In hand specimen, *hongtu* is a brown-red, fine grained, and homogeneous clay (referred to herein as the 'red clay'). *Baitu* is a crystalline rock composed of various mineral phases of differing size, friability, and hardness, supported by a matrix of fine laminated minerals, with a coarse crystalline fraction as discrete minerals. The bulk material was presented in sample Ly-0014. However, ethnographic evidence from interviews with a local descendant of one of the terminal period potters (Fu personal communication 2010) suggests that a refined material composed of the friable fraction (*baitu-hou*) was used in the manufacture of the glazed tiles.

Seven subsamples, including three refined samples, were prepared from the *baitu* material with the purpose of emphasising its different fractions. These subsamples are denoted herein as 'washed friable', 'washed soft block', 'washed hard block', and 'B1' through to 'B4'. Washed samples were mechanically separated using distilled water,

**Table 1.** Description of geological materials.

| Sample name | Visual appearance | Colour |
|---|---|---|
| **Argillaceous Material** | | |
| Red clay | Clay | Brown-red |
| **White Metamorphic Materials** | | |
| Ly-0014 | Fine friable fraction with some blocks | Gray-white |
| Washed friable | Fine powder | Gray-white |
| Washed soft block | Discrete minerals | White with layered structure |
| Washed hard block | Discrete minerals | Yellow-white |
| B1 | Discrete minerals | Green-white |
| B2 | Discrete minerals | White with gray-black intercalated layer |
| B3 | Discrete minerals | White with yellow intercalated layer |
| B4 | Discrete minerals | White |
| **Ethnographically Sourced Material** | | |
| Baitu-hou | Fine friable fraction, may have been washed | Pink-white |

with the resulting non-friable fraction separated into 'washed-soft block' and 'washed-hard block' according to hardness. The 'B' sequence of samples was labelled in order of increasing hardness. The raw material samples are described in Table 1.

## 3. Analytical methods

Major and minor elemental compositions of the archaeological tiles were obtained using wavelength dispersive X-Ray Fluorescence (WDXRF) on fused beads. These analyses were conducted on a PANalytical AXIOS instrument using a rhodium target and SuperQ software. Energy dispersive X-Ray Fluorescence (EDXRF) data were obtained for simulated experimental materials using an EDAX Eagle III XXL operated under the following conditions: tube voltage 25 kV, tube current 600 μA, live time 450 s, and a fundamental parameters standardless calibration with a spot size of 300 μm. Published results show a mean average composition of three spot analyses.

Archaeological samples were prepared by removal of the glazes using an abrasive bladed saw, followed by the crushing and milling of a representative body of material, which was then sieved through a 74 μm mesh. Of the resulting powder, 0.5 g of each sample were mixed with 4.5 g $Li_2B_4O_7$, 0.5 g LiF, and 0.2 g $NH_4NO_3$, and prepared as a fused bead. The Loss on Ignition (LOI) was determined to constant weight using an electrical furnace running at 1000 °C and added to major and minor element results. Where the total values did not fall between 99.5% and 100.3%, they have been normalised to 99.8%, but original totals are also presented. The accuracy, precision, and detection limits of this method have been discussed elsewhere (Duan *et al.* 2009).

X-Ray Diffraction (XRD) analyses were performed on raw materials and fired mixtures. For the raw materials analyses, powdered samples were analysed using a Rigaku D/max 2550PC diffractometer running between 3° and 90° 2θ at 40 kV and 100 mA, using Cu Kα radiation, graphite monochromator, and a scan speed of 8° $min^{-1}$. For the analysis of fired mixtures, including the domestic products, a tube voltage of 40 kV with a current of 150 mA was applied, while other conditions remained constant.

## 4. Results of analysis

### *Chemical analysis*

Sample information and the chemical composition of the 34 tile bodies and the raw materials 'red clay' and 'Ly-0014' are listed in Table 2. It can be seen that the bodies fall into the $MgO$-$Al_2O_3$-$SiO_2$ ternary system, with ranges of 15-25 wt% MgO, 13-17 wt% $Al_2O_3$, 50-56 wt% $SiO_2$, 3-4 wt% $Fe_2O_3$, and less than 2 wt% CaO.

The unusually high MgO and $Al_2O_3$ contents in these ceramics render them unique in the China region among the materials analysed to date. The closest published comparable compositions come from a sequence of Mg-rich Neolithic pottery from southern China (Li 1996), although the much earlier date of these ceramics indicates they are the result of a different tradition. Southern Chinese ceramics show a lower $Al_2O_3$ content, and may derive from raw materials composed of talc-rocks lacking an aluminium-bearing mineral.

Bivariate analysis of bulk compositions suggested covariance into at least two groups of elements. The oxides of iron, aluminium, potassium, and titanium all positively co-varied, suggesting the variable concentration of a

**Table 2.** Sample information and chemical compositions of bodies of architectural glazed tiles and raw materials, analysed by XRD.

| Sample ID | Source of material | Period | $Na_2O$ (%) | MgO (%) | $Al_2O_3$ (%) | SiO2 (%) | $P_2O_5$ (%) | $K_2O$ (%) | CaO (%) | $TiO_2$ (%) | MnO (%) | $Fe_2O_3$ (%) | L.O.I. (%) | Un-normalised Sum (%) |
|---|---|---|---|---|---|---|---|---|---|---|---|---|---|---|
| **Archaeological Tiles** | | | | | | | | | | | | | | |
| WLLM-0025 | Tomb/palace | I | 0.52 | 19.1 | 15.7 | 53.9 | 0.11 | 1.86 | 1.53 | 0.63 | 0.06 | 3.46 | 3.38 | 100.2 |
| WLLM-0044 | Tomb | I-II | 0.64 | 18.7 | 15.2 | 53.8 | 0.13 | 1.60 | 2.01 | 0.66 | 0.06 | 3.43 | 3.61 | 97.9 |
| WLLM-0010 | Tomb | II | 0.53 | 19.3 | 15.8 | 53.2 | 0.12 | 2.10 | 1.91 | 0.63 | 0.05 | 3.47 | 3.36 | 100.4 |
| WLLM-0013 | Palace | II | 0.47 | 20.4 | 15.2 | 53.4 | 0.13 | 1.41 | 1.54 | 0.62 | 0.06 | 3.37 | 3.71 | 100.3 |
| WLLM-0014 | Palace | II | 0.58 | 18.6 | 15.2 | 54.1 | 0.12 | 1.61 | 1.66 | 0.63 | 0.06 | 3.55 | 3.93 | 100.0 |
| WLLM-0033 | Palace | II | 0.62 | 17.9 | 15.6 | 54.5 | 0.09 | 1.94 | 1.19 | 0.63 | 0.06 | 3.63 | 3.64 | 99.8 |
| WLLM-0039 | Tomb | II | 0.45 | 19.9 | 15.4 | 53.0 | 0.12 | 1.68 | 1.60 | 0.62 | 0.05 | 3.25 | 4.03 | 100.0 |
| WLLM-0045 | Palace | II | 0.82 | 19.0 | 15.2 | 53.6 | 0.09 | 1.88 | 1.15 | 0.63 | 0.06 | 3.64 | 3.79 | 98.8 |
| WLLM-0046 | Tomb | II | 0.65 | 16.3 | 15.9 | 54.4 | 0.14 | 2.11 | 1.83 | 0.66 | 0.07 | 4.15 | 3.70 | 98.8 |
| WLLM-0047 | Tomb | II | 0.61 | 20.4 | 14.9 | 53.8 | 0.11 | 1.51 | 1.21 | 0.62 | 0.06 | 3.49 | 3.20 | 98.6 |
| WLLM-0048 | Tomb | II | 0.64 | 19.5 | 15.4 | 54.3 | 0.10 | 2.01 | 0.95 | 0.62 | 0.05 | 3.41 | 2.80 | 98.6 |
| WLLQ-0056 | Tomb | II | 0.46 | 22.3 | 14.2 | 50.4 | 0.10 | 1.41 | 0.96 | 0.59 | 0.05 | 3.22 | 6.08 | 98.3 |
| WLLQ-0057 | Tomb | II | 0.45 | 21.6 | 15.2 | 51.3 | 0.11 | 1.60 | 1.30 | 0.60 | 0.05 | 3.29 | 4.34 | 99.5 |
| WLLQ-0058 | Tomb | II | 0.56 | 19.7 | 15.5 | 53.4 | 0.11 | 1.69 | 1.29 | 0.64 | 0.05 | 3.48 | 3.37 | 97.7 |
| WLLQ-0059 | Tomb | II | 0.46 | 24.3 | 13.3 | 50.1 | 0.09 | 0.92 | 1.13 | 0.57 | 0.05 | 2.94 | 6.02 | 99.5 |
| WLLQ-0061 | Tomb | II | 0.51 | 16.9 | 16.4 | 54.8 | 0.09 | 1.92 | 1.37 | 0.67 | 0.06 | 3.88 | 3.18 | 99.4 |
| WLLM-0008 | Tomb | II-III | 0.58 | 20.0 | 15.2 | 54.3 | 0.10 | 1.61 | 1.18 | 0.63 | 0.06 | 3.53 | 2.71 | 99.1 |
| WLLM-0015 | Tomb/palace | II-III | 0.63 | 20.2 | 15.3 | 54.5 | 0.11 | 1.59 | 1.09 | 0.64 | 0.06 | 3.54 | 2.73 | 100.4 |
| WLLM-0029 | Tomb/palace | II-III | 0.60 | 18.3 | 15.5 | 54.5 | 0.12 | 1.92 | 1.56 | 0.64 | 0.07 | 3.75 | 3.28 | 100.2 |
| WLLM-0030 | Tomb | II-III | 0.51 | 20.6 | 15.6 | 52.3 | 0.10 | 2.03 | 1.05 | 0.62 | 0.05 | 3.33 | 3.89 | 100.0 |
| WLLM-0031 | Tomb | II-III | 0.53 | 18.7 | 16.2 | 54.5 | 0.12 | 2.00 | 1.36 | 0.65 | 0.06 | 3.56 | 2.73 | 100.4 |
| WLLM-0038 | Palace | II-III | 0.54 | 17.6 | 15.8 | 54.9 | 0.12 | 2.07 | 1.25 | 0.64 | 0.06 | 3.55 | 3.21 | 99.7 |

(*Continued*)

**Table 2** - *Continued*

| Sample ID | Source of material | Period | $Na_2O$ (%) | MgO (%) | $Al_2O_3$ (%) | SiO2 (%) | $P_2O_5$ (%) | $K_2O$ (%) | CaO (%) | $TiO_2$ (%) | MnO (%) | $Fe_2O_3$ (%) | L.O.I. (%) | Un-normalised Sum (%) |
|---|---|---|---|---|---|---|---|---|---|---|---|---|---|---|
| WLLQ-0062 | Tomb | II-III | 0.68 | 16.1 | 15.4 | 55.5 | 0.10 | 1.82 | 1.91 | 0.69 | 0.07 | 3.93 | 3.66 | 97.1 |
| WLLM-0011 | Tomb | III-IV | 0.64 | 16.4 | 16.9 | 55.8 | 0.12 | 2.33 | 1.11 | 0.68 | 0.06 | 3.98 | 1.79 | 100.5 |
| WLLM-0012 | Tomb | III-IV | 0.44 | 22.0 | 15.0 | 51.7 | 0.11 | 1.44 | 1.57 | 0.60 | 0.05 | 3.20 | 3.72 | 99.8 |
| WLLM-0018 | Tomb | III-IV | 0.46 | 19.6 | 14.9 | 52.7 | 0.13 | 1.67 | 1.60 | 0.61 | 0.06 | 3.54 | 4.50 | 99.3 |
| WLLM-0021 | Tomb | III-IV | 0.60 | 18.4 | 15.8 | 54.6 | 0.10 | 1.98 | 0.93 | 0.64 | 0.06 | 3.65 | 3.00 | 100.7 |
| WLLQ-0055 | Tomb | III-IV | 0.56 | 20.0 | 15.8 | 54.1 | 0.10 | 1.70 | 0.94 | 0.65 | 0.05 | 3.44 | 2.43 | 98.6 |
| WLLE-0049 | Palace | IV | 0.62 | 21.2 | 15.2 | 51.3 | 0.10 | 1.35 | 1.79 | 0.63 | 0.05 | 3.04 | 4.55 | 98.8 |
| WLLE-0050 | Temple | IV | 0.58 | 15.2 | 16.4 | 54.8 | 0.10 | 2.65 | 1.20 | 0.67 | 0.06 | 4.62 | 3.56 | 98.0 |
| WLLQ-0051 | Temple | IV | 0.60 | 18.4 | 15.6 | 54.2 | 0.09 | 1.70 | 1.65 | 0.64 | 0.06 | 3.48 | 3.41 | 98.5 |
| WLLE-0052 | Temple | IV | 0.59 | 18.4 | 15.7 | 54.5 | 0.10 | 1.78 | 1.38 | 0.64 | 0.06 | 3.49 | 3.26 | 98.0 |
| WLLE-0053 | Tomb | IV | 0.55 | 17.5 | 15.9 | 54.1 | 0.09 | 1.95 | 1.44 | 0.66 | 0.06 | 3.87 | 3.67 | 98.6 |
| WLLQ-0054 | Tomb | IV | 0.57 | 18.2 | 16.3 | 55.1 | 0.09 | 1.96 | 0.89 | 0.66 | 0.05 | 3.54 | 2.41 | 98.7 |
| **Modern Raw Materials** | | | | | | | | | | | | | | |
| Ly-0014 | White raw material | Modern | 0.08 | 33.2 | 8.4 | 31.1 | 0.05 | 0.75 | 0.49 | 0.27 | 0.01 | 1.12 | 24.38 | 99.8 |
| Red clay | Red raw material | Modern | 1.39 | 4.3 | 15.0 | 63.4 | 0.08 | 2.57 | 0.96 | 0.75 | 0.10 | 5.10 | 6.15 | 99.8 |

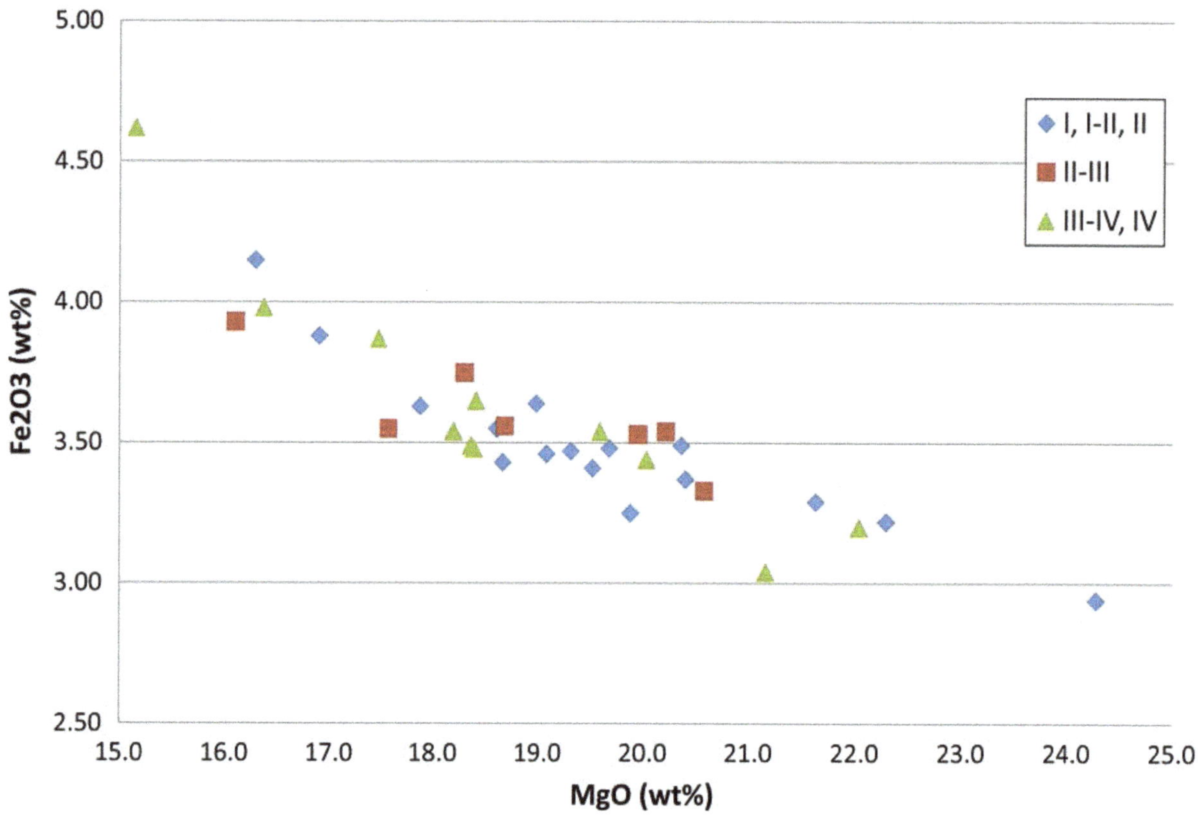


**Figure 2.** Scatterplot of MgO against $Fe_2O_3$ for all tile bodies ($R^2 = 0.78$).

ceramic component that retained a constant relative abundance of these elements. These elements negatively correlated with magnesia, while the correlation with iron gave a 0.78 $r^2$ value (Fig. 2). Such linear relationships typically indicate mixing, and therefore relative dilution between two types of raw material, in this case one supplying the major input of MgO and the other supplying the $Fe_2O_3$ component. Calcium and phosphorus oxides are notable for showing no clear relationship with any other elements and groups of elements. These groupings are consistent with the final tiles' compositional relationship to the two candidates for source materials, with the red clay showing proportionately higher iron (5.5 wt% vs 1.5 wt% after normalisation) and alkali earth oxide contents, and the white rock showing significant magnesia contents (44.0 wt% vs 4.6 wt% after normalisation). The sample Ly-0014 has a very high loss on ignition (24.4%), and this is likely to be a result of the high proportion of hydrous and carbonate minerals, such as chlorite and dolomite, which decompose at high temperatures.

### *Mineralogy analysis*

Potential raw materials and the early domestic pottery bodies were subjected to XRD to determine the crystalline content (Table 3). In all samples, common geological minerals were found to be present, in the form of quartz, microcline feldspar, olivine (forsterite), and enstatite. Accessory minerals, such as muscovite, albite, and spinel, were also found to be common.

Within the red clay, results indicated the presence of quartz, microcline and albite feldspar, chlorite (clinochlore), and muscovite with minor talc and magnesite. This mineralogy is indicative of common ferruginous secondary clays, and is typical of archaeological ceramic sources. As for the nine samples of white raw material, MgO-bearing minerals such as talc, chlorite (clinochlore), and magnesite were predominant, reflecting the regional metamorphic geological context (Zhang 1988; Chen and Huang 1994).

Two groups of material were interpreted, comprised of friable soft fractions and discrete mineral fractions. The first type includes Ly-0014, washed-friable, B1, B2, and *baitu-hou*, all of which showed similar mechanical properties. Among these, clinochlore, talc, quartz, magnesite, and muscovite predominated, mechanically acting as soft grains and plastic minerals. The second type contained the washed-soft block, washed-hard block, B3, and B4, and formed a less friable and denser material. These were mainly composed of magnesite, talc, quartz, and small amounts of clinochlore, all relatively depleted in aluminium content and acting mechanically as more brittle discrete minerals.

### *Petrographic analysis*

An opportunistic sub-sample of archaeological tiles (tiles WLLM-0046, WLLM-0048, WLLQ-0061, WLLQ-0058, WLLE-0049 and WLLQ-0055) was studied petrographically using a variant of Whitbread's well established methodology (Whitbread 1995).

**Table 3.** Mineralogical composition of raw materials and domestic pottery bodies, analysed by XRD.

| Sample name | XRD results |
|---|---|
| **Geological Materials** | |
| Red clay | Quartz, microcline, albite, clinochlore, muscovite, talc, magnesite |
| Ly-0014 | Clinochlore, talc, muscovite, magnesite, quartz, dolomite |
| Washed-friable | Clinochlore, talc, muscovite, magnesite, quartz |
| Washed-soft block | Magnesite, quartz, talc |
| Washed-hard block | Magnesite, quartz, talc |
| B1 | Clinochlore, talc, muscovite, magnesite, quartz |
| B2 | Clinochlore, quartz, talc, muscovite |
| B3 | Magnesite, clinochlore, quartz, talc |
| B4 | Magnesite, clinochlore, quartz, talc |
| Baitu-hou | Clinochlore, talc, muscovite, magnesite, quartz |
| **Pre-Imperial Production Domestic Pottery** | |
| Cugangtai-5 | Quartz, microcline, albite, forsterite, enstatite, talc, muscovite |
| Xigangtai-5 | Quartz, enstatite, cordierite |
| H-106 | Quartz, microcline, forsterite, enstatite |

All samples comprise one fabric class, with the same basic components identified in all of them, along with a sub-group of samples, WLLE-0049 and WLLQ-0055, which are described as variants on the theme. In all cases, the fabric is poorly mixed, with two distinct components forming concentration features. There is a red, argillaceous component (Fig. 3) that forms subrounded and lenticular pellets. Primary inclusions within this component are, in decreasing order of abundance: monocrystalline quartz, biotite and muscovite mica, and opaques, with a 30% total coarse to fine ratio. The colour and birefringence indicate a ferruginous composition, which exhibits linear birefringent streaks. The second component is crystalline with subhedral and anhedral minerals. Where composite rock inclusions are present, these exhibit a schistose texture (Fig. 4), suggesting a metamorphic origin for this lithology. Minerals present in this fraction include, in decreasing order of abundance: chlorite, polycrystalline quartz, talc, muscovite, and magnesite. The groundmass is composed of talc and chlorite. If this comprises a single metamorphic rock-type, this would be consistent with a high pressure, low temperature unit of pelitic composition, possibly of blue-schist facies (Yardley 1989).

The gross tile groundmass is relatively information-poor, with a weak birefringence that suggests high firing. Across the assemblage, the material shows a variable coarse to fine ratio, with coarse material forming between 40% and 20% of each total section, and these components grading down to matrix size. A preferred orientation is often visible, seen among linear fragments, voids, and plastic inclusions, and this is likely to have formed during the tile manufacturing process. Assuming this is the case, we suggest the tile groundmass is formed of a well mixed clay, chlorite, talc, and muscovite paste, with the quartz and less friable phases forming the coarse fraction.

The terminal period samples that form a sub-group (WLLE-0049 and WLLQ-0055) present the common features of the gross assemblage. However, the argillaceous content is darker, with a brown/black appearance. The metamorphic inclusion content is texturally similar, but contains less composite rock inclusions and

**Figure 3.** The presence of plastic, fabric-following argillaceous inclusions bands in tile specimen WLLQ-0061 (field of view = 2.5 mm).

**Figure 4.** From tile specimen WLLE-0049, an elongate schistose inclusion of quartz and talc, likely to represent the white raw material (field of view = 1 mm).

proportionately more talc, magnesite, and chlorite to quartz. The paste of these samples exhibits a paler groundmass, darker brown. Our sample size is currently too limited to fully understand these differences from the main assemblage. We suggest that, as the latest period samples studied, these represent a later manufacturing refinement, or reflect a minor change in the selection of otherwise consistent raw materials.

## 5. Experimental simulation of ceramic bodies

In order to determine the relative proportions and manufacturing behaviour of the two proposed components, a series of ceramic briquettes was produced experimentally. The resulting material was then analysed using EDXRF for comparison with the combined WDXRF and EDXRF data of the archaeological assemblage. Analysis of WLLM-0010 with both techniques shows broadly comparable results, with EDXRF data providing lower concentrations for the lighter elements Mg, Al, and Si, and elevated for the heavier ones: K, Ca, Ti, Mn, and Fe. Phosphorus was not analysed for with EDXRF. WDXRF results should be regarded as fully quantitative, while EDXRF is semi-quantitative.

In devising these recipes, three different combinations of materials were mixed, producing experimental samples composed of: M I) red clay and washed-friable; M II) red clay and the mineral fraction B2; and M III) red clay and *baitu-hou*. The last option was the most consistent with evidence from the ethnographic interview. Of these, M I was tried with varying proportions to investigate the degree of mixing. To compare these results with the assemblage of ceramic tiles, an archaeological sample with a representative MgO and $Fe_2O_3$ composition was chosen: WLLM-0010. The results of the EDXRF analyses are provided in Table 4.

For the variations on M I, the proportion of MgO decreased and $Fe_2O_3$ increased as the proportion of red clay in the mixture increased, a relationship that mirrors the one seen in the archaeological materials in Fig. 2. The proportions of $SiO_2$ and $Al_2O_3$ in the mixtures increased with the proportion of red clay, which may contain more $SiO_2$ and $Al_2O_3$ than the white raw material. The chemical composition produced by a 2:1 ratio of red clay to metamorphic rock in M I was the most similar to our archaeological sample, WLLM-0010.

The XRD spectral comparison of tile sample WLLM-0010 and M I (2:1) is shown in Fig. 5. These samples have comparable phases, i.e. quartz, microcline, enstatite, and forsterite, except for the presence of minor periclase, albite, and hematite in M I (2:1). These components may represent retention of unreacted tile-forming minerals and exotic alteration products, with the residual presence of albite and hematite consistent with unreacted material, and the periclase resulting from the thermal decomposition of magnesite in the washed-friable.

The mineralogical composition of *baitu-hou*, the ethnographically supported material, was comparable to that of washed-friable (Table ). The most significant comparison was between the compositions of the 2:1 mixture of red clay and *baitu-hou* and WLLM-0010 (Table 4), suggesting a viable recipe for synthesising the tiles.

**Table 4.** Compositions of tile sample WLLM-0010 and fired mixtures of red clay and white raw material by EDXRF (wt%).

| Sample name | $Na_2O$ | MgO | $Al_2O_3$ | $SiO_2$ | $K_2O$ | CaO | $TiO_2$ | MnO | $Fe_2O_3$ | Sum |
|---|---|---|---|---|---|---|---|---|---|---|
| **Experimental recipes** | | | | | | | | | | |
| M I (1:1) | 1.2 | 20.2 | 16.6 | 57.5 | 1.3 | 0.5 | 0.4 | 0.0 | 2.4 | 100.1 |
| M I (2:1) | 0.0 | 18.3 | 16.9 | 59.8 | 1.2 | 0.5 | 0.4 | 0.0 | 2.8 | 99.9 |
| M I (3:1) | 0.9 | 13.7 | 17.4 | 62.5 | 1.4 | 0.5 | 0.4 | 0.1 | 3.1 | 100.0 |
| M II (2:1) | 0.0 | 12.3 | 17.7 | 64.4 | 1.7 | 0.6 | 0.5 | 0.0 | 2.9 | 100.1 |
| M III (2:1) | 0.8 | 18.4 | 17.0 | 58.7 | 1.3 | 0.6 | 0.4 | 0.0 | 2.9 | 100.1 |
| **Archaeological comparative material** | | | | | | | | | | |
| WLLM-0010 | 0.5 | 18.1 | 17.6 | 58.1 | 1.5 | 1.4 | 0.4 | 0.0 | 2.4 | 100.0 |

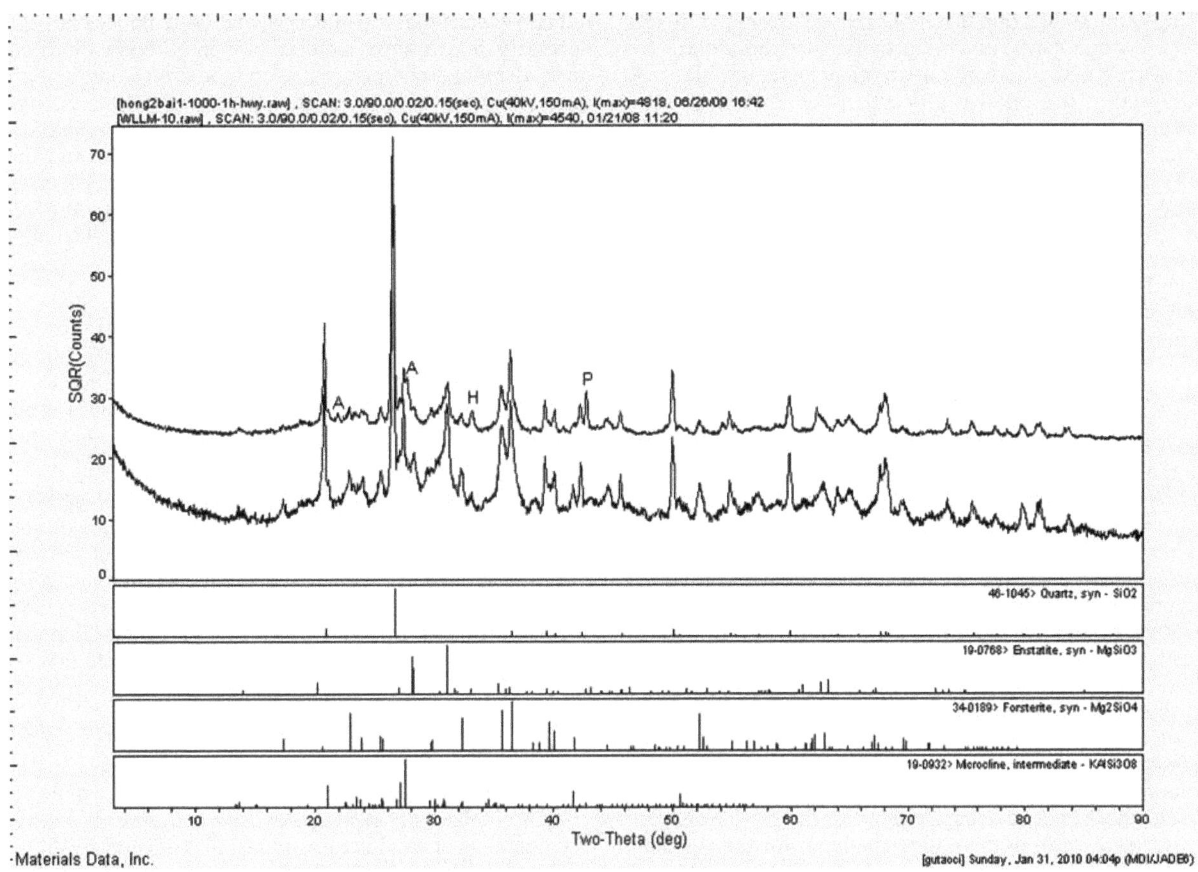


**Figure 5.** XRD patterns of tile sample WLLM-0010 and mixture M I (2:1). Top: M I (2:1); bottom: WLLM-0010; P, periclase; H, hematite; A, albite.

Beyond the mineralogical properties, the experimental synthesis provided insights into the mechanical properties of the ceramics after firing. Samples were fired at 1200 °C in oxidising conditions for an hour. Of the geological samples, red clay altered to brown and fused, while white raw materials sintered poorly. After firing, three white raw material samples mainly containing talc, magnesite, and quartz (Table ), i.e., washed soft block, washed hard block and B4, were all found to be easily mechanically disaggregated. As for the samples with muscovite content, i.e. Ly-0014, washed-friable, B1, *baitu-hou*, these showed some limited mechanical cohesion. Only B2, with the largest muscovite content, sintered and showed high strength. However, its high $SiO_2$ and low MgO (Table 4) are inconsistent with typical ceramics. As a whole, the poor sintering of the white raw materials led to low strength after firing.

Alteration of the colour of the raw materials during firing is significant in comparing the experimental tiles to the archaeological products. In the fired red clay, a strong orange brown to brown colour is observed, likely due to the presence of high iron contents. This hue renders the unmixed red clay unfit for producing tile bodies with light-coloured glazes. Conversely, the white raw materials fired to a white to light yellow colour. With regard to shrinkage after firing, the red clay deformed significantly, whereas the fired washed-friable was far less altered. This suggests that the white raw materials exhibit more refractory behaviour.

It appears that both the aesthetic and physical properties of these materials were improved by the mixing of these components relative to use in their pure state. A comparable pottery technology has been documented ethnographically by Sillar (2000) for the talc-rock tempered cooking ware of Machaca, Peru. In the Peruvian case, the talc was prepared by crushing primary rock material with wooden paddles, then sieving, followed by mechanical mixing with wet clay. The resulting pottery has been compared microscopically with this assemblage and shows broad similarities, with comparable gross proportions and shape amongst the coarse fraction. The key difference observed is in the greater crystal size and abundance of mineral phases in the material from the Huangwa Kiln. There are clear parallels between the two technological traditions in the choice of materials and resulting form.

## 6. Discussion

Firstly, it is clear from the low variability of the archaeological tiles' chemical composition that, throughout 300 years of tile production, the chemical and mineralogical composition remained relatively consistent. This suggests the ceramic technological recipe in use at the Huangwa Kiln, with a common exploitation of raw materials, was stable for an extended period of time.

When considering this recipe, petrological data supports the interpretation provided by primary historical evidence

(Ting *et al.* 1924), of a material produced from the mixing of two components, and we propose that these materials are a clay and a locally outcropping metamorphic rock. Interpreting the relative proportions of these two components in the finished product, we argue that the relationship of MgO (rich in the metamorphic component) to $Fe_2O_3$ (rich in the argillaceous component) forms a valid proxy. Our experimental study indicated a composition by weight of two parts argillaceous clay to one part metamorphic rock. In terms of variation on this recipe, from Fig. 2, the MgO content in most samples falls into a range between 17.5 wt% and 20.5 wt%, while $Fe_2O_3$ content varies from 3.2 wt% to 3.8 wt%. These samples (Fig. 2) cover period I, period I-II, period II, period II-III, period III-IV, and period IV. All of this suggests that the relative proportions of our components vary little with time. In periods III-IV and IV, the iron to magnesium ratio of the tiles exhibits more variability, which may reflect recipe fluctuations during this period.

In interpreting why the ancient potters used more than one type of raw material, we are faced with technological questions. From the point of view of chemical composition, relatively high fluxes existed in the red clay and lower fluxes in the white raw material (Table ), making red clay essential for ceramic-forming. Experimentally, the addition of the white raw material appeared to lead to two key changes in the properties of the material - an increase in the refractory qualities of the ceramic body, and a change in the colour of the resulting tile to an attractive pale white. The first of these may have implications for the glaze layer that is the key feature of these tiles, the latter for the aesthetic properties desired for high status tiles.

## 7. Conclusions

It is clear from the composition of local productions that this unusual tradition of using magnesium-rich pottery source materials continued from the earliest product deposits of the local kiln through three centuries of official tile production. Across four distinct periods of tile production, the chemical compositions of the tile bodies were consistent and comprised a relatively narrow range, indicating that the exotic properties of this raw material selection had been optimised to the point of a stable industry.

This industry was reliant on the mixing of two components, and analysis of locally outcropping materials suggest that all of these could have been sourced locally to the kiln before consumption at sites in a radius of 200 km from the production site. This is consistent with both historical documentation and ethnographic evidence. Red clay and white metamorphic rocks collected from deposits located to the northwest and northeast of the kiln site were viable candidates for the raw components described in the historical record. Amongst several white raw materials outcropping locally, and using evidence from ethnographic interviews, the samples mainly containing clinochlore (chlorite) and talc were the most comparable with the petrographic evidence of the archaeological tiles. Experimental evidence suggested that a recipe of two parts of red clay and one part of white raw material produced a material with the desired properties.

In terms of technological properties, there is clear evidence that there are benefits to this recipe: the deep colour of the red clay means that it could rarely be used alone, since the necessary appearance required a lighter body. Furthermore, the white raw material cannot be used alone, since it does not produce a ceramic when fired under normal conditions. Thus, by combination of the two, a suitably refractory, aesthetically appealing ceramic could be produced that provided the desired base for the covering glaze. This product of unusual raw material selection was then distributed at significant distances across the Liaoning Province.

## References


Chen M. and Huang Z., 1994, The study on the sequence of metamorphism and deformation in metaphilites from Liaohe Group in Dashiqiao-Haicheng area, Liaoning Province. *Liaoning Geology*, 1/2, 44-53.

Duan H., Liang G. and Miao J., 2009, The WDXRF quantitative analysis method research on the major and minor element compositions of the bodies of ancient building glazed tiles. In *Proceedings of the 2009 International Symposium on Ancient Ceramics (ISAC 2009)*, (eds. H. Luo and X. Zheng), 119-124, Shanghai Science and Technology Documents Press, Shanghai.

Fu P. and Lu S., 2007, Field work report of Huangwa Kiln site, Haicheng city, Liaoning Province. *Shenyang Palace Museum Journal*, 4, 6-22.

Kang B., Duan H., Ding Y., Li H., Miao J., Zhao C. and Fu P., 2009, Preliminary research on raw materials and firing technology of architectural glazed tiles of Huangwa Kiln. *Nangfang Relics*, 3, 116-122.

Li W., 1996, *A study of the pottery-making technology in ancient China*, Science Press, Beijing.

Liu D., 1984, *A history of ancient Chinese architecture*, China Architecture and Building Press, Beijing.

Pang T., 1998/1999, Some Chinese glazed tiles of the 6th Century, *Oriental Art*, 44, 61-68.

Sillar B., 2000, *Shaping culture: making pots and constructing households. An ethnoarchaeological study of pottery production, trade and use in the Andes*. British Archaeological Reports International Series 883, Archaeopress, Oxford.

Ting R., Sun S. and Zhang F., 1924, Haicheng county annals in the Republic period, Section 7 Human Resources and Industry. In *Comprehensive Work of Chinese Local Records, Section County annals in Liaoning, Part 5-6*, 304-305, Phoenix Press, Nanjing.

Wang G., 2004, Research on management system of official kilns for architectural glazed tiles in the Yuan, Ming and Qing dynasties. In *Research on management system of ancient official kilns in China*, (ed. G. Wang), 102-115, Forbidden City Press, Beijing.

Whitbread I.K., 1995, *Greek Transport Amphorae: A Petrological and Archaeological Study*, British School at Athens, Athens.

Yardley B., 1989, *An introduction to metamorphic petrology*, Longman Scientific and Technical, Harlow.

Zhang Q., 1988, *Early crust and mineral deposits in the East Liaoning Peninsula*, Geology Press, Beijing.

## ACKNOWLEDGEMENTS

We are grateful to Professor Thilo Rehren and Professor Marcos Martinón-Torres for helpful discussions. Our thanks also go to Mr Changming Zhao and Mrs Pinying Fu of the Anshan Museum, for offering archaeological information and useful help. Expert insights into the petrography were provided by Mr Steve Hirons and Mr Hank Sombroek of the Birkbeck Earth Sciences Department. Ms Lindsay Groom and Mrs Xiaoji Fang made helpful comments on a draft of this paper. This research was supported by the fund of Key Scientific Base Project of State Administration of Cultural Heritage, PRC.

# 15

# Petro-mineralogical and geochemical characterisation of Middle Neolithic Bükk Culture fine ware from Garadna, NE Hungary

**V. Szilágyi[a], K.T. Biró[b], P. Csengeri[c], Gy. Szakmány[d], H. Taubald[e], J. Mihály[f], C. Berthold[e], J.S. Koós[c] and J. Zöldföldi[e]**

[a]Centre for Energy Research, HAS, Budapest, Hungary, szilagyi.vera@gmail.com
[b]Hungarian National Museum, Budapest, Hungary
[c]Herman Ottó Museum, Miskolc, Hungary
[d]Eötvös Loránd University, Budapest, Hungary
[e]University of Tübingen, Germany
[f]Research Centre for Natural Sciences, HAS, Budapest, Hungary

**Abstract** - As a continuation of previous research carried out by the authors, further characterisation and provenancing of Middle Neolithic Bükk pottery (NE Hungary) were undertaken for the purposes of this study. In the context of the current project, a systematic and representative study of Bükk pottery products was initiated at the core of the tribal area. Our basic aim was to compare Bükk fine ware from different sites and to identify local or regional raw materials and production technology. In addition, characteristic incrustation of this fine ware is currently under observation.

The current paper summarises the first results obtained at an important site related to the Bükk Culture in NE Hungary, Garadna - Elkerülő út, site No. 2, the object of a recent large scale rescue excavation. At this site, material belonging to the typology of Bükk fine ware of consecutive chronological phases showed strong similarities, and indicted that the artefacts might have been made from local raw material. Additional conclusions on the collective features of Bükk fine ware were also added. On the one hand, it was proved once again that raw materials of specific, delimited physical properties were utilised by potters making Bükk fine ware at different sites. This suggests the technological preparedness of craftsmen when selecting raw material. On the other hand, the formerly published hypothesis concerning a restricted chemical composition of this raw material appears controversial.

## 1. Introduction - overview of the project

The Bükk Culture is one of the 'industrial' prehistoric cultures in the Carpathian Basin (Middle Neolithic, ca. 5200-4800 BC). The core of the distribution area is located in present-day NE Hungary and SE Slovakia (Fig. 1). The culture is famous for its intensive role in lithic trade, and even more famous for its pottery of artistic perfection. Bükk fine pottery has thin (1-5 mm) walls made from very fine-grained paste and decorated with incised geometric motifs, infilled with white through yellow to red paste (so-called 'incrustation'). Bükk pottery is claimed to have been traded over large areas, much beyond the limits of the tribal quarters. The authors of the present study also wrote the first essay (Szilágyi *et al.* 2008) on the subject of Bükk pottery distribution, carrying out a petroarchaeological analysis of the ceramic in the framework of a German-Hungarian collaboration project focussed on Early Neolithic pottery. As a valuable by-product, the characterisation and provenancing of Bükk pottery was emerging, providing a basis for the continuation of the collaboration.

Based on the results of a previous collaboration (2005-2006) between Tübingen University and numerous research centres in Hungary in the framework of bilateral research projects between Germany and Hungary, we are addressing the complex topic of the prehistoric long

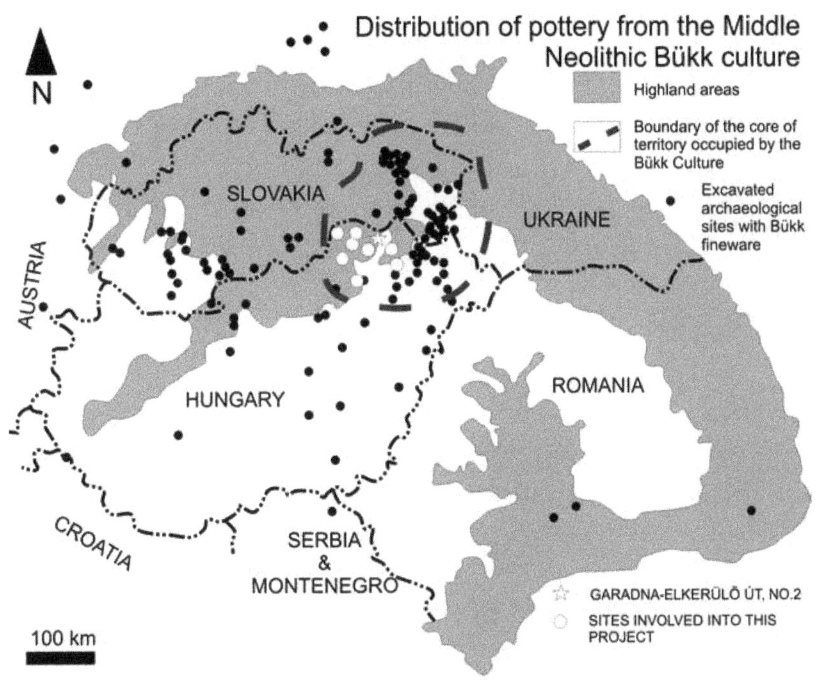


**Figure 1.** Distribution area of Bükk Culture on the map of the Carpathian Basin (dashed lines). The sites included in our study are marked with white circles, Garadna is marked with an asterisk. The presence of Bükk import ware (according to archaeological argumentation) is marked with black dots.

distance trade of ceramics. The previous project was focussed on the investigation of Early Neolithic pottery, potential raw materials, and the characterisation of site-specific features in different geographic environments in Hungary. Neolithic pottery manufacturing was found to have been a typically local activity, with potters utilising raw materials obtained from local or nearby sources. Typical examples in support of this argument are supplied by the case studies of Vörs (Gherdán 2009) and Szarvas-Endrőd (Szakmány and Starnini 2007; Starnini and Szakmány 2009). Long distance trade of pottery can be hypothesised only for some outstandingly high quality ceramics. In Hungary, the first vessels produced also for 'export' as objects of art were claimed to be the fine ware of the Middle Neolithic Bükk Culture (Tompa 1929). The distribution of Bükk pottery was mapped, based on archaeological arguments (style, form, decoration, etc.), by N. Kalicz and J. Makkay (Kalicz and Makkay 1977, Abb. 1). The Bükk Culture is a descendant of the local facies of the Linear Band Pottery Culture, dated around 5000 BC. A limited set of Bükk fine ware was already investigated in the framework of our previous project (Szilágyi *et al.* 2008).

In the present project, we carried out a detailed archaeometric investigation which may be of help in determining whether different occurrences of Bükk fine ware with the same quality, shapes, and decorations were produced using the same raw material, or whether it was the knowledge of the craftsmen that was adapted in other cultural regions using the local raw materials. In order to answer these questions, samples of Bükk fine ware have been systematically examined together with coarse pottery and comparative geological samples (local clay/soil) from each archaeological site. Applying a combination of petro-mineralogical and geochemical investigations (PM, XRF, μ-techniques for incrustated decorations of vessels), complex information regarding the material of the pottery (fabric, tempering, surface decoration) could be obtained.

The aims of the project are (1) to reconsider the distribution map of the Bükk Culture fine ware by scientific investigations and to understand better the civilisation that produce these archaeological objects in light of the obtained results; (2) to determine and gather those (chemical, physical, petro-mineralogical) characteristics which specifically describe the Bükk Culture fine ware using mainly non-destructive methods; and (3) to investigate the surface treatments and materials employed in the making of the Bükk Culture fine ware.

## 2. Notes on the chronology of the Bükk culture

To date, the internal chronology of the Bükk Culture has not been determined with precision. Temporal divisions have been based on the changes of the engraved motifs. According to Lichardus (1974), there were four consecutive phases of Bükk Culture which he denominated as 'A', 'AB', 'B', and 'C' (also Šiška 1995). Hungarian scholars (Kalicz and Makkay 1977) distinguished three phases of Bükk Culture, namely I, II, and III. In the present study, Lichardus' classification is followed. The early phase of the Bükk Culture, i.e., phase 'A', can only be distinguished from the late phase of the preceding Tiszadob Culture (end of the $6^{th}$ millennium BC) with difficulty. The major characteristic feature of its fine ware is the engraved linear and wavy pattern with white incrustation on the

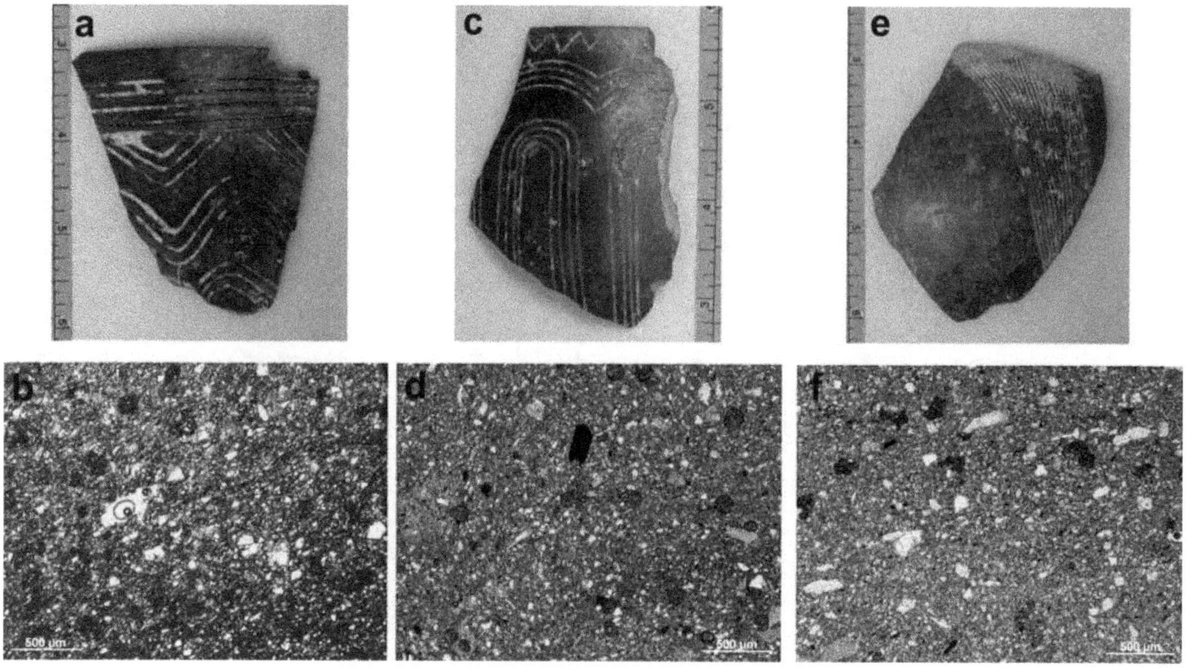

**Figure 2.** Macro- and microscopic scale outlook of fine ware of subsequent Bükk Culture phases from Garadna: (a-b) early 'A' phase; (c-d) (pre)classical 'AB' phase; (e-f) late 'C' phase.

vessels (Fig. 2a). These motifs show a connection to the decorations of the earlier Tiszadob Culture. The subsequent phases 'AB' and 'B' can be interpreted as the pre-classical and classical phases of Bükk Culture, when the development and refinement of the ceramic fashioning technique can be observed, with more dense decoration and common arched motifs. The fine ware of phase 'AB' has dense engraving with white incrustation (zigzag and arched pattern) (Fig. 2c). The late, so-called 'C' phase of Bükk Culture shows an extremely high quality fine ware with refined decoration of dense-filigree lines and plain sectors (Fig. 2e). Its characteristic feature is the appearance of yellow-red incrustation.

It has to be emphasised that there are presently no radiocarbon dates available for associating the coincidence of these 'ceramic phases' with time intervals.

## 3. Samples and methods

The framework of the project involves detailed petrographic, mineralogical, and geochemical characterisation. Following the collection of these data on both archaeological

**Table 1.** Localities of Bükk Culture sampled so far (HOM: Herman Ottó Museum, HNM: Hungarian National Museum).

| Locality | Bükk Culture phase | Museum | No. of archaeological samples | No. of soil samples |
|---|---|---|---|---|
| Garadna - Elkerülő út, No. 2 | A, AB, C | HOM | 40 | 21 |
| Felsoővadász-Várdomb | B, C | HOM | 28 | 18 |
| Sajószentpéter-Kövecses | AB | HOM | 20 | 27 |
| Szendrő-Ördöggáti-barlang | AB | HOM | 20 | 17 |
| Edelény-Borsod-Derékegyháza | B, C | HOM | 30 | 27 |
| Boldogkőváralja-Tekeres-patak | A, AB | HOM | 22 | 10 |
| Boldogkőváralja-Leányvár | C | HNM | 11 | 4 |
| Aggtelek-Baradla-barlang | A, AB, B | HNM | 23 | 23 |
| Bodrogkeresztúr-Tokaji országút | AB, B, C | HNM | 10 | 19 |
| Miskolc-Szeleta-barlang | AB, B, C | HNM | 13 | - |

**Table 2.** Samples analyzed from Garadna-Elkerülő út, site No. 2 (collection of HOM).

| Inv. nr/ Lab nr | Description | Type of analysis | | |
|---|---|---|---|---|
| | | Thin section | XRF | μ-techniques |
| GRDN-01 | Coarse ware, pot, line of holes under the rim | + | | |
| GRDN-02 | Coarse ware, pot, line of holes under the rim | + | | |
| GRDN-03 | Coarse ware, pot, line of holes under the rim | + | | |
| GRDN-04 | Coarse ware, bowl | + | | |
| GRDN-05 | Coarse ware, bowl | + | | |
| GRDN-06 | Coarse ware, bowl | + | + | |
| GRDN-07 | Coarse ware, storage vessel | + | + | |
| GRDN-08 | Coarse ware, amphorae | + | | |
| GRDN-09 | Coarse ware, amphorae | + | | |
| GRDN-10 | Coarse ware, decorated | + | | |
| GRDN-11 | Coarse ware, decorated | + | | |
| GRDN-12 | Coarse ware, decorated | + | | |
| GRDN-15 | Fine ware, AB phase, bit coarser, not polished | + | | |
| GRDN-16 | Fine ware, AB phase, bit coarser, not polished | + | | |
| GRDN-17 | Fine ware, AB phase, “Sajószentpéter type” | + | + | |
| GRDN-18 | Fine ware, AB phase, “Sajószentpéter type” | + | | |
| GRDN-19 | Fine ware, AB phase | + | | + |
| GRDN-20 | Fine ware, AB phase, fine, polished | + | + | |
| GRDN-21 | Fine ware, AB phase, fine, polished | + | + | |
| GRDN-22 | Fine ware, AB phase, fine, polished | + | | |
| GRDN-23 | Fine ware, AB phase, fine, polished | + | | |
| GRDN-25 | Fine ware, AB phase, fine, polished | + | | |
| GRDN-26 | Fine ware, C phase, fine, polished, red-yellow incrus. | + | | |
| GRDN-27 | Fine ware, C phase, fine, polished, red-yellow incrus. | + | + | |
| GRDN-28 | Fine ware, C phase, fine, polished, red-yellow incrus. | + | + | + |
| GRDN-29 | Fine ware, C phase, “Sajószentpéter type”, incrus., not polished, coarser | + | + | |
| GRDN-30 | Fine ware, C phase, “Sajószentpéter type”, incrus., not polished, coarser | + | | |
| GRDN-31 | Coarse ware, C phase | + | | |
| GRDN-32 | Coarse ware, C phase | + | | |
| GRDN-34 | Fine ware, Tiszadob phase, fine, polished, white incrus. | + | + | |
| GRDN-35 | Fine ware, Tiszadob phase, fine, polished, white incrus. | + | | |
| GRDN-36 | Fine ware, Tiszadob phase, fine, polished, white incrus. | + | + | |
| GRDN-37 | Fine ware, Tiszadob phase, fine, polished, white incrus. | + | + | |
| GRDN-38 | Coarse ware, Tiszadob phase | + | | |
| GRDN-39 | Coarse ware, Tiszadob phase | + | | |
| GRDN-40 | Coarse ware, Tiszadob phase, decorated | + | | |

finds and comparative soil/clay samples from the field, the determination of the applied special raw materials and/or special manufacturing techniques of Bükk fine ware becomes possible.

For the purposes of the current project, a number of 10 archaeological sites were already sampled, for a total of 214 samples (Table, Fig. 1). The archaeological specimens were obtained from the collections of the Herman Ottó Museum (HOM) and the Hungarian National Museum (HNM). The sample set contained both fine-grained, thin-walled and coarse-grained, thick-walled pottery. In addition to the archaeological samples, we collected 165 local clayey sediments from the environs of the sites, by 2.0 - 2.2 m deep hand borer, typically at 5 - 10 cm intervals (Table, Fig. 1). The most suitable (plastic, clayey) soil samples were selected and small brick-form test objects were made and fired at 700°C in an electric kiln. Both the archaeological pieces and the experimental objects were sampled and petrographic thin sections were obtained from them. After the evaluation of the thin sections, a further selection was made for geochemical and mineralogical investigation.

As the basic descriptive part of the research, the petro-mineralogical and chemical composition is essential. For this purpose, petrographic microscopic investigations (Department of Petrology and Geochemistry, Eötvös University of Budapest) on a Nikon-ALPHAPHOT-2 polarising microscope and X-ray fluorescence analysis (Department of Geochemistry, University of Tübingen) by a wavelength dispersive X-ray fluorescence instrument (Bruker AXS S4 Pioneer X-ray spectrometer, Rh tube at 4 kW; Hahn-Weinheimer *et al.* 1984) were applied. In order to observe the similarities and differences from an archaeometric point of view, the same methodology was applied on both the archaeological and the comparative raw material samples.

The investigation of the incrusted decoration of ceramics was also part of the research. This was undertaken by the application of completely non-destructive micro-analytical techniques, such as μ-X-ray diffraction (Bruker D8 Discover), μ-X-ray fluorescence analysis (custom made μ-XRF instrument at the Department of Geochemistry, University of Tübingen) and μ-Raman (DILOR LabRam)/Fourier Transform Raman (Digilab FT-Raman) spectroscopy, and by using the quasi-nondestructive attenuated total reflection Fourier Transform Infrared (ATR-FTIR) spectroscopy (Varian 2000 Scimitar with 'Golden Gate' diamond micro-ATR accessory).

For this paper, we have chosen the Garadna - Elkerülő út site No. 2, with good internal stratigraphy and recent excavations covering all the chronological phases of the Bükk Culture (Fig. 1). Thus, in this study, we present a representative site for further comparisons.

## 4. Garadna- Elkerülő út, site No. 2: a prolonged habitation associated with the Bükk culture

As an ideal example for tracking changes in the pottery manufacturing of the Bükk Culture, the Garadna - Elkerülő út, No. 2 site was selected for this presentation of our results. Garadna is situated in the valley of Hernád river, Northern Mountain Range, NE Hungary. The excavations were led by P. Csengeri in 2003, and extended to 2000 $m^2$ on the top and slope of a slightly raised hill near the river. According to the archaeological study of the excavated finds, the settlement was inhabited from the earliest Bükk 'A' phase (or the latest Tiszadob Culture). Evidence of continuous settlement comes from the ceramic finds of the pre-classical 'AB', the classical 'B', and the late 'C' phases. The most significant feature of the earliest phase of settlement is a large-sized pit complex. The later phases are associated with storage and refuse pits, hearths, kilns, and 13 burials.

About half of the excavated 15,000 ceramic pieces associated with the Bükk Culture represented fine ware with thin walls, incised or incrustated decoration, and polished surface. From this ceramic assemblage, a total of 37 samples were analysed (Table 2), among which phase 'A' was represented with 3 coarse and 4 fine pottery specimens, phase 'AB' with 12 coarse and 11 fine (partly white incrustated) ceramics, while 2 coarse and 5 fine (partly yellow-red incrustated) pots were selected from phase 'C'. Almost all the selected samples were investigated by petrography and 11 representative samples were further analysed by XRF.

## 5. Petrography and geochemistry of the fine ceramics

The results of the petrographic microscopic investigations are briefly summarised here. Concerning the fine pottery belonging to the subsequent phases of the Bükk Culture, the ceramic groups proved to be quite homogeneous from a petrographic point of view. Their descriptions will be presented phase by phase. Ceramics of the early 'A' phase have serial, fine-grained (fine silt) fabric. The clay paste is very dense, pure, and not oriented (which suggests that the raw material was not too plastic). The major aplastic constituents are monocrystalline quartz with undulatory extinction, muscovite, and some phyllite fragments (Fig. 2b) which indicate the metamorphic derivation of the sediment. Samples of the (pre)classical 'AB' phase are characterised by a weakly hiatal, fine-grained (fine-coarse silt) fabric. The very dense, pure and less oriented paste of the ceramics emphasises the weak plasticity of the raw material (Fig. 2d). Concerning the mineralogical composition of the aplastic inclusions, there are two types of pottery in this group. On the one hand, ceramics containing monocrystalline quartz with undulatory extinction and muscovite together with phyllite rock fragments suggest a raw material of metamorphic origin. On the other hand, other samples have plagioclase, orthopyroxene, and volcanic lithofragments (effusive rock and tuff related glass shards) grains as aplastic inclusions. The latest 'C' phase of the Bükk Culture is represented by ceramic fragments of hiatal, less fine-grained (coarse silt) fabric. The clay paste is still very dense, pure, and not oriented (which suggests that the raw material was not too plastic) (Fig. 2f). The aplastic composition (undulatory extinguishing monocrystalline quartz, muscovite, and phyllite clasts) supports the idea of a metamorphic-derived raw material.

The chemical composition of the Garadna ceramic samples is listed in Table 3. In addition to the Garadna samples, geochemical data of archaeological ceramics from other Bükk Culture sites were also used for the interpretation ( Fig. 3a). As previously suggested in the study of Bükk fine

**Table 3.** Bulk chemical composition of Bükk pottery from the Garadna site by XRF.

| Name | GRDN-06 | GRDN-07 | GRDN-17 | GRDN-20 | GRDN-21 | GRDN-27 | GRDN-28 | GRDN-29 | GRDN-34 | GRDN-36 | GRDN-37 |
|---|---|---|---|---|---|---|---|---|---|---|---|
| Major elements (w%) | | | | | | | | | | | |
| $SiO_2$ | 62.49 | 63.93 | 65.44 | 65.32 | 68.46 | 69.82 | 69.90 | 66.37 | 67.95 | 68.75 | 69.56 |
| $TiO_2$ | 0.79 | 0.93 | 0.85 | 0.89 | 0.81 | 0.90 | 0.90 | 0.85 | 0.87 | 0.86 | 0.85 |
| $Al_2O_3$ | 17.20 | 20.07 | 17.26 | 18.76 | 14.63 | 15.67 | 15.88 | 16.35 | 15.26 | 14.88 | 13.55 |
| $Fe_2O_3$ | 5.56 | 5.35 | 5.54 | 6.70 | 5.34 | 5.46 | 5.49 | 4.68 | 3.89 | 3.80 | 4.22 |
| MnO | 0.07 | 0.05 | 0.06 | 0.09 | 0.04 | 0.08 | 0.08 | 0.04 | 0.05 | 0.05 | 0.03 |
| MgO | 2.07 | 2.23 | 1.84 | 1.95 | 1.54 | 1.44 | 1.51 | 1.80 | 1.55 | 1.36 | 1.37 |
| CaO | 2.21 | 1.27 | 1.65 | 1.06 | 1.52 | 1.24 | 1.14 | 1.38 | 1.83 | 1.66 | 2.15 |
| $Na_2O$ | 0.58 | 1.15 | 1.10 | 0.63 | 1.17 | 0.84 | 0.84 | 1.11 | 0.87 | 0.83 | 0.88 |
| $K_2O$ | 3.13 | 3.36 | 2.80 | 3.14 | 2.45 | 2.54 | 2.57 | 2.84 | 2.78 | 2.59 | 2.37 |
| $P_2O_5$ | 1.27 | 0.16 | 1.54 | 0.20 | 0.55 | 0.32 | 0.24 | 0.64 | 0.36 | 0.45 | 0.54 |
| LOI | 5.43 | 2.01 | 2.73 | 0.98 | 3.45 | 2.71 | 1.92 | 4.34 | 4.68 | 5.16 | 5.09 |
| Trace elements (μg/g) | | | | | | | | | | | |
| Ba | 1173 | 730 | 1202 | 613 | 881 | 590 | 557 | 783 | 1146 | 1194 | 910 |
| Co | 2 | u.d. | 26 | 4 | 2 | 9 | 8 | u.d. | 4 | u.d. | 1 |
| Cr | 120 | 98 | 113 | 114 | 101 | 108 | 111 | 107 | 94 | 88 | 92 |
| Ni | 34 | 77 | 88 | 114 | 66 | 101 | 95 | 50 | 33 | 24 | 37 |
| Rb | 130 | 171 | 128 | 148 | 112 | 117 | 122 | 123 | 120 | 118 | 102 |
| Sr | 153 | 122 | 177 | 92 | 127 | 117 | 104 | 121 | 177 | 160 | 154 |
| V | 132 | 131 | 114 | 141 | 100 | 113 | 112 | 105 | 94 | 89 | 105 |
| Y | 31 | 45 | 40 | 35 | 38 | 37 | 39 | 38 | 35 | 33 | 40 |
| Zn | 94 | 121 | 72 | 84 | 58 | 58 | 58 | 89 | 82 | 138 | 75 |
| Zr | 174 | 244 | 260 | 194 | 283 | 315 | 312 | 259 | 256 | 259 | 321 |
| Ce | 98 | 121 | 110 | 120 | 96 | 110 | 114 | 103 | 98 | 98 | 110 |
| Eu | 0.8 | 0.8 | 0.8 | 0.6 | 0.9 | 0.7 | 0.8 | 0.7 | 0.8 | 0.8 | 0.9 |
| La | 25 | 39 | 27 | 29 | 27 | 27 | 30 | 28 | 20 | 19 | 30 |
| Nb | 16 | 21 | 18 | 20 | 21 | 21 | 23 | 20 | 19 | 17 | 22 |
| Nd | 35 | 50 | 42 | 34 | 29 | 42 | 41 | 35 | 47 | 46 | 34 |
| Sm | 4.7 | 6.8 | 5.3 | 5.1 | 7.0 | 5.5 | 6.2 | 5.3 | 5.5 | 5.2 | 7.4 |
| Yb | 3.4 | 4.8 | 4.1 | 3.7 | 3.8 | 3.6 | 3.8 | 4.0 | 3.7 | 3.4 | 3.9 |
| Sum | 101.01 | 100.71 | 101.04 | 99.89 | 100.15 | 101.19 | 100.63 | 100.59 | 100.31 | 100.61 | 100.81 |

ware by our research group (Szilágyi *et al.* 2008; 2011), a compositional range was identified for the high quality Bükk fine ware, indicating the usage of a certain raw material type. However, placing our new data points (among which Garadna pottery is also present) into the formerly applied Y/V vs. $TiO_2/Al_2O_3$ discrimination diagram (these major and trace element ratios proved to be useful for provenance investigations due to their resistance under conditions of soil forming and ceramic making processes), it seems that the chemical composition of fine pottery is more varied than it was suggested before. Even if samples from Garadna scatter on the higher Y/V ratio side (right edge) of the former compositional category of Bükk fine ware, samples from other sites show significant deviation.

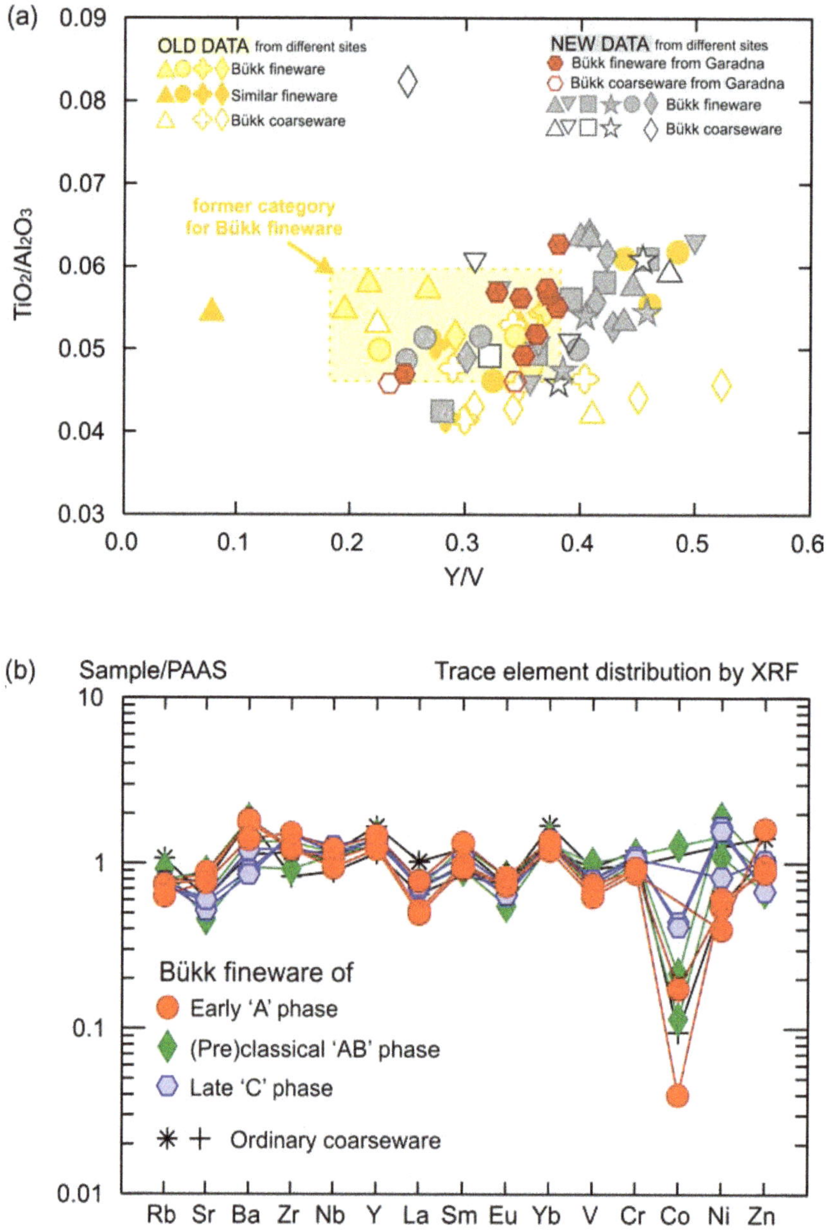


**Figure 3.** Geochemical data on Bükk fine ware from Garadna: (a) Y/V vs. $TiO_2/Al_2O_3$ diagram of both new and former (Szilágyi et al. 2008) data from Garadna and other Bükk Culture sites; (b) trace element distribution (normalised to PAAS) of Bükk fine ware from the Garadna site.

In the observation of fine ware collected exclusively from the Garadna-Elkerülő út site No. 2, the trace element distribution shows strong similarities (Fig. 3b). This feature is true for both mobile (e.g., Rb, Sr, Ba) and immobile (e.g., Zr, Y) elements. In addition, the ceramic composition is quite similar to the reference composition (average value for Post Australian Average Shale = PAAS) used for the normalisation (Nance and Taylor 1976; Taylor and McLennan 1985; McLennan 1989; 2001). The only exceptional elements are Co and Ni, but this can be related to the measuring peculiarities of XRF for these elements (quite high detection limits). The homogeneous chemical behaviour of Garadna fine ware suggests their production from similar raw materials.

## 6. Mineralogy of incrustations

For the detailed investigation of both white and yellow incrustations, some sherds with well-preserved decoration were selected (the complete number of such samples in Table). Applying non-invasive techniques was an essential requirement for the analysis of these covers of paste, some tens of micrometres thick. In the case of the Garadna site, we illustrate our findings with two examples - a white incrustated one, GRDN-24, as representative of the (pre)classical 'AB' Bükk phase, and a yellow incrustated one, GRDN-28, as representative of the late 'C' phase.

The μ-XRF technique was used for fast qualitative screening of the elements of the incrustations. The major

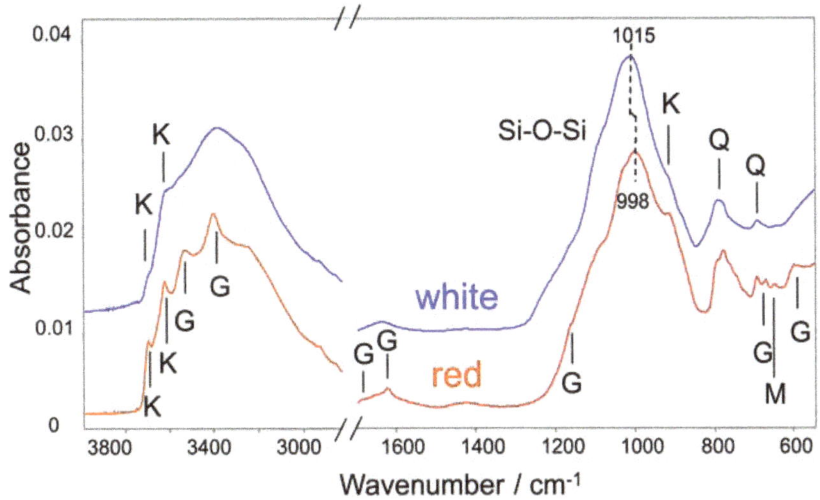


**Figure 4.** The μ-ATR-FTIR spectra of white (phase 'AB') and yellow (phase 'C') incrustations of Bükk fine ware from Garadna (Q indicates quartz).

elements were Fe, Ca, Si, and K. Neither P - corresponding to bone grit (the main component of bone being apatite: $Ca_{10}(PO_4)_6(OH)_2$) -, nor an increased concentration of Ca - indicating the presence of calcite ($CaCO_3$) - as compared to the ceramic matrix was identified.

The μ-XRD measurements revealed quartz, feldspar, and an amorphous phase (presumably heat-treated kaolinite/metakaolinite) as the main minerals of the incrustations. According to our tests, pure kaolinite resulted in the same amorphous-like diffraction pattern after heating up to 500°C for 30 minutes. No further information could be gained by the μ-Raman technique, which indicated the presence of an amorphous phase (possibly carbon?).

The μ-ATR-FTIR spectra of the two selected samples, GRDN-24 and GRDN-28 (Fig. 4), resemble typical spectra of clays. The spectra are dominated by the absorption bands of quartz (symmetric Si-O stretching and bending at 795 and 698 $cm^{-1}$, respectively) and by the Si-O-Si stretching bands of alumino-silicates around 1200-1000 $cm^{-1}$. The broad bands at around 3400 and 1630 $cm^{-1}$ can be attributed to adsorbed water. The well separated -OH bands at 3690 and 3620 $cm^{-1}$, however, are diagnostic for kaolinite, belonging to interlayer and inner O—H groups. When studying the thermal behaviour of clay artefacts, Maniatis *et al.* (1982) observed that clay minerals begin to lose their crystalline hydroxyls and start disorganising at 500-600°C. For some iron-rich minerals, the weak band around 3620 $cm^{-1}$ can persist even up to 800°C, indicating an incomplete dehydroxylation.

The μ-XRD and μ-ATR-FTIR results are summarised in Table 4. All the above results reinforce the assumption of the use of a clay (kaolinite) grit for the incrustated decorations, probably followed by a (re)firing of the incrustated pottery at a temperature higher than 500-600°C. As for the GRDN-28 sample with yellow incrustation, no goethite as colouring mineral was found. The increased relative intensities of the OH stretching bands and the shift of the main Si-O-Si band towards lower wavenumbers (Fig. 4) suggest a lower firing temperature as compared to GRDN-24.

## 7. Discussion and conclusion

According to our research, the Bükk fine ware at the Garadna site proved to be generally made of fine-grained, very dense, and not very plastic raw material — which was probably washed for obtaining such a paste — in the subsequent phases of the Bükk Culture. However, some differences can be outlined. Finer-grained raw materials were used in the earlier periods. In the latest Tiszadob Culture (~ early 'A' phase of Bükk Culture), silt of metamorphic provenance was utilised. Later on, during the pre-classical period ('AB' phase), similarly fine raw materials were applied, but with two kinds of provenance, volcanic and metamorphic. In the latest ('C') phase, the used fine sediment became coarser-grained and more homogeneous, of metamorphic origin. These fabric similarities suggest similar paste preparation techniques, although the raw materials were derived from more sources during the pre-classical period. The observed

**Table 4.** Comparison of mineralogical data on white and red incrustated decoration gained by different μ-techniques.

| | Non-destructive methods | | | Quasi-destructive (2-3 μg) |
|---|---|---|---|---|
| **Sample** | **μ-XRD** | **μ-Raman/FT-Raman** | **μ-XRF** | **μ-ATR-FTIR** |
| White incrustation (phase AB) | amorph | amorphous carbon | Fe, Ca, Si, K | clay (disordered kaolinite) |
| Red incrustation (phase C) | amorph | amorphous carbon | Fe, Ca, Ti, K | clay (disordered kaolinite), magnetite |

similarity of the raw materials is also supported by the similar chemical element distributions.

The limitations in the scope of this paper exclude a detailed analysis of the geological setting. In short, the geographical position of Garadna provides access to a Palaeozoic metamorphic sedimentary source in the form of eroded sediments transported by the Hernád River. This extended alluvial deposit could be appropriate for making fine ware of metamorphic provenance, which was identified throughout the entire Bükk era. The closest source for the occasionally observed volcanic type of fine ware might be located on the opposite side of Hernád River, at the edge of the volcanic Tokaj Mountains.

In the present study, we found new ways of interpreting the chemical data on the fine pottery. In contrast to our previous suggestions based on material from other sites (Aggtelek, Felsővadász, Borsod-Derékegyháza; Szilágyi *et al.* 2008), the chemical composition of Bükk fine ware samples from the Garadna site did not fit into the same range. This indicates that it is not possible to determine one distinct source of raw material for all vessels from different sites. It is more probable that they were manufactured using different raw materials, site by site. This aspect supports the hypothesis of local raw material utilisation rather than that of the use of a regionally accessible source.

Concerning only the fine ceramic from Garadna, the investigation of samples of metamorphic derived aplastic inclusions indicated strong similarities to coarse pottery and to each other (regardless of the different chronological phases of the Bükk Culture) in trace element composition.

As a result of the investigation of the incrusted decorations of the subsequent phases, we observed no significant difference between the white and yellow pastes. The utilisation of kaolinitic clay (probably followed by a (re) firing of the incrusted pottery) could be identified.

Former and new petrographic data clearly indicate that raw materials of similar and relatively delimited physical properties were used for the manufacture of Bükk fine ware at different sites. This means that a clayey silt of less plasticity was preferred for the production of Bükk fine ware. This aspect proves that Bükk Culture potters acted on a special technological 'trick' when selecting raw material for the manufacture of their unique fine ware. With more data added to our database, it has to be stated that a more diverse chemical composition than previously assumed characterises this special ceramic type. In the case of Garadna, Bükk fine ware of consecutive chronological phases show strong similarities both petrographically and geochemically. As they are geochemically similar to the coarse ware, it is very probable that they were made from local raw material.

## References


Gherdán K., 2009, *7000 év kerámiái - Vörs, Máriaasszony-sziget őskori kerámialeleteinek archeometriai összehasonlító vizsgálata területi kitekintéssel.* (Pottery sequence of 7000 years: archaeometric study of pottery finds from Vörs, Máriaasszony-sziget), PhD Thesis, Eötvös Loránd University, Budapest.

Hahn-Weinheimer P.F., Hirner A.V. and Weber-Diefenbach K., 1984, *Grundlagen und praktische Anwendung der Roentgenfluorezenzanalyse (RFA)*, Vieweg-Verlag, Braunschweig.

Kalicz N. and Makkay J., 1977, *Die Linienbandkeramik in der Grossen Ungarischen Tiefebene*, Studia Archaeologica 7, Akadémiai Kiadó, Budapest.

Lichardus J., 1974, *Studien zur Bükker Kultur*, Saarbrücker Beiträge zur Altertumskunde Band 12, R. Habelt, Bonn.

Maniatis Y., Katsanos A.A. and Caskey M.E., 1982, Technological examination of low-fired terracotta statues from Ayia Irni, Kea. *Archaeometry*, 24, 191-198.

McLennan S.M., 1989, Rare Earth Elements in Sedimentary Rocks: Influence of Provenance and Sedimentary Processes. In *Geochemistry and Mineralogy of Rare Earth Elements*, (eds. B.R. Lipin and G.A. McKay), 169-200, Reviews in Mineralogy 21. Mineralogical Society of America, Washington DC.

McLennan S.M., 2001, Relationships between the trace element composition of sedimentary rocks and upper continental crust. *Geochemistry, Geophysics, Geosystems*, 2, 1-24.

Nance W.B. and Taylor S.R., 1976, Rare earth patterns and crustal evolution - I. Australian post-Archean sedimentary rocks. *Geochimica et Cosmochimica Acta*, 40, 1539-1551.

Šiška S., 1995, Zur Problematik des Untergangs der Bükker Kultur. *Slovenská Archeológia*, 43, 5-24.

Starnini E. and Szakmány Gy., 2009, Besides vessels: investigating Early Neolithic fired clay artefacts from Hungary. In *Vessels: inside and outside, Proceedings of the conference EMAC'07, 9th European meeting on ancient ceramics*, (eds. K.T. Biró, V. Szilágyi and A. Kreiter), 165-172, Hungarian National Museum, Budapest.

Szakmány Gy. and Starnini E., 2007, Archaeometric research of the first pottery production in the Carpathian Basin: manufacturing traditions of the Early Neolithic, Körös Culture ceramics. *Archaeometriai M hely*, 4, 5-19.

Szilágyi V., Biró K.T., Csengeri P., Koós S.J., Szakmány Gy., Tóth M. and Taubald H., 2008, Előzetes eredmények a bükki kultúra finomkerámiájának nyersanyag azonosítási és technológiai vizsgálatából. *Archeometriai Műhely*, 3, 27-40.

Szilágyi V., Taubald H., Biró K.T., Koós S.J., Csengeri P., Tóth M. and Szakmány Gy., 2011, Preliminary archaeometric data on fine ware from the Middle Neolithic Bükk culture. In *Proceedings of the 37th International Symposium on Archaeometry, 13th-16th May 2008, Siena, Italy, Part 1*, (ed. I. Memmi), 159-167, Springer, Berlin, Heidelberg and New York.

Taylor S.R. and McLennan S.M., 1985, *The Continental Crust: its Composition and Evolution*, Blackwell Scientific Publications, Oxford.

Tompa F., 1929. A szalagdíszes agyagművesség kultúrája Magyarországon: A bükki és tiszai kultúra. (Die Bandkeramik in Ungarn. Die Bükker- und die Theiss-kultur), Archaeologica Hungarica, 5-6, Budapest.

# 16

# Ceramic technology and the global world: First technological assessment of the Romita ware of colonial Mexico

**Javier G. Iñañez**[a,b]

[a]Grupo Investigación en Patrimonio Construido (GPAC), University of the Basque Country (UPV/EHU), CIEA Lascaray, Avda. Miguel de Unamuno, 3, 01006, Vitoria-Gasteiz, Álava, Spain, javier.inanez@ehu.es

[b]IKERBASQUE, Basque Foundation for Science, 48011, Bilbao, Spain

**Abstract** - Romita Ware is a unique ceramic found in Colonial archaeological contexts throughout Mexico. The ceramic features a white slip and a very thin transparent lead-silica glazed outer coating on top. The characterisation of the technological choices adopted in the production of Romita ware is assessed by means of scanning electron microscopy.

## 1. Introduction

Romita Ware, also known as Indígena ware or Loza Indígena, has been considered by many scholars to be an indigenous imitation of Spanish tin glazed ceramics in colonial Mexico. The pottery is found in abundance in colonial contexts in Mexico City, and, as such, it has interested scholars for decades (Lister and Lister 1982). Because this ware is believed to be most likely an indigenous imitation of European tableware, it is considered important in the study of technological change in colonial situations, the adoption of European aesthetics among indigenous people, and competition between colonisers and indigenous people in the colonial market.

In an effort to identify the origin of this ware, scholars have variously assigned it to Italy (Lister and Lister 1976), Mexico (Lister and Lister 1982, 1987; Maggetti *et al.* 1984; Fournier *et al.* 2007; Iñañez *et al.* 2010), and Spain (Rodríguez-Alegría 2002; Rodríguez-Alegría *et al.* 2003).

Romita ware is an earthenware covered with a white glazed coating, found in many typological forms, such as porringers with leaf-shaped handles, compound-silhouette plates, bowls, and other forms similar to those of European and Mexican majolica. Lister and Lister (1976) named the two main variants of this ware after Rome: Romita Plain for the undecorated variant, and Romita Sgrafitto for the decorated variant, since they first believed that the origin of this ceramic had to be sought in Italy (Lister and Lister 1976). Romita Sgrafitto has decoration outlined probably by carving through the white slip to expose the red colour of the paste, and then filling the areas of the glaze with green and orange pigments, giving the ware its characteristic bright colours. Decorations visible on the Romita Sgraffito vessels include characteristic decorative elements that may derive from pre-Hispanic indigenous traditions, such as corn or eagles, as well as others more similar to the European Renaissance style, including wavy valances, spirals, and circular motifs (for further discussion of the decorations and images of these ceramics, Lister and Lister 1982; Rodríguez-Alegría 2002; Rodríguez-Alegría *et al.* 2003; Fournier *et al.* 2007; Iñañez *et al.* 2010).

Archaeological and historic evidence suggest a prolonged period of production and consumption for these ceramics that would have begun in the 16$^{th}$ century and continued until the middle or even the end of the 17$^{th}$ century (Fournier *et al.* 2007). The geographical distribution of these ceramics is not restricted to Mexico City, where Lister and Lister found mostly the plain variant (Lister and Lister 1982). This ware shows a wide distribution throughout central and northern Mexico and the southern United States, especially of the Sgraffito variant ( Fournier *et al.* 2007, and references therein).

## 2. Archaeometric background

An archaeometric study conducted by Maggetti *et al.* (1984) provided the first chemical and petrographic analyses of the ceramic pastes of Romita ware. Although limited in the number of samples, this research concluded that the petrography exhibited by these samples was compatible with a volcanic geology contrasting with the most common sedimentary tempers used in Spanish majolica, and suggesting that the production area could be a location outside the Valley of Mexico (Maggetti *et al.* 1984) (Fig. 1). More recently, Rodríguez-Alegría conducted a stylistic analysis and a chemical characterisation study of the ceramics from the Metropolitan Cathedral in Mexico City and concluded that it was a European import (Rodríguez-Alegría 2002; Rodríguez-Alegría *et al.* 2003).

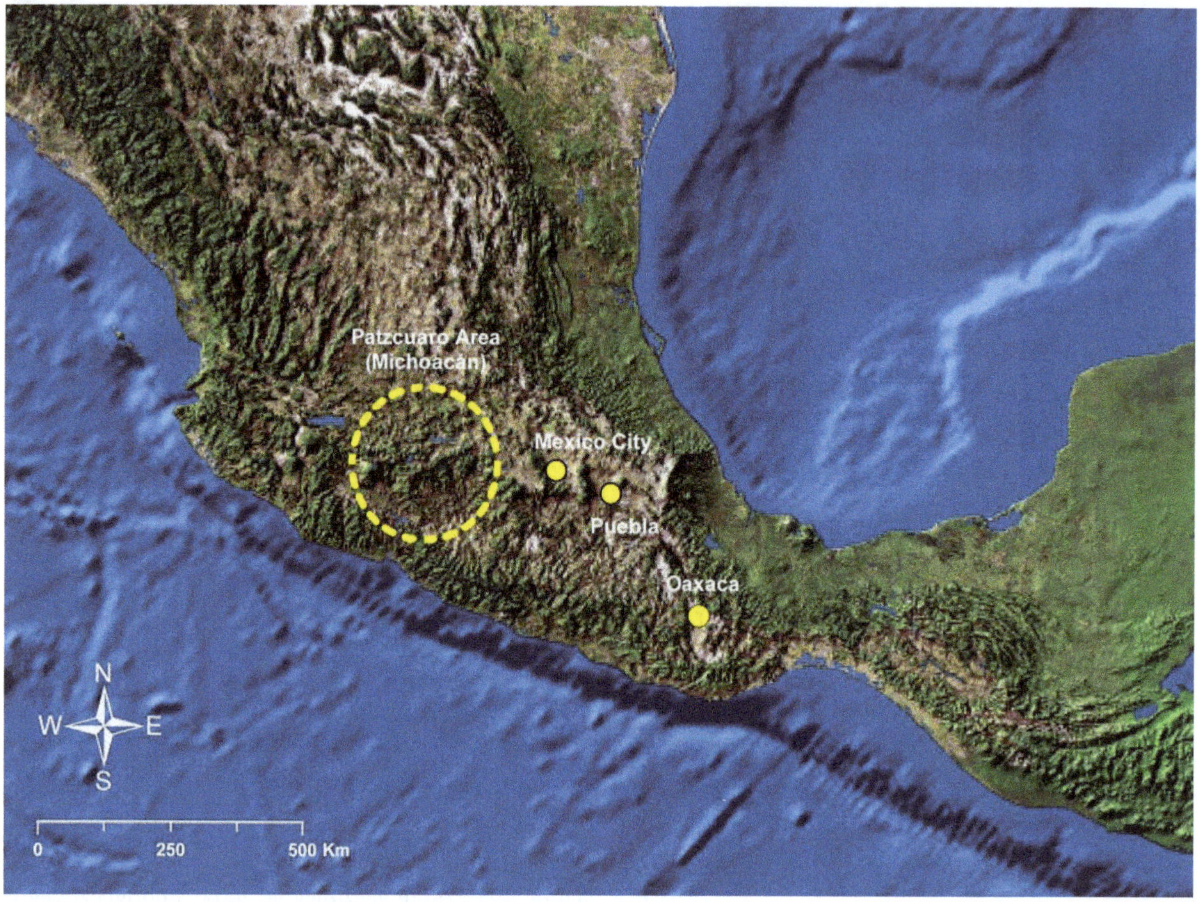


**Figure 1.** Map showing the main locations mentioned in the text and purportedly the Romita Ware production area, represented by a dashed-lined circle.

The chemical composition of the 94 Romita Ware sherds that were analysed by INAA did not match the composition of more than 4,000 samples from Mesoamerica that had been analysed at the Research Reactor of the University of Missouri (MURR). However, Rodríguez-Alegría's study also failed to match the composition of Romita Ware to any European ceramics that had been previously analysed by INAA; nevertheless, the aesthetic features of this ware led him to the conclusion that it had to be European, most likely from Spain. More recently, Fournier *et al.* (2007) carried out an INAA study that included numerous Romita Ware specimens from the Michoacán region, not only archaeological, but also from ethnographic contexts. The authors concluded that the Romita Ware chemical fingerprint identified by INAA suggested the existence of two different groups, very similar in their chemical composition, that were likely produced in the Pátzcuaro Basin area (Fournier *et al.* 2007). Additionally, the statistical reassessment of the Romita Ware INAA data and the comparison with the fingerprint of the main majolica and glazed ware manufactured in the Iberian Peninsula and Western Mediterranean during the colonial period allows ruling out such a provenance (Fig. 2).
Lastly, Iñañez *et al.* (2010) provided the last piece of evidence supporting the hypothesis of a Mexican origin for Romita Ware ceramics after assessing the provenance of the lead sources employed for their glazed coats, following a study of their glazed coatings by Lead Isotope Analysis (LIA). This study confirmed that multiple lead ore sources were exploited in relation to the manufacturing of this ware, all exclusively located within the borders of present-day Mexico. Therefore, after a consideration of all the archaeometric evidence available to date from both INAA and LIA, we can conclusively argue that Romita pottery is a Mexican ware that was likely produced at multiple sites in the Michoacán area.

In the present paper, we bring together evidence pertaining to the chemical composition and the technological features of these ceramics as assessed by SEM-EDS, in order to address a previously unexplored aspect of Romita Ware: the interesting characterisation of the glazing technology of this ceramic. Revealing the technological characteristics of Romita Ware is of paramount importance for assessing the degree of technological 'influence' of European potting traditions to indigenous societies during the first periods of contact. Surely, as identified in the studies of other aspects of life under contact situations in other regions and by other populations, from food to social relationships, the struggles, dominations, and cultural influences occurring during these contacts resulted in many new expressions that agglutinated to various extents cultural and technological influences from the different agents that took part in them (e.g., Deagan 1996; Ortiz 2002 [1940]; Armstrong and Hauser 2009).

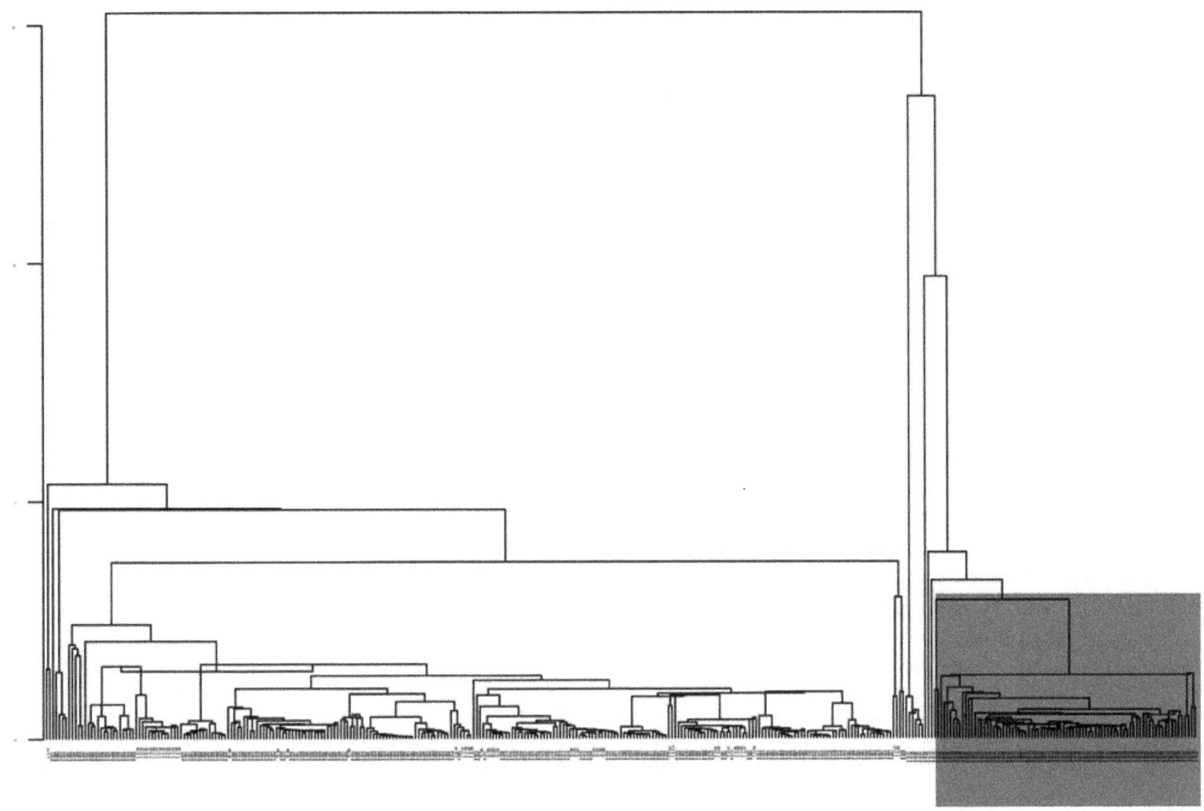

**Figure 2.** Dendrogram resulting from the cluster analysis of INAA data of the ceramics from the main majolica and glazed ware production centres in the Iberian Peninsula and Western Mediterranean compared to the Romita Ware ceramics using square Euclidian distance and the centroid algorithm on the sub-composition: La, Lu, Nd, Sm, U, Yb, Ce, Cr, Cs, Eu, Fe, Hf, Sc, Ta, Tb, Th, Zn, Zr, Al, Ca, Dy, Mn, Ti, and V, after central log-ratio transformation. Dashed area represents the group formed by Romita Ware.

Along these lines, indigenous pottery production also seems to have been affected by these transculturation processes, since, according to the current archaeological knowledge, these pots probably began to be produced around the mid-16th century in West Mexico. Therefore, the Spanish colonisation of this area may have represented an important cultural and technological influence, which might be traceable by analytical means.

## 3. Analytical methodology

Chemical analyses of Romita Ware and Spanish majolica and utilitarian ware ceramic pastes were conducted by neutron activation analysis (NAA) at the Archaeometry Laboratory of the University of Missouri Research Reactor (MURR)[1]. Exterior surfaces of the ceramics were mechanically removed using a silicon-carbide abrader to remove slip, glaze, and other surface contamination. Samples were then ground and homogenised in agate mortars or using an automatic mill with a tungsten carbide cell, and the resulting powdered material was transferred to into glass vials. Prior to weighing, the powdered pottery samples were oven-dried at 100 °C for at least 24 h. Approximately 150 mg of sample were weighed in small polyvials used for short irradiations. At the same time, 200 mg of each sample were weighed into high-purity quartz vials used for long irradiations. Along with the ceramic samples, reference standards of SRM-1633a (coal fly) and SRM-688 (basalt rock) were prepared, as well as quality

**Table 1.** Average, standard deviations, and range concentrations for SEM-EDS data obtained for the glazes and slips of the Romita Ware ceramics.

| wt% | Average | St.dev. | Average | St.dev. |
|---|---|---|---|---|
| | Glaze | | Slip | |
| $Na_2O$ | 0.6 | 0.3 | 1.3 | 0.6 |
| MgO | 0.2–1.0 | | 0.5 | 0.3 |
| $Al_2O_3$ | 9.9 | 4.4 | 36.8 | 7.4 |
| $SiO_2$ | 31.6 | 5.2 | 52.2 | 5.6 |
| PbO | 56.0 | 8.6 | 0.0–7.9 | |
| $K_2O$ | 0.0–0.6 | | 0.8 | 0.4 |
| CaO | 0.1–0.9 | | 0.1–4.0 | |
| $Fe_2O_3$ | 0.0–5.1 | | 4.4 | 2.7 |
| $TiO_2$ | 0.4 | 0.2 | 1.2 | 0.2 |
| $SnO_2$ | 0.0–0.8 | | 0.0–0.4 | |

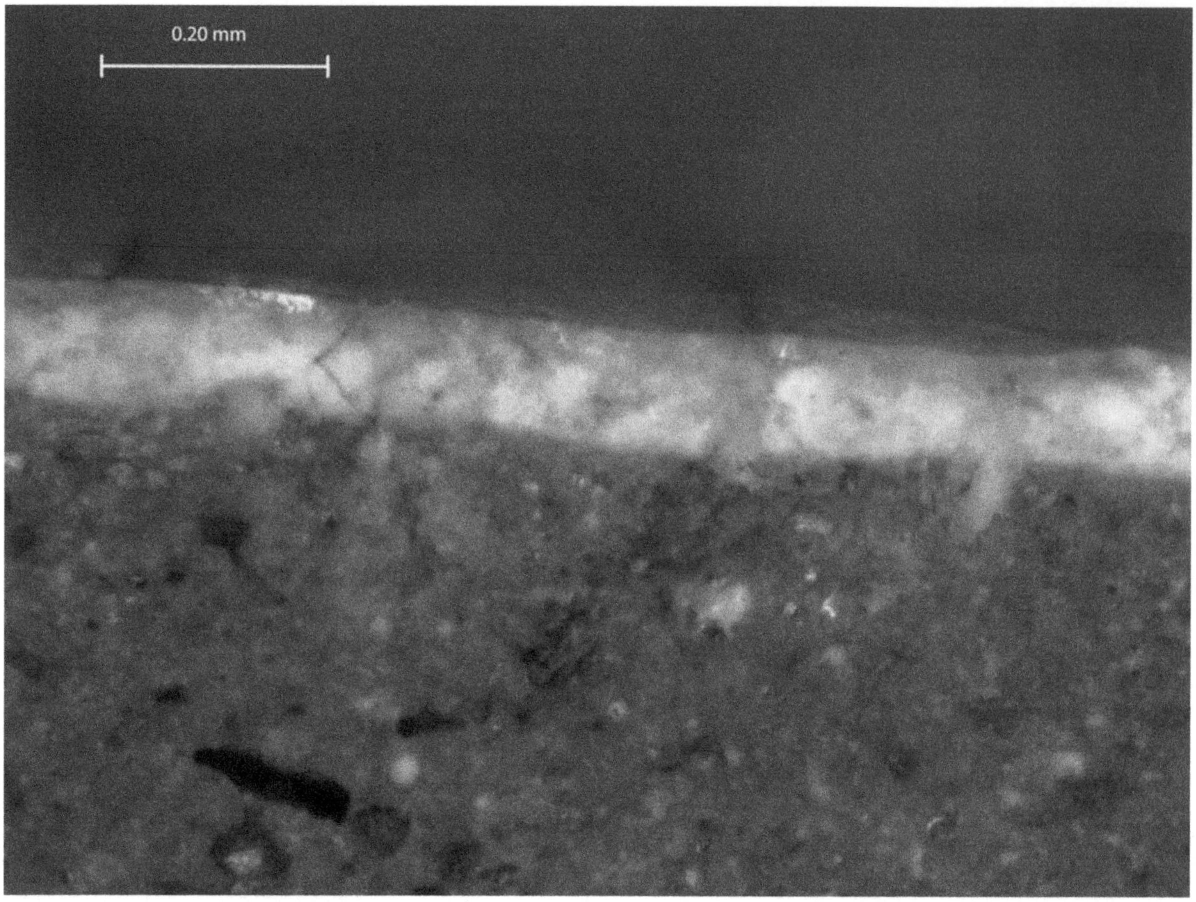


**Figure 3.** Extended Depth of Field (EDF) microphotography of a Romita Ware ceramic sample.

control samples of SRM-278 (obsidian rock) and Ohio Red Clay. At MURR, the NAA of pottery consists of two irradiations and a total of three gamma counts. Short irradiations involve a pair of samples being transported through a pneumatic tube system into the reactor core for a 5 s neutron irradiation using a thermal flux of $8x10^{13}$ n $cm^{-2}$ $s^{-1}$. After 25 min of decay, the samples are counted for 720 s using a high-resolution germanium detector. This count yields data for nine short-life elements: Al, Ba, Ca, Dy, K, Mn, Na, Ti, and V. For the long irradiation, bundles of 50 or 100 of the encapsulated quartz vials are irradiated for 24 h at a flux of $5x10^{13}$ n $cm^{-2}$ $s^{-1}$. Following the long irradiation, samples decay for seven days, and then are counted for 1800 s (known as 'middle count') on a high-resolution germanium detector coupled to an automatic sample changer. This middle count yields determination of seven medium half-life elements: As, La, Lu, Nd, Sm, U, and Yb. After an additional two-week decay, a second count for 9000 s is carried out on each sample. This final measurement allows quantification of 17 long-life elements: Ce, Co, Cr, Cs, Eu, Fe, Hf, Ni, Rb, Sb, Sc, Sr, Ta, Tb, Th, Zn and Zr (for a detailed discussion of the analytical conditions, Glascock *et al.* 2007).

Scanning electron microscopy (SEM) with energy dispersive spectroscopy (EDS) was conducted on a subset of 10 Romita Ware ceramics. Samples were cross-sectioned using a Buehler Isomet slow speed saw before casting and embedding them in epoxy resin. Metallographic polishing was conducted using diamond paste and alumina in different steps in order to achieve an optimal specular surface for SEM microanalysis. After each polishing step, specimens were sonicated in distilled water baths for 30 s in order to remove any adhering polishing compound. Polished specimens were subsequently mounted on aluminum stubs and carbon coated for SEM examination. SEM examinations were conducted at the Conservation Institute of the Smithsonian Museum using a Hitachi S-3700N variable pressure SEM and Bruker EDS. Chemical microanalyses were acquired for 200 live seconds in areas of interest (ca. 50 $\mu$m or larger at 500x magnification). The operational conditions were kept at an optimal working distance for microanalyses of 10 mm, and an acceleration voltage of 20 kV under full vacuum.

## 4. Results and discussion

The characterisation of the glazing technology is of great importance for the study of Romita Ware, in order to assess technological features that this pottery may or may not share with European majolica. Therefore, SEM-EDS examinations appear crucial to understanding the materials and technical traditions that potters used to manufacture Romita Ware ceramics.

SEM-EDS examinations reveal the nature of the technology employed by ancient Mexican potters for achieving a certain degree of opacification in Romita Ware glazes. The purpose of opacifying the intrinsically transparent lead

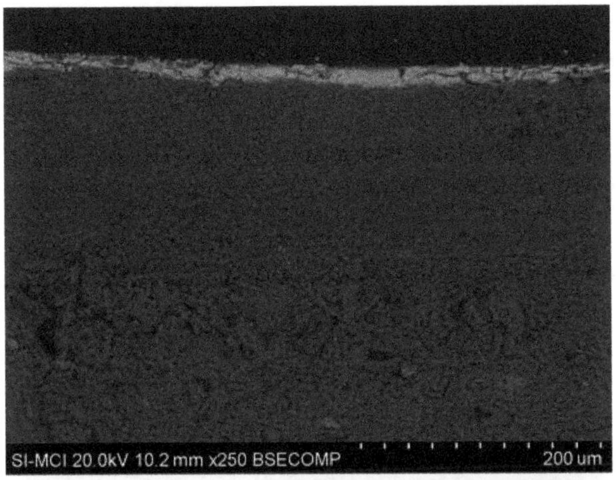


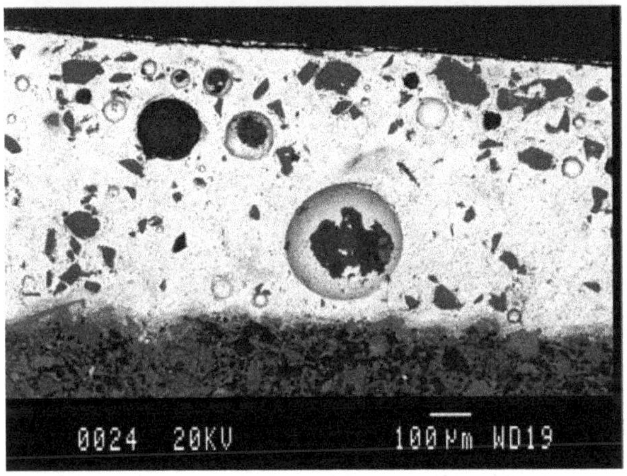


**Figure 4.** (upper) SEM-BS microphotograph of a Romita ware ceramic, featuring the glaze layer (in bright white), the white slip, and the clay matrix; (lower) SEM-BS microphotography of a majolica sample from Seville, featuring the opaque glaze layer and inclusions, as well as part of the clay matrix.

glaze is to achieve a white 'canvas' where chromatic decoration could be subsequently applied, and also to block the colour of the paste. This latter aspect is of high significance, since it may introduce an undesired darker coloration in the glaze. Tin is typically the agent used in European majolica for achieving opaque glaze coatings, given that it recrystallises in the vitreous matrix of the glaze during firing, forming small-size cassiterite crystals, $SnO_2$, which increase the dispersion and scattering of incident light in the glaze. Therefore, the glaze acquires its desired characteristic white colour (for further discussion, Tite *et al.* 1998; Vendrell *et al.* 2000; Iñañez 2007, and references therein). In contrast to European majolica technology, Romita Ware does not have $SnO_2$ in the glaze (Table 1). Instead, opacification in Romita Ware ceramics was achieved by the use of a white mineral or clay slip that was applied to the external surfaces of the ceramic, which was subsequently covered by a transparent thin lead glaze layer (Figs. 3 and 4). SEM examinations corroborate the existence of two different layers of material applied on top of the ceramic matrix. In Figure 4, the layer immediately above the clay matrix can be identified as the white slip from the previous photograph, due to its different textural characteristics. The thickness of the Romita white slip varied slightly from one sample to another and also depending on the area of the pot that was analysed. Accordingly, the slip was estimated to range from 100 to 150 μm, although slightly thicker and thinner slips were also found (up to 200 μm and as thin as 50 μm). On top of the white slip layer, a very thin and transparent lead glaze layer was observed (ranging from 5 to 25 μm). This thin lead glaze layer showed no apparent inclusions or crystals, which may have affected its visual transparency (Fig. 4).

The chemical composition of the slip and glazes provides some insights into the features of Romita Ware ceramics. In Figure 5, chemical data acquired by SEM-EDS are plotted using the most relevant analytes as axes: $Al_2O_3$, $SiO_2$, PbO. As such, it becomes possible to identify the chemical composition of the white slip and glazes, which are clearly identified as separate entities. In this case, the white slip was made out of a rich siliceous and alumina matrix (Table 1). Unfortunately, the current state of the investigations does not allow identifying the nature of this compound, and future studies will appropriately address this question. Additionally, the glaze coating of Romita Ware is mainly composed of lead (> 50 wt%) and silica (around 30 wt%) (Table 1). Therefore, and also taking into consideration the low amounts of $Na_2O$ and $K_2O$ (< 1 wt% or absent), it should be considered a high lead glaze (Tite *et al.* 1998). This last consideration can be directly related to the Spanish majolica glazing technology tradition, which overwhelmingly used high lead glazes for coatings during the early and late Renaissance (Iñañez 2007). Finally, as mentioned above, tin is not present in either the white slip or the transparent lead glaze coat.

When compared to a true majolica, Romita Ware lacks most of the technological features that are characteristic of majolica ceramics. Thus, the opacification is not achieved by the use of tin in the glaze mixture, but taking advantage of the use of a clay-like slip dominated by silica and alumina. Furthermore, the thickness of the glaze is unequivocally different. On the one hand, the glaze of Romita Ware consists of a very thin transparent layer, with no apparent inclusions. In contrast (Fig. 4), majolica glaze consists of a thick opaque layer, whose opacity is achieved by the use of tin oxides, usually ranging from 200 to 500 μm in thickness, and often presenting different inclusions, such as quartz and feldspar grains. Finally, while the majority of the majolica bodies are consistent with calcareous pastes, with lime content usually ranging between 15 and 25 wt% (Iñañez 2007; Tite 2009), Romita bodies were manufactured from non-calcareous clays, with lime contents around 1 wt%, according to the NAA data (Rodríguez-Alegría 2002). Therefore, the coloration of Romita pastes is dark reddish or brownish, likely due to the presence of hematite ($Fe_2O_3$), contrasting with the buff light coloured pastes of Spanish majolica (Iñañez *et al.* 2008), in which the process of incorporation of iron oxides into calcium iron silicates has already been established (Molera *et al.* 1998).

## 5. Conclusion

As presented in other studies ( Fournier *et al.* 2007; Iñañez *et al.* 2010), chemical data from INAA of the ceramic pastes demonstrates a New World origin for Romita Ware ceramics, with the Michoacán region (Mexico) being the likely production locale. Moreover, these results allow us to unequivocally rule out a Spanish provenance for these ceramics.

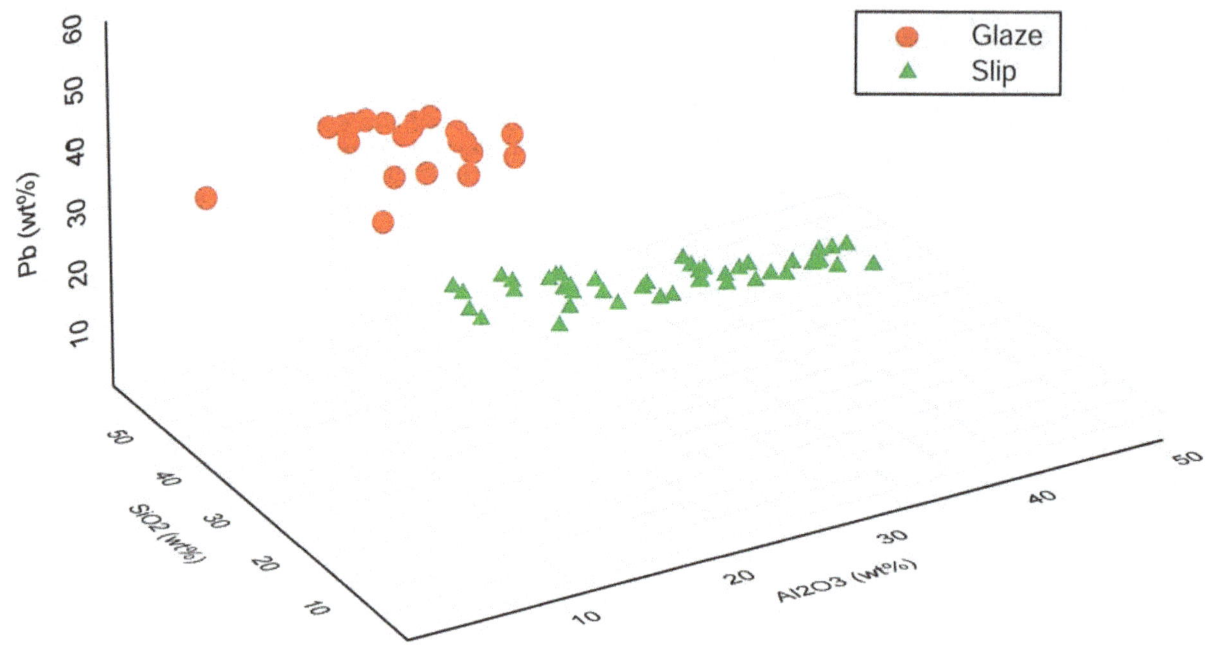


**Figure 5.** Three-dimensional plot featuring multiple readings of SEM-EDS compositions for $Al_2O_3$, $SiO_2$, and PbO of the glaze and slip areas of Romita Ware ceramics.

The present study represents the first attempt to characterise the glazing technology of the Romita Ware ceramics. Its results indicate that Romita Ware shows a very characteristic technological choice, i.e., the use of white slips under a thin transparent lead glaze layer. A relationship between these ceramics and products of European manufacture, majolica in this case, clearly exists. The similarities between Romita and majolica ceramics relate primarily to the external appearance of the vessels; Romita artefacts clearly resemble some of the forms and shiny metallic surfaces that are characteristic of Spanish majolica. However, the technological choices of the craftsmen who produced the Romita pottery were substantially different from Spanish majolica producers. Thus, the choice of non-calcareous clays, as well as the use of slip for achieving white opacification, can be seen as a continuation of traditional potting traditions as well as of constraints on the availability of local clays and other materials (e.g., tin), although in an effort to imitate the new products brought by Europeans colonisers. Therefore, Romita Ware ceramics can be interpreted as the successful combination of two different technological traditions that, as a consequence of acculturation processes, resulted in the emergence of a new pottery technology.

### Note

1. For additional information see Rodríguez-Alegría (2002); data is published on-line at http://archaeometry.missouri.edu/datasets/datasets.html.

## References

Armstrong D.V. and Hauser M.W., 2009, A Sea of Diversity: Historical Archaeology in the Caribbean. In *International Handbook of Historical Archaeology*, (eds. D. Gaimster and T. Majewski), 583-612, Springer, New York.

Deagan K.A., 1996, Colonial transformations: Euro-American cultural genesis in the early Spanish-American colonies. *Journal of Anthropological Research*, 52, 135-160.

Fournier P., Blackman M.J. and Bishop R.L., 2007, Los alfareros Purépecha de la Cuenca de Pátzcuaro: producción, intercambio y consumo de cerámica vidridada durante la época virreinal. In *Arqueología y Complejidad Social*, (eds. P. Fournier, W. Wiesheu and T. Charlton), 195-221, Instituto Nacional de Antropología e Historia, Ciudad de México.

Glascock M.D., Speakman R.J. and Neff H., 2007, Archaeometry at the University of Missouri Research Reactor and the provenance of obsidian artefacts in North America. *Archaeometry*, 49, 343-357.

Iñañez J.G., 2007, *Caracterització arqueomètrica de la ceràmica vidrada decorada de la Baixa Edat Mitjana al Renaixement dels principals centres productors de la Península Ibèrica*, PhD Thesis, Universitat de Barcelona, Barcelona.

Iñañez J.G., Speakman R.J., Buxeda i Garrigós J. and Glascock M.D., 2008, Chemical characterization of majolica from 14th-18th century production centers on the Iberian Peninsula: a preliminary neutron activation study. *Journal of Archaeological Science*, 35, 425-440.

Iñañez J.G., Bellucci J.J., Rodríguez-Alegría E., Ash R., McDonough W. and Speakman R.J., 2010, Romita pottery revisited: a reassessment of the provenance of ceramics from Colonial Mexico by LA-MC-ICP-MS. *Journal of Archaeological Science*, 37, 2698-2704.

Lister F.C. and Lister R.H., 1976, Italian presence in tin glazed ceramics in Spanish America. *Historical Archaeology*, 10, 28-41.

Lister F.C. and Lister R.H., 1982, *Sixteenth century maiolica pottery in the Valley of Mexico*, Anthropological Papers

Series of the University of Arizona 39, The University of Arizona Press, Tucson.

Lister F.C. and Lister R.H., 1987, *Andalusian ceramics in Spain and New Spain: A Cultural Register from the Third Century B.C. to 1700*, The University of Arizona Press, Tucson.

Maggetti M., Westley H. and Olin J.S., 1984, Provenance and technical studies of Mexican majolica using elemental and phase analysis. In *Archaeological Chemistry III*, (ed. J.B. Lambert), 151-191, Advances in Chemistry Series 205, American Chemical Society, Washington DC.

Molera J., Pradell T. and Vendrell-Saz M., 1998, The colours of Ca-rich ceramic pastes: origin and characterization. *Applied Clay Science*, 13, 187-202.

Ortiz F., 2002 [1940], *Contrapunteo cubano del tabaco y del azúcar*, Cátedra, Madrid.

Rodríguez-Alegría E., 2002, *Food, Eating, and Objects of Power: Class Stratification and Ceramic Production and Consumption in Colonial Mexico*, PhD Thesis, University of Chicago, Chicago. (dataset available on http://archaeometry.missouri.edu/datasets/datasets.html).

Rodríguez-Alegría E., Neff H. and Glascock M.D., 2003, Indigenous ware or Spanish import? The case of Indígena ware and approaches to power in Colonial Mexico. *Latin American Antiquity*, 14, 67-81.

Tite M.S., 2009, The production technology of Italian maiolica: a reassessment. *Journal of Archaeological Science*, 36, 2065-2080.

Tite M.S., Freestone I., Mason R., Molera J., Vendrell-Saz M. and Wood N., 1998, Lead glazes in antiquity - methods of production and reasons for use. *Archaeometry*, 40, 241-260.

Vendrell M., Molera J. and Tite M.S., 2000, Optical properties of tin-opacified glazes. *Archaeometry*, 42, 325-340.

## ACKNOWLEDGEMENTS

This work is supported by the Marie Curie International Outgoing Fellowships program, endorsed by the European Commission (PIOF-GA-2008-223319). SEM analyses were conducted at the Museum Conservation Institute of the Smithsonian Institution INAA analyses of the 94 Romita Ware samples from the Metropolitan Cathedral were conducted by Enrique Rodríguez-Alegría; and INAA analyses of majolica and glazed ware from the Iberian Peninsula and Western Mediterranean were conducted by Javier G. Iñañez, both datasets acquired at the Missouri University Research Reactor (SBR-9802366; National Science Foundation grant 0504015). Thanks are extended to Enrique Rodríguez-Alegría for providing the Romita ware samples included in this technological study and to Melvin Wachowiak for assistance with the EDF image. The author is also indebted to Robert J. Speakman for his support and advice.

# 17

# Archaeometric investigation of Buda white ware ($12^{th}$–$14^{th}$ century AD, North Hungary)

**V. Szilágyi[a], T.Á. Rácz[b], V. Gál-Mlakár[c], E. Simonyi[d] and I.E. Sajó[e]**

[a]Centre for Energy Research, Budapest, HAS, Hungary, szilagyi.vera@gmail.com
[b]Tragor Ignác Museum, Directorate of Pest County Museums
[c]Herman Ottó Museum, Directorate of Borsod-Abaúj-Zemplén County Museum
[d]Hungarian National Museum
[e]Research Centre for Natural Sciences, HAS, Budapest, Hungary

**Abstract** - This paper summarises the main archaeological questions regarding the medieval ($12^{th}$–$14^{th}$ century) 'Buda white ware' (BWW), based on our knowledge accumulated from the last 50 years of research. Considering the known facts and contradictions concerning typology, chronology, raw material sources and workshops, and distribution of this peculiar pottery group, the necessity of a scientific characterisation of its composition becomes clear. Our initial results are based on the investigation of a limited set of archaeological samples, but they confirm that the group of BWW, formerly described as homogenous, can be classified into at least three types based on differences in the raw materials used. In addition, there is evidence of contemporary lower quality imitations. These preliminary results establish the need for further analyses of a larger set of the BWW finds.

## 1. Introduction

The first appearance of the white-body pottery and red-on-white painted table ware was a unique event in the history of ceramics of early medieval Hungary, and can be dated to the $12^{th}$ century AD. Its utilisation very quickly became widespread, and the white table ware was already a popular product during the $13^{th}$ century. The largest amount of this pottery was excavated at Buda, the capital of the medieval Hungarian Kingdom, and in its vicinity. This aspect resulted in the name 'Buda white ware' (BWW), and the description of these ceramic finds from Buda provided a basis of characterisation for archaeologists carrying out research around the middle of the last century. Since at Buda and the surrounding sites the BWW appeared to be a homogeneous group, this idea has been accepted as an axiom in most of the subsequent publications. However, in excavations carried out in NE Hungary, white ceramic assemblages with white table ware types different from BWW were found. A meaningful description of the spatial and temporal distribution of the different white ware types cannot be fulfilled without the investigation of their raw materials. The present research was undertaken because no information about the raw materials used was yet available for reaching archaeological conclusions.

The aim of this paper is to describe the problems concerning BWW from an archaeological point of view, and to report the results of the first preliminary investigations on a limited sample set. For this purpose, the principal facts of the archaeological research on BWW are provided first. Subsequently, the archaeometric data that we gained by petro-mineralogical investigations of a limited white ware sample set from some selected archaeological sites is presented.

## 2. Summary of the previous 'Buda white ware' research

In 1930, Höllrigl was the first researcher who noted the preponderance of yellowish white to pale rose coloured pottery among Arpadian age (1000–1301 AD) vessels (Höllrigl 1930). Later on, Holl published some basic statements about this ceramic group in the framework of an investigation of the potting techniques of the $13^{th}$ century (Holl 1956). It was his observation that the white ware was manufactured using the coil technique on a turntable. This process was in use even until the $15^{th}$ century, but, at that time, only for the white pottery. Holl (1963) concluded that the first appearance of BWW can be estimated to either the end of the $12^{th}$ century or the beginning of the

13$^{th}$ century. As this type of white pottery was produced continuously at various locations until the 19$^{th}$ century, he assumed that more than one source of raw material was used.

Due to the recent excavations at Buda Castle, the predominance of the BWW in the pottery of the royal palace during the 13$^{th}$ and 14$^{th}$ centuries was established by finds of cultural layers dated by coins. White vessels found with coins that were hidden from the Tartar invasion (1241–1242) have a special importance from the point of view of dating (Parádi 1963, 206–8, Figs. 2/7 and 9). An early occurrence of BWW (at Budaújlak – Kárpáti 2002, 509, comment no. 8, and 603, Fig. 10) and a ceramic assemblage with a high proportion of BWW dated before the 1240s were noted (at Kána – Terei 2010) in the local rural surroundings of the royal court. BWW was commonly utilised in this region (Pest County) from the 13$^{th}$ century. White pottery forms an easily identified group in the chronologically overlapping ceramic assemblages of the Arpadian age settlements dated from the 10$^{th}$–11$^{th}$ until the 14$^{th}$ century (Rácz 2010). The highest amount of BWW was detected in ceramic assemblages dated to the 13$^{th}$–14$^{th}$ century (e.g., Üllő-7, Maglód-1 sites). However, BWW also appeared at sites (e.g., Vecsés-67) with only a few pits dated to the 13$^{th}$ century. As an earlier example, white pottery fragments were discovered in a smaller quantity in a 12$^{th}$ century context (e.g., Üllő-1, Üllő-2 sites).

It is very probable that the white ware became common soon after its first appearance, and replaced the former red or brown ware almost entirely. There is strong evidence for that from the 13$^{th}$–14$^{th}$ century objects of the Maglód-1 site, where more than 700 ceramic fragments were excavated, with only 10 remains of the formerly known handicraft and seven pieces of 14$^{th}$ century grey pots (interpreted as imported). All the other pottery finds proved to be BWW. However, among the 10$^{th}$–11$^{th}$ century objects of the same site, there were no white ceramics at all. At the Üllő-7 site (dated to the 13$^{th}$ century and deserted before the 14$^{th}$ century), the ceramic assemblage contained 85% BWW. This phenomenon has to be connected to the high quality and usefulness (for cooking) of BWW. It was better fired, and hence less porous and liable to break than the former red and brown ceramics. Another general trend was that, paralleling the appearance of BWW, the enrichment (5–6 times) of the quantity of pottery at the 13$^{th}$–14$^{th}$ century settlements could be observed.

In addition to the region near the capital, the white ware also spread to the northern and north-eastern part of present-day Hungary (Nógrád, Heves, and Borsod-Abaúj-Zemplén Counties). It could be dated to a period anterior to the late 13$^{th}$ – early 14$^{th}$ century at Salgó Castle (Feld 1984, 228–9, 261, Fig. 32). The spirally engraved BWW could be detected from the first part of the 13$^{th}$ century (e.g., at the castle of Füzér – Cabello and Feld 1980; at the castle of Komlósla-Pusztavár – Gál-Mlakár 2007), while the red decorated BWW (at Felsőzsolca – Simonyi 2003, 126) appeared at the turn of the 14$^{th}$ century. Archaeological sites of this eastern region provided ceramic assemblages with significant (sometimes predominant) amounts of BWW (e.g., at Fehérkő Castle at Sámsonháza – Rácz 2006; at Sárospatak, Roman Catholic cemetery – Gömöri 1970; at Mezőnyárád – Wolf 1986; at Hejőkeresztúr – Wolf 1999, 169).

Although the presence of BWW can also be detected at early medieval sites located west of Buda, their occurrence is sporadic (at Győr-Homokgödör – Takács 2000, 268, Fig. 10/1; at Kajárpéc-Pokolfadomb – Takács 1993, 219, Fig. 12/5-6). These finds were interpreted as representative of a non-local ceramic production.

The distribution of the 13$^{th}$–14$^{th}$ century white ware in present-day Slovakia was observed by Hosso, who provided a map of regions with different ceramic types (Hosso 1983, 220, and Fig. 1) ( the boundary indicated by Hosso on our Fig. 1). According to this author, two territories (an eastern – south-eastern and a north-western) can be said to be characterised by the presence of medieval white body pottery in present-day Slovakia. He estimated its first appearance to have taken place during the 13$^{th}$–14$^{th}$ centuries. The eastern boundary of Hosso's eastern – south-eastern, so-called 'painted and white pottery' territory was drawn at Trebišov (Terebes) and the Košice (Kassa) basin. Its northern limit is associated with the sites of Obišovce (Castle of Abos), Jelšava (Jolsva), and Filakovo (Fülek), while the region extended till Chl'aba (Helemba) at the Duna-Ipoly junction to the south-west.

The later history of the BWW from NE Hungary was partly recorded by Tomka (2004, 82–95). He indicated that – three centuries after the general extension of red decoration on liquid container vessels – red painting of white cooking pots became common in the 16$^{th}$ century. White pottery seemed to be less abundant in the late Middle Ages: from 40% of the total in the 13$^{th}$–14$^{th}$ centuries, its proportion decreased to 10% in the 16$^{th}$–17$^{th}$ centuries.

## 3. The question of the raw material of 'Buda white ware'

Clayish raw material displaying a white colour after firing is not known among the geological formations around Buda. This means that either the raw material or the finished product had to be traded from the source to the potters or consumers. However, this hypothesis presupposes a very intensive exchange of goods throughout the region around Buda, since BWW could be detected at every settlement of this era in the vicinity of Buda and in Pest County, even in the smallest and most inaccessible villages (MRT 9, XIII/2). The large amount of white pottery suggests that the raw material used was easily accessible for most potters.

Ethnographic studies of the 19$^{th}$ century found that white pottery was manufactured at former Gömör and Kis-Hont County (in present-day Slovakia), in the vicinity of Rimavska Sobota (Rimaszombat) and in villages south of Jelšava (Jolsva) during the 19$^{th}$ century, and its trade extended especially to the Great Hungarian Plain and partly to other regions. The direction of the trade was determined by the geographic circumstances of the territory, namely closed in from the north by mountains, and with southward oriented valleys. The distribution of this 19$^{th}$ century pottery overlaps with the occurrences of BWW in the 13$^{th}$ century. It has long been an assumption in Hungarian ethnography that the white ware of the 13$^{th}$ century was connected to the Gömör pottery handicraft (Kresz 1960, 315; Kiss 2006, 158). However, archaeologists did not yet take a stand on the source of the raw material for the 12–13$^{th}$ century products. Accepting Holl's

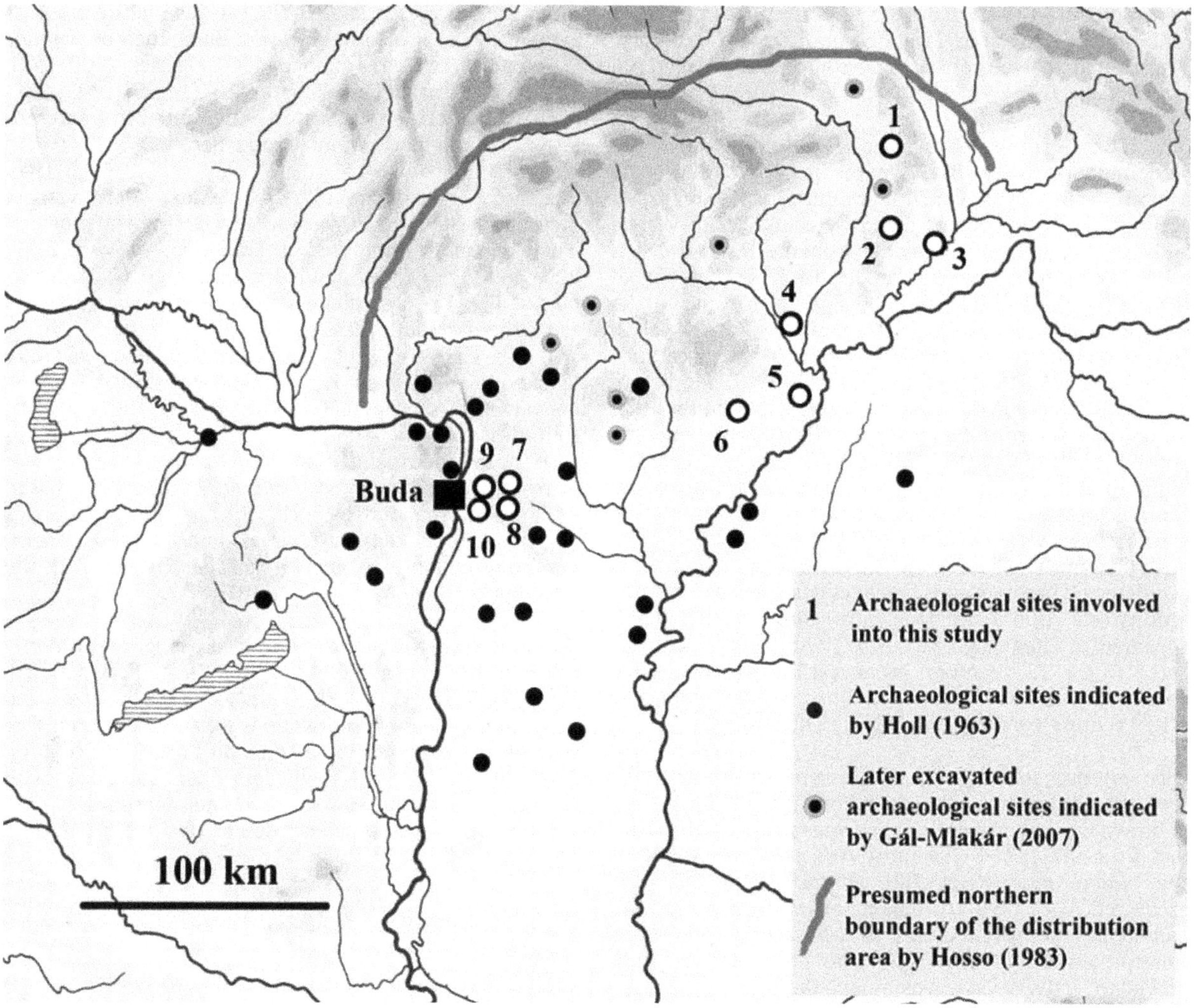


**Figure 1.** Distribution of 'Buda white ware' in Hungary and Slovakia. The map follows the compilation of Gál-Mlakár (2007). Archaeological sites included in this study are marked with numbers: (1) Castle of Füzér; (2) Castle of Komlóska-Pusztavár; (3) Sárospatak - parish church cemetery; (4) Felsőzsolca-Várdomb; (5) Hejőkürt-Cifrahát; (6) Mezőkeresztes-Lucernás; (7) Maglód-1; (8) Üllő-1, Üllő-2, and Üllő-7; (9) Vecsés-36; (10) Gyál-9.

first idea about a domestic production of BWW in settlements around Buda, typology and dating became the more emphasised topics. A hindering factor in the archaeological research to date is that no kiln or firing pit producing BWW has been as yet excavated anywhere.

As a result of ethnographic data collected by Kresz (1960), potential regions of white ceramic production could be identified in the former Gömör, Nógrád, and also Bihar (in present-day Romania) Counties. She presumed that in these counties, significant clay sources were employed for pottery production in medieval times. In the northern region (Gömör and Kis-Hont and Nógrád Counties, present-day SE Slovakia), this source was identified as the Poltár Formation of the Košice (Kassa) basin by Hosso (1983, 222). New ethnographic studies reported this type of raw material in the middle parts of the Gömör and Kis-Hont Counties in the valleys of the Rimava (Rima), Blh (Balog), Murán (Murány), and Turiec (Túróc) rivers, and described it as high quality, white firing colour clay rich in kaolinite. A similar clay outcrop is known near Csákvár at the Vértes Hills, Hungary, from the written sources (Kiss 2006, 156). Likewise, in the eastern region (in present-day Romania), a clay type resulting in a white colour after firing is known near Vadu Crişului (Körösrév) in former Bihar County, and in the former Hunyad County, in the middle of Transylvania (Kiss 2006, 156). It is an ethnographic suggestion that the clay exploitation was more significant at the northern sites, but an identification of raw material source sites for various BWW has yet to be achieved.

The 13$^{th}$ century proved to be a turning-point in the Hungarian potter's craft from the point of view of raw materials, and also of vessel shapes and types. Some features of the BWW were known from former ceramic products, e.g., spiral decoration and bottom stamp. However, some esentially new characteristics also occurred, e.g., standardisation of the raw material and the shaping of the rim. In addition, the simultaneous qualitative and quantitative improvement of the vessels suggests the introduction of a new social group highly specialised in pottery manufacturing. It is also presumable that the handicraft traditions of this new group did not derive directly from the former regional practices.

Investigating the raw material of the BWW, some comprehensive questions concerning the archaeological era are raised: (1) Which social group started to make this type of pottery? (2) What was the region of production? (3) Where did the raw material come from? (4) How could this ware become so popular and disperse so densely in the central part of the Hungarian Kingdom? (5) Is it realistic to assume such an intensive exchange of raw material or finished products already in the 13$^{th}$ century? These questions have no answer in the written sources, but the solutions to them are hidden in the ceramic fragments.

## 4. Methods and samples

Preliminary archaeometric data on the BWW was provided by microscopic petrographic observations and basic mineralogical investigations (XRPD carried out at the Chemical Research Center, Hungarian Academy of Sciences). The Nikon ALPHAPHOT-2 polarising microscope was used with the permission of the Department of Petrology and Geochemistry, Eötvös Loránd University. XRPD scans were taken on a Phillips PW3710/PW1050 diffractometer with Cu K$\alpha$ radiation, graphite monochromator, and a proportional counter.

Our first attempts to characterise the BWW and identify its raw material sources concentrated on the ceramic assemblages of rural settlements and fortresses of the 13$^{th}$–14$^{th}$ century in NE Hungary. The white ware finds from Komlóska-Pusztavár Castle were selected as a reference material since its BWW assemblage is one of the best described and seems to be very heterogeneous. By the characterisation of the diversity of this white pottery assemblage, we can obtain a comparative reference material for the classification of the BWW sherds from other 13$^{th}$–14$^{th}$ century NE Hungary sites (Castle of Füzér, Sárospatak – cemetery of the parish church, Felsőzsolca-Várdomb), and the contemporary BWW pots from villages close to Buda (Üllő-2, Üllő-7, Maglód-1, Vecsés-36), which were also included in our study (Fig. 1). In this first stage of the research, a limited number (27 pieces) of ceramic finds was investigated from 12 archaeological sites dating to the 12$^{th}$–14$^{th}$ centuries, located at different distances from Buda. A list of samples and a short classification determined by archaeological and archaeometric research are provided in Table 1. The selected samples cannot represent the ceramic assemblages of the selected sites in their entirety. However, as easily distinguished components of the pottery collections of each site, these sherds can be compared to each other and help us understand the white ware supply of the Hungarian Kingdom in the period between the 12$^{th}$ and the 14$^{th}$ century.

## 5. Results

### *Determining diversity of BWW by macroscopic archaeological observations*

BWW was basically described as white pottery (table ware and cooking pots) without decoration, or characterised by engraved and/or red painting of strips, spots, or spiral motifs around them (Höllrigl 1930). This description suggested that the BWW is a homogeneous, easily recognisable group of pottery. However, recent excavations in NE Hungary identified BWW artefacts with a more heterogeneous macroscopic appearance.

The ceramic assemblage of the medieval Castle of Komlóska-Pusztavár (Gál-Mlakár 2007) contains white ceramics in proportion of 20% of the total, a number which can be further divided into three distinguishable types (Fig. 2) based on their macroscopic features (Gál-Mlakár 2009 and Table 2). The first type of BWW is the so-called 'white type', which is interpreted as the *sensu stricto* 'Buda white ware'. Its pure white colour, both on the surface and in section, suggests a well controlled atmosphere during the firing. The second type of BWW is denominated as 'greyish white type', since its colour is greyish white both on the surface and in section, which also indicates well controlled firing. The third type of BWW is generally referred to as the 'yellowish white type', and includes potsherds differing greatly from the classical BWW in their colour (sandwich structured, which can be interpreted as a result of less controlled firing), quality (thin-walled), and material. All of the three types of BWW have – probably – washed paste and sand temper.

Although it is very probable that – because of their decoration and fragment shapes – the majority of the BWW consisted of table ware (bottles, jugs, etc.), the fragmented condition of the finds makes it impossible to verify this statement. Nevertheless, there are also vessels with other functions in all three types of BWW (e.g., cooking pots of greyish white colour, mugs of white colour).

In order to comprehensively investigate the characteristics of BWW in the eastern part of the distribution area, further archaeological sites were included in this research. The major point of interest was whether the above distinguished types of BWW could also be identified at these sites. White ceramic remnants from the early 13$^{th}$ century stratum of the castle of Füzér (Simon 2006) could be classified into all three types of BWW. One 13$^{th}$ century pit and some strata of the cemetery at Sárospatak (Gömöri 1970) also contain white ceramic fragments belonging to all three types of BWW.

Other eastern Hungarian archaeological sites (Hejőkürt-Cifrahát, Mezőkeresztes-Lucernás, Felsőzsolca-Várdomb) with a smaller quantity of white ware in their ceramic assemblages were also included in this preliminary study. Although represented by only one or a few samples, their investigation contributes to the understanding of BWW production and distribution. The 12$^{th}$–13$^{th}$ century ceramic collection of Hejőkürt-Cifrahát is quite homogeneous (Simonyi 2005). It contains almost exclusively beige (rosy white), spirally engraved pots, mugs, and cauldrons of plastic but grainy (sand-tempered?) material which were fashioned on a turntable. Based on the physical appearance (compact, dense body), we can ascertain that the vessels were well fired. The excavation at Mezőkeresztes-Lucernás identified 12$^{th}$–13$^{th}$ century dated beige cauldrons and vessel fragments with spirally engraved decoration and coarse sand tempering (Simonyi 2005). The ceramic material of the third archaeological site, Felsőzsolca-Várdomb, contained greyish white cauldrons, a pure white pot, and rosy-white ceramics (Simonyi 2003; 2005). In addition, some fragments of white cooking pots made of sandy material, a ribbed-rimmed mug, and red painted

**Table 1.** List of samples included in this study with the archaeological classification of ceramics and grouping on the basis of archaeometric analysis. Abbreviations: W BWW = 'white' Buda white ware; GW BWW = 'greyish white' Buda white ware; YW BWW = 'yellowish white' Buda white ware.

| Arch. site | Inv. No. | Lab. No. | Age | Arch. class. | Petrogr. class. |
|---|---|---|---|---|---|
| Castle of Füzér | 1998.2.17.7.22. | FUZ-1 | $13^{th}$ c. | 'yellowish white' pot | High quality YW BWW |
| | 1998.2.8.2.7. | FUZ-3 | | 'white' pot | High quality W BWW |
| | 1998.2.13.2.1. | FUZ-2 | | 'greyish white' pot | High quality GW BWW |
| Komlóska-Pusztavár (Castle of Solymos) | 2007.3.294. | KOM-1 | $13^{th}$ /$14^{th}$ c. | 'yellowish white' pot | High quality GW BWW |
| | 2005.118.121. | KOM-3 | | 'white' pot | High quality W BWW |
| | 2007.3.69. | KOM-2 | | 'greyish white' pot | Silty clay |
| Sárospatak - cemetery | 71.3.45. | SP-1 | $13^{th}$ /$14^{th}$ c. | 'yellowish white' pot | High quality YW BWW |
| | 71.3.15. | SP-3 | | 'white' pot | High quality W BWW |
| | 71.2.55. | SP-2 | | 'greyish white' pot | High quality GW BWW |
| Felsőzsolca-Várdomb | 1992/ditch No.1/ white pot | FZS-19 | $13^{th}$ c. | Buda white ware | High quality W BWW |
| | IV$^{th}$ oven | FZS-21A | $12^{th}$ - $13^{th}$ c. | 'beige' pot | Lower quality W BWW |
| | 1992/ditch No.1/ cauldron | FZS-21B | $13^{th}$ c. | 'beige' cauldron | Lower quality YW BWW |
| | Painted bottle | FZS-10 | $13^{th}$ /$14^{th}$ c. | 'beige' painted bottle | Fine silt |
| | 1993/ditch No.1 | FZS-20 | $15^{th}$ c. | white pot | High quality YW BWW |
| | Obj. 1994/1, well | FZS-9 | $13^{th}$ c. | Buda white ware | Lower quality W BWW |
| Hejőkürt-Cifrahát | 1/12 | HC-8 | beg. $13^{th}$ c. | 'beige' pot | Lower quality GW BWW |
| Mezőkeresztes-Lucernás | S6/7 | ML-7 | $12^{th}$ /$13^{th}$ c. | 'beige' pot | Hiatal fabric + sand temper |
| Maglód-1 | 2008.2.409.8. | M1-8 | $13^{th}$ - $14^{th}$ c. | Painted, white bottle | High quality W BWW |
| | 2008.2.409.2. | M1-2 | | White pot (szilke) | High quality W BWW |
| Üllő-1 | 2006.8.36.3. | Ü1-3 | end $12^{th}$ c. | White cooking pot | High quality GW BWW |
| Üllő-2 | 2006.9.9.2. | Ü2-2 | $12^{th}$ c. | White cooking pot | High quality W BWW (bit ferrous) |
| Üllő-7 | 2006.10.38.2. | Ü7-2 | beg. $13^{th}$ c. | White cauldron 'white' | High quality W BWW (bit ferrous) |
| | 2006.10.3.5. | Ü7-5 | | Painted bottle | Lower quality W BWW |
| | 2006.10.40.1. | Ü7-1 | | White pot | High quality W BWW (bit ferrous) |
| | 2006.10.40.3. | Ü7-3 | | White cauldron 'beige/rose' | Lower quality W BWW |
| Vecsés-36 | 2006.11.31.2. | V36-2 | $13^{th}$ c. | White ceramic | High quality W BWW (bit ferrous) |
| Gyál-9 | 2007.10.206.1. | GY9-1 | $13^{th}$ - $14^{th}$ c. | White caldron | High quality W BWW (bit ferrous) |

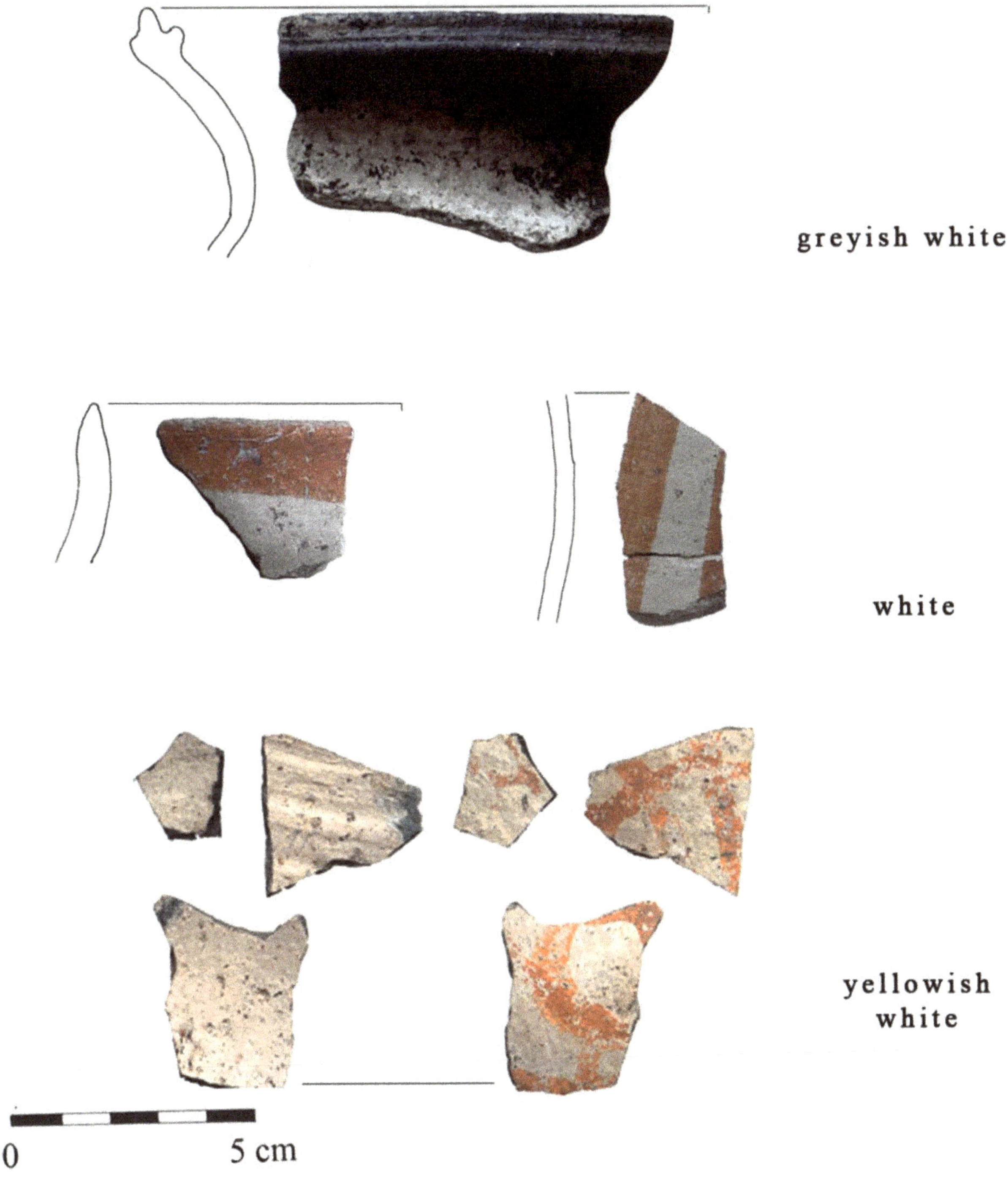


**Figure 2.** Macroscopically distinguished 'Buda white ware' types at Komlóska-Pusztavár.

ceramics were also found. The pure white pottery seemed to be different from other white ware found at Felsőzsolca and similar to the white ware of Buda Castle. A red painted, beige (rosy)-white bottle manufactured on a turntable with potter-mark was also mentioned from the material of the ditch (a similar bottle is published from Muhi – Laszlovszky *et al.* 1997, 147, Fig. 151), which was dated to the end of the 13$^{th}$ – beginning of the 14$^{th}$ century.

Similarly to their eastern relatives, representatives of BWW at sites near Buda were manufactured with a coil technique on a turntable from gravelly, sometimes micaceous raw material. The colour of the vessels is esentially white, even if some greyish hues appear, due to the use of open fire. At sites Üllő-7 and Maglód-1, there are some vessels (mainly bottles and cauldrons) of bright orange or rose colour. However, this latter colour is more possibly related to a less controlled firing technique than to a different raw material. A majority of the ceramic assemblage consists of cooking pots (or small pots), but new, formerly unknown vessel shapes also occur (e.g., bottle, cup, jug) (Fig. 3). As part of their uniform appearance, homogeneous raw material,

**Table 2.** Macroscopic characteristics of the three main types of Buda white ware found at Komlóska-Pusztavár (abbreviations in Table).

| **Feature** | **W BWW** | **GW BWW** | **YW BWW** |
|---|---|---|---|
| *Surface colour* | Pure white | Greyish white | Yellowish white - beige (rosy) |
| *Section colour* | Homogeneous, pure white | Homogeneous, greyish white | Sandwich structured, light-dark-light grey |
| *Paste* | Fine-grained | Fine-grained | Very fine-grained |
| *Temper* | Micaceous fine-grained (<1mm) sand | Medium-grained (≤1mm) sand | Medium to coarse-grained sand (>1mm, even up to gravel size) |
| *Surface* | Smooth, dry touch | Grainy, greasy touch, matt | Smooth, uneven and grainy, greasy touch, shiny (sometimes porous, blistered) |
| *Wall thickness* | 5-6 mm | 4-6 mm | 3-4 mm |
| *Decoration* | Engraved/painted red | Engraved | Engraved/painted red |

double rim, and spiral engraving can be observed. Nevertheless, the vessels' sizes and rim diameters are unique, piece by piece, with the exception of the cups. While bottles and jugs have variable fashioning, the cups seem to be more consistent, with a wide, bent out rim, less curved bulge, and narrow bottom. Bottles have spiral decorations, simple motifs of red pigment are visible on jugs, while parallel engraving on the shoulder or bulge of small pots are common. Although a white cauldron was a rare vessel type, fragments of nine white and four red cauldrons were discovered at the Üllő-7 site.

### *Differentiation by petro-mineralogical composition*

In the first stage of the archaeometric research, the aim was to investigate whether the macroscopically distinguished white ware types from the eastern part of the Kingdom can also be differentiated by material analysis. The petrographic-mineralogical characterisation concentrated on the classified samples from three archaeological sites (Füzér, Komlóska, and Sárospatak) of the eastern region. After obtaining instrumental data pertaining to the three provisional categories of BWW, the further identification of white ceramic fragments from sites closer to Buda became easier.

According to our petrographic investigations, the 'white' type of the BWW has a very pure, plastic paste deriving from iron-free, micaceous clay. The fragments have hiatal fabric, from artificially added medium to coarse-grained, metamorphic (granitoid) sand. The aplastic constituents are quartz, muscovite, K-feldspar, and quartzite fragments (Fig. 4b). The XRD measurements proved that these vessels were derived from mixed, (15) illitic-smectitic clay, and contain quartz, feldspars, and an amorphous phase (Fig. 4a). The petro-mineralogical characteristics summarised here

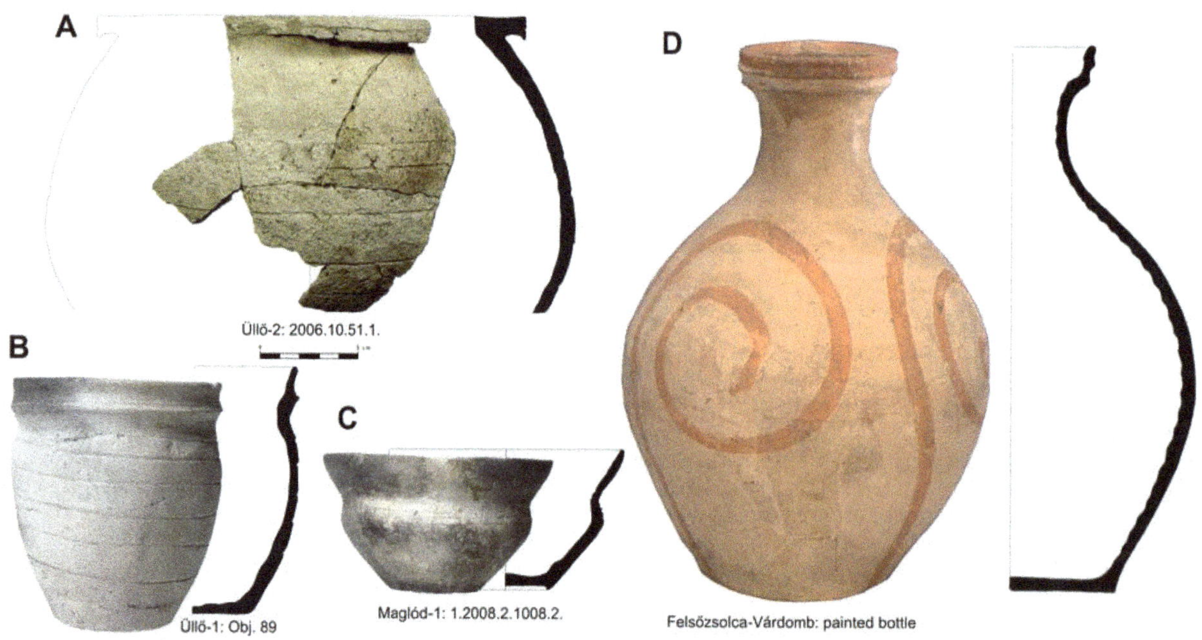


**Figure 3.** Characteristic vessel shapes of 'Buda white ware': (a) cauldron; (b) cooking pot; (c) cup; and (d) bottle.

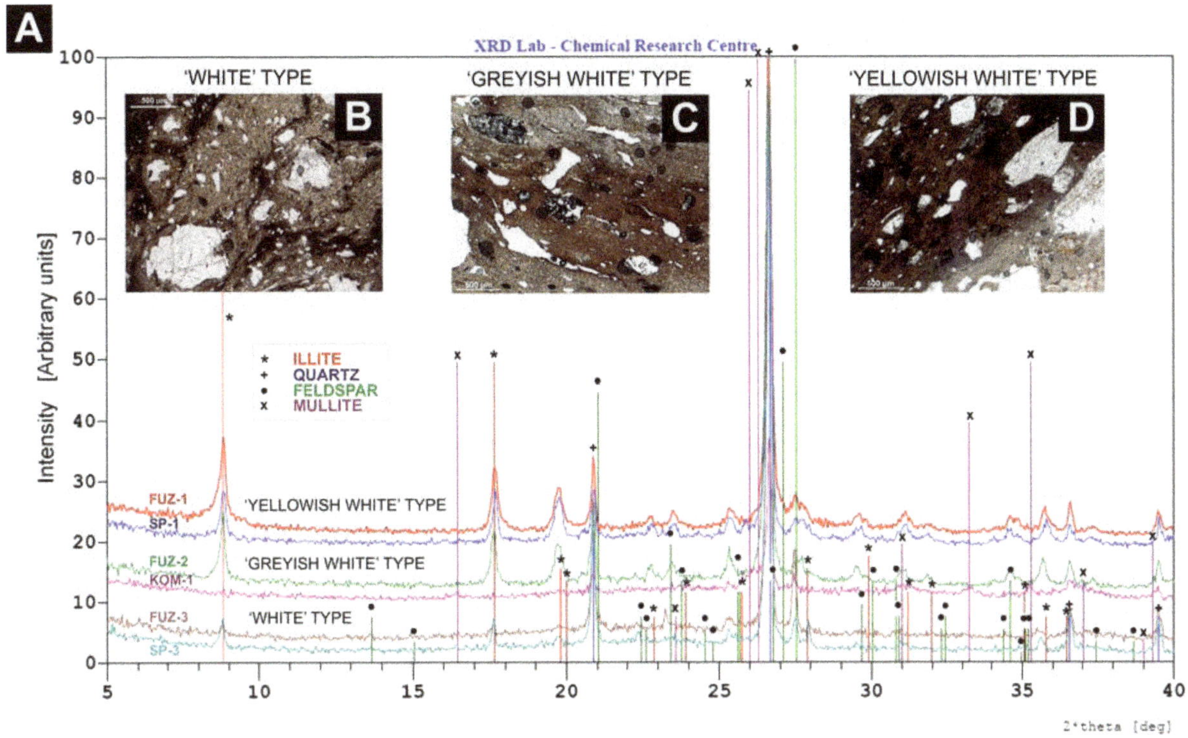


**Figure 4.** Mineralogical composition (a) and fabric features (b-d) of the three raw material types of 'Buda white ware' obtained as a result of microscopic petrographic and XRPD investigations. Two samples of each BWW type were measured by XRD.

could be associated to samples FUZ-3, SP-3, and KOM-3, which represented the 'white' type of BWW in the initial archaeological sample set.

Ceramic fragments belonging to the 'greyish white' type have a very pure, plastic paste which derives from ferrous clay. Their hiatal fabric contains artificially added, coarse-grained, metamorphic (granitoid) derived sand. The average mineralogical composition is quartz, muscovite, K-feldspar, and quartzite lithofragments (Fig. 4c). Based on XRD analysis, the plastic paste of the pottery is illitic clay. Quartz, amorphous phase, and feldspar relate to the aplastics. In one case, absence (decomposition) of the primary phyllosilicate and neoformation of the high-temperature mullite phase could be detected, indicating firing at a temperature consistently above 900°C (Fig. 4a). The neoformation of mullite indirectly attests the (at least partly) kaolinitic composition of the raw clay. These features correspond to samples FUZ-2 and SP-2, which were archaeologically described to belong to the 'greyish white' type of BWW. However, a ceramic fragment of the same archaeological classification from Komlóska (KOM-2) did not fit into this group. Furthermore, another sample (KOM-1), macroscopically determined to belong to the 'yellowish white' type of ceramic proved to belong to this petrographic group.

The 'yellowish white' type of BWW contains hiatal fabric ceramics which have very pure, very plastic paste of pleochroic clay and artificial tempering with coarse-grained sand of metamorphic (granitoid) origin. The mineralogical composition of the aplastics is quartz, muscovite, K-feldspar, and phyllite lithofragments (Fig. 4d). XRD analysis showed an illitic-smectitic clay mineral composition of the clay paste. In addition, quartz, feldspar, and a lower amount of amorphous phase could be identified (Fig. 4a). This description is valid for samples FUZ-1 and SP-1, which were macroscopically classified into the 'yellowish white' type of BWW. The above detailed mixing of selected samples from Komlóska has to be mentioned here too.

## *Utilisation of grouping*

Since it is suggested above that the provisional three categories of BWW can often be distinguished according to petro-mineralogical characteristics, white ceramic finds from archaeological sites closer to Buda can be described in a similar way and classified into one of the three groups (or differentiated from them). For this purpose, white ware types from Komlóska, Füzér and Sárospatak were chosen as reference samples. The comparison was based on petrographic microscopic investigations, since instrumental analysis of the mineralogical composition could not be carried out at this stage of the research for these western samples (for brief descriptions, Table).

The reference samples for the 'white type' of BWW were FUZ-3, SP-3, and KOM-3. This group proved to be the most abundant type of BWW in the western part of the investigated region, since vessels of the same raw material quality (sometimes with patchy ferrous clay) were described from the Felsőzsolca, Üllő-2, Üllő-7, Vecsés-36, Gyál-9, and Maglód-1 sites (8 pieces). In addition, four potsherds of similar but lower quality raw material (less pure, grainy clay) were also found at Felsőzsolca and Üllő-7.

The reference samples of the 'greyish white' type of BWW were FUZ-2 and SP-2. One vessel sherd of the same raw material quality was analysed from Üllő-1, and another sherd from Hejőkürt showed similar but lower quality (less pure) raw material. Hence, this group seems to have had a limited distribution in the western part of the investigated region.

The only sample that proved to be of the same raw material quality to the 'yellowish white' type of BWW (reference samples FUZ-1, SP-1) originated from Felsőzsolca, and was described as a fragment of a white slipped vessel from the 15th century (Simonyi 2003; 2005). Another fragment from the same site showed similar but lower quality raw material composition (less pure, patchy anisotropic clay). The 'yellowish white' type of BWW was also found less frequently at archaeological sites closer to Buda.

Some of the white vessel fragments proved to be made from a raw material different from all of the above distinguished compositions. These fragments originated from the ceramic assemblages of Komlóska, Fels´o´zsolca, and Mezo´´keresztes. The samples had a less pure clay component than the BWW, and thus have to be excluded from the category of BWW. The fact that in spite of their macroscopic appearance, some white ceramic fragments cannot be classified into any of the here described BWW types suggests the complexity of the raw material supply and the manufacturing sectors of this pottery during the period between the 12th and 14th century.

## 6. Discussion

Based on our results, it was possible to differentiate 12th–14th century BWW into subgroups according to microscopic petrographic and XRD analyses. A common feature of BWW potsherds is the usage of very fine-grained, very plastic clay tempered with sand derived from low-grade metamorphic and granitoid rocks. Our identification of measurable physical differences between the representatives of BWW which were formerly considered as a homogeneous group by the archaeological researchers represents a new archaeological and archaeometric insight. According to the differing mineralogical compositions, three archaeological types of BWW could be distinguished: 'white', 'greyish white', and 'yellowish white' types. The main difference lies with the quality of the clay (iron-free and mixed clay mineral; ferrous and illitic/kaolinitic?; ferrous, pleochroic and mixed clay mineral, respectively) and partly with the sometimes variable firing conditions. These collective and distinctive features suggest that during the manufacturing of all three types of BWW, potters tried to follow the same fashion (style of vessel shapes, surface quality and treatment), but the raw material supply differed from place to place. According to the suggestions of the archaeologists, the sensu stricto, i.e., highest quality 'Buda white ware' is the 'white' type described above.

The above created categories are the result of the investigation of BWW finds from NE Hungary (Castle of Füzér, Komlóska-Pusztavár, Sárospatak cemetery), where the three types also seem to be distinguishable macroscopically. This region was considered as a 'reference territory' for the purposes of our research. White ceramics from other medieval sites closer to Buda show a heterogeneous distribution among these three BWW categories, as well as other material types.

One group of the archaeological sites included in this study is situated close to our 'reference territory', e.g., Felsőzsolca-Várdomb, Hejőkürt-Cifrahát, and Mezőkeresztes-Lucernás. Macroscopically, potsherds of BWW and also a so-called 'beige' type were distinguished in the ceramic assemblage of these sites. The latter type was interpreted as a local version (Simonyi 2005). Following the archaeometric investigation, the BWW fragments were identified as belonging to the 'white' type, and the beige sherds as belonging to the 'yellowish white' and 'greyish white' types of BWW.

Another group of the archaeological sites included in this study is located close to Buda, e.g., the Üllő-1, Üllő-2, Üllő-7, Maglód-1, Gyál-9, and Vecsés-36 sites. White ceramic fragments found at these sites can be determined almost entirely as belonging to the 'white' type of BWW. The only exception is Üllő-1, which has white pottery of the 'greyish white' type.

Our suggestion is that the 'beige' group is identical with the 'yellowish white' type of BWW. According to the former interpretation of the so-called 'beige' ceramics, their first appearance was (18) estimated to be in the first half of the 14th century (Wolf 1986, 157). However, the earliest 'yellowish white' specimens from Komlóska-Pusztavár and Castle of Füzér are dated to the first half of the 13th century. A continuous production of the 'yellowish white' type can be assumed to have been in place until the 15th century. In addition to the extension of the dating of the 'beige' or 'yellowish white' ware, another suggestion is that this type of BWW was common exclusively in the eastern part of the Kingdom.

At this stage of our research, it was not possible to collect comparative raw material samples from the regions suggested by the ethnographic sources. However, a kaolinitic clay source of white firing colour is mentioned by Kiss (2006). This is partly in contradiction with our results regarding the rather illitic clay component of the investigated BWW.

## 7. Conclusion

The present paper summarises the main questions concerning the medieval (12th–14th centuries) 'Buda white ware' raised during the last five decades of archaeological research in Hungary. Including the archaeological records of the recent excavations, research on BWW – which was formerly considered as a homogeneous group – reached the stage where further scientific analyses appeared necessary to answer questions concerning the source(s) of the raw material(s). Our preliminary results indicate that BWW can no longer be considered as a homogeneous ceramic category. At least three types of raw materials – and probably of pottery-making techniques – have to be assumed for the understanding of the manufacturing and distribution processes of the Arpadian age BWW. Some of the investigated ceramic fragments can be classified into the three material types suggested by our research, while in some cases the sherds look like lower quality imitations made from almost the same raw clays.

These results commend further study of this peculiar white ware of the medieval Hungarian Kingdom.

Further and detailed petro-mineralogical and geochemical investigations of more BWW samples will be essential. It is necessary to obtain more evidence concerning the classification of BWW reported here both from a macroscopic and a microscopic-instrumental point of view. For comparison, the study of excavated materials from Buda, the capital of the Kingdom, other sites (e.g. Muhi, Ónod), and of potential raw materials will contribute to the better understanding of the observed archaeological context. In addition, white pottery from the 18th–19th centuries should also be investigated. This can be useful since the market region of the 18th–19th century white pottery known from Hungarian ethnography and the distribution area of the Arpadian age white ware are highly overlapping. Correlating these data by tracking back the ethnographic information further to the past might render these two similar pottery groups comparable. For this purpose, it is our future task to collect written sources and to observe the chronology of BWW in detail (from the point of view of vessel shapes), a task which requires a standardised description of the archaeological records of BWW. This could be followed by the revision of the distribution map of the Arpadian age white pottery. In addition, previous and recent archaeological data on the Slovak sites of the era could be included in this study.

## References

Cabello J. and Feld I., 1980, Jelentés a füzéri vár 1977. évi kutatásáról. Bericht über die Erforschung der Burg von Füzér im Jahre 1977. *Archeológiai Értesítő*, 107, 214-225.

Feld I., 1984, Jelentés a salgói vár 1981–83. évi régészeti kutatásáról. Bericht über die Ausgrabung in der Burg Salgó in den Jahren 1981–1983. *Nógrád Megyei Múzeumok Évkönyve*, 10, 213-264.

Gál-Mlakár V., 2007, Komlóska-Pusztavár régészeti feltárásának eredményei. Archaeological investigation at Komlóska-Pusztavár. *Herman Ottó Múzeum Évkönyve*, 46, 87-113.

Gál-Mlakár V., 2009, A 13–14. századi kerámia kutatástörténete Északkelet-Magyarországon. *MARISIA (Maros Megyei Múzeumok Évkönyve), Studiişi materiale, Arheologie*, 29, 165-185.

Gömöri J., 1970, Beszámoló a sárospataki róm. kat. templom mellett 1968-ban végzett ásatásról. Excavations beside the Sárospatak Gothic Church in 1968. *Herman Ottó Múzeum Évkönyve*, 9, 109-128.

Holl I., 1956, Adatok a középkori magyar fazekasság munkamódszereihez. Beiträge zu den Arbeitsmethoden der ungarischen Töpferei des mittelalters. *Budapest Régiségei*, 17, 177-194.

Holl I., 1963, Középkori cserépedények a budai várpalotából. Mittelalterliche Keramik aus dem Burgpalast von Buda (13–15 Jahrhundert). *Budapest Régiségei*, 20, 335-394.

Hosso J., 1983, Prehlad vyvoja stredovekej keramiky na Slovensku. *Archeologica Historica*, 1983/8, 215-231.

Höllrigl J., 1930, Árpádkori keramikánk I. Fenékbélyeges edények. *Archeológiai Értesítő*, 44, 142-169.

Kárpáti Z., 2002, Árpád-kori majorság Budaújlak területén. Árpádian Age Manor in the Area of Budaújlak. *Budapest Régiségei*, 35, 587-617.

Kiss E., 2006, A gömöri fazekasok által használt nyersanyagokról. About Raw Materials Used by Potters in County Gömör. In *Az agyagművesség évezredei a Kárpát-medencében*, (eds. Sz.A. Holló and J. Szulovszky), 155-166, Budapest and Veszprém.

Kresz M., 1960, Fazekas, korsós, tálas. Néhány szempont fazekas-központjaink kutatásához és összehasonlításához. *Ethnographia*, 71, 297-379.

Laszlovszky J., Pusztai T. and Tomka G., 1997, Muhi –Templomdomb. In *Utak a Múltba: az M3-as autópálya régészeti leletmentései=Paths into the past: rescue excavations on the M3 motorway*, (eds. T. Racky, T. Kovács and A. Anders), 144-150, Magyar Nemzeti Múzeum, Eötvös Loránd Tudományegyetem Régészettudományi Intézet, Budapest.

MRT 9: Dinnyés I., Kővári K., Kvassay J., Miklós Zs., Tettamanti S. and Torma I., 1993, Magyarország Régészeti Topográfiája 9. Pest megye régészeti topográfiája XIII/2. A szobi és a váci járás.

Parádi N., 1963, Magyarországi pénzleletes középkori cserépedények (XI-XVII. század). Münzfunde Hortende Mittelalterliche Gefässe in Ungarn (11–17. Jh.). *Archeológiai Értesítő*, 90, 205-251.

Rácz M., 2006, A sámsonházi Fejérkő régészeti kutatása (2004–2005). Archäologische Untersuchung in der Burg Fejérkő von Sámsonháza (2004–2005). In *"Gondolják, látják az várnak nagy voltát..." Tanulmányok a 80 éves Nováki Gyula tiszteletére*, (eds. Gy. Kovács and Zs. Miklós), 237-246, Budapest.

Rácz T.Á., 2010, Árpád-kori települések szerkezetének sajátosságai Pest megyében. Kutatások az M0-s autópálya és a 4-es számú elkerülő főút nyomvonalában. Characteristics of the Structure of Árpádian-era Settlements. Investigations Along the Line Followed by the M0 Motorway and the Highway 4 Bypass. In *A középkor és a kora újkor régészete Magyarországon. Archaeology of the Middle Ages and the Early Modern Period in Hungary*, (eds. E. Benkő and Gy. Kovács), 69-79, Budapest.

Simon Z., 2006, *Füzér vára*, Budapest.

Simonyi E., 2003, Előzetes jelentés a felsőzsolca-várdombi ásatásról (1992–2001). Report on the excavations at Felsőzsolca-Várdomb (1992–2001). *Herman Ottó Múzeum Évkönyve*, 42, 109-133.

Simonyi E., 2005, Adatok a 10–13. századi kerámiakészítéshez Északkelet-Magyarországról. Data on Pottery from the 10th–13th Century in North-East Hungary. *Herman Ottó Múzeum Évkönyve*, 44, 37-55.

Takács M., 1993, Árpád-kori településrészlet Kajárpéc-Pokolfadombon. Arpadenzeitlicher Siedlungsteil von Kajárpéc-Pokoldomb. *Communicationes Archaeologicae Hungaricae*, 201-228.

Takács M., 2000, A győr-homokgödröki 10–11. századi temetőrészletek és középkori település. Fragment of cemeteries from the 10–11th centuries and medieval settlement at Győr-Homokgödrök. In *A népvándorláskor kutatóinak kilencedik konferenciája*, (eds. T. Petercsák and A. Váradi), 253-309, Heves Megyei Régészeti Közlemények 2, Eger.

Terei Gy., 2010, Az Árpád-kori Kána falu. Kána, a Village from the Arpadian Era (Twelfth–Thirteenth Centuries). In *A középkor és a kora újkor régészete Magyarországon. Archaeology of the Middle Ages and the Early Modern Period in Hungary*, (eds. E. Benkő and Gy. Kovács), 81-111, Budapest.

Tomka G., 2004, *Északkelet-Magyarország kora újkori kerámiája*, PhD Thesis, ELTE BTK, Budapest.

Wolf M., 1986, Középkori kerámia leletek Mezőnyárádon. Mittelalterliche Keramische Funde in Mezőnyárád. *Communicationes Archaeologicae Hungaricae*, 147-159.

Wolf M., 1999, Árpád-kori település Hejőkeresztúr határában. A Népvándorláskor Fiatal Kutatói 8. Találkozójának Előadásai, Veszprém., 166-179.

# Ceramic technology between the Final Bronze Age and the First Iron Age in NE Italy: The case of Oppeano (Verona)

**Massimo Saracino**[a], **Lara Maritan**[b] **and Claudio Mazzoli**[b]
[a]Independent researcher, Verona, Italy
[b]Department of Geosciences, University of Padova, Via G. Gradenigo 6, 35131 Padova, Italy

**Abstract** - The archaeometric study of pottery from the settlement of Oppeano (Verona, NE Italy) indicates important changes in production technology taking place between the 10th and 5th centuries BC. Ceramic production in the Final Bronze Age was characterised by the use of grog, as in other coeval sites of the southern Adige and eastern Po plains. From the beginning of the Iron Age, the number of recipes increased considerably, the use of grog gradually decreased, and cases of importation are also attested. These changes were probably due to increasing socio-economic complexity and the introduction of new production techniques.

## 1. Introduction

The production technology of pottery from the Final Bronze and First Iron Ages in the Veneto region (NE Italy) has been only partially investigated from an archaeometric viewpoint, as a result of studies carried out on both micro-regional areas, particularly in the southern Verona and Rovigo areas (Jenkins *et al.* 1999; Jones *et al.* 2002; Saracino 2006; Saracino *et al.* 2006), and isolated settlements and specific ceramic classes, as in the case of Concordia Sagittaria (Bertelle *et al.* 2001), Montebelluna (Bianchin Citton *et al.* 2000), Montagnana (Crivellari *et al.* 2007), and Este (Maritan 2001) (Fig. 1a). These studies indicate that the coarse ware (known as '*impasto*') of this period was locally produced with raw materials available within the hydrographical basin on which the settlement lies, and that the paste was often tempered with grog.

Chemical analyses also covered a set of 'Italo-Mycenaean' potsherds from several settlements in the region (Frattesina, Montagnana, Fabbrica dei Soci, Castello del Tartaro, Fondo Paviani), dated to the Late Helladic III C, revealing probable importation from both the Aegean and southern Italy (Jones *et al.* 2002).

In contrast, during the First Iron Age, the temper was almost completely substituted by rock inclusions collected from coarse alluvial deposits or locally available outcrops, such as the Euganean Hills in the case of Este (Maritan 2001). As for fine ware, only a few archaeometric studies have been carried out in this area (Maritan 1998, 2004; Maritan *et al.* 2005), investigating *ceramica zonata*, Etruscan-Padan and grey pottery, respectively. The studies report that these ceramic classes were locally produced, and that short-distance exchanges took place between regional centres, such as those of grey and Etruscan-Padan pottery between Padua and Este. Recent archaeometric studies carried out using LIBS and Raman spectroscopy on *ceramica zonata* from several tombs of the Este necropolis dated to the 6th–4th centuries BC have focused on the provenance of sherds and the nature of the black and red colouring (Lenzi 2008), partially confirming older analyses (Biscontin *et al.* 1984).

In this framework, the settlement of Oppeano is an interesting case, due to the presence of Final Bronze Age to First Iron Age pottery (Guidi 2008), the archaeometric analysis of which can supply important information regarding production technology and possible exchanges with other regional and extra-regional centres. The settlement of Oppeano is located about 20 km south-east of Verona, on an alluvial plateau 82 hectares across and elliptic in shape, surrounded by alluvial Holocene plain (Fig. 1a). The terrace on which the settlement lies is south of the spring-water line, on an ancient alluvial cone of the Adige river, composed of sandy-, silty- and clayey-sized deposits and pit lenses on the south-eastern side of the plateau (Fig. 1b). The terrace rises between 26 and 30 m above sea level, about 2 to 5 m from the surrounding plain (Figs. 1c, d), and is composed of the steps which are very common in the high and middle Verona plain, and which are especially common where river erosion was particularly incisive (Balista 2004; Rioda 2008). The southern slope of the plateau, which formed during the late Pleistocene (Sorbini *et al.* 1985) is of lower altitude than the northern one, and was shaped by the

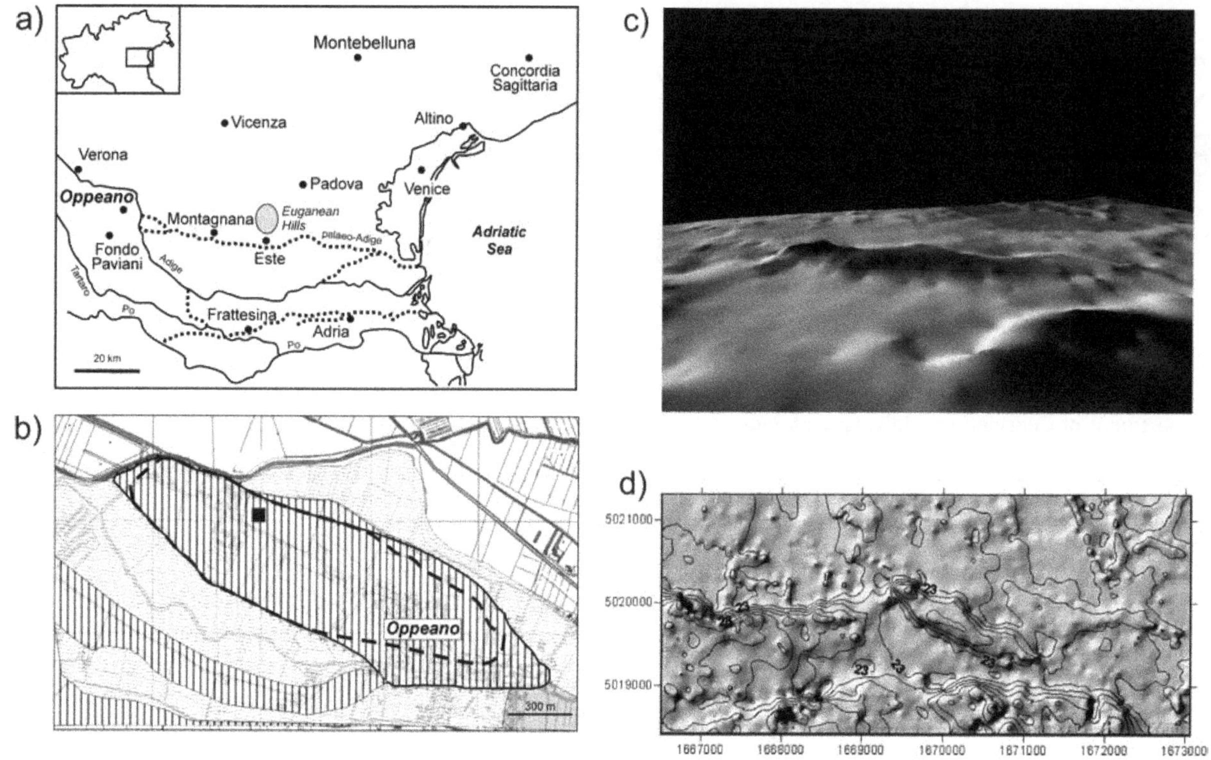


**Figure 1.** a) Geographic location of Oppeano in the eastern Po plain; main rivers and palaeo-rivers also indicated; b) geomorphological sketch of the alluvial plateau of Oppeano (Rioda 2008, modified); black outline: settlement area; black box: excavation from which the analysed potsherds were collected; c-d) digital terrain model and contour line of the Oppeano settlement, processed by ITABC – CNR Rome.

Adige during the early Holocene. A 'defensive system' (articulated in three main phases), originating in the First Iron Age, has been identified next to the older slope (Balista 2004; Saracino 2004). The discovery of the settlement and of several necropoles at Oppeano was documented since the 18th century, but it is only from the last decades of the 19th century that the area has been archeologically investigated (Ferrari 2008). Intensive archaeological research carried out since 2000 by the University of Verona in collaboration with the Veneto Archaeological State Antiquity Office, both through surveys on the settlement and necropoles areas and excavation campaigns in the north-western part of the plateau (Fig. 1b), indicates that the area was continuously settled from the Final Bronze Age to the First Iron Age, from the 10th to the 5th centuries BC (Guidi and Salzani 2008). The distribution of pottery in the Final Bronze and First Iron Ages, according to 'leopard-spotted' scattering, corresponds to that detected in many other proto-urban centres in central and northern Italy (Guidi 2008; Guidi and Saracino 2008). During the First Iron Age (from the 6th century BC), the occupation expanded over most of the plateau, assuming the dimensions of a 'town'. The importance and the socio-cultural vitality of the centre during this period are also attested by the increasingly richer funeral sets, the construction of a complex defensive system along the southern slope, and the increasingly frequent occurrence of Attic, black-glazed, and Etruscan-Padan pottery, inscriptions borrowed from Northern Etruscan models, and the higher concentration of manufacturing (i.e., metallurgy and pottery).

The ceramic repertoire found during the excavation of 2003 (Guidi *et al.* 2008; Saracino and Sboarina 2010), the subject of this research, is mainly composed of common ware, showing various surface treatments and colours (Saracino 2009), and of a few findings of *ceramica zonata*, a wheel-made type of pottery produced in the ancient Veneto region from the 6th century BC onwards and characterised by alternating black and red coloured bands on the outer surface. Rare fragments of fine ware, typical of various Italian and Mediterranean cultures, were also found and archaeologically interpreted as commercial products and/or gifts for the aristocracy (Saracino 2009).

The present research is an archaeometric study of a set of 37 potsherds found at the site, with the main aims of determining the production technology and possible cases of importation during the period between the Final Bronze and Second Iron Ages, over a timespan of six centuries. This study represents preliminary research on the Oppeano pottery, which needs to be integrated with future work on a larger sample to increase its representativeness. The selection of samples presented in this paper was based on the effective frequency and identification of archaeological structures in this and several other parts of the settlement. The older layers (Final Bronze Age) are less well preserved and are disturbed by later structures (i.e., Early, and mainly First Iron Age) (Morandini and Saracino 2008; Saracino 2008). Fragments of common ware, represented by open and closed shapes such as bowls, *ollae*, biconical pots and *dolia* (Figs. 2 and 3), and of fine ware, represented by Etruscan-Padan pottery, *ceramica zonata*, a small amphora

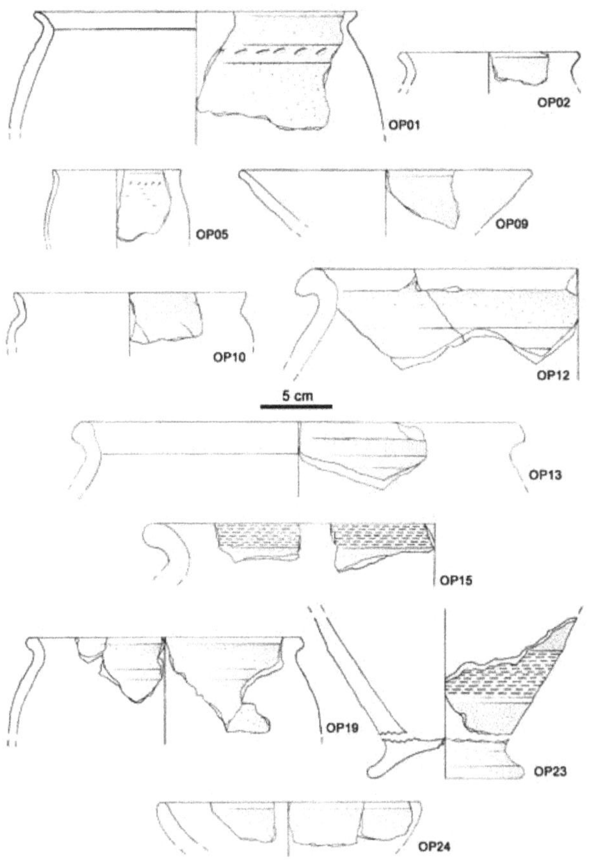


**Figure 2.** Archaeological drawing of some of the studied samples: fragments dating to the Early Iron Age (OP01, OP02, OP05, OP09, OP10) and First Iron Age (OP12, OP13, OP15, OP19, OP23, OP24).

with a horizontal grey-brownish band and a wavy line motif, of uncertain origin, and a black-glazed *kylix* (Fig. 3) were studied from the petrographic and mineralogical viewpoints. A secondary deposition sherd of final Neolithic Age with scratched/impressed decoration (end of the fourth millennium BC) was also studied in order to define provenance and understand the possible oldest human occupation of the site. In addition, four samples of clay, representing locally available material, were collected immediately outside the plateau, from the Holocene alluvial deposits of the Adige, and subsequently analysed (Fig. 1b).

Petrographic analysis was carried out following the terminology and descriptive scheme proposed by Whitbread (1986, 1989, 1995). The surfaces of the potsherds were removed with a micro-drill in order to eliminate possible surface contamination. Samples were then reduced to powder in an agate mortar. The mineralogical composition was obtained by X-ray powder diffraction (XRD) on a PANalytical X'Pert PRO diffractometer in Bragg-Brentano geometry, equipped with a Cu X-ray tube (40 kV and 30 mA, Cu-K$\alpha$ radiation) and an X'Celerator detector, in the 3–70° 2$\theta$ range with 0.02° step size and counting 1 s per step. The results were compared with archaeometric data from other coeval sites referable to the same culture, in order to define analogies in pottery production and reconstruct possible trade routes.

## 2. Results and discussion

The petrographic analysis shows that most of the samples are characterised by a homogeneous groundmass, optically active, with a striated birefringent-fabric (*b-fabric*), sometimes with clay pellets, and few pores, mainly represented by some vughs with a size ranging from dozens to hundreds of microns and scarce millimetric channels. According to the textural features and mineral-petrographic nature of the inclusions, they were divided into seven main petrographic groups (Fig. 4), a brief description of which is provided below (Table 1).

Samples belonging to *group 1* (*alluvial sand-rich potsherds*) contain abundant inclusions of quartz, showing single-space distribution, angular to sub-rounded shape, and occurring as single crystals, polycrystalline aggregates of metamorphic origin, and chert, associated with scarce crystals of plagioclase, K-feldspar, biotite, muscovite, epidote, a few opaque minerals, and occasionally fragments of micritic limestone, rhyolite, crystals of rutile, tourmaline, garnet, zircon, pyroxene, chlorite, and/or amphibole (Fig. 4a). On the basis of the abundance and maximum grain size, two subgroups were distinguished, with a 50:50 c:f (coarse:fine) ratio and inclusions up to approximately two millimetres (samples: OP01, OP08, OP09, OP30), and 30:70–40:60 c:f ratio and inclusions up to approximately one millimetre (samples: OP05, OP23, OP24, OP34, OP35, OP37), respectively.

The samples of *group 2* (*alluvial sand- and calcite-rich potsherds*) (samples: OP07, OP16, OP18, OP20) are characterised by abundant (40:60 c:f ratio) micron-sized,

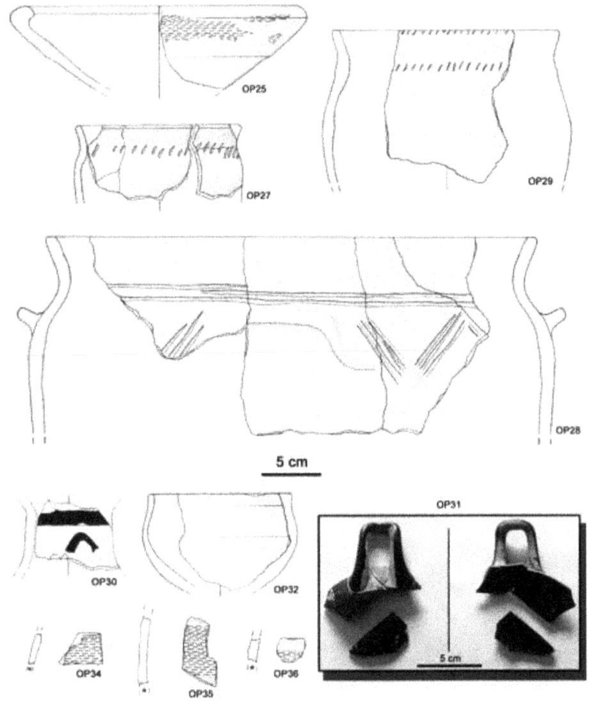


**Figure 3.** Archaeological drawing of some of the studied samples: fragments dating to the Second Iron Age (OP25) and Final Bronze Age (OP27, OP28, OP29), small amphora (OP30), black-glazed *kylix* (OP31), and *ceramica zonata* (OP34, OP35, OP36).

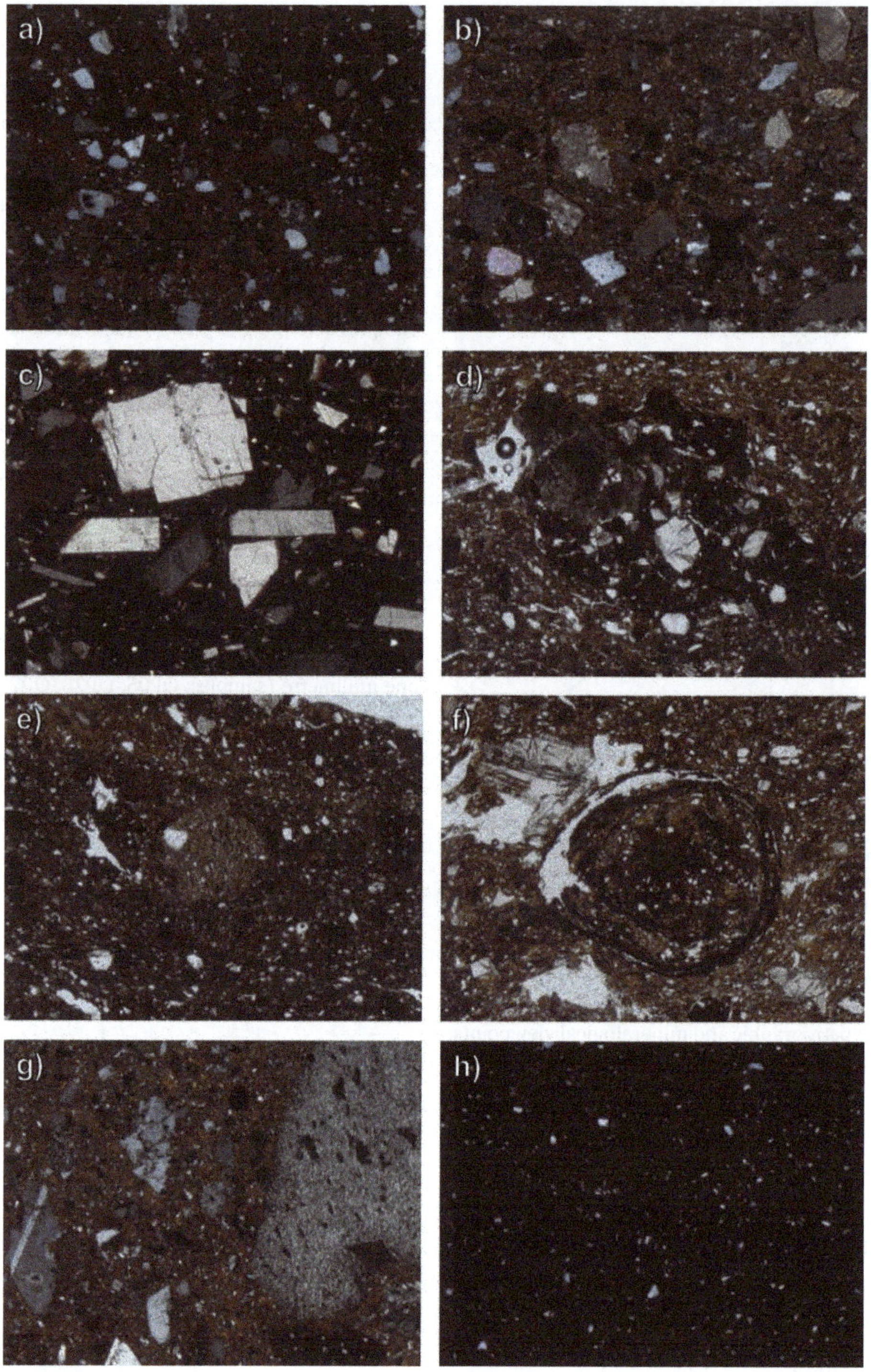


**Figure 4.** Microphotographs of representative samples from the petrographic groups identified: a) group 1 (sample OP34); b) group 2 (sample OP16); c) group 3 (sample OP04); d) group 4 (sample OP13); e) group 5 (sample OP10); f) clay pellet in sample OP12; g) group 6 (sample OP15); h) group 7 (sample OP12). Image width: 3.5 mm. Crossed polars: images a, b, c, g, h; parallel polars: images d, e, f.

angular fragments of calcite (Fig. 4b), as well as the same type of inclusions as group 1.

The quantity of calcite fragments increases in *group 3* (*calcite-rich potsherds*) (samples: OP04, OP19, OP21, OP22), and the alluvial sand component decreases drastically in comparison with the previous groups (Fig. 4c).

Potsherds belonging to *groups 4 and 5* are characterised by fragments of grog up to millimetric size (Figs. 4d, e). This type of temper was distinguished from other possible

**Table 1.** Schematic petrographic description of samples. Groundmass: homogeneity: HOM = homogeneous; Optical state: A= active, IN = inactive; *b-fabric*: ST = striated; OR = oriented, IS = isotropic; Pores: shape: Ch = channels, PV = planar voids, Ve=vesicles, Vu = vughs; Inclusions: c:f related distribution: CS = close-spaced, SS = single-spaced, DS = double-spaced, OP = open-spaced; grain-size distribution: B= bimodal, U= unimodal; shape: A= angular, SA = sub-angular, SR = sub-rounded, R=rounded; composition: Qtz = quartz, Cc = calcite, Pl = plagioclase, Kfs =K-feldspar, Op = opaque minerals, Bt = biotite, Ms = muscovite, Rt = rutile, Ep = epidote, Zr = zircon, Grt = garnet, Px = pyroxene, Am = amphibole, Tor = tourmaline, Chl=chlorite, Qtz met = metamorphic quartz, Lim = limestone, Ry = rhyolite, Tr = trachyte, Met = fragments of metamorphic rocks; abundance: xxxxx = predominant (>70%), xxxx = dominant (50–70%), xxx = frequent (30–50%), xx = common (15–30%), x= few (5–15%),++ = very few (2–5%), + = rare (< 5%). Ceramic types also listed.

| | | | Groundmass | | | Pores | | | | | | | Inclusions | | | | | | | | Inclusion composition | | | | | | | | | | | | | | | | | | | | | | |
|---|---|---|---|---|---|---|---|---|---|---|---|---|---|---|---|---|---|---|---|---|---|---|---|---|---|---|---|---|---|---|---|---|---|---|---|---|---|---|---|---|---|---|---|
| | | | | Optical state | | | | Shape | | | | | | | | | Shape | | | | | | | | | | | | | | | | | | | | | | | | | | |
| Group | Sample | Ceramic typology | Homo-genity | (b-fabric) | Preferred orientation | % | Max size (μm) | Ch | PV | Ve | Vu | c:f ratio | c:f related distribution | Grain-size distribution | Max size (μm) | A | SA | SR | R | Sample | Qzt | Cc | Pl | Kfs | Op | Bt | Ms | Rt | Ep | Zr | Grt | Px | Am | Tor | Chl | Chert | Qtz met | Lim | Ry | Tr | Met | Grog |
| 1 | OP01 | Ovoid pot | HOM | A (ST) | OR | 5 | 2000 | * | | | * | 50:50 | CS | U | 900 | * | * | | * | OP01 | xxx | xx | ++ | + | x | + | + | + | + | + | + | | | | + | x | xx | xxxx | + | | | |
| 5 | OP02 | Ovoid pot | HOM | A (ST) | OR | 3 | 700 | * | | | * | 30:70 | SS | B | 2000 | * | * | | * | OP02 | xxx | | + | | ++ | + | ++ | | + | | + | + | | + | + | xx | xx | xx | ++ | | | x |
| 5 | OP03 | End-conical bowl | HOM | A (ST) | OR | 5 | 500 | * | | | * | 40:60 | SS | B | 2000 | * | * | | | OP03 | xxx | | ++ | + | x | + | x | | + | | | | + | + | + | x | x | | | | | xxx |
| 3 | OP04 | Ovoid pot | HOM | A (ST) | OR | 3 | 400 | * | | | | 50:50 | CS | B | 2000 | * | * | | | OP04 | x | xxxxx | | | ++ | + | ++ | + | + | | | + | + | + | + | | + | | | | | |
| 1 | OP05 | Small ovoid pot | HOM | A (ST) | OR | 3 | 700 | * | | | * | 50:50 | SS | U | 700 | * | * | | * | OP05 | xxx | | ++ | ++ | x | + | ++ | | | | | | | | | xxx | xxx | | | | | |
| 2 | OP07 | Ovoid pot | HOM | A (ST) | IS | 10 | 1000 | * | | | * | 50:50 | CS | B | 1200 | * | * | * | | OP07 | xx | | + | ++ | x | + | + | + | + | | + | | | | | x | xx | xxx | | | | |
| 1 | OP08 | Decorated sherd | HOM | A (ST) | OR | 10 | 1000 | * | | | * | 40:60 | SS | U | 3000 | * | * | * | * | OP08 | xxx | xx | + | x | x | + | ++ | + | | | + | + | | | | xx | xxx | | x | | | |
| 1 | OP09 | End-conical bowl | HOM | A (ST) | OR | 5 | 600 | | | * | * | 40:60 | SS | B | 2500 | * | * | * | * | OP09 | xxx | | ++ | | x | | x | | + | | + | + | + | + | + | xxx | xxx | x | + | | | |
| 5 | OP10 | Bowl | HOM | A (ST) | OR | 10 | 1200 | * | * | | * | 50:50 | SS | B | 3000 | * | * | | | OP10 | xxx | | + | + | xx | + | x | | + | | + | | | | + | x | x | ++ | ++ | | | xxxx |
| 5 | OP11 | Ovoid *dolium* | HOM | A (ST) | IS | 10 | 1000 | * | | | * | 40:60 | SS | B | 1500 | * | * | | * | OP11 | xx | xx | | + | x | + | ++ | | + | + | + | | | | | x | xxx | x | | | | x |
| 4 | OP12 | Ovoid *dolium* | HOM | A (ST) | OR | 3 | 600 | | | | * | 30:70 | SS | B | 2500 | * | * | | * | OP12 | xx | xxx | | | x | + | ++ | + | ++ | | | | | + | | x | x | xx | ++ | | | x |
| 4 | OP13 | Ovoid *dolium* | HOM | A (ST) | OR | 10 | 1800 | * | | | * | 40:60 | SS | B | 2300 | * | * | | * | OP13 | xx | xxx | | ++ | x | + | x | + | | + | | | | + | | x | x | x | | | x | xx |
| 4 | OP14 | Ovoid *dolium* | HOM | A (ST) | IS | 5 | 1600 | * | * | | * | 40:60 | SS | B | 2100 | * | * | | * | OP14 | xxx | xxx | | | x | + | ++ | + | + | | | | | + | | ++ | xx | x | | | | x |
| 6 | OP15 | Ovoid *dolium* | HOM | A (ST) | IS | 3 | 1000 | | | | * | 40:60 | SS | B | 2400 | * | * | | * | OP15 | xxx | | xx | x | x | ++ | ++ | + | + | + | | | + | + | + | | x | | | xxx | | ++ |
| 2 | OP16 | Ovoid *dolium* | HOM | A (ST) | OR | 10 | 1800 | * | | | * | 40:60 | SS | B | 2200 | * | * | | * | OP16 | xx | xx | ++ | ++ | ++ | ++ | x | + | | | + | + | | + | + | ++ | xx | xx | ++ | | x | |
| 4 | OP17 | Ovoid *dolium* | HOM | A (ST) | OR | 10 | 2000 | * | * | | * | 30:70 | SS | B | 1500 | * | * | | * | OP17 | xxx | xxx | ++ | ++ | xx | | x | + | | | | + | | | | ++ | x | x | | | | xx |
| 2 | OP18 | Ovoid *dolium* | HOM | A (ST) | OR | 10 | 2500 | * | | | * | 40:60 | SS | B | 2100 | * | * | | * | OP18 | xx | xx | ++ | ++ | ++ | + | ++ | | + | | | + | | | | xx | x | x | x | | x | ++ |
| 3 | OP19 | Ovoid pot | HOM | A (ST) | OR | 3 | 800 | * | | | * | 40:60 | SS | B | 2000 | * | * | | * | OP19 | xx | xxxx | + | | x | + | ++ | + | + | | | | | + | | | | + | | | | |
| 2 | OP20 | Ovoid pot | HOM | IN | OR | 5 | 600 | * | | | * | 40:60 | SS | B | 1800 | * | * | | * | OP20 | xxx | xxx | ++ | + | x | + | ++ | | | | + | + | | | | x | x | x | ++ | | | |
| 3 | OP21 | End-conical pot | HOM | A (ST) | OR | 20 | 2200 | * | | | * | 40:60 | SS | B | 1800 | * | * | | | OP21 | xx | xxxxx | | | x | + | ++ | + | + | + | | + | | + | | | | x | | | | |
| 3 | OP22 | Ovoi pot | HOM | A (ST) | OR | 10 | 500 | * | * | | * | 40:60 | SS | B | 1400 | * | * | | | OP22 | xx | xxxx | | | + | | + | + | | | | | | + | | + | + | x | | | | |

*(Continued)*

**Table 1.** - *continued*

| | | | Groundmass | | | Pores | | | | | | Inclusions | | | | | | | | | Inclusion composition | | | | | | | | | | | | | | | | | | | | | |
|---|---|---|---|---|---|---|---|---|---|---|---|---|---|---|---|---|---|---|---|---|---|---|---|---|---|---|---|---|---|---|---|---|---|---|---|---|---|---|---|---|---|---|
| | | | | Optical state | | | | Shape | | | | | | | | Shape | | | | | | | | | | | | | | | | | | | | | | | | | | |
| Group | Sample | Ceramic typology | Homo-genity | (b-fabric) | Preferred orientation | % | Max size (μm) | Ch | PV | Ve | Vu | c:f ratio | c:f related distribution | Grain-size distribution | Max size (μm) | A | SA | SR | R | Sample | Qzt | Cc | Pl | Kfs | Op | Bt | Ms | Rt | Ep | Zr | Grt | Px | Am | Tor | Chl | Chert | Qtz met | Lim | Ry | Tr | Met | Grog |
| 1 | OP23 | *Ceramica zonata* | HOM | A (ST) | OR | 10 | 600 | * | * | | * | 40:60 | SS | B | 900 | * | * | | * | OP23 | xxx | xx | + | + | x | + | ++ | + | + | + | + | | + | + | | xx | x | ++ | x | | | |
| 1 | OP24 | End-conical bowl | HOM | A (ST) | OR | 15 | 1800 | * | | | | 40:60 | SS | U | 500 | * | * | * | * | OP24 | xxx | x | ++ | ++ | x | ++ | ++ | | + | | + | | | | | xx | xx | | x | | ++ | |
| 1 | OP25 | End-conical bowl | HOM | IN | OR | 15 | 1800 | * | * | | * | 30:70 | SS | U | 600 | * | * | * | * | OP25 | xxx | | + | ++ | x | + | ++ | | + | | + | | | + | | xx | xx | | x | | ++ | |
| 5 | OP26 | Biconic pot | HOM | A (ST) | OR | 15 | 2000 | * | * | | * | 40:60 | SS | U | 1500 | * | * | | * | OP26 | xxx | | + | + | x | + | ++ | | + | | | | | | | xx | x | xx | + | | + | xx |
| 5 | OP27 | Ovoi pot | HOM | A (ST) | OR | 10 | 1200 | * | * | | * | 30:70 | DS | B | 2000 | * | | | * | OP27 | xxx | | + | + | + | | x | | + | + | | + | | + | | x | xx | | | | | xxx |
| 5 | OP28 | Ovoi pot | HOM | A (ST) | IS | 20 | 3000 | * | * | | * | 30:70 | DS | B | 2000 | * | * | | * | OP28 | xxxx | | + | + | xx | + | x | + | | | | | | + | | | + | | | | | xx |
| 5 | OP29 | Ovoi pot | HOM | A (ST) | IS | 10 | 1000 | * | | | * | 30:70 | SS | B | 1500 | * | | | * | OP29 | xxxx | | ++ | + | x | ++ | x | + | + | + | + | | | | | x | xx | | | | + | xx |
| 7 | OP30 | Small amphora | HOM | IN | OR | 5 | 600 | | | | * | 30:70 | DS | B | 2000 | * | * | | | OP30 | xxxx | | + | + | x | ++ | xx | + | | + | + | | | | | ++ | x | | | | | |
| 7 | OP31 | Black-glaze *kylix* | HOM | IN | IS | 1 | 200 | | | | * | 10:90 | OS | U | 50 | * | | | | OP31 | xxxx | | | | x | + | xx | + | | | | | | | | | ++ | | | | | |
| 3 | OP32 | Globular bowl | HOM | IN | IS | 20 | 2000 | | | | * | 30:70 | DS | U | 1200 | * | * | | | OP32 | xx | xxxx | + | | x | | ++ | | | | | | | | | | + | | | | | |
| 1 | OP34 | *Ceramica zonata* | HOM | A (ST) | OR | 20 | 800 | * | * | | * | 30:70 | DS | U | 1200 | * | * | * | * | OP34 | xx | | + | + | ++ | + | ++ | | + | + | + | + | + | | | xx | xx | | xx | | ++ | |
| 1 | OP35 | *Ceramica zonata* | HOM | A (ST) | OR | 15 | 1500 | * | | | * | 30:70 | DS | U | 1200 | * | * | * | * | OP35 | xx | + | + | + | ++ | + | ++ | | + | | + | + | | | | xx | xx | ++ | x | | ++ | |
| 6 | OP36 | *Ceramica zonata* | HOM | A (ST) | OR | 10 | 1100 | * | | | * | 30:70 | DS | U | 1300 | * | * | * | * | OP36 | xxx | | x | ++ | x | + | ++ | + | + | + | | + | | | | ++ | ++ | | | xx | | |
| 1 | OP37 | *Ceramica zonata* | HOM | A (ST) | OR | 20 | 1000 | * | | | | 30:70 | DS | B | 700 | * | * | * | * | OP37 | xx | x | | ++ | x | + | + | | | | + | + | | | | xx | x | ++ | xx | | + | |

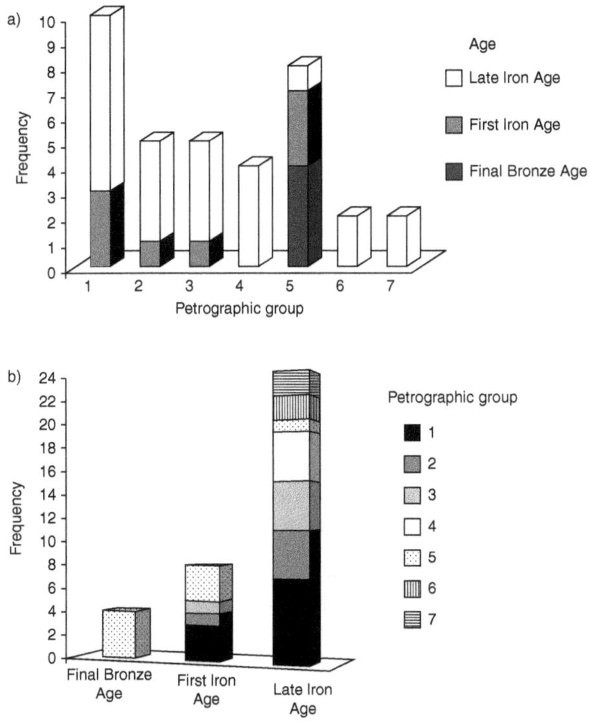


**Figure 5.** Frequency distribution of potsherds according to: a) petrographic group; b) age.

argillaceous inclusions, in particular clay pellets, according to microscopic features, as described by Whitbread (1986). Examined in more detail, the grog is angular in shape, has sharp boundaries, and in many cases shows preferred orientation and neutral optical density, whereas the clay pellets are rounded, with merging boundaries, concentric internal structure, and equant shape (Fig. 4f). In *group 4 (calcite and grog-bearing potsherds)*, recycled pottery is associated with abundant fragments of calcite (samples: OP12, OP13, OP14, OP17); in *group 5 (grog-rich potsherds)* (OP02, OPO3, OP10, OP11, OP26, OP27, OP28, OP29), it constitutes the only type of added temper, associated with more or less abundant sand, representing the naturally occurring coarser fraction of the clay materials. Only two potsherd samples (OP15, OP36) also contain fine sandy-sized inclusions, with a mineral-petrographic composition analogous to that of the previous groups, with sub-rounded fragments of trachyte of some hundreds of microns (*group 6, trachyte-bearing potsherds*) (Fig. 4g). Lastly, although samples OP30 (*oinochoe*) and OP31 (*kylix*) contain different quantities of inclusions, 20% and 10% respectively, they were classified together in *group 7 (fine potsherds)*, as they are both fine-grained and mainly composed of quartz and subordinate opaque minerals, crystals of muscovite, biotite, rutile and, in the case of OP30, also by rare plagioclase, K-feldspar, zircon, garnet and chert (Fig. 4h).

The results of petrographic analysis were used to define the main technological features of this pottery, as well as to constrain its provenance. As such, the textural features of the groundmass, and in particular the oriented *b-fabric*, according to the results of experimental shaping studies by Courty and Roux (1995), indicate that the pastes were wheel-turned. Fragments of grog in the potsherds of two of the groups (4 and 5) indicate that these ceramic pastes were deliberately tempered by the potters. The samples of groups 2 and 3, containing angular sand-sized crystals of calcite, were also produced by mixing base clays with crushed calcite. The potsherds of groups 1 and 6 show a continuous grain-size distribution, and the sub-rounded shape of the coarser fraction is not useful for identifying features that would allow interpreting production recipes. The possible addition of sand to the base clay is very hard to detect and demonstrate. A particular case is represented by fine potsherds (OP30, OP31), which were probably produced with depurated clay materials.

As regards provenance, the presence in many of the samples of groups 1, 2, and 5 of fragments of rhyolite, referable to the Permian volcanites which constitute the porphyric platform of the Trentino Alto Adige region, as well as of metamorphic quartz and minerals related to metamorphic rocks such as garnet, epidote, tourmaline, and zircon, indicate that the clay materials were locally collected from the Adige plain deposits (Jobstraibizer and Mallesani 1973). The lack of any mineral-petrographic marker in the samples of groups 3, 4 and 7 do not allow a delimitation of the provenance of these ceramic objects, although the presence of marble and sparitic limestone in the coarser Adige deposits does not exclude the possibility that the pottery of groups 3 and 4 was locally produced. According to this petrographic evidence, only the samples of group 6 were certainly imported from another production centre. The presence of small fragments of trachyte, a rock outcropping in the area of the Euganean Hills – Padua (Piccoli *et al.* 1981), about 50 kilometres downriver from Oppeano (Fig. 1) constrain these samples

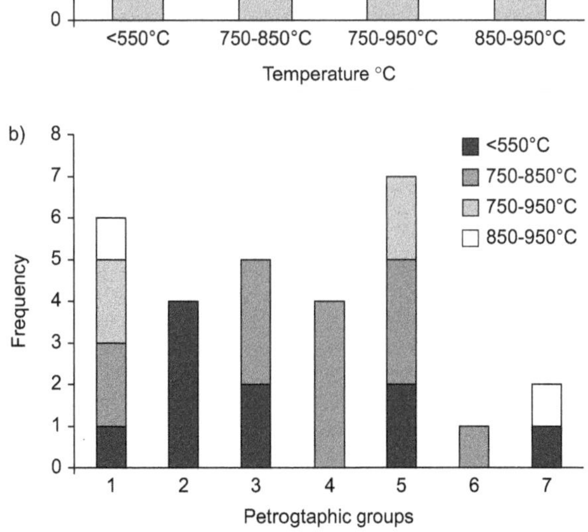


**Figure 6.** Firing temperatures of potsherds, according to: a) mineralogical assemblage; b) petrographic group.

to this area. In particular, the ceramic paste of these samples, which is very similar to the one described by Maritan (1998) for the *ceramica zonata* production at Este, indicates that these objects, a *ceramica zonata* (OP36) and a *dolium* with internal and horizontal red and black stripes (OP15), were imported from there. It is very interesting that other objects, the stylistic features of which correspond to those of the *ceramica zonata* (OP34, OP35), were produced with a different recipe, petrographically corresponding to group 1 (*alluvial sand-rich potsherds*), which may be considered as locally produced at Oppeano.

When examining the distribution of petrographic groups over the three main chronological phases, Final Bronze Age, First Iron Age, and Second Iron Age, it is very interesting to note that, despite the poor statistics, the number of production recipes drastically increased from the Final Bronze Age to the First and Second Iron Ages (Fig. 5). In particular, pottery using exclusively grog temper was produced during the Final Bronze Age, whereas in the Iron Age other types of temper were also used, and many new recipes were introduced in the ceramic production technology at Oppeano.

Petrographic groups are generally not correlated with specific ceramic types, with the exception of *ceramica zonata*, produced according to recipe 1, with the exception of the imported object OP36, which belongs to group 6. This suggests that, at least for the more abundant ceramic types, little care was taken in the selection of the raw materials.

The mineralogical composition of the ceramic sherds is closely related to the nature of the raw materials, maximum firing temperature, and oxygen fugacity during firing (Bimson 1969; Philpotts and Wilson 1994). Therefore, the presence or absence of specific minerals is essential in defining the technological features of production. The differing mineral assemblages identified in the potsherds are listed below:

- quartz, plagioclase, K-feldspar, illite/muscovite, chlorite, ± calcite, ± dolomite;
- quartz, plagioclase, K-feldspar, illite/muscovite, calcite ± haematite;
- quartz, plagioclase, K-feldspar, illite/muscovite;
- quartz, plagioclase, K-feldspar, illite/muscovite, pyroxene.

Quartz is generally the most abundant phase, followed by plagioclase, K-feldspar, and illite/muscovite. Calcite, dolomite, pyroxene, haematite and/or chlorite occur only in some samples.

The XRD analysis of four samples of clay, collected from the surroundings of the Adige plain site and representing locally available raw materials, indicated that they are composed of quartz, dolomite, calcite, illite/muscovite, chlorite, plagioclase and K-feldspar, in differing proportions.

The maximum firing temperatures of the Oppeano potsherds were determined according to the mineralogical association of samples and after comparisons with firing experiments in both oxidising and reducing conditions, on illitic-chloritic calcareous and non-calcareous clays similar in terms of mineralogical composition to locally available clay materials (Maritan 2003; Maritan *et al.* 2006; Nodari *et al.* 2007) (Fig. 6). About one third of the samples were fired at a temperature lower than 550°C, and half of them between 750°C and 850°C (Fig. 6a). These temperatures are determined by the occurrence of chlorite peaks in the former (lower temperature) group and the lack of chlorite and presence of calcite and haematite in the latter (high temperature). Where calcite was lacking, the firing temperature was identified on the basis of the absence of chlorite peaks and the occurrence of those of illite/muscovite. Only in a few cases did the maximum firing temperature fall in the interval 850–950°C, associated with the occurrence of both pyroxene and illite/muscovite. When considering firing temperatures in relation to petrographic groups, it is interesting to note that some correspondences were found (Fig. 6b), despite the poor statistical representativeness of most of the petrographic groups. In particular, the samples of group 2 were fired at a temperature below 550°C, while those of groups 4 and 6 were fired at temperatures between 750 and 850°C. Pottery produced according to recipes 1, 3, 5 and 7 displays a wide range of firing temperatures, from below 550°C to 950°C, probably as the result of poor control of the firing technology. All this evidence suggests specific and differentiated production technologies, in terms of both production recipes and control of firing conditions.

## 3. Conclusions

The mineral-petrographic analysis of a selection of pottery samples from Oppeano (Verona, NE Italy) has provided very important data to define ceramic trade with other coeval regional and distant centres, and to characterise the production technology of this site. Although only preliminarily, the production recipes identified at Oppeano indicate in one case a link between pottery paste and type, as in the case of the *ceramica zonata* locally produced with the recipe corresponding to group 1. In view of these data, a wider investigation of the Oppeano pottery may provide important indications pertaining to the interpretation of functional pottery in relation to its paste type.

This technological change observed at Oppeano cannot be easily associated with other regional and extra-regional situations, due – in terms of the statistical diachronic and synchronic representativeness of regional sites – to the limited number of archaeometric studies on pottery from this period (Final Bronze Age to Second Iron Age). During the Italian Bronze Age, grog was commonly used in those settlements located in sedimentary areas, because of the lack of rock outcrops nearby (Jenkins *et al.* 1999; Jones *et al.* 2002; Levi and Sonnino 2003; Saracino 2006; Saracino *et al.* 2006). For this reason, it is possible that grog was considered as an easily available raw material of good quality.

During the First Iron Age, grog was less frequently used, as it possibly represented a temper related to the past, and from the 6$^{th}$-5$^{th}$ centuries BC onwards it was replaced by sand as well as spathic calcite, probably collected from alluvial sand and pebble deposits. During the same period, important technological changes were introduced, such as wheel throwing and through or up-draught kilns, both of which have been archaeologically confirmed in Oppeano and other settlements in Italy (Iaia 2009; Saracino 2009). The increase in the number of recipes in the First and especially Second Iron Age (Fig. 5b) is interpreted as due to the developments related to socio-economic complexity and increased trade.

During the Final Bronze and First Iron Ages, ceramic production was carried out in household workshops, whereas from the Second Iron Age onwards, the emergence of a well-established aristocracy regulated settlement growth and led to a population increase (Guidi 2008), both of which gave rise to greater demand for products for local use or for trade, as attested in other parts of Europe (Gosden 1985). This demand could be satisfied by setting up more workshops in special areas of the settlement (a kind of artisan district, as, for example, in the settlements of Padua and Oppeano, or in those at Trebbio-Sansepolcro, in central Italy – De Min *et al.* 2005; Guidi and Salzani 2008; Iaia and Moroni Lanfredini 2009), where several full-time potters could practise their technological and cultural know-how. Nevertheless, we must also take into account the factor of importation. Petrographic and stylistic analyses indicate the importation of both *dolia* and *ceramica zonata* pottery from the Euganean Hills area, probably from Este, and of fine ware, a black-glazed *kylix* and a fine small decorated amphora from unidentified localities. The lack of any petrographic marker prevents us from assigning a particular area/district of provenance for these objects. Only a detailed analysis of the chemical composition, which was not ascertained due to the small quantity of material available, can provide information on the origin of these materials. In particular, the *kylix* may be related to a central Italic production or more probably to the local Greek-Etruscan *emporium* of Adria (F. Wiel-Marin personal communication). The small amphora seems to be made with a naturally fine clay, without any depuration process. From a typological point of view, it recalls similar products found at the Ca' Cima necropolis (Adria, NE Italy) (S. Bonomi personal communication) but its provenance cannot at present be confirmed.

## References


Balista C., 2004, Oppeano 2001–2002: la geomorfologia e le difese meridionali dell'area insediativa dell'età del Ferro. In *Ricerche archeologiche in Italia e in Siria: atti delle Giornate di studio, Verona, 6–7 maggio 2002*, (eds. A. Guidi and S. Ponchia), 27-36, Sargon, Padova.

Bertelle M., Calogero S., Leotta G., Stievano L., Salerno R. and Segnan R., 2001, Firing Techniques of the Impasti from the Protohistoric Site of Concordia Sagittaria (Venice). *Journal of Archaeological Science*, 28, 197-211.

Bianchin Citton E., Levi S.T. and Magrini M., 2000, Osservazioni archeometriche sul vasellame di scarto della prima età del Ferro di Montebelluna (TV) – Via Monte Civetta. In *Proceedings of Conference on Le scienze della terra e l'archeometria, Este 26-27/02/1999*, (eds. C. D'Amico and C. Tampellini), 31-38, Patron Editore, Este.

Bimson M., 1969, The examination of ceramics by X-ray powder diffraction. *Studies in Conservation*, 14, 85-89.

Biscontin G., Driussi G., Leonardi G., Moretti G., Quartarone G., Valle A. and Zingales A., 1984, Indagine chimico-fisiche su alcuni manufatti della necropoli paleoveneta del Piovego-Padova. *Archeologia Veneta*, 7, 321-340.

Courty M.A. and Roux V., 1995, Identification of wheel throwing on the basis of ceramic surface features and microfabrics. *Journal of Archaeological Science*, 22, 17-50.

Crivellari F., Bianchin Citton E., De Vecchi G., Fabbri B., Gualtieri S. and Rigoni A.N., 2007, Alcune classi di manufatti non vascolari dell'abitato di Montagnana-Borgo S. Zeno. *Materiali argillosi non vascolari: un'occasione in più per l'archeologia. Atti della IX Giornata di Archeometria della Ceramica (Pordenone 18–19 Aprile 2005)*, 87-94, Consiglio Nazionale delle Ricerche, Faenza.

De Min G., Gamba M., Gambacurta G., Ruta Serafini A. (eds.), 2005, *La città invisibile. Padova preromana. Trent'anni di scavi e ricerche*, Tipoarte Industrie Grafiche, Ozzano Emilia.

Ferrari C., 2008, Cronistoria degli scavi e delle scoperte archeologiche. In *Oppeano: vecchi e nuovi dati sul centro protourbano*, (eds. A. Guidi and L. Salzani), 10-15, Quaderni di Archeologia del Veneto Serie Speciale 3, Giunta regionale del Veneto, Venezia.

Gosden C., 1985, Gifts and kin in Early Iron Age Europe. *Man*, 20, 475-493.

Guidi A., 2008, Per un tentativo di sintesi dei dati sul popolamento. In *Oppeano: vecchi e nuovi dati sul centro protourbano*, (eds. A. Guidi and L. Salzani), 165-172, Quaderni di Archeologia del Veneto Serie Speciale 3, Giunta regionale del Veneto, Venezia.

Guidi A., Salzani L. (eds.), 2008, *Oppeano: vecchi e nuovi dati sul centro protourbano*, Quaderni di Archeologia del Veneto Serie Speciale 3, Giunta regionale del Veneto, Venezia.

Guidi A. and Saracino M., 2008, New data on the proto-urban site of Oppeano (Verona, North-Eastern Italy). *Ritus und Religion in der Eisenzeit, Beiträge zur Sitzung der AG Eisenzeit während der Jahrestagung des Mittel- und Ostdeutschen Verbandes für Altertumsforschung e.V. in Halle an der Saale* 2007, (ed. C. Eggl, et al.,), 135-142, Beiträge zur Ur- und Frühgeschichte Mitteleuropas 49, Beier & Beran, Langenweißbach.

Guidi A., Candelato F. and Saracino M., 2008, Il popolamento del territorio veronese nell'età del Ferro. *I Veneti antichi: novità ed aggiornamenti*, 15-45, Cierre Edizioni, Sommacampagna.

Iaia C., 2009, Una panoramica sugli impianti di cottura per ceramica nella protostoria italiana, fino al VI seolo a.C. In *L'Età delferro a Sansepolcro. Attività produttive e ambiente nel sito di Trebbio*, (eds. C. Iaia and A. Moroni Lanfredini), 55-72, Aboca Museum Edizioni, Perugia.

Iaia C., Moroni Lanfredini A. (eds.), 2009, *L'Età delferro a Sansepolcro. Attività produttive e ambiente nel sito di Trebbio*, Aboca Museum Edizioni, Perugia.

Jenkins D., Williams J.L. and Levi S.T., 1999, Petrographic analysis in provenancing and classification of sherds: the lower Po valley – a case study. In *Proceedings of the 4th European Meeting on Ancient Ceramics on Estudis arqueològics i arqueomètrics, Andorra 12-15/11/1997*, (eds. M. Maggetti and M. Vendrell-Saz), 175-181, Govern d'Andorra Ministeri de Turisme I Cultura, Andorra.

Jobstraibizer P. and Mallesani P., 1973, I sedimenti dei fiumi veneti. *Memorie della Società Geologica Italiana*, 12, 411-452.

Jones R.E., Vagnetti L., Levi S.T., Williams J., Jenkins D. and De Guio A., 2002, Mycenaean pottery from northern

Italy: archaeological and archaeometric studies. *Studi Micenei ed Egeo-Anatolici*, 44, 221-261.

Lenzi A., 2008, *Studio della ceramica veneta con decorazione a fasce rosse e nere*, Unpublished PhD thesis, University of Pisa.

Levi S.T. and Sonnino M., 2003, Archaeogeology Project: a Tool for the Study of Provenance and Technology of Ancient Pottery. In *Ceramics in the society: proceedings of the 6th European Meeting on Ancient Ceramics, Fribourg, Swtizerland 3–6 October 2001*, (eds. S. Di Pierro, V. Sernels and M. Maggetti), 197-208, Department of Geosciences, Mineralogy and Petrography. University of Fribourg, Fribourg.

Maritan L., 1998, Studi petrografici, mineralogici e chimici di materiali ceramici di età protostorica provenienti da Este e Padova. *Plinius*, 20, 147-149.

Maritan L., 2001, Coarse Iron age pottery from Este (Padova), Italy. In *Proceedings of the 3rd International Congress on Science and Technology for the Safeguard of Cultural Heritage in the Mediterranean Basin, Alcalà de Henares, Spagna 9–14 July 2001*, (eds. J. Alpuente, I. de Bustamante, P. López and J. Sanz), 1-6, Elsevier, Paris.

Maritan L., 2003, *Studio archeometrico di ceramiche di tipo etrusco padano dell'area veneta: indagini petrografiche, chimico-fisiche e confronto con i risultati ottenuti da prove sperimentali di cottura di materiali argillosi*, Unpublished PhD thesis, University of Padova.

Maritan L., 2004, Archaeometric study of Etruscan-Padan type pottery from the Veneto region: petrographic, mineralogical and geochemical-physical characterisation. *European Journal of Mineralogy*, 16, 297-307.

Maritan L., Mazzoli C., Nodari L. and Russo U., 2005, Second Iron Age grey pottery from Este (north-eastern Italy): study of provenance and technology. *Applied Clay Science*, 29, 31-44.

Maritan L., Nodari L., Mazzoli C., Milano A. and Russo U., 2006, Influence of firing conditions on ceramic products: experimental study on clay rich in organic matter. *Applied Clay Science*, 31, 1-15.

Morandini A. and Saracino M., 2008, Considerazioni preliminari. In *Oppeano: vecchi e nuovi dati sul centro protourbano*, (eds. A. Guidi and L. Salzani), 137-139, Quaderni di Archeologia del Veneto Serie Speciale 3, Giunta regionale del Veneto, Venezia.

Nodari L., Marcuz E., Maritan L., Mazzoli C. and Russo U., 2007, Hematite nucleation and growth in the firing of carbonate-rich clay for pottery production. *Journal of the European Ceramic Society*, 27, 4665-4673.

Philpotts A.R. and Wilson N., 1994, Application of petrofabric and phase equilibria analysis to the study of a potsherd. *Journal of Archaeological Science*, 21, 607-618.

Piccoli G., Sedea R., Bellati R., Di Lallo E., Medizza F., Girardi A., De Pieri A., De Vecchi G.P., Gregnanin A., Piccirillo E.M., Norinelli A. and Dal Prà A., 1981, Note illustrative della carta geologica dei Colli Euganei alla scala 1:25.000. *Memorie di Scienze Geologiche dell'Università degli Studi di Padova*, 34, 523-566.

Rioda V., 2008, Inquadramento geologico e geomorfologico. In *Oppeano: vecchi e nuovi dati sul centro protourbano*, (eds. A. Guidi and L. Salzani), 16-20, Quaderni di Archeologia del Veneto Serie Speciale 3, Giunta regionale del Veneto, Venezia.

Saracino M., 2004, Oppeano Veronese (VR): i materiali archeologici del saggio esplorativo n. 3 nei pressi della località Montara. In *Ricerche archeologiche in Italia e in Siria: atti delle Giornate di studio, Verona, 6–7 maggio 2002*, (eds. A. Guidi and S. Ponchia), 37-39, Sargon, Padova.

Saracino M., 2006, Archeologia della produzione ceramica nell'alto e medio Polesine tra il Bronzo Medio 2 e l'inizio della prima età del Ferro. *Padusa*, 41, 57-76.

Saracino M., 2008, Le schede della ricognizione sistematica. In *Oppeano: vecchi e nuovi dati sul centro protourbano*, (eds. A. Guidi and L. Salzani), 120-129, Quaderni di Archeologia del Veneto Serie Speciale 3, Giunta regionale del Veneto, Venezia.

Saracino M., 2009, *Archeologia della produzione ceramica durante l'età del Ferro: il caso di Oppeano*, Unpublished PhD thesis, University of Verona, Verona.

Saracino M. and Sboarina M., 2010, Analisi dei materiali ceramici provenienti dal centro protourbano di Oppeano-ex Fornace (campagna di scavo 2003). *Bollettino Museo Civico di Storia Naturale*, 34, 37-49.

Saracino M., Angelini I., Artioli G., Bellintani P. and Chiaffoni B., 2006, Frattesina – fase tre: Archeologia ed Archeometria della produzione ceramica. In *La ceramica in Italia quando l'Italia non c'era: atti della 8a Giornata di archeometria della ceramica (Vietri sul Mare, 27–28 aprile 2004)*, (eds. B. Fabbri, S. Gualtieri and M. Romito), 63-74, Edipuglia, Bari.

Sorbini L., Accorsi C.A., Bandini Mozzanti M., Forlani L., Gandini F., Meneghel M., Rigoni A. and Sommaruga M., 1985, *Geologia e geomorfologia di una porzione di pianura a Sud-Est di Verona*, Memorie del Museo Civico di Storia Naturale di Verona, II serie, Sezione Scienze della Terra 2, Museo Civico di Storia Naturale di Verona, Verona.

Whitbread I.K., 1986, The characterisation of argillaceous inclusions in ceramic thin sections. *Archaeometry*, 28, 79-88.

Whitbread I.K., 1989, A proposal for the systematic description of thin sections towards the study of ancient ceramic technology. In *Archaeometry: Proceedings of the 25th International Symposium*, (ed. Y. Maniatis), 127-138, Elsevier, Amsterdam.

Whitbread I.K., 1995, *Greek Transport Amphorae: A Petrological and Archaeological Study*, Fitch Laboratory Occasional Paper 4, British School at Athens, Athens.

## ACKNOWLEDGEMENTS

The authors would like to thank Prof. Alessandro Guidi (University Roma Tre) and Dr. Luciano Salzani (Veneto Archaeological State Antiquity Office) for their help and contribution to this research. They are very grateful to Gabriel Walton for revising the English text.

# 19

# Computerised documentation of painted decoration on pottery vessels using 3D scanning

**David Ben-Shlomo[a], Avshalom Karasik[a,b] and Uzy Smilansky[a]**
[a]Department of Physics of Complex Systems, The Weizmann Institute of Science, Rehovot 76100, Israel
[b]The Institute of Archaeology, The Hebrew University, Mount Scopus, Jerusalem 91904, Israel

**Abstract** - The article examines the possibility of the incorporation of painted decoration on pottery into the computerised documentation of ceramics, using 3D scanning. A new method is described for the recording of standard decorations painted on the surfaces of pottery vessels as 2D images, viewed in various projections. The diverse images are captured directly from the 3D model after it was accurately positioned. The output is automatically incorporated in the final publication together with the final drawing of the vessel. Several archaeological examples are presented and discussed.

## 1. Introduction

Computerised recording and analysis of ancient pottery vessels has advanced considerably in the past few years (e.g., Sablatnig and Menard 1996; Gilboa *et al.* 2004; Mara and Sablatnig 2005; Karasik and Smilansky 2008, and references therein; Mara *et al.* 2009), and its importance to archaeological research has been amply demonstrated and acknowledged. Several studies have succeeded both in recording large quantities of pottery sherds efficiently and accurately, and in analysing their morphology and typology in a mathematical manner (Gilboa *et al.* 2004; Karasik *et al.* 2004; Karasik and Smilansky 2008, 1151-6). A prerequisite for such studies is a robust and accurate algorithm which identifies the rotation axis of wheel-produced ceramics. This algorithm has recently been improved, tested, and used as a highly efficient practical tool for pottery analysis (Karasik and Smilansky 2008). The profile of the axially symmetric vessel, as well as the reconstruction of its diameter can be obtained, recorded, and analysed using this method. The details of the shape of the profile are most commonly used for typological analysis of pottery in archaeological studies. Nevertheless, some of the attributes of the pottery vessels have not yet been addressed by these procedures. These include non-axially-symmetric morphological parts such as handles and spouts applied to the vessel body (Karasik and Smilansky 2008, 1149), as well as the painted decoration on the surfaces of the vessel. This article aims to consider the latter aspect.

Painted decoration on pottery, when present, is one of its significant components in relation to archaeological research and interpretation. The contents, style and technique of the decoration can reflect cultural or ethnic aspects of the society using the pottery, as well as interactions between various pottery workshops (e.g., Friedrich-Hardin 1970; Hardin 1984; Rice 1987, 249-69). This applies both to stylistic and cultural interpretations of pottery, as well as to chronological and technological analyses. In some cases of a specific pottery decoration style or 'ware', the decoration can be the primary classification factor discussed in archaeological research (such as for the Late Bronze Age Mycenaean pottery, e.g., Mountjoy 1999, or Iron Age 'Black on Red' pottery, e.g., Gjerstad 1948; Schreiber 2003). In other cases, the geometric attributes of the decoration and its symmetry may reflect certain innovative technological aspects, such as the fast wheel and multiple-brush (e.g., Eiteljorg 1980). In any case, the decoration is an organic component of the archaeological record related to pottery vessels.

The documentation and illustration of decorated pottery are often acknowledged to be a more complex issue in archaeological research (e.g., Becker 1985; Olin and Dillon 1985, 103-8; Adkins and Adkins, 1989, 173, Fig. 8.11; Orton *et al.* 1993, 91). Such documentation may require particular solutions for specific vessel forms and decoration layouts (e.g., Becker 1985, 116-20). To date, very few studies tackle the issue of computerised analysis of painted decoration on archaeological artefacts, whether with regard to 2D or 3D acquisition (Tanahashi *et al.* 1995; Lock 2003, 156, Fig. 4.11; Kammerer *et al.* 2005; Mara *et al.* 2007; Kolomenkin *et al.* 2008 for the analysis of *relief* decoration on archaeological artefacts). Usually, these studies only make initial suggestions for further research rather than supply complete practical analytical procedures for archaeological data.

The computerised documentation of pottery decoration entails several methodological issues that will be evaluated and discussed below. The first is of a geometrical nature:

the projection of a drawing on a curved surface (the vessel surface) to a flat surface - the page where the pottery is illustrated. This is a problem dealt with in detail in many fields, such as cartography, for example, where geographic features on the surface of the earth are projected to form planar maps. The second issue, which is more relevant in the present context, concerns transforming the digitised texture signal obtained from the 3D scan to a depiction which conforms to the conventions and standards of archaeological publications. A process of scanning the pottery sherds with texture data and the recording of the texture on the potsherd surface is suggested, and several examples of actual pottery sherds are presented. The procedure presented complements the 3D scanning method as an archaeological recording system for painted pottery, as well as facilitating the presentation of this pottery in archaeological reports.

## 2. Decorations on curved surfaces and their projections to the plane

The issue of transforming or projecting a curved surface onto a flat surface belongs to the ancient and vast discipline of cartography, where the spherical surface of the earth is projected onto planar maps (for example, McDonnell 1979; Snyder and Steward 1988). Naturally, this field cannot be reviewed in detail here; rather, it will be shown how such projections may be used effectively to depict decorations on the surfaces of pottery vessels. Any given type of projection from a curved to a flat surface introduces a certain amount of distortion. Hence, finding the optimal projection involves a compromise which should be decided upon by the archaeologist. In the present note, we describe three simple projections which offer a range of alternatives amongst which an optimal representation can be reached. Their suitability for the effective and accurate depiction of decorations on various types of pottery sherds will be illustrated by applying them to some pottery fragments.

These projections are:

A. *Parallel Projection:* This projection is completely determined by the choice of the direction from which the object is viewed. The viewing plane is perpendicular to the viewing direction, and the points on the surface are projected to the flat surface by lines which are parallel to the viewing direction. This is illustrated in Figure 1, where projections corresponding to two different viewing directions of a schematic decorated surface are shown. Assuming that the potsherd is positioned with its symmetry axis along the $z$-axis, the 'horizontal' projection with viewing direction along the $x$-axis provides a distorted image because of the inclination of the fragment. This can be improved by tilting the viewing direction to the perpendicular to the surface, as shown for the second projection (Fig.1). However, if the surface is curved, and its size is of the order of the radius of curvature, a single viewing direction cannot be perpendicular to the entire surface. Thus, the parallel projection which is very easy to implement is restricted to small, almost flat fragments.

B. *'Radius-dependent' cylindrical projection:* In a cylindrical projection, the surface is mapped by projecting the curved surface on an imaginary cylinder (whose axis coincides with the vessel's axis of rotation) wrapped around it, and subsequently unfolded unto a flat rectangular surface; in this case, the 'meridians' are equally spaced (e.g., Snyder 1987, 5). As pottery vessels usually have a highly varying radius in various points of their profile, such a projection could significantly distort the size and area of patterns located on pottery vessel surfaces. Therefore, instead of the regular cylindrical projection, in our study we have used a 'radius-dependent' cylindrical projection. The 2D outcome is therefore not rectangular (Fig. 2b; this projection is somewhat similar, yet not identical, to the 'Albers equal-area conic projection'; Snyder 1987, 98-103). Here, the transformation of a point $P_i = (x_i, y_i, z_i)$ which has an angle $\theta_i$ with the positive $x$-axis onto the radius-dependent cylindrical projection is given by the following formula:

$$\text{Transformation 1}: \hat{x} = \sqrt{(x_i^2 + y_i^2)} \cdot \theta_i \left[\theta_i = \arctan\left(\frac{x_i}{y_i}\right)\right]$$
$$\hat{y} = z_i$$

This projection can be useful for larger and more curved pottery sherds, or for complete vessels, where a parallel projection provides only a partial and distorted representation (Fig. 2b); it is especially accurate if the surface on which the decoration is painted bears a resemblance to a

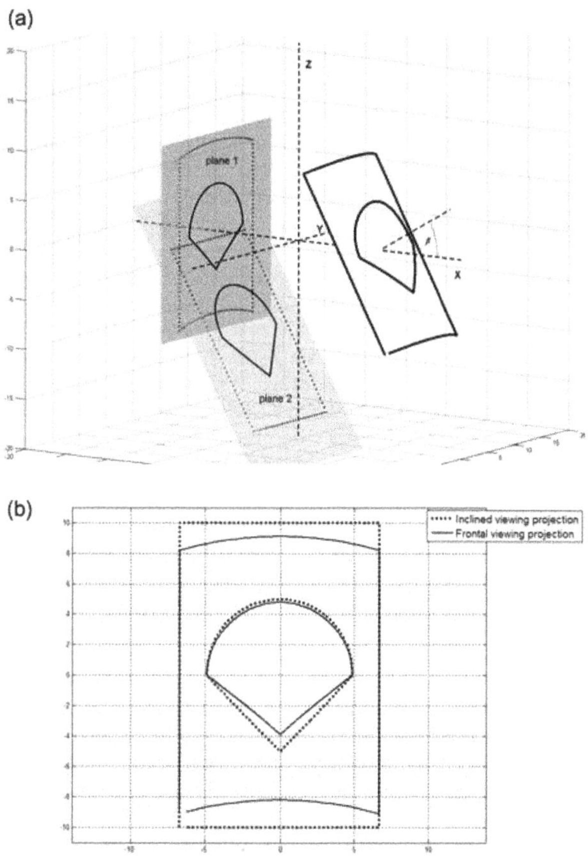


**Figure 1.** (a) Synthetic sherd inclined by angle β according to the location of decoration; Plane 1: frontal parallel projection; Plane 2: inclined parallel projection; (b) Plane 1 and 2 overlapped.

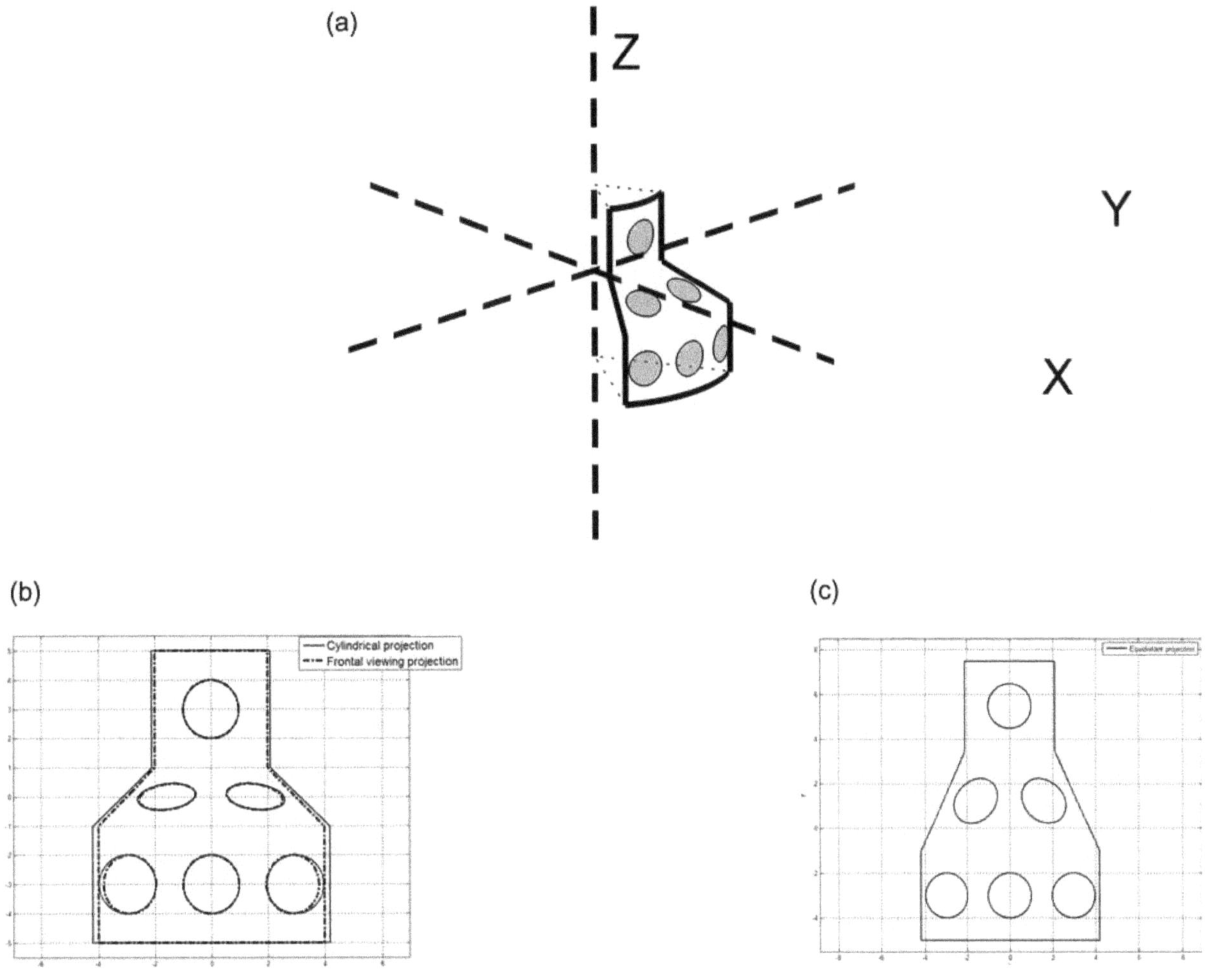


**Figure 2.** Various projections of a decoration pattern on a synthetic vessel: (a) An isometric view of the vessel; (b) Parallel and radius-dependent cylindrical projections of the decoration pattern overlapped; (c) Equidistant projection of the decoration pattern conserving the distances along the vessel's profile.

cylindrical shape. Obtaining this projection manually is rather difficult and time-consuming in manual documentation of pottery (a somewhat similar outcome is generated from a 'rollout drawing', e.g., Becker 1985, 118-121, Figs. 8.3-8.5). A similar projection of texture derived from 3D scanning of objects was presented in the past, yet it was based on a single view (e.g., Colombo *et al.* 2002; also Tanahashi *et al.* 1995).

C. *Equidistant projections:* Equidistant projections maintain various metrical relationships existing on the curved surface when depicted on the flat surface. Such a depiction may illustrate the decoration in the exact manner it was drawn on the surface of the vessel, as it retains the distances between the points on the surface of the vessel. There are various equidistant projections (e.g., Snyder 1987, 191-202), and in practice one should adapt the projection used to the estimated body shape of the pottery vessel in the area in which the relevant decoration is located. Such projections are virtually impossible to obtain accurately in manual drawing of pottery vessels. The transformation used below transforms the 3D data into a 2D image preserving the distances between the points along the profile of the surface of the vessel (Fig. 2c). The $\hat{x}$ remains as with the radius-dependent cylindrical projection, but for the $\hat{y}$ we need to define a new variable which is the 'arc-length of the profile' (denoted by '$s$'). The computation of $s$ in each point on the surface of the sherd is carried out along the mean profile of the entire sherd (as described in Karasik and Smilansky 2008, 1153):

$$\text{Transformation 2}: \hat{x} = \sqrt{(x_i^2 + y_i^2)} \cdot \theta_i \left[\theta_i = \arctan\left(\frac{x_i}{y_i}\right)\right]$$
$$\hat{y} = s_i$$

Such a projection may be compared to an equidistant projection in which distances along vertical meridians are conserved. While this may not be an ideal equidistant projection for any type of body shape (such as perfectly spherical or conical shapes), we have chosen this projection as it naturally derives from the treatment of actual pottery vessel profiles already developed and used in computerised recording and analysis.

## 3. Documentation of actual pottery sherds with decoration utilising the texture signal obtained from the 3D scanning

The aim of the procedure described below is to automatically integrate the painted decoration onto a conventional 2D image of the profile for the documentation and publication of ancient pottery. Such an image, common in illustrations of pottery sherds in archaeological reports, shows the mean profile and contour of the sherd with the symmetrical reconstruction of the vessel. This depiction usually uses a parallel projection. Nevertheless, as shown above, it is evident that in the case of large sherds or complete vessels, in order to represent the decoration accurately, various projections from 3D into 2D, as well as several viewing angles may be needed. Thus, once the scanning procedure for the texture is complete, the different projections described above can be automatically applied in order to obtain an optimal view of the decoration. The decision on which one to use depends on the user and on the nature of the vessels and their decoration. In all projections we assume that the rotation axis of the fragment was already computed and is chosen as the $z$-axis in space. Moreover, the 3D model of the fragment is arbitrarily rotated so that its centre of mass is on the positive $x$-axis.

The decoration appearing on the surfaces of the vessels is represented by the 'texture' acquisition of the 3D scanning. To enable texture acquisition together with 3D information, we use a third camera alongside the other two that are used to capture the geometry of the fragment (as described in Karasik and Smilansky 2008, 1150, Fig. 1). The colour camera captures 2D colour snapshots together with the 3D information obtained by the other two cameras. Consecutive views of the object are recorded simultaneously by all three cameras. The on-line analysis is performed and controlled with the QTSculptor software (produced by Polygon Technology, Darmstadt, Germany).

Several points regarding 3D scanning with texture should be noted. A reasonable quality and resolution could be obtained only when a single object was scanned each time. If several objects are scanned together within a frame (Karasik and Smilansky 2008, 1150), the resolution of the acquired texture decreases significantly. The texture can also be imported from regular digital pictures of the object and attached to the 3D geometry after a careful segmentation. The size of the typical output file for a sherd with texture is between 5 and 15 MB; in most cases, a complete vessel would usually require about 30-60 MB. The final file comprises a '.wrl' file with the 3D geometry of the object, as well as references to '.jpg' colour images files. Each point on the scanned vessel is assigned with a RGB (red-green-blue) texture value (a 3-dimensional vector of integer values in the range of 0-255). As several colour images are obtained (one from each view taken, usually six), many points can be seen from more than one view; in these cases we use an averaged value. The '.wrl' file is transformed into a '.mat' file that can be further manipulated by the 3D geometry and the average profile of the vessel (Karasik and Smilansky 2008), as well as the average texture of each point on the vessel. For the sake of simplicity, from this point onwards the average RGB values are transformed into 'grey-scale' values (yet, RGB values can be similarly used and coloured images will be obtained):

$$T_{gray} = 0.2989T_{red} + 0.5870T_{green} + 0.1140T_{blue}$$

It should be emphasised that the 'texture' acquired by the 3D scanning is far from a straightforward representation of the painted decoration on the vessel surface. The issue of computerised analysis of texture is a complex subject that is dealt with in various fields of computerised acquisition technology, pattern recognition, artificial and 'machine intelligence', involving methods such as 'edge detection' and 'colour segmentation' (Cheng *et al.* 2001 for a review of some 2D methods). Some studies deal with 3D images and their analysis (e.g., Sturm and Priese 2003), yet most developed tools are for the analysis of two-dimensional images. Analysis of archaeological objects, such as painted pottery sherds, is clearly more complicated and difficult, as the paint is most often only partly preserved.

In general, several methodological difficulties arise in the analysis of the texture of ancient pottery (Kammerer *et al.* 2005, 154-5):

1. The texture data collected by the scanning camera includes other information aside from that pertaining to the painted decoration. This can include: various stains, irregularity of the fabric colour, soot marks, registration marking with ink, patina, dirt, etc. Often these elements, which can be considered as 'noise', give a

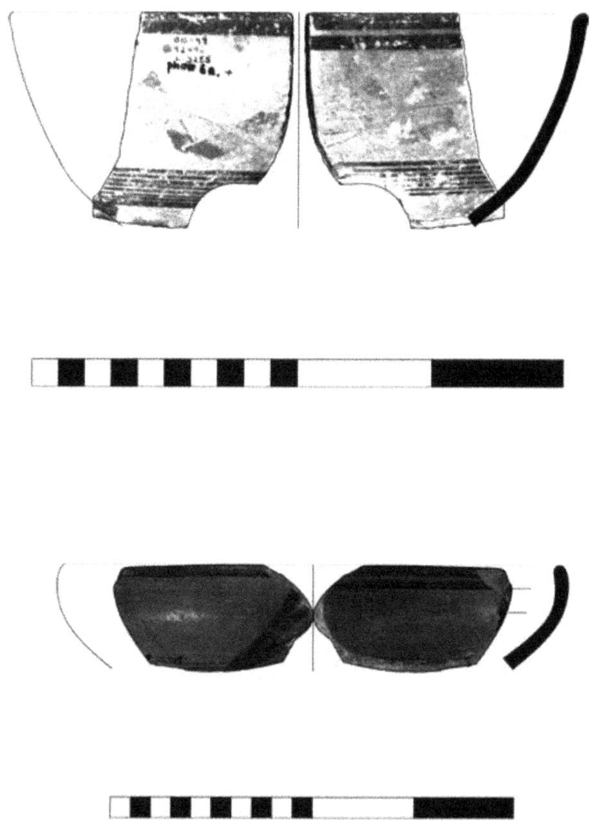

**Figure 3.** Profile and frontal parallel projection of two 'Black on Red' bowls from Tel Dor and Horvat Rosh Zayit (lower), Israel.

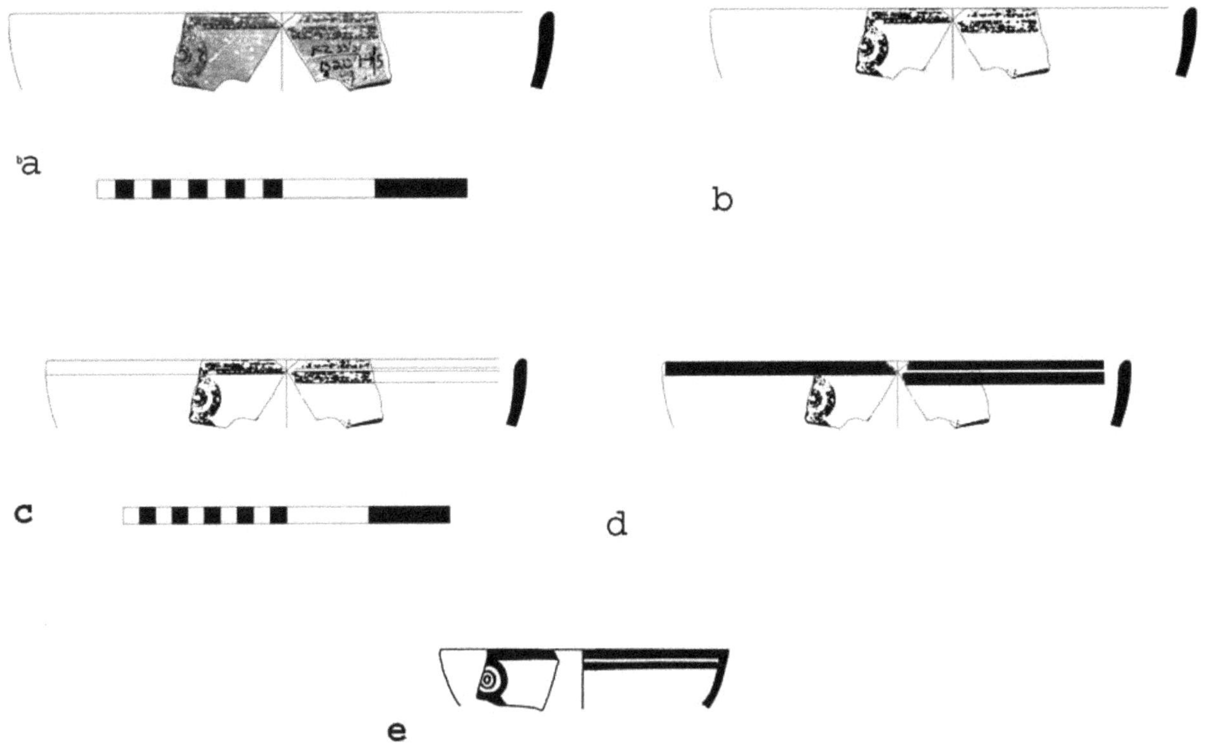


**Figure 4.** A 'Black on Red' sherd from Horvat Rosh Zayit: (a) Original scanning; (b) After transformation into a black and white image and cleaning; (c) After marking of the reconstructed limits of the horizontal decoration bands; (d) After filling reconstructed areas in black; (e) An illustration of the same bowl from the archaeological excavation report (Gal and Alexandre 2000, Fig. VI.13:21).

stronger signal than the painted decoration itself. This is in contrast to the analysis of the geometry of the vessel, in which such 'noise' or irregularity of the signal appears in a relatively inferior magnitude to that of the 'signal' and can thus be 'smoothed' by various methods.

2. The painted decoration on the ancient pottery sherds is often worn or chipped (and very often a large portion of the decoration motifs presented in archaeological reports is reconstructed). In certain cases the decoration is in very faint or light-coloured paint, or in a hue relatively close to that of the background colour of the sherd. This aspect, together with the one presented at No. 1 may highly complicate, or in certain cases can render impossible, the automatic and clear distinction between painted and unpainted zones on many ancient pottery vessels.

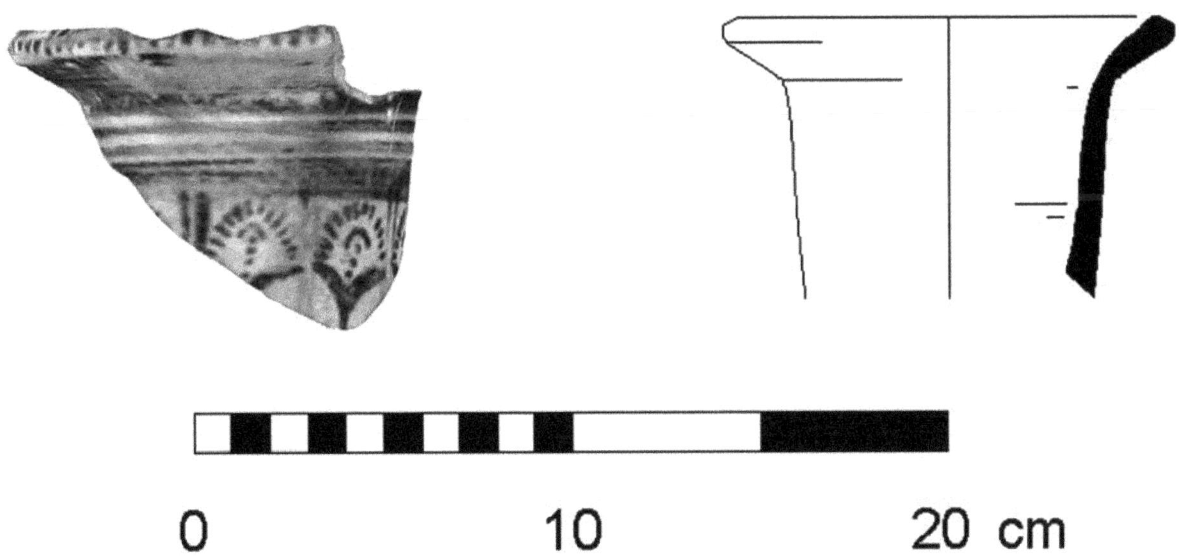


**Figure 5.** Profile and outer parallel projection of a Mycenaean rhyton sherd from Tell Abu Hawam, Israel.

**Figure 6.** Radius-dependent cylindrical projection of outer surface decoration on a Mycenaean rhyton sherd.

3. Mathematical analysis of painted decoration is problematic as most decorative motifs are abstract and do not bear any intrinsic or other systematic pattern that can be conveniently described using mathematical formulas. Thus, they cannot be incorporated and described in 2D along the arc-length of the profile, as is the case for axially symmetric decorations (e.g. Iron Age 'Black on Red' ware).
4. If more than one decorative shade of colour appears on the sherd (as black and red for example), distinguishing between the shades is necessary in order to perform archaeological documentation and analysis.

In order to create a 2D image from the projection of the scanned surfaces of the vessel, the plane is divided into pixels, where the average colour of the points in each pixel set its colour in the final image. Obviously, two independent images are available for the inner and the outer surfaces of the fragment, and the user should choose which one to create each time the projection is applied. The distinction between inner and outer points is carried out very easily by using the × value of their normal vectors, since the normals to the inner and outer surfaces point in opposite directions. Positive values belong to points on the outer surface and the negative ones to those on the inner surface. In the computation of the precise colour/grey-scale value of each pixel we use the information from the original triangulated mesh that connects between the points. For each pixel, we look on the triangle in which its centre of gravity is located, and averaging the weighed values of each vertex according to its distance from the centre of the pixel. In order to create an image with reasonable resolution, a pixel size of about 0.1 mm was selected (equivalent to around 300 dpi). The value is rounded in order to fit a discrete 0-255 range of grey-scale output. The procedure can involve a relatively long computational time (about 15-100 minutes per sherd, depending on the size of the sherd and the resolution of the scanning), as all triangle indices are examined consecutively. The obtained matrix is saved as a '.jpg' or '.tiff' file and is combined with the contour line of the sherd and its profile on one image (as in Fig. 3).

We have scanned several bowl fragments of decorated 'Black on Red' ware (Schreiber 2003), dated to the Iron Age II (ca. 1000-600 BC). The decoration consists of well-executed horizontal concentric circles drawn on both sides of the sherd (three examples are presented in Figs. 3-4). As these sherds are relatively small and flat, they can be quite accurately documented using the parallel projection, which can be integrated with the depiction of the profile of the vessel (Fig. 3).

One example (Fig. 4) illustrates how the grey-scale image can be easily transformed into a standard black and white drawing, by changing of contrast, for example (Fig. 4b; Fig. 7); in a further stage, the decoration bands can be outlined (Fig. 4c), and thence reconstructed by filling the outlined areas in black (Fig. 4d). This would result in a drawing such as those used in archaeological reports (Fig. 4e; note the proposed diameter of the bowl as drawn in the archaeological report is smaller than the one computed by the algorithm). Various irrelevant details on the drawing, such as registration marking with ink, patina, dirt, and soot marks can be easily erased manually with a graphic software such as 'Photoshop' (Fig. 4b). The drawing illustrates only the decoration as captured by the camera; as often the motifs are very faint and incomplete, restoration in dotted lines can also be easily done by graphic software. This may be carried out manually by the person preparing the drawings for publication (conferring with the archaeologist). If a more minimalistic and accurate account is desired, the original grey-scale or black and white image can be used (as in Figs. 3, 4a, 7). The great advantage of the documentation method described here is that it uses the colour information of the pattern together with the accurate alignment of the 3D model. Then, the pixelisation process is precise, easy to implement and does not require any manual measurements.

Another example is that of a large sherd of a Mycenaean rhyton (a conical vessel usually used for ceremonial drinking) from Tell Abu Hawam, dated to the Late Bronze Age (ca. 1550-1200 BC) (Figs. 5-8). The decoration here is in a dark and fairly well-preserved paint, and appears on the outer surface both as horizontal lines and a more elaborate reccurring motif below. As this is a large and much more curved sherd than the first example, and the decoration more complex, it was documented in the three different projections suggested above. Figure 5 illustrates the sherd with a parallel projection integrated with the computerised illustration of the profile of the vessel. This is a standard archaeological documentation, yet it does not depict the entire decoration on the sherd, as its extreme margins cannot be viewed (especially the third occurrence of the motif on the right side). A radius-dependent cylindrical projection of the decoration on the outer surface of the vessel is shown in Figure 6, transformed into a standard black and white drawing in Figure 7. In this case the radius-dependent cylindrical projection seems to be quite useful for presenting the entire decoration in one view. An equidistant projection of the decoration on the outer surface of the Mycenaean rhyton is given in Figure 8. The difference between the radius-dependent cylindrical (Fig. 6) and equidistant (Fig. 8) projections is evident, especially in the area where

**Figure 7.** Black and white image derived from the radius-dependent cylindrical projection of a decoration on a Mycenaean rhyton sherd.

**Figure 8.** An equidistant projection conserving the distances along the vessel's profile of the outer surface decoration on the Mycenaean rhyton.

the surface of the sherd curves outwards (as the upper part in Fig. 8; this area is 'stretched' in the equidistant projection and can also result in a different illustration of the decoration.

In another example, an upper part of a decorated Phoenician Bichrome jug from Tel Dor was scanned (dated to the Late Iron Age I; for example, Gilboa 1999). In this case, the vessel was preserved almost in its entire perimeter in the neck, shoulder and upper body area. The main decoration consists of black and red painted motifs of concentric circles (in black, on the sides) and a schematic tree design in the middle (in red). In Figure 9, the profile of the vessel and the parallel projection can be seen. In this projection, the image can be easily pasted on the drawing, creating a standard illustration of an excavated report. Note also that the resolution in this case is sufficient for distinguishing the two colours, black and red, in this grey-scale image. However, only a certain portion of the decoration can be seen in this projection, and, moreover, the motif of the schematic tree is not very visible in this angle. In Figure 10, the same projection is shown but viewed in various inclinations (of 30 and 45 degrees). This substantially improves the visibility of the motif, and the archaeologist can choose any of these images for better documentation. An equidistant projection depicting only the frontal part of the vessel (Fig. 11) can also be used instead of a parallel projection with the vessel profile. This could be another way to achieve a better documentation of the upper part of the 'tree' motif, which is not very visible in the regular parallel projection.

Illustrating the same vessel in radius-dependent cylindrical projection (Fig. 12, upper) and equidistant projection (Fig. 12, lower) enables us to see the entire decoration on the vessel in one image. The fact that the concentric circles on both sides can be observed in full is especially useful. Once again, one can see how the equidistant projection 'stretches' the pattern in the *y*-axis, giving a different view of the decoration on the surface of the vessel.

## 4. Discussion and suggestions for further research

We have presented a procedure that can complement the 3D technology for pottery recording and analysis (Karasik and Smilansky 2008), previously only dealing with the geometry of the vessels, with the documentation of the decoration as well. The sherds can be recorded and depicted in the correct positioning with the profile of the vessel, together with the texture obtained from the 3D scanning, which represents the painted decoration on the pottery. By using the computerised depiction of the decoration, various different projections can be easily and accurately attained to meet the request and requirements of archaeological recording and research. In turn, these projections can be easily transformed into publication-quality, correctly positioned illustrations of decorated pottery sherds. This is a new tool for the documentation of ancient pottery; previously, the decoration on such sherds would have been added only manually, which was time- consuming and inaccurate ( for example, Adkins and Adkins 1989, 173; Becker 1985).

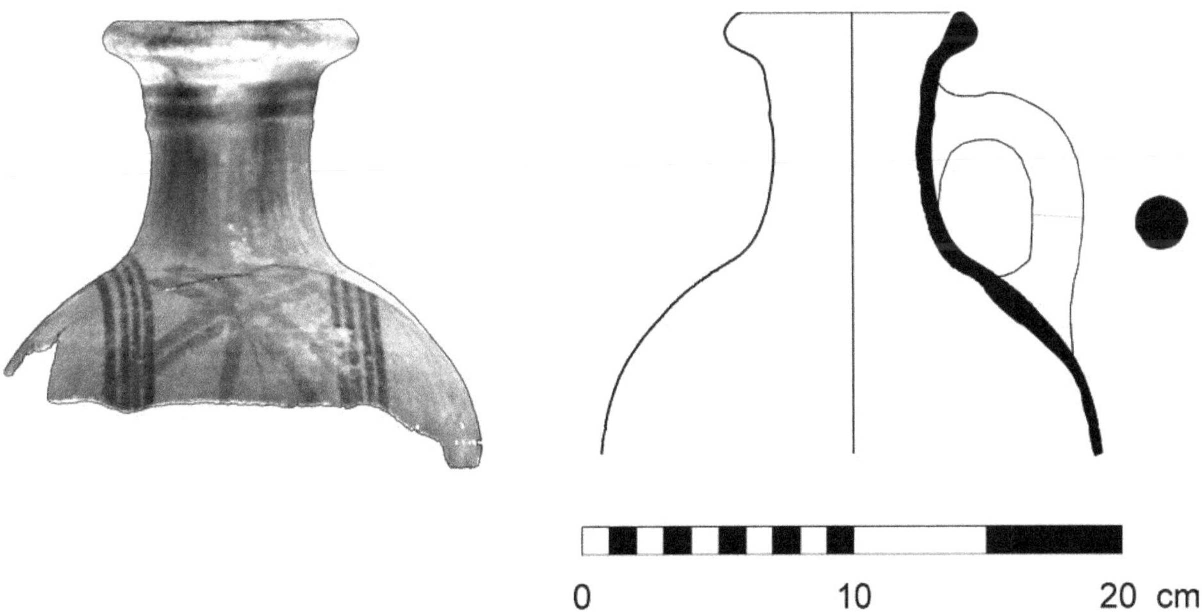


**Figure 9.** Profile and parallel projection of a Phoenician Bichrome jug from Dor.

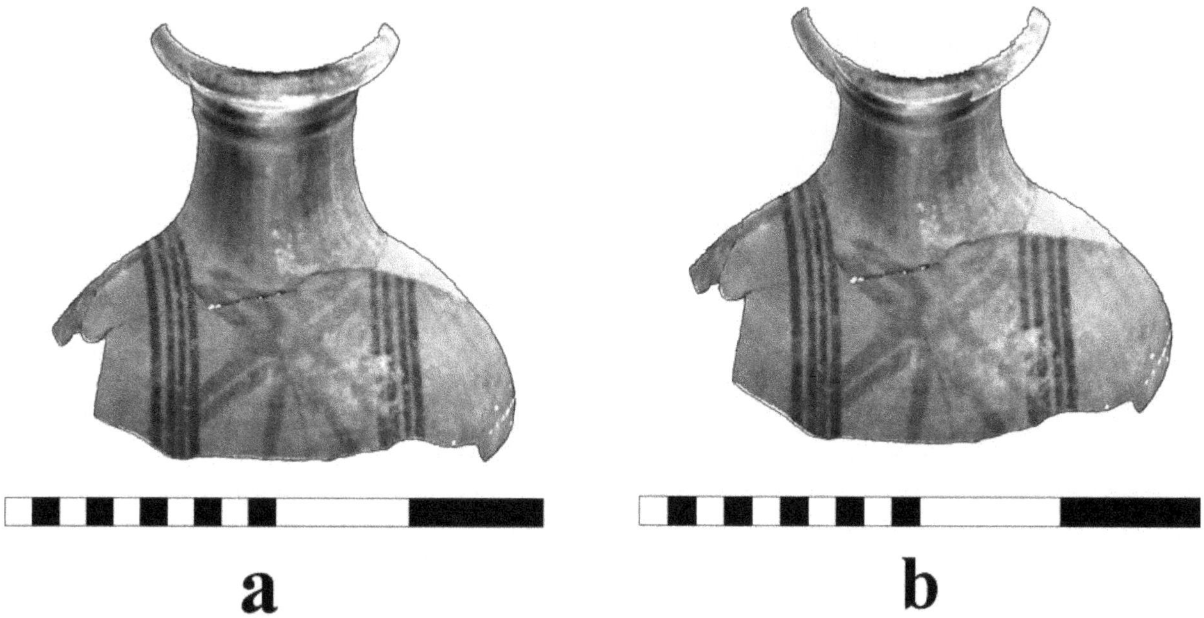

Figure 10. Parallel projections of the jug from Dor with various inclinations: (a) 30 degrees; (b) 45 degrees.

The ability to obtain various projections of the decoration motifs painted on the curved surface of the pottery vessel in a 2D image may also enable us to investigate various artistic aspects of ancient pottery decoration. Possibly, by comparing parallel and equidistant projections, one can evaluate the relationships between how the motifs were drawn on the surface of the vessel and the way they were perceived by the viewer, and thus better understand methods used by potters to execute such motifs successfully.

So far, the texture treatment deals only with rendering and recording, while a quantitative analysis of this information in combination to the geometry of the vessel is still to be developed in the future. Several methodological difficulties, especially relating to the texture signal in 3D rendering have been noted. Two main avenues for this research may be suggested. A) The analysis of axially symmetric decoration (most commonly in the form of horizontal bands), which can be incorporated into the vessel's average profile. Once a complete assemblage with axially symmetric decorations has been scanned successfully, comparisons between line decoration characteristics of different vessels can be made mathematically. In this process, it will be important to minimise the 'noise', enhance the signal, and define the boundaries on the line decoration more clearly. B) Analysis of non-axially symmetric decoration, which cannot be combined into the average profile of the vessel. Here, probably, each decorative motif should be isolated, defined and analysed separately. The main initial difficulty is to clearly define the decorated and undecorated zones on the pottery sherd according to the characteristics of the signal (hue intensities, distribution, etc.).

Edge detection and segmentation methods may be used and modified for 3D analysis (e.g. Mara *et al.* 2007, Fig. 8), to assist in the resolution of some of these methodological problems. In these methodology, cluster analysis, 'mean shift', 'density analysis', and edge detection methods are used to define boundaries, usually on 2D images, of coloured segments on the object (e.g., Christoudias *et al.* 2002, and references therein). In the analysis of axially symmetric decoration, it may be useful to apply certain 'archaeological constraints' together with these pattern recognition methods. Examples of such constraints would be a minimal and maximal thickness of the decoration band, the fact that the bands are relatively parallel to

Figure 11. Equidistant projection of the frontal part of the outer surface of the jug from Dor conserving the distances along the vessel's profile.

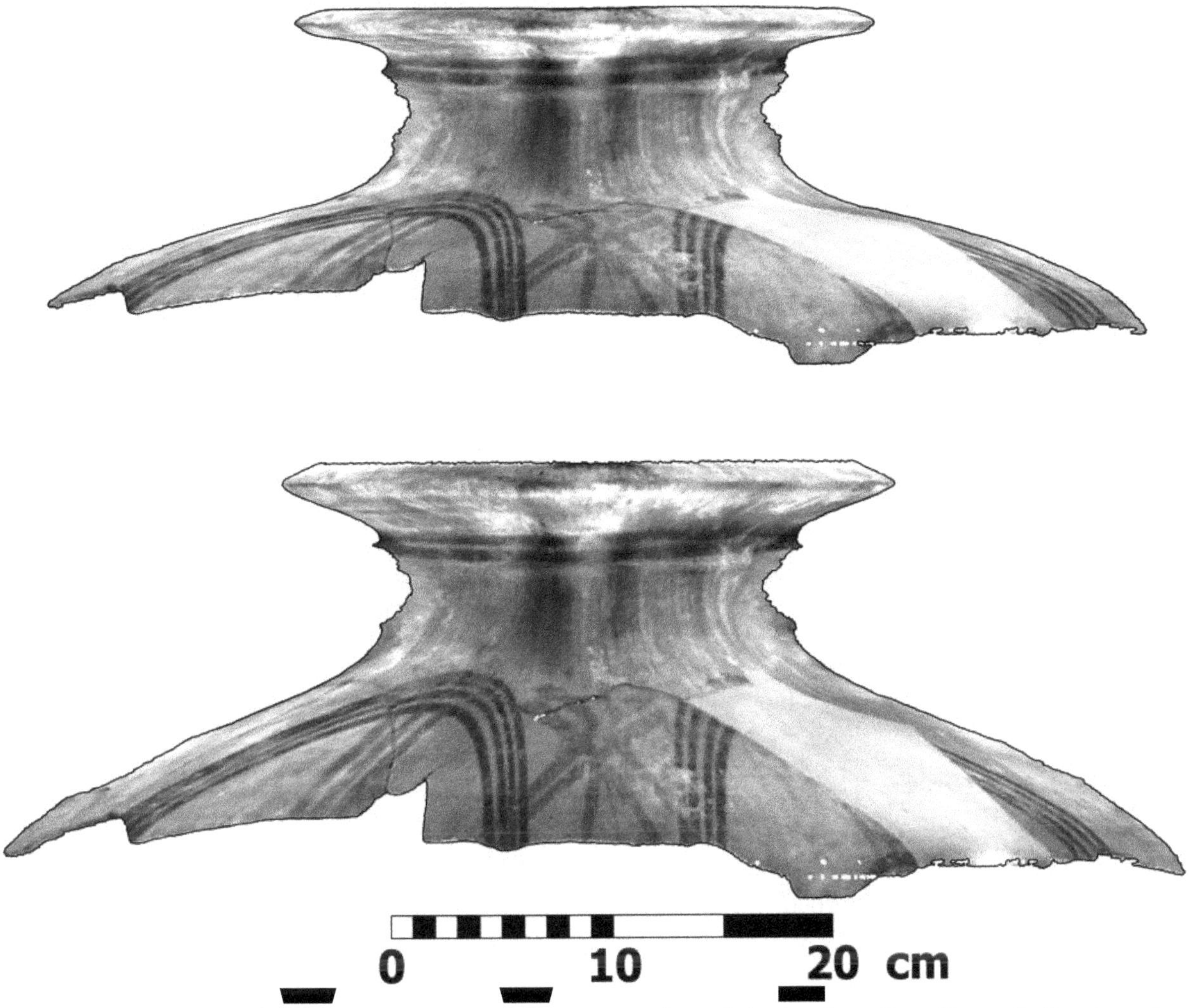


**Figure 12.** Projections of the entire perimeter of the outer surface of the jug from Dor: Upper: Radius-dependent cylindrical projection. Lower: Equidistant projection conserving the distances along the vessel's profile.

each other, and the assumption that the inner and outer boundaries of the painted band are parallel (this assumption could possibly be also made for non-axially symmetric decorations). Another issue concerns the identification and recording of more than one hue of decoration on the pottery, since pottery is often decorated in two or more colours (as red and black, black and white, etc.).

These issues require additional and expanded research, while the specifications of the texture capture of the scanner camera used should also be more optimally adapted for this purpose. In the future, if painted decoration on vessels represented by 3D images can be computerised, this could result in a greater potential for the analysis of decoration, including the analysis and comparison of decorative motifs and compositions, and their integration within a computerised typology of pottery vessels. In the meantime, any type of texture can be accurately and automatically incorporated into the final publication procedure of ceramics according to the methodology presented in this study.

## References

Adkins L. and Adkins R.A., 1989, *Archaeological Illustration*, Cambridge University Press, Cambridge.

Becker J., 1985, Special Problems in Ceramic Illustration. In *A Student's Guide to Archaeological Illustrating*, (ed. B.D. Dillon), 112-121, University of California, Los Angeles.

Cheng H.D., Jiang X.H., Sun Y. and Wang J., 2001, Colour Image Segmentation: Advances and Prospects. *Pattern Recognition*, 34, 2259-2281.

Christoudias C.M., Georgescu B. and Meer P., 2002, Synergism in Low Level Vision. *16th International Conference on Pattern Recognition Vol. IV*, 150-156, Institute of Electrical and Electronics Engineers Computer Society.

Colombo C., Bimbo A-D. and Pernici F., 2002, Uncalibrated 3D Metric Reconstruction and Flattened Texture Acquisition from a Single View of a Surface of Revolution. *Proceedings of the First International*

*Symposium on 3D Data Processing Visualization and Transmission*, 277-284, Institute of Electrical and Electronics Engineers.

Eiteljorg H., 1980, The Fast Wheel, the Multiple Brush Compass and Athens as Home of the Protogeometric Style. *American Journal of Archaeology*, 84, 445-452.

Friedrich-Hardin M., 1970, Design Structure and Social Interaction: Archaeological Implications of an Ethnographic Analysis. *American Antiquity*, 35, 332-343.

Gal Z. and Alexandre Y., 2000, *Horbat Rosh Zayit: An Iron Age Storage Fort and Village*, Israel Antiquity Authority Reports, Jerusalem.

Gilboa A., 1999, The Dynamics of Phoenician Bichrome Pottery: A View from Tel Dor. *Bulletin of the American Schools of Oriental Research*, 316, 1-22.

Gilboa A., Karasik A., Sharon I. and Smilansky U., 2004, Towards Computerized Typology and Classification of Ceramics. *Journal of Archaeological Science*, 31, 681-694.

Gjerstad E., 1948, *The Cypro-Geometric, Cypro Archaic and Cypro-Classical Periods. The Swedish Cyprus Expedition Vol. IV Part 2*, Swedish Cyprus Expedition, Stockholm.

Hardin N.A., 1984, Models of Decorations. In *The many Dimensions of Pottery: Ceramics in Archaeology and Anthropology*, (eds. S.E. van der Leeuw and A.C. Pritchard), 573-614, CINGULA 7, Institute of Pre- and Proto-History, Amsterdam.

Kammerer P., Mara H., Kratzmueller B. and Zolda E., 2005, Detection and Analysis of Lines on the Surface of Archaeological Pottery. In *Electronic Imaging and the Visual Arts*, (eds. V. Cappellini and J. Hemsley), 154-159, Pitagora Editrice, Bologna.

Karasik A. and Smilansky U., 2008, 3D Scanning Technology as a Standard Archaeological Tool for Pottery Analysis: Practice and Theory. *Journal of Archaeological Science*, 35, 1148-1168.

Karasik A., Sharon I., Smilansky U. and Gilboa A., 2004, Typology and Classification of Ceramics Based on Curvature Analysis. In *Enter the Past. The E-way into the Four Dimensions of Cultural Heritage. CAA 2003, Computer Applications and Quantitative Methods in Archaeology*, (eds. K.F. Ausserer, W. Bo rner, M. Goriany and L. Karlhuber-Vo ckl), 472-475, Archaeological Reports International Series. Archaeopress, Oxford.

Kolomenkin M., Shimshoni I. and Tal A., 2008, Demarcating Curves for Shape Illustration. *Proceedings of ACM SIGGRAPH Asia 2008*, ACM Transactions on Graphics 27, No. 5, Association of Computing Machinery, New York.

Lock G., 2003, *Using Computers in Archaeology*, Routledge, New York.

Mara H. and Sablatnig R., 2005, Semiautomatic and automatic profile generation for archaeological fragments. In *Proceedings of 10th Computer Vision Winter Workshop*, (eds. A. Hanbury and H. Bischof), 123-134, CVWW05.

Mara H., Kampel M., Niccolucci F. and Sablatnig R., 2007, Ancient Coins and Ceramics - 3D and 2D Documentation for preservation and retrieval of lost Heritage. In *Proceedings of the 2nd ISPRS International Workshop on 3D Virtual Reconstruction and Visualization of Complex Architectures (3D-ARCH 2007)*, (eds. F. Remondino and S. El-Hakim), 10-18, International Society for Photogrammetry and Remote Sensing, Zurich.

Mara H., Trinkl E., Kammerer P. and Zolda E., 2009, 3D-Acquisition of Attic Red-Figured Vessels and Multi-Spectral Readings of White Primed Lekythoi of the Kunsthistorisches Museum Vienna for the new CVA Volume. In *Proceedings of 9th European Meeting on Ancient Ceramics - "Vessels: Inside and Outside" (EMAC'07)*, (eds. K.T. Biró, V. Szilágyi and A. Kreiter), 21-28, Hungarian National Museum, Budapest.

McDonnell P.W., 1979, *Introduction to Map Projections*, Marcel Dekker, Inc., New York.

Mountjoy P.A., 1999, *Regional Mycenaean Decorated Pottery*, Marie Leidorf, Rahden/Westfalen.

Olin J. and Dillon B.D., 1985, Ceramics. In *A Student's Guide to Archaeological Illustrating*, (ed. B.D. Dillon), 96-109, University of California, Los Angeles.

Orton C., Tyers P. and Vince A., 1993, *Pottery in Archaeology*, Cambridge University Press, Cambridge.

Rice P.M., 1987, *Pottery Analysis: A Sourcebook*, Chicago University Press, Chicago.

Sablatnig R. and Menard C., 1996, Computer-based Acquisition of Archaeological finds. In *Proceedings of the 3rd International Symposium on Computing and Archaeology, Vol I*, (eds. P. Moscati and S. Mariotti), 429-446, Archeologia e Calcolatori, Firenze.

Schreiber N., 2003, *The Cypro-Phoenician Pottery of the Iron Age*, Culture and History of the Ancient Near East 13, Brill, Leiden and Boston.

Snyder J.P., 1987, *Map Projections - A Working Manual*, US Geological Survey Professional Paper 1395, US Government Printing Office, Washington DC.

Snyder J.P. and Steward H., 1988, *Bibliography of Map Projections*, US Geological Survey Bulletin 1856, US Government Printing Office, Washington DC.

Sturm P. and Priese L., 2003, 3D-Colour-Structure-Code - A Hierarchical Region Growing Method for Segmentation of 3D-Images. In *Image Analysis: 13th Scandinavian conference, SCIA 2003, Halmstad, Sweden, June 29 - July 3, 2003: proceedings*, (eds. J. Bigun and T. Gustavsson), 603-608, Lecture Notes in Computer Science 2749. Springer, Berlin and New York.

Tanahashi H., Sakaue K. and Yamamoto K., 1995, Recovering Decorative Patterns of Ceramic Objects from a Monocular Image Using a Genetic Algorithm. *Proceedings of the Third International Conference on Document Analysis and Recognition, 14-16 August 1995, Montreal, Canada*, 339-342, Institute of Electrical and Electronics Engineers Computer Society Press, Los Alamitos, CA.

## ACKNOWLEDGEMENTS

This research was supported by the Center of Excellence (Grant No. 300/06) of the Israel Science Foundation and by the ISF Grant No. 168/06. We wish to thank Talia Goldman for her assistance and useful remarks. The sherds that were scanned in Haifa University were kindly provided by Dr. Ayelet Gilboa of Haifa University and the Israel Antiquities Authority.

20

# Insights into manufacturing techniques of archaeological pottery: Industrial X-ray computed tomography as a tool in the examination of cultural material

**Stephan Karl[a], Daniel Jungblut[b], Hubert Mara[c], Gabriel Wittum[b] and Susanne Krömker[c]**

[a]Department of Archaeology, University Graz, Universitätsplatz 3/II, 8010 Graz, Austria
[b]Goethe Center for Scientific Computing (G-CSC), University Frankfurt, Kettenhofweg 139, 60325 Frankfurt am Main, Germany
[c]Interdisciplinary Center for Scientific Computing (IWR), University Heidelberg, Im Neuenheimer Feld 368, 69120 Heidelberg, Germany

**Abstract** - The application of X-radiography in ceramic studies is becoming an increasingly valued method. Using the potential of industrial X-ray computed tomography (CT) for non-destructive testing as an archaeometric or archaeological method in pottery studies, especially regarding aspects such as manufacturing techniques or pottery fabrics, requires controlled data-acquisition and post-processing by scientific computing adjusted to archaeological issues. The first results of this evaluation project show that, despite the difficulties inherent in CT technology, considerable information can be extracted for pottery analysis. The application of surface morphology reconstructions and volumetric measurements based on CT data will open a new field in future non-invasive archaeology.

## 1. Introduction

For as long as archaeometric methods have been well established in the study of archaeological pottery, there have been many efforts to reduce their impact on the object and to look for less-destructive or non-destructive methods. X-radiography and computed tomography (CT) have played an important role within this type of non-invasive archaeology. These techniques were generally used for the visualisation of manufacturing details and to provide more accurate vessel profiles, particularly of closed shapes. The further development of the medical CT to an industrial X-ray computed tomography for non-destructive testing (NDT) established a technology that enables a much closer examination of the object due to high local resolution. Industrial CT generates a complete 3D digital model of the object with all features of its shape and its ceramic paste, which are recorded in different grey levels due to their local material density (X-ray absorption). Various, mostly medical, software programs can be used for 3D visualisations or virtual cross-sections of these objects (Fig. 1).

The successful use of the large volume of CT data for pottery analysis requires controlled data acquisition and new methods of post-processing specifically adjusted to archaeological ceramic studies (Rice 1987; Orton *et al.* 1993). This article presents preliminary results of a pilot project that demonstrates how industrial CT can be an important tool in the examination of museum objects, as well as providing an overview of the potential of its application in ceramic studies. It also offers an outlook on an integrated approach by which recent improvements in scientific computing pave the way for the use of these large data volumes for analytical purposes and help to evaluate the segmented volumetric data with surface reconstruction methods.

## 2. Historical review

The application of X-radiography in ceramic studies dates back to the 1930s (Berg 2008). It was used for the visualisation of manufacturing details and for the reconstruction of pottery-making techniques, but to a lesser degree also for the investigation of pottery fabrics. In the late 1980s various medical and industrial radiographic

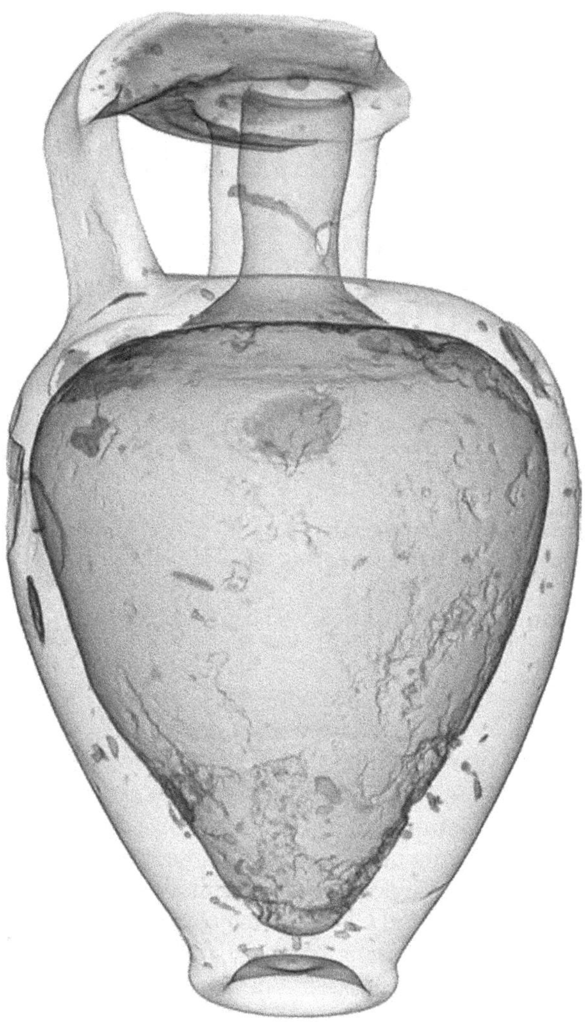

**Figure 1.** Protocorinthian Aryballos UMJ ARCH inv. 4214 (CT01184: 79 kV, 197 μA, voxel size 159 μm, Al 0.5mm), visualisation using the volume renderer developed at the VRVis Research Center, supported by ÖGI.

techniques were comprehensively reviewed and evaluated for their usefulness with regard to archaeological and anthropological case-studies (Carr 1990; Carr and Riddick 1990). Despite these, there are only a few recent ceramic projects that use X-radiography and approaches for a systematic digital radiography of archaeological pottery (Middleton 2005; Greene and Hartley 2007; Berg 2008).

Computed tomography, a three-dimensional imaging system developed in the early 1970s, has had an immense impact on medical diagnosis. Although most common in medicine, CT has also played an important role within non-invasive archaeology (Applbaum and Applbaum 2005). However, the lack of medical CT scanners available for archaeologists to use prevented investigations on a larger scale until recently. Nevertheless, a group of the Allard Pierson Museum and the Academic Medical Centre of the University of Amsterdam has shown the potential of systematically using CT in studies of Greek ceramics (Jansen and Koens 1996; Van de Put 1996; 2006; Jansen *et al.* 2001; Borgers and Brijder 2007). The technique was generally used for two purposes: to provide more accurate vessel profiles, particularly of closed shapes, and for insights into ancient pottery techniques. A very interesting aspect of the Amsterdam project was the attempt to use the variations of the grey tones of a CT scan, which are largely due to different petrographical compositions of the ceramic material, for the comparison and determination of pottery fabrics (Koens and Jansen 1999; Van Duivenvoorde 2000). Closer inspection of inner structures and the detection of inclusions and voids are greatly restricted by medical scanners due to their low resolution and contrast, which are adjusted to the investigation of *in vivo* human bodies.

In recent decades, industrial X-ray computed tomography has opened up a new field of NDT. Industrial CT has created a procedure in which the finest interior and exterior structures of mainly manufactured castings and components can be examined for the industry with high resolution and high penetration rates according to certain quality requirements. This technological progress transformed industrial CT into a method most suitable for the non-destructive investigation of archaeological objects. Some of the archaeological issues in pottery analysis benefit from experiences in industrial applications, such as defect detection of NDT (Eisele 2002). Nevertheless, industrial CT is rarely applied in the study of archaeological objects. The reason for this can be found in the cost of such an investigation, and sometimes in the lack of mobility of museum objects, because the specimens must be transported to the site where the CT scanner is located. However, the main problem until recently was the huge datasets generated. Within the research on Greek pottery, industrial CT has recently been applied to single objects, but the results have mainly been presented in terms of their visualisation capabilities, whereas their potential for further applications in material analysis has only been touched upon (Dimitrov *et al.* 2006; Friedmann and Bente 2007).

## 3. Method and data acquisition

Cooperation with the Austrian Foundry Research Institute (ÖGI) in Leoben made it possible to evaluate industrial CT as an archaeometric or archaeological method in pottery studies (Karl 2010). For the present study, a homogenous group of small closed vessels (maximum size about 15 cm), representing mainly Corinthian ceramic, and a sample group of ceramic fragments of different origins was selected from the archaeological collections of the Universalmuseum Joanneum in Graz (UMJ) and the Department of Archaeology at the Karl-Franzens-University Graz (KFUG) (Lehner *et al.* 1993; Karl *et al.* 2009). All objects were acquired in the $19^{th}$ century and have not undergone any subsequent conservation. From an archaeological point of view, the objects can be attributed confidently to well-known pottery types of the Archaic time ($7^{th}$–$6^{th}$ century BC). Minimal contact with the vessels during the entire scanning procedure and carriage from the museum and back to it has proven ideal.

The objects were scanned using a Phoenix X-ray v|tome|x cone beam CT-system equipped with a 240 kV micro-focus tube with a variable focal spot size of 4 to 200 μm and a 16 bit flat panel detector (512×512 px). The maximum resolution of the scanner is 5 μm, depending on the size of the object. The material ceramic turned out to be appropriate for this type of radiography because it absorbs hardly any X-rays due to its low specific gravity that requires low energy. The requirement of low energy is very important because, in handling museum objects, the issue of preservation of integrity and the degree of non-destructiveness have to be addressed. It is well known that the objects were exposed to ionising radiation, which influences the luminescence signal (Haustein *et al.* 2003; Castaing and Zink 2004). Although dating of museum objects with unknown post-antique transformation processes by thermoluminescence (TL) often provides very inaccurate results, in the present study the effects of the added radiation dose on the TL age were investigated. By means of the analysis of different loaded series of ceramic samples, limiting values – as in medicine – were introduced in order to minimise this impact. These main limits for the Phoenix v|tome|×CT-system were defined as follows: time of scanning=720 sec, acceleration voltage ≤150 kV, and tube current ≤300 µA. In brief, the first preliminary results of the TL analysis show that the impact of CT within these limiting values falls within an acceptable range and displaces the TL age by no more than 20% of the age of the investigated ceramic samples of ancient times (Rudolf Erlach personal communication), which is well within the accuracy range for museum objects (Dittberner 2009).

This low impact is made possible because the Phoenix v|tome|×CT-system has a fast data acquisition time of 720 sec. Within the defined limiting values for radiation, these parameters are directed to achieving the best possible image information, with high contrast and resolution. The determination of consistent parameters, even just for the one CT scanner used in this project, is a difficult undertaking, because these mainly depend on the complexity of the object (size, material density, wall-thickness, metallic applications, etc.) and on the aim of achieving the best range of optical densities. Using a 1 mm aluminium filter together with a 1 mm copper filter and a high acceleration voltage (but within the limit of 150 kV) has proven optimal for the quality of the CT data. This has reduced the softer, more easily scattered X-rays with longer wavelengths, and corrected some of the beam-hardening artefacts, which are a common phenomenon in CT technology (Eisele 2002, 6–8). The next step in CT scanning is the reconstruction process, which lasts about the same amount of time as scanning and generates a volumetric record which is represented in so-called voxels, the 3D-equivalent of a pixel, with a certain edge length – in the present study always isotropic. The 16 bit system detector has the ability to record a wide spectrum of grey values, between 0 and 65535. Finally, one record consists of $512^3$ voxels and is c. 260 MB in size, which conventional PCs can easily cope with. However, there are still beam-hardening effects, like dark bands or streaks and cupping artefacts, which require further attempts at an adequate beam filtration and a better calibration correction in the reconstruction process.

## 4. Volumetric data

Volumetric data of CT scans have a well-known additional advantage over all other surface-based data acquisition methods, such as the 3D scanner based on structured light (DePiero and Trivedi 1996). The documentation is efficient and the scanning time is not determined by the complexity of an object. CT provides a complete 3D data set of an object and allows archaeologists to look into the body of a vessel. Inaccessible parts of objects become visible and can therefore be evaluated. Measurements can be taken at every position. In comparison to a conventional archaeological drawing, the inner profile line of an object with a narrow mouth can be accurately determined for the first time (Fig. 2).

In addition, there are also ways to use CT data for comparative analysis. The volume of the ceramic can be measured using a so-called isosurface. For this purpose, a grey level that is in between the grey levels representing the ceramic and the air is chosen. This way, all voxels with values higher than that threshold are counted as material. The iso-grey level can either be calculated automatically by software programs like VGStudio, or set manually. In addition to the dry weight of an object under defined conditions such as temperature (21–23°C) and humidity (40–50%), it also becomes possible to calculate bulk density, i.e., the density including the ceramic matrix with inter-particle voids and internal pore volume.

Various voids can be identified in the CT data and show differences in the preparation of the clay paste. One possibility involves obtaining this information about porosity from two-dimensional cross-sections and image processing (Lang *et al.* 2005). Comparable grey values for further processing have been achieved using the histogram peaks for air and ceramic. The isosurface was defined at the mean of these two peaks. The maximum grey value for the ceramic matrix has been placed symmetrically to the ceramic peak, at the same distance as to the isosurface. The next step was the reduction of the CT grey values to grey-scale images of 256 values in such a way that the lowest CT value (the isosurface) obtains the new value 255 and the highest (the densest material in the ceramic matrix) the value 0. This mapping allows an approximated evaluation of the visible voids, depending on the CT resolution (Pierret *et al.* 1996), with software programs (Adobe Photoshop, Scion Image <http://scion-image.software.informer.com>) using filtering tools for enhancing the edges, correcting the cumulative effect and reducing the number of grey values to binary images (Fig. 3).

Nevertheless, the results obtained in this first attempt with the use of standard software programs were still unsatisfactory in terms of their accuracy and the lack of a three-dimensional analysis. However, through the application of scientific computing, the ceramic volume could already be obtained with a new method of surface reconstruction, and significantly improved the accuracy of the data because of its verifiability. This surface reconstruction also enables a non-contact measurement of the capacity of the vessels, which represents important information for the standardisation of ancient pottery production (outlook for more details).

## 5. Forming techniques

Much useful information about the production process can be obtained by looking into the interior of objects and into their ceramic material. Forming techniques are clearly visible in the specific texture of the ceramic matrix and in the orientation of the voids and the inclusions. A characteristically helical structure from the base to the rim (Tellenbach 2002), in the present study always in a clockwise direction, shows the rotation of the potter's

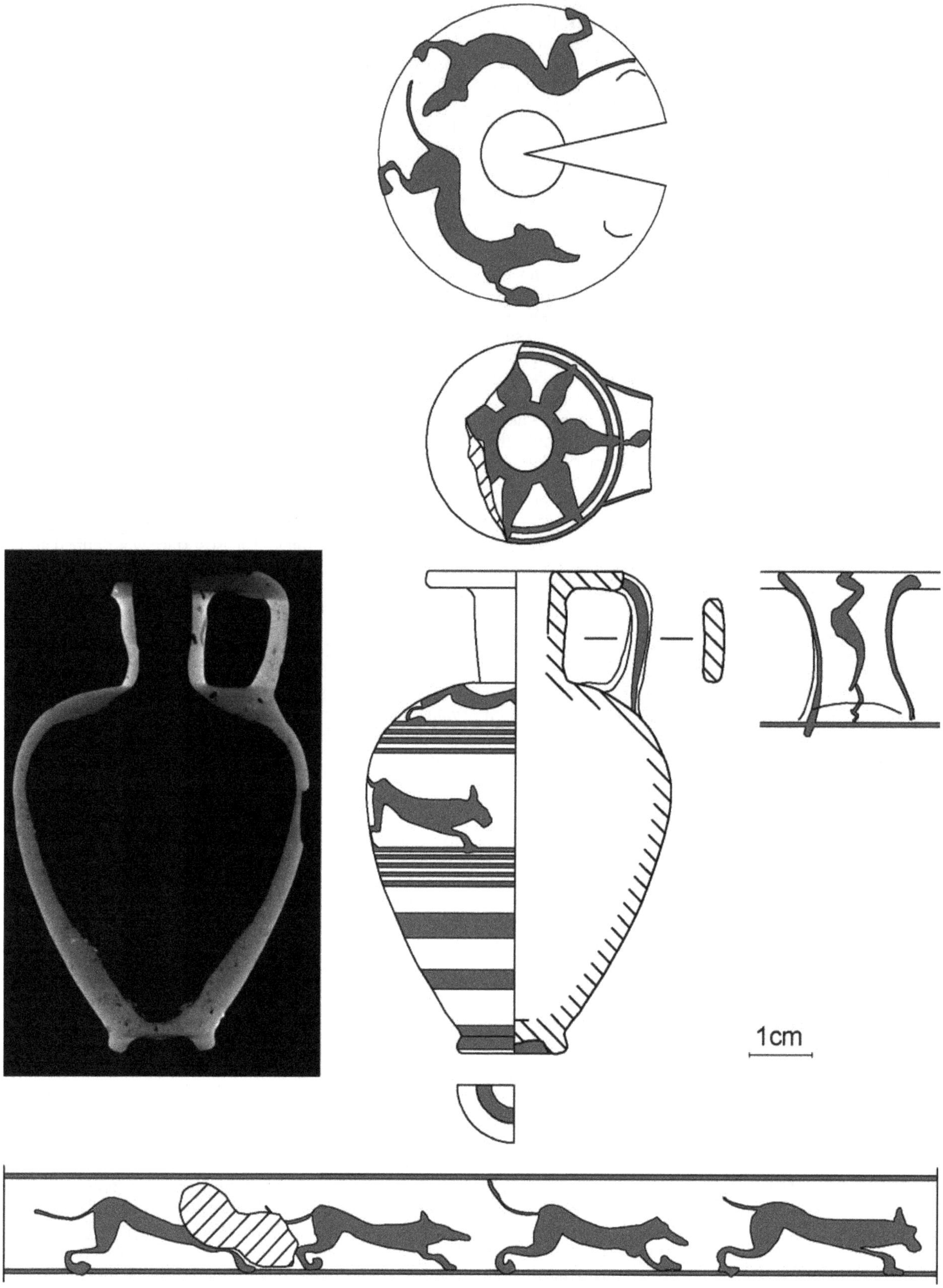


**Figure 2.** Protocorinthian Aryballos UMJ ARCH inv. 4214 (CT01184:), conventional drawing and CT cross-section.

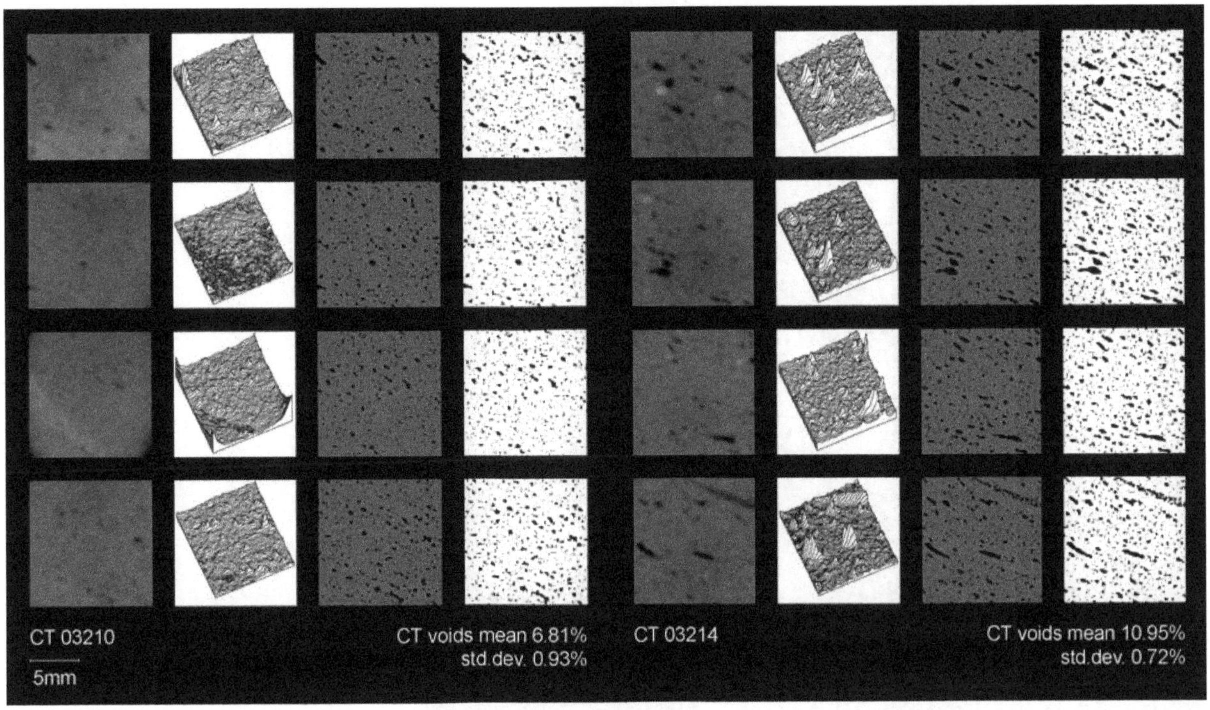


**Figure 3.** CT cross-sections of a Corinthian (left) and Protocorinthian (right) example (CT03210/CT03214: 150 kV, 225 μA, voxel size 184μm, Al+Cu 1.0+1.0mm) for image analysis.

wheel (anti-clockwise turning), and probably gives an indication of a right-handed potter (Schreiber 1983). Due to the high resolution of the CT, wheel-thrown vessels and those that are coiled and then wheel-shaped can be distinguished (Berg 2008). Turning marks seen inside at the bottom depending on traces left by the hand of the potter, or spiral stress lines at the shoulder running up into the neck, enables the reconstruction of the entire process of production from a lump of clay on the wheel to the finished piece (Rye 1988; Schreiber 1999). The studies show that, despite their outward similarities, vessels present differences in their construction and in the usage of additional shaping tools. CT reveals the composition of different parts, such as necks or bases, which are recognised most readily because of the alteration in the texture.

Scans of the Protocorinthian and Corinthian aryballoi indicate different manufacturing techniques of these small globular vessels with narrow mouths, which were changed over the years. They were all thrown on the wheel, but

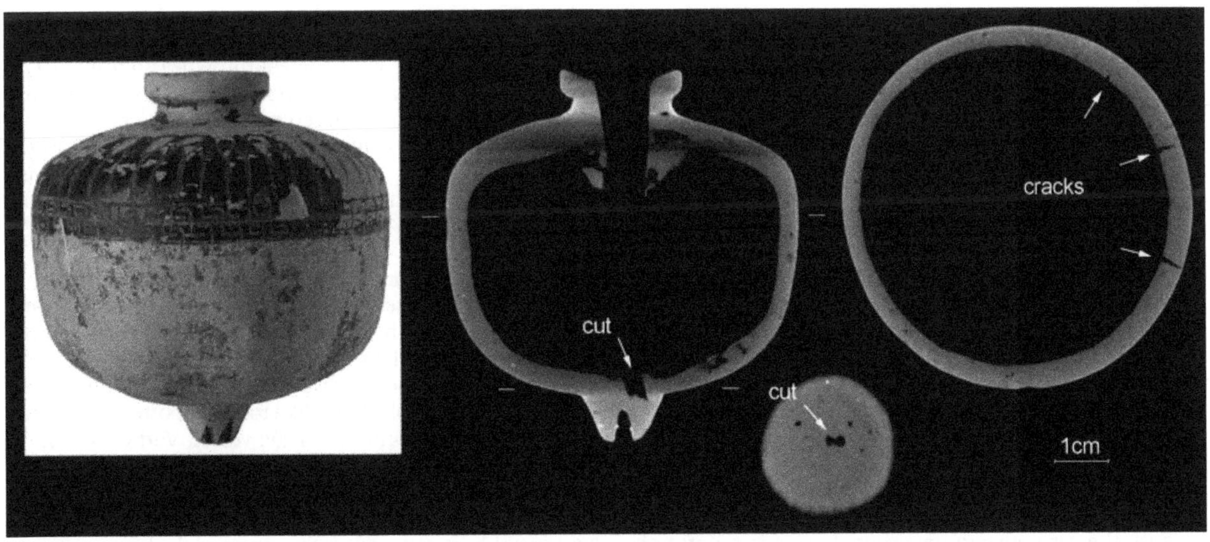


**Figure 4.** East-Greek pomegranate-vessel KFUG Dept. Arch. inv. G 56 (CT03047: 85 kV, 225 μA, voxel size 190 μm, Al 1.0mm), image and CT cross-sections.

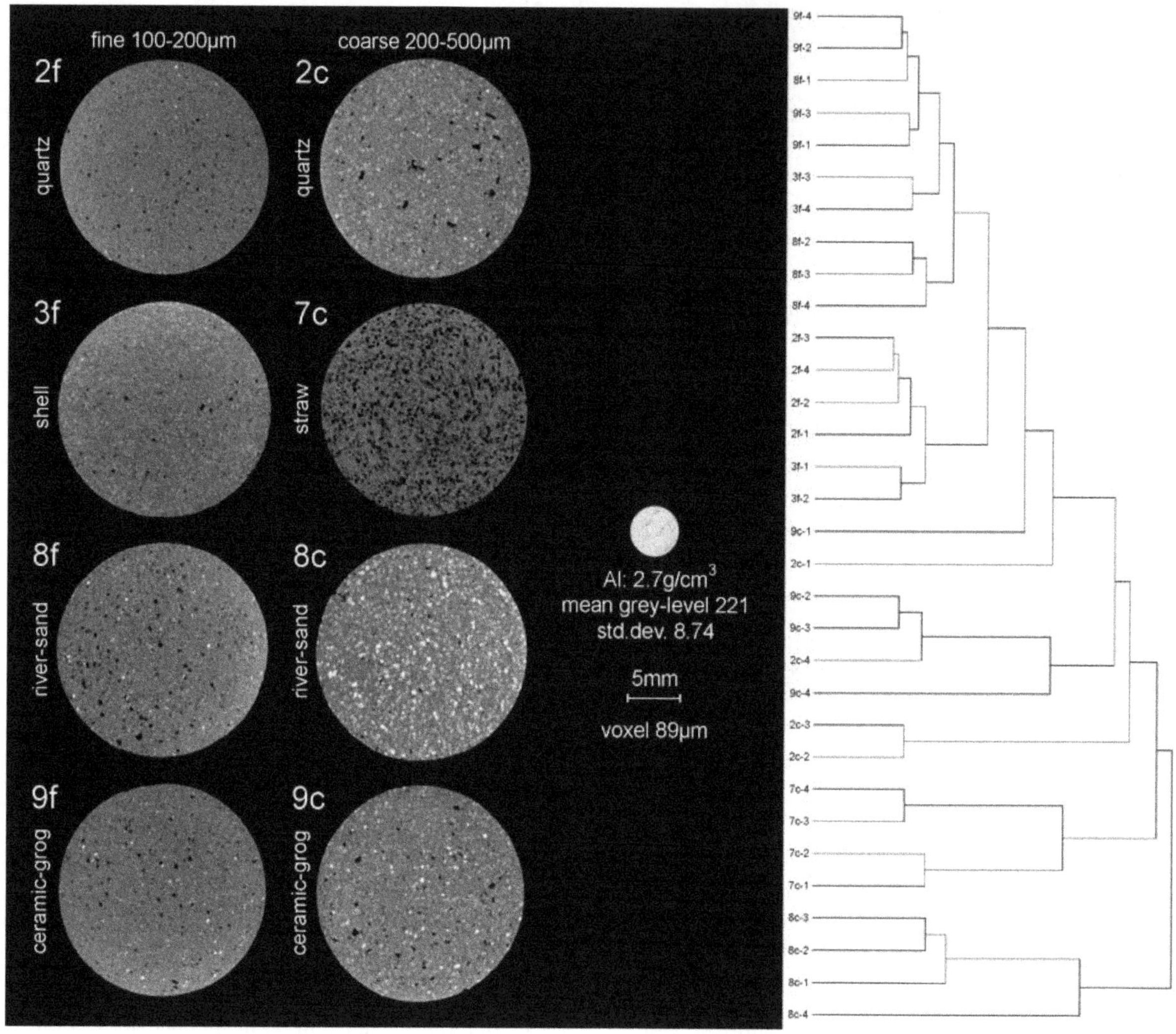


**Figure 5.** Test tiles 3f, 7c, 8f, 8c, 9f, and 9c (CT03071: 110 kV, 132 µA, voxel size 89 µm, Al 2.0 mm), CT cross-sections and classification dendrogram resulting from cluster analysis performed using Euclidian distance.

display differences in the making of the opening, with a very narrow neck and a wide mouth. The most difficult task in the production of these small vessels was angling the wall inward towards the centre and narrowing the opening to an inside diameter of around 5 mm. Sharp breaks and edges demonstrate the variety of potters' tools used for the manufacture of these shapes. Characteristic of the Early Corinthian aryballoi is a type of dowel that is used for this process of collaring-in and has left its mark on the break to the shoulder. The data raise a number of important questions concerning the development of this shape and the introduction of the globular aryballos. The results of each individual piece will be published in detail in a forthcoming Corpus Vasorum Antiquorum (CVA) containing the collection of Greek vessels of the Universalmuseum Joanneum in Graz.

One piece proved exemplary for the application of the CT method for the identification of forming techniques, because the results were unexpected. This is a pomegranate-shaped vessel which belongs to a well-known group of so-called East Greek plastic vases of the late 7th to 6th century BC (Ducat 1966, 142–4). The *communis opinio* is that these vessels were wheel-thrown (Boldrini 1994, 46–7). The investigation has shown the opposite; there are no turning marks at all, moreover the wall thickness varies, and the vessel was made of three separate sections joined together: the body, the mouth, and the cone end, which was originally part of the remaining sepals of the pomegranate (Fig. 4).

Due to the vertical orientation of the pores and inclusions, it is likely that this vessel was partially formed by using a one-piece mould into which the potter spread and pressed the clay wall more or less evenly with the fingers. After the formation of the body, a disc-shaped spout was appended, and, with the help of a knife-like tool, the potter opened the body wall into the interior of the vessel. Vertical cracks in the belly area were caused by physical stresses during the contraction of the wall. The possibility that perhaps all East Greek pomegranate-shaped vessels were made in moulds and could be compared to the group of Archaic terracotta figurines opens up new vistas for the archaeological understanding of this ceramic type.

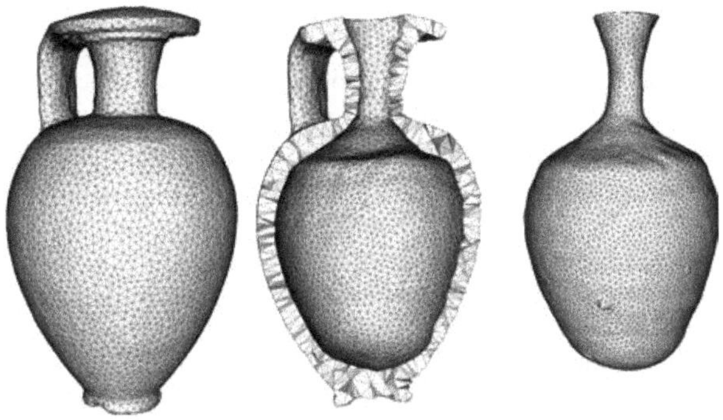

**Figure 6.** Aryballos UMJ ARCH Inv. 25277 (CT03145: 77 kV, 285 μA, voxel size 182 μm, Al 2.0 mm), surface reconstruction method for visualisation (left), ceramic volume (middle), and capacity (right).

## 6. Textural analysis

Although the emphasis has been on the different manufacturing techniques and on clarifying the hidden inner structures, the results of the present study show potential for the visual classification of pottery fabrics. A large variation of grey values is visible in the ceramic body and shows enough distinguishing criteria for such a classification. This is due to various inclusions and particles that have different radiation-absorbing rates from the ceramic matrix. CT is no substitute for petrographical investigation, but can detect these higher density particles. Perhaps the addition of reference materials with known specific gravity would enables a form of calibration (e.g., aluminium with 2.7 g/cm$^3$) with a view to measuring the volumetric proportion of these different particles within the vessel and using it to determine fabric groups.

The use of a series of test tiles with different tempering might help to determine the efficacy of this approach. For example, test tiles were made of modern fine levigated clay (Westerwald stoneware) with a 10% addition of one kind of tempering (quartz, shell-sand, straw, river sand, ceramic grog, etc.), each of these in two different particle sizes (100–200 μm and 200–500 μm), and fired in a modern electric kiln at a temperature of 960°C. Although some of these particles have similar specific gravity to the clay (mean 2.65 g/cm$^3$), for example quartz (mean 2.63 g/cm$^3$), they can be easily recognised because of their homogeneity as compared to that of the clay matrix, in which the inter-particle micro-voids decrease the radiodensity of the clay due to the partial volume effect (cf. Berg 2008, 1186) (Fig. 5).

Depending on the resolution, these particles could be distinguished as diagnostic features that can be analysed in terms of frequency, size, elongation, etc. by standard methods such as the optical classification of pottery fabrics normally applied at fresh breaks of ceramic sherds (Gassner 2003, 23–34; Gassner and Schaller 2009). In the further procedure, four CT cross-sections were made through each of the test tiles and grey-scale images were generated according to the procedure already described above. Each image was segmented by thresholding into the regions 0–85 (less dense, air) and 170–255 (dense), and analysed regarding area, elongation, grey-level, and major axis of these features with the use of the imaging program (UTHSCSA ImageTool <http://compdent.uthscsa.edu/dig/itdesc.html>). Analysis programs like RapidMiner (<http://rapid-i.com>) help to examine these data and to classify them into groups based on these sets of values and a similarity measure. Such an agglomerative clustering (visualised by a classification dendrogram) shows strong correlation to the initial fabrics. In spite of this two-dimensional approach, the results are encouraging and show that textural analysis using CT data is possible. Nevertheless, petrological examination by thin-sectioning of these test tiles is envisaged in order to test the reliability of this type of analysis.

## 7. Outlook

The reconstruction algorithm NeuRA2 (Jungblut *et al.* 2013) generates a triangular surface mesh of the scanned objects (Fig. 6). The application of a volumetric mesh generator (TetGen <http://wias-berlin.de/software/tetgen>; Si 2008) transforms the surface mesh into a tetrahedral volumetric mesh. Summing up the volumes of each single tetrahedron yields an approximation of the volume of the scanned ceramics, and, since the mass of those objects is usually known, the bulk density of the objects can be calculated. Reconstructing the inner parts of the vessels also enables one to calculate an approximation of their capacity. However, an error estimation for the volume and capacity calculation has to be evaluated, and this will represent the objective of future research.

## 8. Conclusions

The above results provide considerable information regarding different manufacturing techniques and also demonstrate the possibility of using CT for textural analysis of pottery fabrics. Despite the difficulties inherent in the data acquisition and image processing, reasonably good results can be achieved by means of CT. These results indicate both a need for better grey-scale calibration as well as a reduction of the artefact-formations and beam-hardening effects. The final outcome has to be the use of the volumetric data of the CT not only for 3D visualisations, but also for textural analysis of the complete object. The application of surface reconstruction methods to the volumetric CT data is a promising approach for this type of 3D analysis. It is to be hoped that the general development of CT in industry will accommodate archaeological applicability in the future. If this occurs, studies using CT scanners will become more and more reasonable and cost-effective, the devices will be used more widely, and, as in the field of digital photography, resolution and data storage features will increase rapidly. A careful examination of the various details of manufacturing techniques relating to the reconstruction of the pottery production process, may lead to a better understanding of the ancient potter's craft.

## Post scriptum

The methodological approach using industrial X-ray computed tomography for non-contact and non-invasive investigation of ancient ceramics is now comprehensively documented:

Karl S., Jungblut D. and Rosc J., 2013, Berührungsfreie und nicht invasive Untersuchung antiker Keramik mittels industrieller Röntgen-Computertomografie. Mit einem Beitrag von Rudolf Erlach. In *Interdisziplinäre*

*Dokumentations- und Visualisierungsmethoden, Corpus Vasorum Antiquorum Österreich Beiheft 1*, (ed. E. Trinkl), 73-114, Verlag der Österreichischen Akademie der Wissenschaften, Wien, <http://hw.oeaw.ac.at/0xc1aa500d_0x002ec069.pdf>.

## References


Applbaum N. and Applbaum Y.H., 2005, The Use of Medical Computed Tomography (CT) Imaging in the Study of Ceramic and Clay Archaeological Artifacts from the Ancient Near East. In *X-rays for Archaeology*, (eds. M. Uda, G. Demortier and I. Nakai), 231-245, Springer, Dordrecht.

Berg I., 2008, Looking through pots: recent advances in ceramics X-radiography. *Journal of Archaeological Science*, 35, 1177-1188.

Boldrini S., 1994, *Le ceramiche ioniche*, Gravisca 4, Edipuglia, Bari.

Borgers O.E. and Brijder H.A.G., 2007, *Attic black-figure amphorae, pelikai, kraters, hydriai, olpai, oinochoai, and tripod kothon, Allard Pierson Museum, University of Amsterdam*, Corpus Vasorum Antiquorum – The Netherlands 11, Allard Pierson Museum, University of Amsterdam, Amsterdam.

Carr C., 1990, Advances in Ceramic Radiography and Analysis: Applications and Potentials. *Journal of Archaeological Science*, 17, 13-34.

Carr C. and Riddick E.B., 1990, Advances in Ceramic Radiography and Analysis: Laboratory Methods. *Journal of Archaeological Science*, 17, 35-66.

Castaing J. and Zink A., 2004, Influence de l'irradiation aux rayons X sur la luminescence et la datation des terres cuites. *Technè: La science au service de l'histoire de l'art et des civiliations*, 19, 130-133.

DePiero F.W. and Trivedi M.M., 1996, 3-D Computer Vision Using Structured light: Design, Calibration, and Implementation Issues. *Advances in Computers*, 43, 243-278.

Dimitrov L.I., Wenger E., Šrámek M., Trinkl E. and Lang-Auinger C., 2006, VISAGE: An Integrated Environment for Visualization and Study of Archaeological Data Generated by Industrial Computer Tomography. In *The e-volution of Information Communication Technology in Cultural Heritage, The 7th International Symposium on Virtual Reality, Archaeology and Cultural Heritage, Nicosia 30.10.-4.11.2006*, (eds. M. Ioannides, D. Arnold, F. Niccolucci and K. Mania), 49-55, Archaeolingua, Budapest.

Dittberner L., 2009, *Die Thermolumineszenzanalyse zur Altersbestimmung von Keramik*, Unpublished diploma thesis, University of Applied Sciences and Arts Hildesheim/Holzminden/Göttingen.

Ducat J., 1966, *Les vases plastiques rhodiens archaiques en terre cuite*, Bibliothèque de l'École Française d'Athènes et de Rome 209, F. de Boccard, Paris.

Eisele H., 2002, *Automated defect detection and evaluation in X-ray CT images*. PhD thesis, University of Heidelberg, <http://www.ub.uni-heidelberg.de/archiv/3106>.

Friedmann G.K. and Bente K.A., 2007, Zu antiken, neuzeitlichen und aktuellen Restaurierungen auf der Grundlage materialwissenschaftlicher Diagnostik. In *Konservieren oder Restaurieren – Die Restaurierung griechischer Vasen von der Antike bis heute*, (eds. M. Bentz and U. Kästner), 129-132, Corpus Vasorum Antiquorum – Deutschland Beiheft 3, Beck, München.

Gassner V., 2003, *Materielle Kultur und kulturelle Identität in Elea in spätarchaischer-frühklassischer Zeit. Untersuchungen zur Gefäß- und Baukeramik aus der Unterstadt*, Velia-Studien 2, Verlag der Österreichischen Akademie der Wissenschaften, Wien.

Gassner V. and Schaller K., 2009, Ein webbasiertes Informationssystem für Keramikscherbentypen (fabrics) im zentralen Mittelmeerraum (FACEM). *Forum Archaeologiae*, 53/XII/2009, <http://farch.net>.

Greene A. and Hartley C., 2007, (Re)discovering ceramic structure: a multiscalar approach to understanding ancient ceramic practices. In *Proceedings of 9th European Meeting on Ancient Ceramics – Vessels: Inside and Outside (EMAC'07)*, (eds. K.T. Biró, V. Szilágyi and A. Kreiter), 5-14, Hungarian National Museum, Budapest.

Haustein M., Krbetschek M.R. and Pernicka E., 2003, Influence of radiation used by the security control at airports on the TL signal of quartz. *Ancient TL*, 21, 7-10.

Jansen R.J. and Koens H.F.W., 1996, Computer Tomography. A Tool in the Study of Ancient Ceramics, contribution to P. Heesen. In *The J.L. Theodor Collection of Attic Black-Figure Vases*, (ed. H.A.G. Brijder), 199-201, Allard Pierson Series 10, Allard Pierson Museum, Amsterdam.

Jansen R.J., Koens H.F.W., Neeft C.W. and Stoker J., 2001, Scenes from the Past. CT in the Archaeologic Study of Ancient Greek Ceramics. *RadioGraphics*, 21, 315-321.

Jungblut D., Karl S., Mara H., Krömker S. and Wittum G., 2013, Automated GPU-based Surface Morphology Reconstruction of Volume Data for Archaeology. In *Scientific Computing and Cultural Heritage, Contributions in Computional Humanities*, (eds. H.G. Bock, W. Jäger and M.J. Winckler), 41-49, Springer, Heidelberg.

Karl S., 2010, Durchleuchtungen griechischer Keramik – Industrielle 3D-Röntgen-Computertomografie als archäometrische Methode. In *Primus conventus austriacus archaeometriae. Scientiae naturalis ad historiam hominis antiqui investigandam MMIX, Tagungsband zum Ersten Österreichischen Archäometriekongress, 15.-17. Mai 2009, Paris Lodron Universität Salzburg, ArchaeoPlus 1*, (eds. J. Kiesslich, F. Lang, K. Schaller, C. Uhlir and M. Unterwurzacher), 99-104, University of Salzburg, Interfakultärer Fachbereich Gerichtsmedizin und Forensische Neuropsychiatrie, Salzburg.

Karl S., Modl D., Porod B. (eds.), 2009, *Katalog Archäologiemuseum. Schloss Eggenberg, Universalmuseum Joanneum*, Schild von Steier 22, Universalmuseum Joanneum, Graz.

Koens H.F.W. and Jansen R.J., 1999, Computed Tomography and Ancient Ceramics: a new approach.

In *Classical Archaeology towards the Third Millennium: Reflections and Perspectives, Proceedings of the XVth International Congress of Classical Archaeology, Amsterdam 12-17.7.1998*, (eds. R.F. Docter and E.M. Moorman), 223-224, Allard Pierson Series 12, Allard Pierson Museum, Amsterdam.

Lang J., Middleton A., Ambers J. and Higgins T., 2005, Radiographic images. In *Radiography of Cultural Material*, (eds. J. Lang and A. Middleton), 20-48, Elsevier Butterworth-Heinemann, Amsterdam and London.

Lehner M., Lorenz Th. and Schwarz G., 1993, *Griechische und italische Vasen aus der Originalsammlung des Instituts für Klassische Archäologie der Karl-Franzens-Universität Graz*, Veröffentlichungen des Instituts für Klassische Archäologie der Karl-Franzens-Universität Graz 1, Graz.

Middleton A., 2005, Ceramics. In *Radiography of Cultural Material*, (eds. J. Lang and A. Middleton), 76-95, Elsevier Butterworth-Heinemann, Amsterdam and London.

Orton C., Tyers P. and Vince A., 1993, *Pottery in Archaeology*, Cambridge Manuals in Archaeology, Cambridge University Press, Cambridge.

Pierret A., Moran C.J. and Bresson L.-M., 1996, Calibration and Visualization of Wall-Thickness and Porosity Distributions of Ceramics Using X-radiography and Image Processing. *Journal of Archaeological Science*, 23, 419-428.

Rice P.M., 1987, *Pottery analysis: A Sourcebook*, University of Chicago Press, Chicago and London.

Rye O.S., 1988, *Pottery technology. Principles and Reconstruction*, Manuals on Archaeology 4, Taraxacum, Washington DC.

Schreiber T., 1983, In *Greek Vases in the J. Paul Getty Museum*, (eds. J. Frel and S. Knudsen Morgan), 149-153, Occasional Papers on Antiquities 1, J. Paul Getty Museum, Malibu, CA.

Schreiber T., 1999, *Athenian Vase Construction. A Potter's Analysis*, J. Paul Getty Museum, Malibu, CA.

Si H., 2008, *Three Dimensional Boundary Conforming Delaunay Mesh Generation*. PhD thesis, University of Berlin, <http://opus.kobv.de/tuberlin/volltexte/2008/1966/>.

Tellenbach M., 2002, Exkurs zur Frage der Drehscheibenkeramik. In *An die Mächte der Natur. Mythen der altperuanischen Nasca-Indianer*, (eds. A. Wieczorek and M. Tellenbach), 54-63, Publikationen der Reiss-Engelhorn-Museen, P. von Zabern, Mainz.

Van de Put W.D.J., 1996, The Use of Computer Tomography for the Study of Greek Ceramics, contribution to P. Heesen. In *The J.L. Theodor Collection of Attic Black-Figure Vases*, (ed. H.A.G. Brijder), 203-205, Allard Pierson Series 10, Allard Pierson Museum, Amsterdam.

Van de Put W.D.J., 2006, *Red-figure and white-ground lekythoi, Allard Pierson Museum, University of Amsterdam*, Corpus Vasorum Antiquorum – The Netherlands 10, Allard Pierson Museum, University of Amsterdam, Amsterdam.

Van Duivenvoorde W., 2000, *Attisch aardewerk gevormd in de CT-scan. Computertomografie in de archeologie: de vervaardiging van Attisch zwartfigurig aardewerk uit de zesde en vijfde eeuw voor Christus*. Unpublished diploma thesis, University of Amsterdam.

## ACKNOWLEDGEMENTS

The present research was supported by a grant of the Styrian government (Department 3 – Science and Research). Our thanks go to the Universalmuseum Joanneum in Graz for the access to its Greek vase collection; to Georg Geier, Daniel Habe and Jördis Rosc of the Austrian Foundry Research Institute in Leoben for all their support in the course of the CT scanning process; to Rudolf Erlach of the Institute of Art and Technology of the University of Applied Arts in Vienna for the TL analysis; to Wendy van Duivenvoorde of the Western Australian Maritime Museum in Fremantle and Lena Dittberner, Hannover, for their unpublished theses; to Robert Fürhacker, Gutenberg a.d. Raabklamm, and Günther Nograsek of the Universalmuseum Joanneum in Graz for their help with preparing and firing the test tiles; to Dietmar Klammer and Isabella Egartner of the Institute of Applied Geosciences of the University of Technology in Graz for thin-sectioning the test tiles; to Patrick Fazioli, University of Buffalo, and Christine Rogl, Spital a.d. Drau, for their archaeological comments.

# 21

# Messages impressed in clay: Scientific study of Iron Age Judahite Bullae from Jerusalem

**Yuval Goren[a], Shira Gurwin[b] and Eran Arie[c]**

[a]Laboratory for Comparative Microarchaeology, Department of Archaeology and Ancient Near Eastern Cultures, Tel Aviv University, ygoren@post.tau.ac.il

[b]The Eretz-Israel Museum, Tel Aviv, and Department of Archaeology and Ancient Near Eastern Cultures, Tel Aviv University

[c]The Israel Museum, Jerusalem, and Department of Archaeology and Ancient Near Eastern Cultures, Tel Aviv University

## 1. Introduction

The application of ceramic petrography in archaeology is celebrating now about seven decades since its introduction. Over the years, it has become one of the most common scientific methods in archaeology, due to its availability, low cost and impressive track record. Even the introduction of other, often more sophisticated methods for provenancing ceramics, first and foremost Neutron Activation Analysis (NAA), has not effected the use of petrography due to its ability to supply a wide range of technological as well as provenance data. However, the method has always been limited by its destructive nature, making the analyses of delicate artifacts practically impossible.

During the last decade, one of us (YG) developed a methodology for petrographic examination of delicate clay objects, such as cuneiform tablets and clay seal impressions (bullae). Amongst other things, it has been used for provenance studies of cuneiform tablets from Amarna (Goren *et al.* 2004), Ugarit (as yet unpublished), Hatti (Goren *et al.* 2011) and Israel (Goren *et al.* 2004; 2007; 2009), and the so-called Knossian Replica-Rings from Crete and Thera (Goren and Panagiotopoulos, as yet unpublished). The present paper presents the methodology, results, and implications of the examination of Iron Age bullae from the City of David, namely Iron Age Jerusalem.

Little has been preserved in the archaeological record from the rich literary material of the Kingdom of Judah (Fig. 1). Despite the discovery of some contemporary written sources, such as ostraca and seals, it may be assumed that many of the documents were apparently written on scrolls or papyri that have not survived. Consequently, most of the scholarly records from this period often referred to in the biblical sources, have been lost forever. Only some meager remains of these texts have been preserved in the form of bullae, namely, the clay sealings that were once attached to them. Bullae are small lumps of clay, often fingernail-sized and shaped as flat disks. They were usually affixed to a cord binding a papyrus-document and then stamped with a seal (Fig. 2). Other bullae were apparently sealing basketry or fabrics, most likely small bags containing some commodities, as evident by the impressions on their reverse sides.

Only a relatively small amount of bullae have been found in the course of over a century of archaeological exploration of the major Iron Age sites of Judah (Avigad 1997, 167-241). The reasons for this situation are as follows: First, due to their small size, such tiny clay objects can easily escape the attention of inexperienced workers who do not search specifically for them. Moreover, careful sifting was not always the common practice in many of the large-scale excavations of the major Judahite tells in the past. Last but not least, the preservation of unfired bullae is often impossible in the sub-humid conditions of the southern Levant, hence most of the bullae found in recorded excavations were discovered as part of very few caches, which survived due to specific depositional conditions.

The bullae sampled in our research include two major groups, dating to two different phases of the Iron Age. The first group, comprising 51 items, was uncovered during the 1982 season of excavations in Area G at The City of David by Shiloh (1984, 19-20; 1986; Shiloh and Tarler 1986) and was published by Shoham (2000) and Brandl (2000). At the base of the famous stepped stone structure, a series of buildings was erected during the seventh century BCE. They spread over two terraces: the 'House of Ahiel' and the 'Burnt Room' were found on the upper terrace while on the lower one stood the 'Bullae House'. The floor of the latter, only partly excavated, was covered by a thick charred destruction layer containing the bullae together with pottery vessels, arrowheads, a scale weight and four limestone altars. The finds are typical of the final stage of the Iron Age and the bullae found in this context clearly date to the last phase before the Babylonian destruction of Jerusalem in 586 BCE. Most of the bullae are in very good state of preservation, hence fully legible (Fig. 2); four of which are unepigraphic. They bear dozens of Hebrew personal names, two of which belong to figures known from the Bible. The first is Gemaryahu son of Shaphan (Shoham 2000, 33), a high official at the court of King Jehoiakim, and the second is 'Azaryahu son of Hilkiyahu (*ibid*, 43), a member of a priestly family appearing in the book of Chronicles (Schneider 1988).

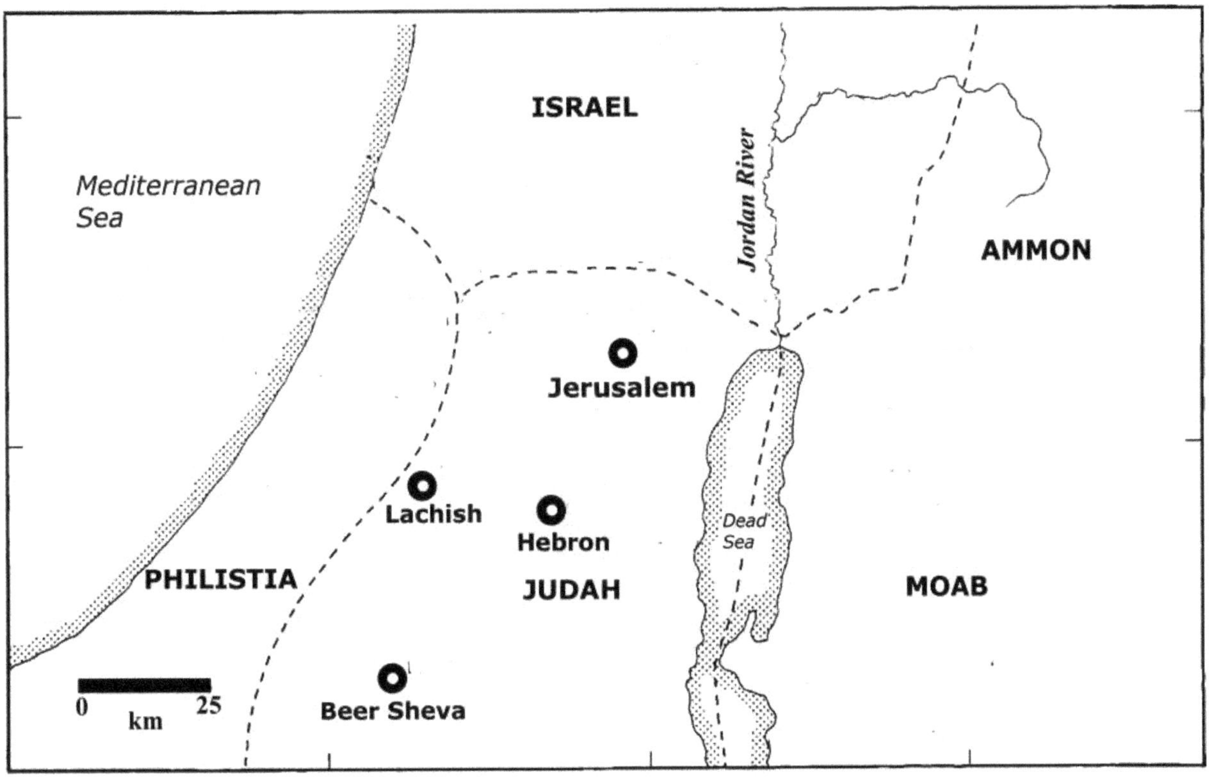


**Figure 1.** Jerusalem and the boundaries of the Kingdom of Judah between the 8th and the 6th century BCE.

The second group of bullae includes over 120 items, dating to the 9th - late 8th Centuries B.C., discovered only recently in the water system of City of David near the Gihon spring (Reich *et al.* 2007, 156-157). Being earlier to the second group, these bullae are not epigraphic and their seal impressions contain only decorative patterns. These bullae will be discussed in here only with regard to the methodology of their examination.

## 2. Method

This study was aimed at providing the opportunity to analyze some as yet undetermined aspects of the Judahite bullae. Since it is widely believed that bullae were used to seal documents or small parcels sent from one authority to another, ensuring the discrete reading of a message or opening of the parcel by the addressee alone, we first attempted to disclose the geographical origin of the bullae through the provenance of their clays, in order to map the network of the administrative correspondence of Judah during the middle to the end of the Iron Age. By doing so, we hoped to reveal the location of several personalities and to draw the network of Judahite bureaucracy. Therefore, our first question was whether the material composition of a given assemblage of bullae would reflect sufficient similarity to justify their assignment to a single site, or whether the analysis would show that they were made of clay from different locations. The answer to this question was the key to the research questions that followed, since if the first option were true, then the following question had to be related with the issue of provenance of the entire lot. However, if greater diversity of raw materials were recognized, then the study would need to focus on the correlation of the clay types with the textual and stylistic characteristics of the bullae and the location of their discovery. As in standard provenance studies of ceramics in archaeology, it would be naïve to expect that the mineralogical or chemical analyses of small samples of clay would enable provenance determinations to the single site level. Moreover, there are obviously some fundamental differences between bullae and pottery, and consequently, between the preferences of potters and scribes or other officials. However, as is the case of pottery, the background of the study of bullae is such that the interpretation of the results can be considerably narrowed by many archaeological and historical considerations. In fact, the distribution of the Judahite Late Iron Age sites where bullae could have been issued and used is such that it leaves only very few possibilities open, if the analysis were to suggest even a general area within the confines of Judah.

As a result, our research project was planned to be made in three stages. In the first stage we examined the structural and technical aspects of the bullae based on surface microscopic observations under a stereomicroscope, with magnifications ranging between 10 to 100 times. This was made in order to record minute details of the papyrus, fabric, or leather, and the cord impressions, the fingerprints and other imprints, and of course the seal impression themselves. These examinations attempted to address some technical questions, such as the general composition of the fabric and the formation process.

In the second stage, minute samples were extracted from the bullae by the peeling technique and examined in thin sections under the petrographic microscope. In this

**Figure 2.** General stereomicroscopic view of one of the inscribed bullae from the "Bullae House" in Area G, the City of David (IAA 84-123). The palaeo-Hebrew seal reads: "(of) Nachum son of Sha'ala". Late $7^{th}$ - early $6^{th}$ century BCE, field width: about 2.5 cm.

method, a shallow lamina, sizing only few millimeters, is peeled off from a broken facet of the bulla or from its reverse side under the stereomicroscope with the aid of a scalpel or a botanical needle. The samples were set in improvised moulds made of small rounded polyethylene test tube cups, and dried at room temperature for a few hours. Then the cups with the samples were put in a dessicator, where the samples were impregnated with Buehler Epo-Thin low viscosity epoxy resin under vacuum conditions. After curing, the resulting pellet was used for the preparation of a standard thin-section and subjected to routine petrologic examination under a polarizing microscope using X40 - X400 magnifications.

The petrographic definition of each sample was then supported by structural and chemical examinations under the analytical Quanta 200 FEG Environmental Scanning Electron Microscope (ESEM), operated by the Wolfson Applied Materials Research Center of the Tel Aviv University. The microscope combines high vacuum, low vacuum and wet-mode to support a variety of material characterization applications. The specific instrument was chosen because it allows for the examination of non-conducting, contaminated, hydrated and even living samples without significant sample preparation, in addition to those samples that have always been viewable under conventional scanning electron microscopes. It allows for user selection of accelerating voltage, magnification, gas type, gas pressure, and detector type. The microscope uses tungsten electron source and allows wide range of accelerating voltages from 200 V to 30 kV. It also has complete set of detectors providing imagining in secondary and backscattered electrons in all the operating modes (high and low vacuum) at the resolution of 3.5 nm. The system also includes energy dispersive spectroscopy (EDS, supplying qualitative and quantitative data) with Ultra-Thin Window for light element detection down and including carbon with spectral resolution better than 132 eV. Hence

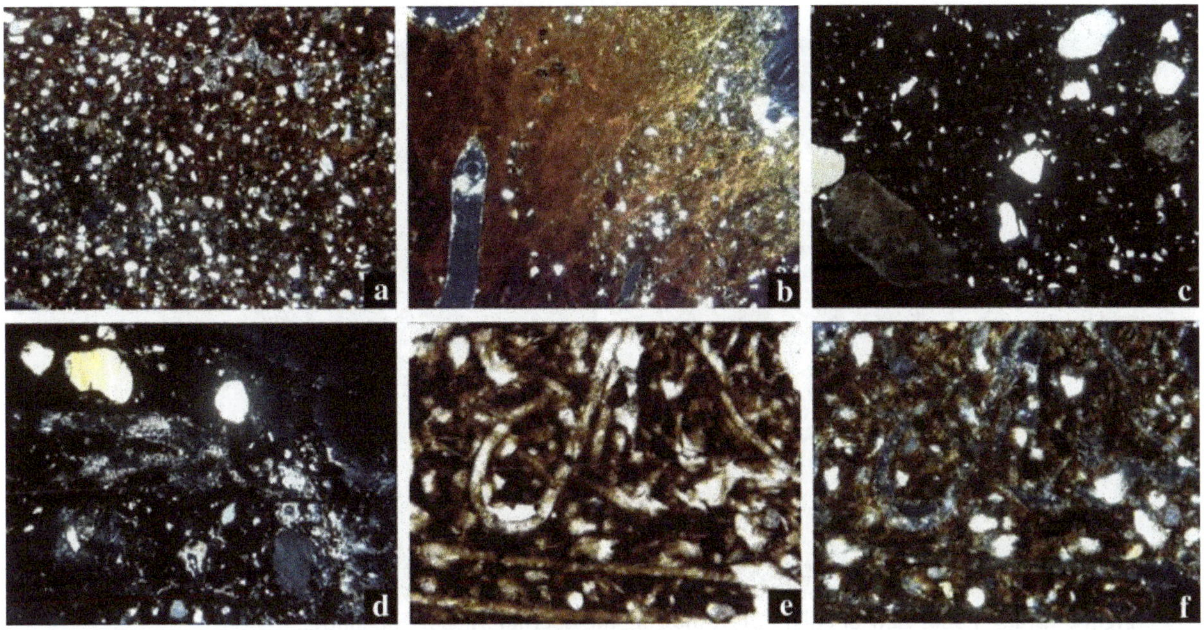

**Figure 3.** Bullae fabrics in thin section: (a)Bulla 20867, XP, Alluviated *Terra Rossa* with high silt contents. Field width: 2.5 mm. (b) Bulla 18692, XP, *Terra Rossa* with low silt contents, note the typical high optical activity with striated b-fabric. The elongated voids indicate vanished vegetal material. Field width: 2.5 mm. (c) Bulla 16771, XP, *Terra Rossa* with nearly isotropic silty matrix and sand inclusions containing spherical quartz and some limestone. Field width: 2.5 mm. (d) Bulla 19314, XP, as 3c but with secondary (post-depositional) crystallization of calcite in the voids. Field width: 2.5 mm. (e) Bulla 22112, PPL, the *Terra Rossa* matrix is mixed with fine fibers. field width: 0.8 mm. (f) As 3e but under XP. Field width: 0.8 mm.

the ESEM enables the examination of the intact bullae at low vacuum, without any process of coating.

## 3. Results

Based on the petrographic data, combined with the ESEM results, the raw material of all the examined specimens are readily identified as derived from soil deposits which are, in fact, Quaternary alluvial beds derived from *Terra Rossa* soils and, in few cases, *Terra Rossa* from *in situ* exposures. It should be emphasized that none of the bullae that we have examined so far were made of clay or marl from older geological formations, such as the local Moza and Teqiye clay formations, even though these were extensively used for pottery production in Judah throughout the ages.

In thin section, most of the bullae appear as non calcareous, ferruginous matrix (Figure 3a). This fabric is typified by reddish-tan to dark matrix in thin section, highly optically active to nearly opaque under crossed polarizers, with silt ranging between 5% (rare) to nearly 20% (common). The silt is mainly quartzitic but it often contains some accessory heavy minerals of which hornblende and zircon are the most common. The inclusions are made of fine sand containing mainly quartz or limestone. Other minerals or rock fragments that rarely appear in the inclusions are chert or chalcedony.

*Terra Rossa* soils occur on hard limestone and dolomite exposures in the semiarid to subhumid Mediterranean climatic zones. This soil material is eroded downslope, forming colluvial-alluvial soils. All the soil materials in Israel include, to varying extents, aeolian dust of desert origin. Carbonate rocks do not contain silt-size quartz grains, but large amounts of such grains occur in the soils that developed on these rocks. The external source of the silt-size quartz grains is considered to be an aeolian contribution to the soil. The largest amount of aeolian dust occurs in soils that developed on hard limestone and dolomitic limestone, in which the residual material released from the dissolution of the rocks is only about 2% (Adan-Bayewitz and Wieder 1992). Only in a few cases nearly non-silty *Terra Rossa* was used (Fig. 3b), indicating the employment of soil from an *in situ* exposure.

The inclusions are sparsely spread and occasional, reflecting opportunistic use of different soil mixtures where often fine sand naturally occurs. This sand is essentially quartzitic, often with the addition of calcareous rock fragments (Fig. 3c). In many cases, very fine splinters of vegetal material were added, as indicated by the void structure, often infilled by secondary re-crystallization of calcite (Fig. 3d). This vegetal material can be, in fact, the result of the inclusion of cords in the bullae. In one case, delicate fibers of unidentified type were densely packed within the matrix of a bulla (Fig. 3 e,f).

The petrographic examinations were enhanced by the ESEM-EDS analyses. The latter were made on the entire surfaces of the bullae rather than on samples which were extracted from them. This necessitated the opening of the vacuum chamber of the ESEM whenever the bulla needed to be tilted, hence in many cases we preferred to define *a priori* the desired surface for scanning. In many cases, this

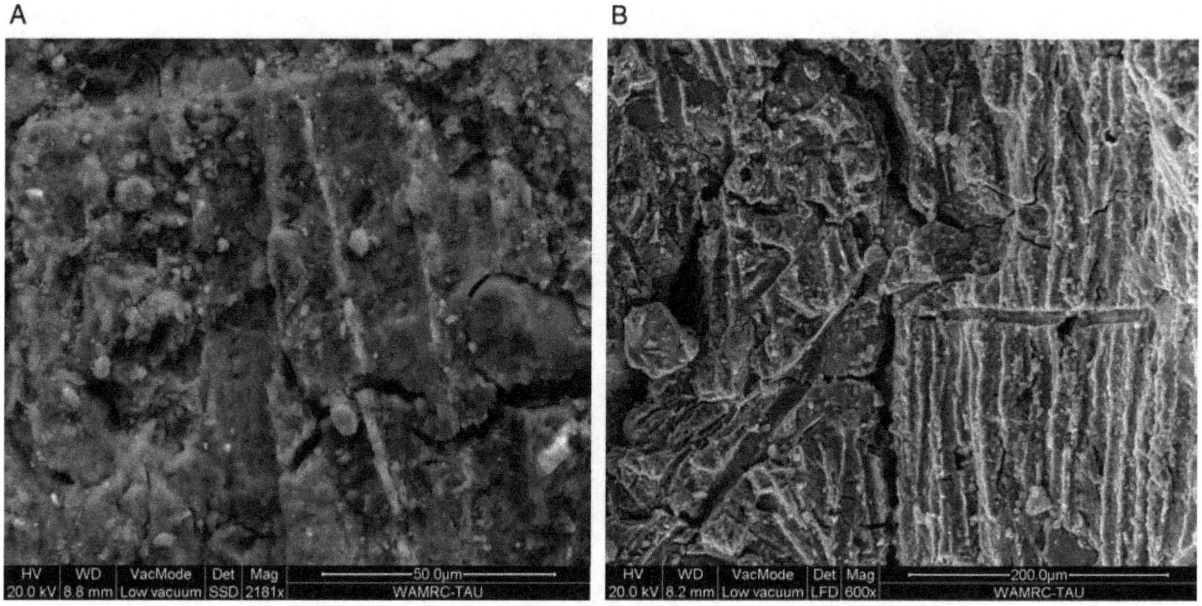


**Figure 4.** (a,b) ESEM view of bulla 22112 showing the fiber imprints in the matrix.

was the surface where the sample for petrography was extracted by the peeling method. This enabled a comparison between the petrographic sample and the surface scanned by the ESEM. A few examples can be seen in Figures 4 and 5: the non plastics (silt and fine sand) could be observed also by the ESEM and identified on the spot with the aid of the EDS system. Structural features, such as the cord impressions, fibers within the clay (Fig. 4), secondary crystallization of calcite (Fig. 5) and other phenomena, could be observed by the ESEM using the secondary and backscattered electron detectors.

To sum up, both the petrographic and the ESEM analyses revealed that the entire two groups of bullae from the City of David in Jerusalem were made of *Terra Rossa* soil, having the same mineralogical composition of silt and temper inclusions. Although *Terra Rossa* soils are quite widespread in the Mediterranean sub-humid parts of Israel, where they develop on hard limestone and dolomite of the Mediterranean climatic zones, the uniform composition of the silt with its accessory minerals, and the coarse fraction, suggests that the assemblage is the product of a single location. Moreover, the distribution of *Terra Rossa* in the confines of Judah is limited to the Judean Anticline, where Jerusalem is the only major site of this period. In addition, this composition is identical to the fabric of the numerous local pillar figurines from the City of David (Goren *et al.* 1996). Therefore, the entire two sets of bullae from the City of David may be regarded as the local production of this site.

## 4. Discussion

The rather unexpected results of this study indicate that the entire assemblage of bullae from the City of David was most likely made locally around Jerusalem. It enables us now to define more precisely the nature of bureaucratic networks of Judah according to the evidence at hand. The fact that all the bullae were found to be made of clay from the major city where they were deposited negates the assumption that the bullae sealed letters that arrived from far away. Instead, it appears that either the bullae sealed locally circulated documents, restricted to the immediate surroundings of the city in which they were found, or that they were used to seal local legal and administrative documents.

The contents of the Lachish ostraca (Torczyner 1938), dating to the same period as the later lot of bullae sampled in our research, indicate that letters written on papyrus were sent together with ostraca. Na'aman's analysis of the archive of Ya'ush points out that the ostraca uncovered in the gate of Lachish were part of a much wider correspondence, most of which was written on papyri (Na'aman, 2003, 175, 179). The assumption that letters were written on papyri is reinforced by the Wadi Murabba'at document, the only provenanced Iron Age

**Figure 5.** ESEM view of secondary re-crystallization of calcite in a void as seen in thin section.

papyrus found so far in the southern Levant (Milik 1961, 93-100). Although it is a palimpsest and its text is fragmented, it is clear that it was a personal letter. Hence it seems that while ostraca are more frequent than papyri in the archaeological record, they substituted for papyri only for economic reasons, while the latter constituted the common writing material.

Can we thus assume that the bullae in question represent a local correspondence of letters? Before jumping to a conclusion, let us check some more data that might help in our interpretation. Two important archives of papyri dated to the Persian period, hence somewhat later than our case, were unearthed in our region. The first one, dated to the 5th century BCE, was discovered at Elephantine in Upper Egypt (Porten 1992; 1996). It consists of almost one hundred Aramaic documents from several private and communal archives that were mostly composed of contracts and letters. The second archive was found in a cave in Wadi Daliyeh, in the desert east of Samaria (Cross 1974; Lapp 1974; Leith 1997; Gropp 2001). It was dated to the 4th century BCE and contained 18 legal documents, all written in Aramaic. In these two groups, found in areas of extreme aridity, some papyri were discovered with their bullae still attached (Gropp 2001, Pls. 1-3, 17). Their perfect state of preservation and their proximity in time to the Late Iron Age may hint at the possible contents of the Iron Age sealed papyri as well.

The papyri from Elephantine and Wadi Daliyeh illustrate the sealing practices used in the Persian period. After the text was written and witnessed, it was rolled from bottom to top, flattened and folded. Strings and pieces of papyrus-fiber were looped around separately to tie the documents, which in turn were sealed by the bullae.

A comparison between the letters and the contracts found in Elephantine made by Porten (1992, 447-448) raises some significant technical differences, which are relevant to our case: letters were shorter than contracts and unlike contracts, they were usually written on both sides of the papyrus. While contracts were rolled up and folded in thirds, letters were rolled up and folded in half. Since contracts were meant to be stored for an extended period of time, a blank space at the top of the contract was an insurance against any external damage obliterating any part of the opening lines. These differences illustrate the special treatment of legal documents, most likely due to their long-term importance, as opposed to letters.

The archives from Elephantine and Wadi Daliyeh demonstrate that most of the sealed documents were formal legal records, which were concerned with the most personal forms of social interaction. The impression of personal seals upon these papyri was thus a highly complex act of signification. The seal must have had a complex power and its imprint was the insurance of the owner for the authority. A seal impression therefore established a set of moral obligations, which bound those who encountered it. To break or forge of a seal was an act which everyone knew was wrong.

Therefore we join the opinion first presented by Avigad (1997, 33-39) and Shiloh (1986, 36-37) and we assume that Judahite bullae were used as sealings of legal documents. In order to support this hypothesis, a brief cross-cultural account of sealing practices of legal documents is required. The most informative description of the preparation of a deed of sale in the days of the monarchy is found in the Book of Jeremiah (32:1-15). It concerns the purchase of a field by Jeremiah just a few days before the Babylonian conquest. The story reveals the technical features of the legal bureaucracy during the same days of the later group of bullae examined in our research. Although Jeremiah's purchase was of symbolic nature, this is most likely an accurate description of the legal process by which deeds were carried out at that time. Two texts, an original and a duplicate copy, were written on two separate sheets of papyrus. The first was termed the "sealed deed" because it was rolled up and sealed with a bulla or bullae; it would be opened before judicial authorities only when absolutely necessary. The second, "open deed" was a copy of the sealed one and was intended for daily use.

## 5. Conclusion

The use of delicate sampling techniques for petrographic analyses, combined with other complementary methods for the examination of extremely small samples, is a rewarding approach. It enables now the study of unique, delicate artifacts which could not be hitherto examined by the highly efficient and informative method of petrography. The results of the present study, which were not predicted by the visual examination of the bullae under discussion, provide significant information about the nature of the Judahite bureaucracy and the historical significance of the few clusters of Iron Age bullae which were discovered in recorded excavations. Since hundreds of contemporary, unprovenanced bullae originating from the antiquities market are found now in several museums and private collections, this methodology can be applied also on them. Besides serving as a tool for authenticity determinations, it may disclose the possible existence of some other Judahite administrative centers where legal administration may have been practiced.

## References

Adan-Bayewitz D. and Wieder M., 1992, Ceramics from Roman Galilee: A comparison of several techniques for fabric characterization. *Journal of Field Archaeology*, 19, 189-205.

Avigad N. (*Revised and completed by Sass, B.*), 1997, *Corpus of West Semitic Stamp Seals*, Israel Exploration Society, Jerusalem.

Brandl B., 2000, Bullae with Figurative Decoration. In *Excavations at the City of David 1978-1985 Directed by Yigal Shiloh. Volume IV. Inscriptions*, Qedem, 41, (ed. D.T. Ariel), 58-74, Jerusalem.

Cross F.M., 1974, The Papyri and Their Historical Implications. In *Discoveries in the Wa dõ ed-Da liyeh*, (eds. P.W. Lapp and N. Lapp), 17-29, AASOR XLI, Cambridge.

Goren Y., Finkelstein I. and Na'aman N., 2004, *Inscribed in Clay: Provenance Study of the Amarna Tablets and Other Ancient Near Eastern Texts*, Institute of Archaeology of Tel Aviv University, Tel Aviv.

Goren Y., Kletter R. and Kamaiski E., 1996, The technology and provenience of the Iron Age figurines from the City of David: Petrographic Analysis. In *City of David Excavations, Final Report IV*, (eds. D.Z. Ariel and A. de Grott), 87-89, Jerusalem.

Goren Y., Na'aman N., Mommsen H. and Finkelstein I., 2007, Provenance study and re-evaluation of the cuneiform documents from the Egyptian residency at Tel Aphek. *Ägypten und Levante*, XVI, 161-171.

Goren Y., Mommsen H., Finkelstein I. and Na'aman N., 2009, Provenance study of the Gilgamesh fragment from Megiddo. *Archaeometry*, 51, 763-773.

Goren Y., Mommsen H. and Klinger J., 2011, Nondestructive provenance study of cuneiform tablets using portable X-ray fluorescence (pXRF). *Journal of Archaeological Science*, 38, 684-696.

Gropp D.M., 2001, *Wadi Daliyeh II: The Samaria Papyri from Wadi Daliyeh*, Discoveries in the Judaean Desert XXVIII. Oxford.

Lapp P.W., 1974, An Account of the Discovery. In *Discoveries in the Wa dõ ed-Da liyeh*, (eds. P.W. Lapp and N. Lapp), 1-6, AASOR XLI, Cambridge.

Leith M.J.W., 1997, *Wadi Daliyeh I: The Wadi Daliyeh Seal Impressions*, Discoveries in the Judaean Desert XXIV, Oxford.

Milik J.T., 1961, Textes He breux et Arame ens. In *Les Grottes de Murabba'a t*, (eds. P. Benoit, J.T. Milik and R. de Vaux), 67-205, Discoveries in the Judean Desert II, Oxford.

Na'aman N., 2003, The Distribution of Messages in the Kingdom of Judah in Light of the Lachish Ostraca. *Vetus Testamentum*, 53, 169-180.

Porten B., 1992, Elephantine Papyri. In *The Anchor Bible Dictionary* (vol. II), (ed. D.N. Freedman), 445-455, Yale University Press, New York.

Porten B., 1996, *The Elephantine Papyri in English. Three Millennia of Cross-Cultural Continuity and Change*, Brill, Leiden.

Reich R., Shukron E. and Lernau O., 2007, Recent Discoveries in the City of David, Jerusalem. *Israel Exploration Journal*, 57, 153-169.

Schneider T., 1988, Azariahu Son of Hilkiahu (High Priest?) on a City of David Bulla. *Israel Exploration Journal*, 38, 139-141.

Shiloh Y., 1984, *Excavations at the City of David I. 1978-1982. Interim Report of the First Five Seasons*, Qedem 19, Jerusalem.

Shiloh Y., 1986, A Group of Hebrew Bullae from the City of David. *Israel Exploration Journal*, 36, 16-38.

Shiloh Y. and Tarler D., 1986, Bullae from the City of David. A Hoard of Seal Impressions from the Israelite Period. *The Biblical Archaeologist*, 49/4, 196-209.

Shoham Y., 2000, Hebrew Bullae. In *Excavations at the City of David 1978-1985 Directed by Yigal Shiloh. Volume IV. Inscriptions*, (ed. D.T. Ariel), 29-57, Qedem 41, Jerusalem.

Torczyner H., 1938, *Lachish I (Tell ed Duweir): The Lachish Letters*, Oxford University Press, London.

## ACKNOWLEDGEMENTS

The preliminary stage of this study was first supported by a grant from the Horowitz Foundation on behalf of the Interdisciplinary Center for the Conservation and Study of Historical Heritage in Israel (ESHMOR), and a grant from the Early Israel framework on behalf of the New Horizons program of Tel Aviv University. The follow-up of this research is supported by Israel Science Foundation (ISF) grant No. 947/12 entitled "The Administration of Judah under Assyrian, Babylonian and Persian Rule". We would like to thank Michal Dayagi-Mendels of The Israel Museum, Fawzi Ibrahim of the Rockefeller Museum, Ronny Reich from the Institute of Archaeology, Haifa University, and Hava Katz, Michael Saban, Eli Shukrun, and Debi Ben-Ami of the Israel Antiquities Authority for enabling us to study the bullae. Oded Lipschits (Tel Aviv University) read the draft of this article and made useful comments.

# The geochemistry and distribution of Archaic, Classical, and Hellenistic wares of the territory of ancient Sagalassos (SW Turkey): A reconnaissance study

**Dennis Braekmans[a,b], Patrick Degryse[c], Jeroen Poblome[d], Bert Neyt[c] and Marc Waelkens[d]**

[a]Laboratory for Ceramic Studies, Faculty of Archaeology, Leiden University, P.O.Box 9515, 2300 RA Leiden, The Netherlands, dennis.braekmans@ees.kuleuven.be

[b]Materials in Art and Archaeology, Laboratory of Materials Science, Delft University of Technology, Mekelweg 2, 2628 CD Delft, The Netherlands

[c]Section Geology, Earth and Environmental Sciences, Katholieke Universiteit Leuven, Celestijnenlaan 200 E - bus 2408, B-3001 Leuven, Belgium

[d]Sagalassos Archaeological Research Project, Katholieke Universiteit Leuven, Blijde-Inkomststraat 21 - bus 3314, B-3000, Leuven, Belgium

**Abstract** - A reconnaissance survey carried out within the territory of Sagalassos yielded vast amounts of unknown ceramic wares, tentatively dated to the Archaic, Classical, and Hellenistic periods. An archaeometric programme including geochemical analysis and applied multivariate statistics was applied in order to grasp the diversity of the wares and consequently identify variability in production. Within the chemical spectrum, it becomes clear that at least two large production groups exist, with a substantial 'internal' distribution of wares. Elemental evidence supports the hypothesis of a local production of ceramics related to the geographical situation of the sites.

## 1. Introduction

Mineral resources within the territory of Roman Sagalassos (Pisidia, SW Turkey) were used for the production of a wide array of artefacts, especially pottery. During several survey campaigns, sites were revisited and discovered within the study area, spanning the Epipaleolithic to late Ottoman times, reconstructing regional occupation (Vanhaverbeke and Waelkens 2003). At each of the surveyed sites, 'grab samples' were taken of the artefacts present. Sites belonging to the Archaic, Classical, and Hellenistic period were identified. These were situated at focal points within the entire territory, seemingly producing a wide array of pottery. Crucial questions that remain pertain to how and when these activities and the associated socio-economic patterns occurred.

The main purpose of the study was to fingerprint the material belonging to six sites (Düver, Seydiköy, Hisar, Kepez Kalesi, Belören, Kozluca and Aykırıkça; Fig. 1) attested in the study area which was later to become the territory of Sagalassos (Vanhaverbeke and Waelkens 2003; Degryse and Waelkens 2008). In this case, the extent of geochemical diversity of the wares is important in order to explain possible variability in production. In addition, determining the scale of distribution and circulation of these wares on a regional level is of importance, as it would allow a synchronic and diachronic socio-economic pottery study, and identifying production in the region from the Archaic period into Classical and Hellenistic times. In addition, the question of whether the growth of Sagalassos as a regional centre in Pisidia can be measured against regional production of the ceramics needs evaluation.

## 2. Materials and methods

Substantial amounts of pottery were found at each site, and samples were thus collected from each assemblage. The collection was carried out after a macroscopic assessment of the main features, following an initial macroscopic classification as described by D.P.S. Peacock (1977). For the purposes of a reconnaissance study, we opted to select

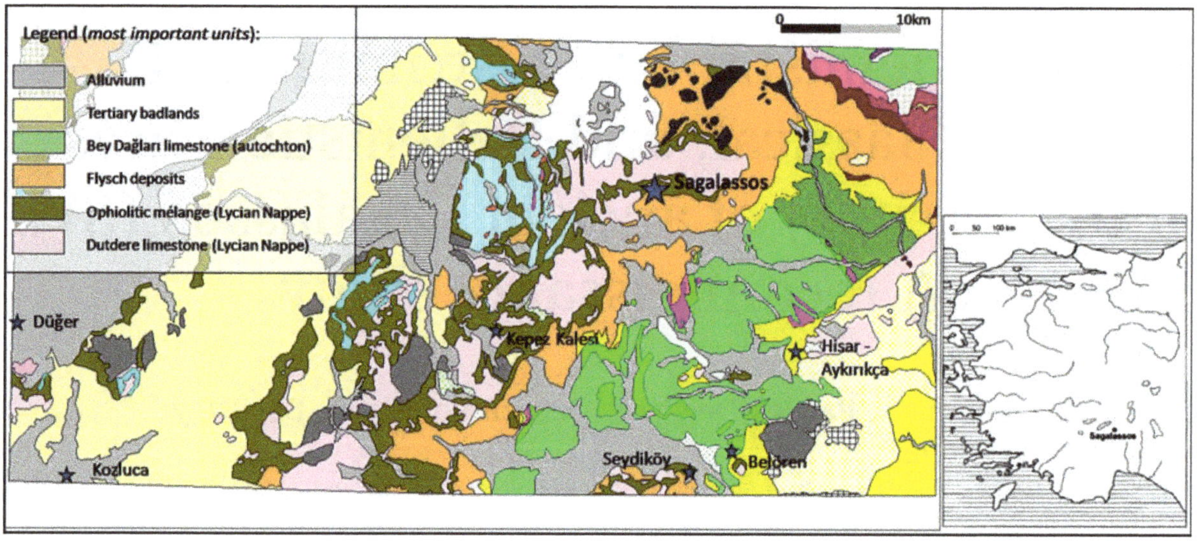


**Figure 1.** Approximate location of the studied sites on a geological map of the study area, indicating the major geological units within the region.

the main fabric groups that are most substantially represented on each site for chemical analysis. In this selection, rare fabric groups, representing only a marginal amount of the sherds encountered, were not sampled. Nevertheless, the characteristics of these ceramics need to be explored in future work.

A classification of ceramics dating from the Early Iron Age (and thus the Archaic period) in Southern Anatolia was proposed by J. Mellaart (1954, 1955), incorporating Phrygian-influenced grey wares, Southwest Anatolian painted wares ('black on red', 'black on buff', and 'bichrome black and red on buff') (Birmingham 1964; Aydal *et al.* 1997), Central Anatolian wares, 'red on buff', and 'black on purplish grey' wares; all incorporating geometric decoration patterns.

As the survey of the area carried out by the Sagalassos archaeological team continued, only occasional sherds could be ascribed to a Southwest Anatolian 'black on red' tradition (Poblome and Degeest 2003; Talloen *et al.* 2006). In this respect, no grey wares, commonly associated with the contemporaneous Phrygian tradition were recorded, nor was any other class of geometric decorated pottery.

Although no immediately identifiable wares belonging to the categories identified above were recognised, most sites in the region exhibited a fairly macroscopically similar assemblage of wares. These included relatively high-fired, wheel-thrown oxidised coarse wares, ranging from light orange to dark reddish brown sherds, as well as a partly slipped finer group of wares. These more levigated wares, although dissimilar in quality from the defined Southwest Anatolian wares, often bear some banded slipped decoration in various forms, representative of the so-called 'matt-painted' wares. After the macroscopic assessment, it has been noted that ware groups seem to circulate and are not restricted to any one site. Also, infrequent amounts of Classical-Hellenistic pottery were retrieved at nearly all sites, suggesting a (partial) occupation of the sites, whether or not continuous, during the 5$^{th}$-2$^{nd}$ centuries BC. This suggestion was based on typological analogies of echinus bowls and sparse Achaemenid bowls, omphalos bowls, fishplates, and more generic technomic types (Robinson 1950; Sparkes and Talcott 1970; Rotroff 1997, 2006; Çokay-Kepçe 2006). In this respect, mostly an 'orange' coloured type of wares seemed to have been used, for common wares, cook wares, and table wares alike, thus suggesting several varieties of levigation and/or inclusions. Less frequently, a bleaker, 'white' type of ware was attested.

Samples (Table 1) include material from the site of Düver, which had been occupied already during the (Late) Bronze Age, as suggested by the few Mycenaean pottery remains in the area (Mee 1978; Mellaart and Murray 1995). Düver is situated on a promontory overlooking the salt lake of Yarısli, surrounded by fertile plains (Waelkens and Loots 2000). At the site, Archaic material was collected, of which some sherds are richly decorated with a geometric pattern, dated to the 8$^{th}$-6$^{th}$ centuries BC, and related to a Southwest Anatolian tradition (Birmingham 1964; Cummer 1970; Waelkens and Loots 2000). Two groups of grey-cored 'Düver' technomic wares were discerned along with generic 'orange' wares, and brown and red slipped wares that were associated with Classical-Hellenistic traditions and 'buff' technomic wares. No Phrygian influence was visible in the ceramic assemblage, despite the presence of Phrygian-influenced architecture at the site (Waelkens and Loots 2000; Talloen *et al.* 2006). The site of Kozluca also procured significant amounts of both Archaic and Classical-Hellenistic pottery (Waelkens and Loots 2000), albeit somewhat different in nature from the material from Düver at first sight. The locations of Düver and Kozluca are of rather significant importance, as both sites controlled part of a road system linking Central and Western Anatolia with the Southern coast of Turkey (Waelkens and Loots 2000). Ceramics at Aykırıkça include a coarse, 'typical' dark grey reduced fabric, abundantly present at the site, and buff-reddish sherds with occasional traces of painted decoration. No sherds dating to the Classical-Hellenistic period were found within the survey

**Table 1.** Sample list, including survey site, statistical code, macroscopic identification, and geochemical group.

| Sample nr. | Site | Site-code | Macroscopic Identification | Geochemical group |
|---|---|---|---|---|
| 1 | Aykırıkça | AYK | Archaic, Aykırıkça technomic ware | A |
| 2 | Aykırıkça | AYK | Archaic, Aykırıkça technomic ware | B |
| 3 | Aykırıkça | AYK | Archaic, Aykırıkça technomic ware | B |
| 4 | Aykırıkça | AYK | 'matt-painted' ware | D |
| 5 | Belören | BEL | Class/Hell, 'orange' technomic ware | C |
| 6 | Belören | BEL | Archaic, technomic ware | E-1 |
| 7 | Belören | BEL | Class/Hell, levigated 'orange' ware | E-1 |
| 8 | Belören | BEL | Archaic, 'matt painted' 'Brown on red' | D |
| 9 | Belören | BEL | Class/Hell, oxidized storage vessel | C |
| 10 | Belören | BEL | Archaic, brown slipped technomic ware | D |
| 11 | Belören | BEL | oxidized tugla | C |
| 12 | Belören | BEL | oxidized tugla | D |
| 13 | Belören | BEL | Archaic, Aykırıkça technomic ware | A |
| 14 | Belören | BEL | Class/Hell, 'white' technomic ware | E-2 |
| 15 | Belören | BEL | Class/Hell, oxidized ware, grey slipped | D |
| 16 | Belören | BEL | Archaic, Aykırıkça technomic ware | A |
| 17 | Düver | DUV | Archaic, Düver technomic ware II | E-3 |
| 18 | Düver | DUV | Archaic, Düver technomic ware I | F |
| 19 | Düver | DUV | Southwest Anatolian, 'Black on Red' | E-4 |
| 20 | Düver | DUV | 'matt-painted' ware | E-4 |
| 21 | Düver | DUV | Southwest Anatolian, 'Black on Buff' | E-2 |
| 22 | Düver | DUV | Archaic, Düver technomic ware I | E-4 |
| 23 | Düver | DUV | Archaic, Düver technomic ware I | F |
| 24 | Düver | DUV | Class/Hell, 'orange' technomic ware | E-4 |
| 25 | Düver | DUV | Class/Hell, 'orange' technomic ware, brown slipped | E-4 |
| 26 | Düver | DUV | Class/Hell, res slipped table ware | E-4 |
| 27 | Düver | DUV | 'matt-painted' ware | E-1 |
| 28 | Düver | DUV | Class/Hell, 'orange' technomic ware, brown slipped | E-3 |
| 29 | Düver | DUV | Archaic, Düver technomic ware I | E-4 |
| 30 | Düver | DUV | Archaic, Düver technomic ware I | E-1 |
| 31 | Düver | DUV | Archaic, Düver technomic ware I | E-3 |
| 32 | Düver | DUV | Class/Hell, 'orange' technomic ware, brown slipped | F |
| 33 | Hisar | HIS | Class/Hell, 'orange' technomic ware | C |
| 34 | Hisar | HIS | Class/Hell, 'buff' technomic ware, grey slipped | E-1 |
| 35 | Hisar | HIS | tugla, oxidized, brown slipped | B |

**Table 1** - *continued*

| Sample nr. | Site | Site-code | Macroscopic Identification | Geochemical group |
|---|---|---|---|---|
| 36 | Hisar | HIS | tugla, 'orange' | 'outlier' |
| 37 | Kepez Kalesi | KEP | Class/Hell, technomic ware | C |
| 38 | Kepez Kalesi | KEP | 'matt-painted' ware | D |
| 39 | Kepez Kalesi | KEP | Class/Hell, 'orange' technomic ware, red slipped | D |
| 40 | Kepez Kalesi | KEP | Class/Hell, 'orange' technomic ware, grey cored | E-1 |
| 41 | Kepez Kalesi | KEP | Cook ware | B |
| 42 | Kepez Kalesi | KEP | Class/Hell, cook ware | 'outlier' |
| 43 | Kepez Kalesi | KEP | 'orange' ware | C |
| 44 | Kepez Kalesi | KEP | Omphalos bowl, 2nd century BC | E-1 |
| 45 | Kozluca | KOZ | Class/Hell, 'white' technomic ware, grey slipped | E-1 |
| 46 | Kozluca | KOZ | Class/Hell, 'orange' technomic ware, red slipped | E-2 |
| 47 | Kozluca | KOZ | Class/Hell, 'orange' technomic ware, red slipped | 'outlier' |
| 48 | Kozluca | KOZ | Archaic, 'buff' technomic ware | 'outlier' |
| 49 | Seydiköy | SEY | Archaic, 'orange' ware, incised decoration | A |
| 50 | Seydiköy | SEY | Class/Hell, 'orange' technomic ware, orange wash | F |
| 51 | Seydiköy | SEY | Class/Hell, 'orange' technomic ware | C |
| 52 | Seydiköy | SEY | Archaic, Seydiköy technomic ware, brown wash | C |
| 53 | Seydiköy | SEY | Archaic, Seydiköy technomic ware, brown wash | A |
| 54 | Seydiköy | SEY | 'matt-painted' ware | C |
| 55 | Seydiköy | SEY | Archaic, Seydiköy technomic ware, brown wash | D |
| 56 | Seydiköy | SEY | Archaic, Seydiköy technomic ware, brown wash | B |
| 57 | Seydiköy | SEY | Class/Hell, 'orange' tableware | E-2 |
| 58 | Seydiköy | SEY | Archaic, Seydiköy technomic ware, incised decoration | B |

assemblage. However, it was suggested that the site was abandoned and related in Hellenistic times to the site of Hisar (Waelkens and Poblome 1997; Vanhaverbeke and Waelkens 2003). The 'typical' dark reduced Aykırıkça fabric was, however, not attested at Hisar; instead, several likely Classical-Hellenistic ware 'buff' and 'orange' ware groups were sampled. The Archaic site of Belören - a hilltop site - developed into a Hellenistic city, Keraitai (Waelkens and Poblome 1995), which is represented in the ceramic assemblage by a collection of Archaic related 'matt-painted' wares, grey and brown slipped table wares, and a wide array of 'white', 'orange', and other oxidised common ware groups, as well as related tile fabrics. In the centre of the study area lies the settlement of Kepez Kalesi, also situated on a promontory overlooking a fertile valley. The main occupation of the site was dated to the early Hellenistic period, although pottery remains suggest an earlier substrate with reference to the matt-painted wares (Waelkens *et al.* 1997). Furthermore, an 'orange' technomic ware, as well as a more unique reddish brown coloured ware, have been documented. Seydiköy, similarly situated, presented a typical brown technomic ware group, Archaic related 'matt-painted' ceramics, as well as substantial amounts of 'orange' technomic wares.

Geochemical analysis is regarded as a robust technique for assessing the general origin and/or similarity of ceramic samples. In this respect, variation within the chemical composition can often be traced back to the clay raw material, the temper material present or technological interventions. ICP-MS analysis (Actlabs Ltd., Canada; standards and quality control: NIST 694, DNC-1, BIR-1, FK-N, NIST 1633b, SY-3, W-2a, NIST 696 and JSD-3; Table 2) was conducted for major elements ($SiO_2$, $Al_2O_3$,

**Table 2.** Measurements of reference samples used in the geochemical procedure. (Mean = measured values; Cert = certified values).

| | $SiO_2$ | $Al_2O_3$ | $Fe_2O_3$(T) | MnO | MgO | CaO | $Na_2O$ | $K_2O$ | $TiO_2$ | $P_2O_5$ | LOI | Total | Sc | Be | V | Ba | Sr | Y | Zr | Cr | Co | Ni | Cu | Zn | Ga | Ge | As |
|---|---|---|---|---|---|---|---|---|---|---|---|---|---|---|---|---|---|---|---|---|---|---|---|---|---|---|---|
| Unit Symbol | % | % | % | % | % | % | % | % | % | % | % | % | ppm | ppm | ppm | ppm | ppm | ppm | ppm | ppm | ppm | ppm | ppm | ppm | ppm | ppm | ppm |
| GXR-1 Meas | | | | | | | | | | | | | | | | | | | | < 20 | 7 | 30 | 1110 | 730 | 12 | | 428 |
| GXR-1 Cert | | | | | | | | | | | | | | | | | | | | 12 | 8.2 | 41 | 1110 | 760 | 13.8 | | 427 |
| WMG-1 Meas | | | | | | | | | | | | | | | | | | | | 780 | 193 | 2500 | 4970 | 120 | 10 | | 9 |
| WMG-1 Cert | | | | | | | | | | | | | | | | | | | | 770 | 200 | 2700 | 5900 | 110 | 10.3 | | 7 |
| NIST 694 Meas | 11.62 | 1.97 | 0.74 | 0.012 | 0.34 | 43.82 | 0.9 | 0.6 | 0.117 | 30.16 | | | | | 1592 | | | | | | | | | | | | |
| NIST 694 Cert | 11.2 | 1.8 | 0.79 | 0.0116 | 0.33 | 43.6 | 0.86 | 0.51 | 0.11 | 30.2 | | | | | 1740 | | | | | | | | | | | | |
| DNC-1 Meas | 45.3 | 17.82 | 9.17 | 0.142 | 9.85 | 10.81 | 1.84 | 0.29 | 0.467 | 0.07 | | | 30 | < 1 | 171 | 104 | 137 | 17 | 36 | 280 | 55 | 250 | 110 | 90 | 14 | 1 | < 5 |
| DNC-1 Cert | 47 | 18.3 | 9.93 | 0.149 | 10.1 | 11.3 | 1.87 | 0.234 | 0.48 | 0.09 | | | 31 | 1 | 148 | 114 | 145 | 18 | 41 | 285 | 54.7 | 247 | 96 | 66 | 15 | 1.3 | 0.2 |
| BIR-1 Meas | 49.55 | 16.11 | 11.26 | 0.177 | 9.99 | 13.57 | 1.87 | 0.04 | 1.002 | 0.03 | | | 45 | < 1 | 325 | 6 | 110 | 13 | 15 | 390 | 51 | 160 | 130 | 80 | 16 | 2 | < 5 |
| BIR-1 Cert | 47.8 | 15.4 | 11.3 | 0.171 | 9.68 | 13.2 | 1.75 | 0.03 | 0.96 | 0.05 | | | 44 | 0.58 | 313 | 7 | 108 | 16 | 16 | 382 | 51.4 | 166 | 126 | 71 | 16 | 1.5 | 0.44 |
| MICA-FE Meas | | | | | | | | | | | | | | | | | | | | 90 | 25 | 30 | < 10 | 1320 | 96 | 3 | < 5 |
| MICA-FE Cert | | | | | | | | | | | | | | | | | | | | 90 | 23 | 35 | 5 | 1300 | 95 | 3.2 | 3 |
| GXR-2 Meas | | | | | | | | | | | | | | | | | | | | 30 | 8 | < 20 | 100 | 570 | 34 | | 18 |
| GXR-2 Cert | | | | | | | | | | | | | | | | | | | | 36 | 8.6 | 21 | 76 | 530 | 37 | | 25 |
| FK-N Meas | 63.52 | 17.88 | 0.08 | 0.002 | 0.01 | 0.1 | 2.41 | 12.37 | 0.004 | 0.02 | | | < 1 | < 1 | < 5 | 193 | 36 | < 2 | 4 | | | | | | | | |
| FK-N Cert | 55 | 18.6 | 0.09 | 0.005 | 0.01 | 0.11 | 2.58 | 12.8 | 0.02 | 0.024 | | | 1 | 1 | 5 | 200 | 39 | 0.5 | 13 | | | | | | | | |
| LKSD-3 Meas | | | | | | | | | | | | | | | | | | | | 80 | 29 | 50 | 40 | 130 | | | 26 |
| LKSD-3 Cert | | | | | | | | | | | | | | | | | | | | 87 | 30 | 47 | 35 | 152 | | | 27 |
| MAG-1 Meas | | | | | | | | | | | | | | | | | | | | 90 | 20 | 50 | 30 | 140 | 22 | | 7 |
| MAG-1 Cert | | | | | | | | | | | | | | | | | | | | 97 | 20.4 | 53 | 30 | 130 | 20.4 | | 9.2 |
| NIST 1633b Meas | 50.54 | 29.22 | 10.95 | 0.017 | 0.82 | 2.21 | 0.28 | 2.48 | 1.324 | 0.58 | | | 42 | | 291 | 725 | 1056 | | | | | | | | | | |
| NIST 1633b Cert | 49.2 | 28.4 | 11.1 | 0.02 | 0.8 | 2.11 | 0.27 | 2.35 | 1.32 | 0.53 | | | 41 | | 296 | 709 | 1040 | | | | | | | | | | |
| SY-3 Meas | 61.71 | 12.19 | 6.09 | 0.328 | 2.69 | 8.08 | 4.35 | 4.53 | 0.142 | 0.46 | | | 7 | 22 | 118 | 467 | 275 | 132 | 302 | | | | | | | | |
| SY-3 Cert | 59.6 | 11.8 | 6.49 | 0.32 | 2.67 | 8.25 | 4.12 | 4.23 | 0.15 | 0.54 | | | 7 | 20 | 50 | 450 | 302 | 718 | 320 | | | | | | | | |
| NIST 696 Meas | 3.68 | 54.05 | 8.28 | 0.005 | 0.02 | 0.02 | | 0.05 | 2.594 | 0.05 | | | | | 371 | | | | 1039 | | | | | | | | |
| NIST 696 Cert | 3.79 | 54.5 | 8.7 | 0.004 | 0.012 | 0.018 | | 0.009 | 2.64 | 0.05 | | | | | 403 | | | | 1040 | | | | | | | | |
| JSD-3 Meas | 78.67 | 10.06 | 4.21 | 0.144 | 1.16 | 0.58 | 0.41 | 2.07 | 0.419 | 0.09 | | | | | | | | | | | | | | | | | |
| JSD-3 Cert | 76 | 9.908 | 4.368 | 0.148 | 1.17 | 0.56 | 0.411 | 1.971 | 0.403 | 0.0817 | | | | | | | | | | | | | | | | | |

**Table 2** - *continued*

| | Rb | Nb | Mo | Ag | In | Sn | Sb | Cs | La | Ce | Pr | Nd | Sm | Eu | Gd | Tb | Dy | Ho | Er | Tm | Yb | Lu | Hf | Ta | W | Tl | Pb | Bi | Th | U |
|---|---|---|---|---|---|---|---|---|---|---|---|---|---|---|---|---|---|---|---|---|---|---|---|---|---|---|---|---|---|---|
| Unit Symbol | ppm | ppm | ppm | ppm | ppm | ppm | ppm | ppm | ppm | ppm | ppm | ppm | ppm | ppm | ppm | ppm | ppm | ppm | ppm | ppm | ppm | ppm | ppm | ppm | ppm | ppm | ppm | ppm | ppm | ppm |
| GXR-1 Meas | 3 | 1 | 18 | 32.7 | 0.8 | 50 | 122 | 2.6 | 9 | 15 | | 7.7 | 2.6 | 0.6 | 3.7 | 0.8 | 4.6 | | | 0.38 | 2.1 | 0.29 | 0.6 | < 0.1 | 164 | 0.4 | 731 | 1380 | 2.3 | 32 |
| GXR-1 Cert | 14 | 0.8 | 18 | 31 | 0.77 | 54 | 122 | 3 | 7.5 | 17 | | 18 | 2.7 | 0.69 | 4.2 | 0.83 | 4.3 | | | 0.43 | 1.9 | 0.28 | 0.96 | 0.175 | 164 | 0.39 | 730 | 1380 | 2.44 | 34.9 |
| WMG-1 Meas | | 5 | 2 | < 0.5 | | 2 | 8.3 | < 0.5 | 9.5 | 17.7 | | 8.7 | 2.2 | 0.77 | | 0.4 | 2.4 | 0.5 | | 0.21 | 1.3 | 0.2 | 1.6 | 0.3 | < 1 | | 19 | | 1.1 | 0.7 |
| WMG-1 Cert | | 6 | 1.4 | 2.7 | | 2.2 | 1.8 | 0.48 | 8.2 | 16 | | 9 | 2.3 | 0.82 | | 0.3 | 2.8 | 0.5 | | 0.2 | 1.3 | 0.21 | 1.3 | 0.5 | 1.3 | | 15 | | 1.1 | 0.65 |
| DNC-1 Meas | 4 | 1 | < 2 | < 0.5 | | | 3 | < 0.5 | 4.5 | 8.8 | 1.09 | 4.6 | 1.4 | 0.63 | 1.9 | 0.4 | 2.8 | 0.6 | 2 | 0.32 | 2 | 0.3 | 1.1 | < 0.1 | < 1 | < 0.1 | 6 | < 0.4 | 0.2 | < 0.1 |
| DNC-1 Cert | 4.5 | 3 | 0.7 | 0.027 | | | 0.96 | 0.34 | 3.8 | 10.6 | 1.3 | 4.9 | 1.38 | 0.59 | 2 | 0.41 | 2.7 | 0.62 | 2 | 0.38 | 2.01 | 0.32 | 1.01 | 0.098 | 0.2 | 0.026 | 6.3 | 0.02 | 0.2 | 0.1 |
| BIR-1 Meas | < 2 | < 1 | < 2 | < 0.5 | | < 1 | 0.6 | < 0.5 | 0.8 | 2.2 | 0.38 | 2.3 | 1.1 | 0.56 | 1.8 | 0.4 | 2.6 | 0.6 | 1.8 | 0.28 | 1.7 | 0.25 | 0.6 | < 0.1 | < 1 | < 0.1 | < 5 | < 0.4 | < 0.1 | < 0.1 |
| BIR-1 Cert | 0.25 | 0.6 | 0.5 | 0.036 | | 0.65 | 0.58 | 0.005 | 0.62 | 1.95 | 0.38 | 2.5 | 1.1 | 0.54 | 1.85 | 0.36 | 2.5 | 0.57 | 1.7 | 0.26 | 1.65 | 0.26 | 0.6 | 0.04 | 0.07 | 0.01 | 3 | 0.02 | 0.03 | 0.01 |
| MICA-FE Meas | > 1000 | 270 | < 2 | | 0.7 | 73 | | 180 | 240 | 400 | 51.4 | 172 | 33.1 | 0.65 | 20.9 | 2.8 | 10.8 | 1.5 | 3.8 | 0.55 | 3.4 | 0.48 | 29.2 | 35 | 11 | 16 | 9 | < 0.4 | 165 | 92.3 |
| MICA-FE Cert | 2200 | 270 | 1.2 | | 0.6 | 70 | | 180 | 200 | 420 | 49 | 180 | 33 | 0.7 | 21 | 2.7 | 11 | 1.6 | 3.8 | 0.48 | 3.5 | 0.5 | 26 | 35 | 15 | 16 | 13 | 1.9 | 150 | 80 |
| GXR-2 Meas | 80 | 8 | < 2 | 14.2 | < 0.2 | 3 | 35 | 5.4 | 29.7 | 54.5 | | 18.3 | 3.5 | 0.75 | 2.9 | 0.5 | 2.9 | | | 0.27 | 1.7 | 0.27 | 6.3 | 0.8 | 2 | 1 | 746 | < 0.4 | 8.1 | 3 |
| GXR-2 Cert | 78 | 11 | 2.1 | 17 | 0.252 | 1.7 | 49 | 5.2 | 25.6 | 51.4 | | 19 | 3.5 | 0.81 | 3.3 | 0.48 | 3.3 | | | 0.3 | 2.04 | 0.27 | 8.3 | 0.9 | 1.9 | 1.03 | 690 | 0.69 | 8.8 | 2.9 |
| LKSD-3 Meas | 75 | 7 | < 2 | 0.6 | | 2 | 3.2 | 2.3 | 55.8 | 94.6 | | 39.6 | 7.3 | 1.46 | | 0.9 | 5 | | | | 2.7 | 0.41 | 4.8 | 0.6 | 1 | | 13 | | 10.1 | 4.5 |
| LKSD-3 Cert | 78 | 8 | 2 | 2.7 | | 3 | 1.3 | 2.3 | 52 | 90 | | 44 | 8 | 1.5 | | 1 | 4.9 | | | | 2.7 | 0.4 | 4.8 | 0.7 | 2 | | 29 | | 11.4 | 4.6 |
| MAG-1 Meas | 149 | 12 | 2 | < 0.5 | < 0.2 | 3 | 1.7 | 8.6 | 48.8 | 90.3 | 9.7 | 34.2 | 6.8 | 1.48 | 5.5 | 1 | 5.1 | 1 | 2.8 | 0.42 | 2.5 | 0.37 | 3.6 | 1.1 | 2 | 0.2 | 24 | < 0.4 | 11.3 | 2.9 |
| MAG-1 Cert | 149 | 12 | 1.6 | 0.08 | 0.18 | 3.6 | 0.96 | 8.6 | 43 | 88 | 9.3 | 38 | 7.5 | 1.6 | 5.8 | 0.96 | 5.2 | 1.02 | 3 | 0.43 | 2.6 | 0.4 | 3.7 | 1.1 | 1.4 | 0.59 | 24 | 0.34 | 11.9 | 2.7 |

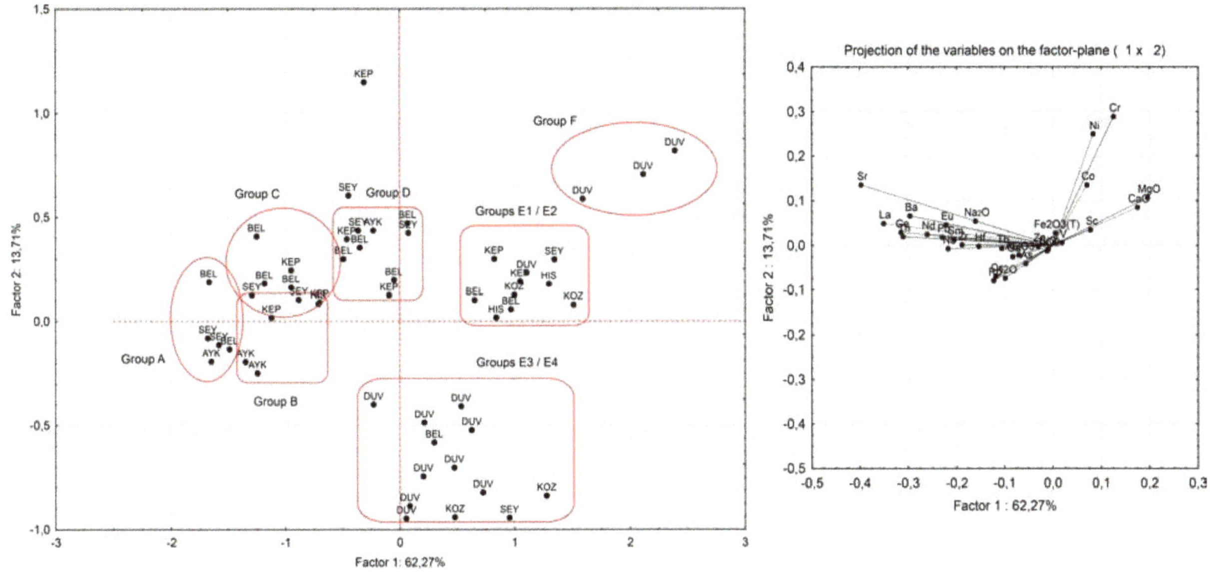


**Figure 2.** Graphical output of Principal Component Analysis and variables on the factor-plane.

$Fe_2O_3$, MnO, MgO, CaO, $Na_2O$, $K_2O$, $TiO_2$ and $P_2O_5$) and trace elements (Be, V, Ba, Sr, Y, Zr, Cr, Co, Ni, Cu, Zn, Ga, Rb, Nb, Cs, La, Ce, Pr, Nd, Sm, Eu, Gd, Tb, Dy, Ho, Er, Tm, Yb, Lu, Hf, Ta, Tl, Pb, Th and U). The accuracy and precision were consequently high (less than 2-3%), as expected from the application of this technique. Statistical procedures were conducted to visualise the raw data, and to document and explain the variability and importance of the elements measured as compared to the local geology. After logarithmic transformation, the data set was analysed to identify outliers by means of robust Mahalanobis distance and standard residuals, resulting in the removal of three samples (36, 47, and 48) from the dataset. A $\log_{10}$ transformation of data is desirable to balance the magnitude between major and trace elements.

## 3. Results and discussion

Following the use of Principal Component Analysis (PCA) as a survey tool for elemental distribution (Fig. 2), the results showed, from a site perspective, a large variation correlating to a certain extent with the geographical situation. The variation is explained to a proportion of almost 76% for the first two components. Sherds collected at Düver and Kozluca occupy a different region of the plot, with a small group from Düver distinguished from the main group by higher values of Cr, Ni, Co, MgO, Sc, and CaO. The rest of the sites group together, spanning different compositional groups differentiated by variations in either MgO, CaO, and Sc; Co, Cr, Ni; Rb, $K_2O$, Cs, Sr, Ba, Th, $Na_2O$ and the REEs.

The issue of discriminating production workshops within the chemically distinguished zones was approached by analysing data from the regions under consideration. Hierarchical cluster analysis using Ward's method (Ward 1963) and squared Euclidean distances was applied (Johnson and Wichern 1992) for both the Düver-Kozluca assemblage and the Aykırıkça, Kepez Kalesi, Belören, Hisar and Seydiköy assemblages to evaluate ware groups. The dendrogram resulting from the hierarchical cluster analysis is reproduced in Fig. 3. Descriptive statistics of the measurements are provided in Table 3.

The cluster analysis shows a similar grouping pattern as previously shown with the PCA, resulting in six cluster groups. Because of large distance values, these were normalised to 100 (DLink/DMax*100) to allow an overview of the groups. When looking at the raw chemical data, cluster E exhibits a wide variety of values; therefore, it should be possible to discriminate between the products from this large cluster. This analytical data was analysed separately from the other groups, in more detail, by a subsequent cluster analysis. This resulted in the discovery of four possible (sub)groups (labelled E-1 to E-4; Fig. 4).

To explain the clusters defined by the statistical procedures and to document variability, multi-element plots ($Al_2O_3$, $Fe_2O_3$ (total), CaO, $Na_2O$, $K_2O$, $TiO_2$, Sc, Ba, Sr, Zr, Cr, Co, Ni, Rb, Cs, REE: La, Ce, Nd, Sm, Eu, Gd, Tb, Dy, Er, Yb & Lu, Hf, Ta, Th) were constructed for each group, allowing for an examination of enrichments or depletions in comparison to an 'average' sediment signal. We opted to use NASC-normalisation (North American Shale Composite) as ceramics are most closely related to sedimentary material. Values for the normalisation are derived from Gromet *et al.* (1984), and are plotted on a logarithmic scale (cf. Fig. 5 for the most significant chemical signals).

Cluster A is characterised by notable enrichments in $Na_2O$, $TiO_2$, Ba, Sr, Zr, Hf, Ta, Th, and LREEs, exposing a positive Eu anomaly. Cluster B exhibits a highly similar pattern, only showing sample 2 as being depleted in Hf. Cluster C, however, exhibits a strong enrichment in Cr and Ni as compared to clusters A and B, and also varies strongly in CaO content. Sample 51 has an anomalous high Cs enrichment for this cluster. Within cluster D, a strong CaO enrichment is visible, as well as a minimal enrichment in $Na_2O$ and depletion in Cs and Rb. Once again, one sample exhibits a high Cs value; in this case, sample 15.

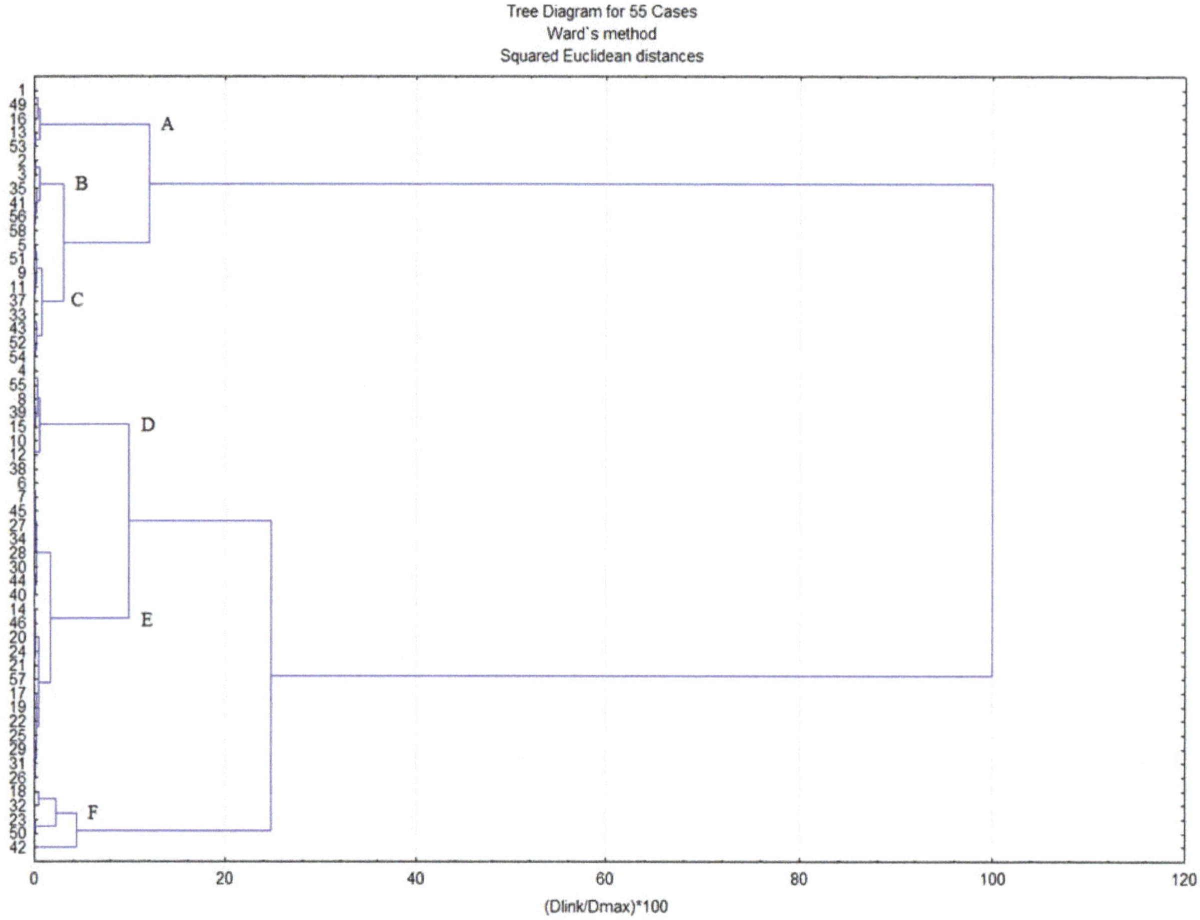


**Figure 3.** Hierarchical clustering dendrogram using Ward's method and squared Euclidean distances of all samples.

Groups E-1 through E-4 show a large variability and do not present consistent signals; their main attribute is the absence of any substantial REE enrichments. Also, no Th, Ta, or Hf enrichment is visible. Instead, this group is mainly distinguished by a substantial CaO, Sr, Cr, and Ni enrichment, and a $Na_2O$ depletion for group E-1. Within group E-2, the Cr and Ni enrichment is absent, and instead a Ba and Co deficiency seems present. In terms of the rare earth elements, no noteworthy Eu anomaly is visible. Groups E-3 and E-4 are highly similar, and no strong high or low values can be indicated apart from a small Cr, Ni, Th, and LREE enrichment as compared to NASC. Variations are noted in CaO, $K_2O$, $Na_2O$, and Sr values.

Two distinct signals are accounted for within cluster F, easily recognised and highly distinctive from the previous clusters mentioned. As such, all samples have a strong enrichment in CaO, $TiO_2$, Sc, and most explicitly in Cr and Ni. Samples 23 and 32 are however more strongly depleted in $K_2O$, Ba, Sr, Hf, Ta, Th, and LREE. In particular, the general pattern of the REEs seems to suggest a different resource usage.

Sample 42, belonging to cluster G, can be described as an 'outlier' and varies substantially from the other signals, as also detected in the PCA. It appears reasonable to describe this sherd as 'non-local'.

It becomes clear that clusters A, B, C, and D mainly contain material from Kepez Kalesi, Belören, Hisar, Aykırıkça, and Seydiköy, whereas clusters E and F are dominated by material found at Düver and Kozluca. When examining the geological substrate of the region, it becomes clear that an intricate network of ophiolite, flysch, and limestone beds is situated in the eastern part of the study area, opposite to the lake sediments in the area of lake Burdur (Düver, Kozluca). This aspect is in agreement with the elemental evidence. High Fe and Ni (and compatible Cr) values may reflect a mafic parent rock of the clay raw material. Moreover, Cr is also commonly accumulated within the clayey matrix itself. This regularly dispersed as an accessory element in zircon, sphene, epidote, and apatite, and thus commonly related to igneous rocks. Sr substitutes for Ca in plagioclase and may be the result of the weathering of carbonates and plagioclase. Na is generally attested in feldspars and Na-rich mica. Highly compatible alkali metals and high field strength elements (Zr, Hf) generally do not show significant variability, an aspect which may reflect the absence of felsic-based raw material. Clusters E-4 and F, on the other hand, show a very strong depletion of alkali metals, arguing for a very different resource. Differences between the two production regions seem easily detectable by examining the REE pattern, which is far less pronounced in clusters E and F. These two clusters are, in addition, lacking a strong positive Eu anomaly under the NASC normalisation of the data. Moreover, La is mainly enriched

**Table 3.** Descriptive statistics of elements measured, including valid N, average, minimum, maximum and standard deviation for each group.

| | | $SiO_2$ | $Al_2O_3$ | $Fe_2O_3$(T) | MnO | MgO | CaO | $Na_2O$ | $K_2O$ | $TiO_2$ | $P_2O_5$ | LOI | Sc | Be | V | Ba | Sr | Y | Zr | Cr | Co | Ni | Cu | Zn | Ga |
|---|---|---|---|---|---|---|---|---|---|---|---|---|---|---|---|---|---|---|---|---|---|---|---|---|---|
| | | wt.% | wt.% | wt.% | wt.% | wt.% | wt.% | wt.% | wt.% | wt.% | wt.% | wt.% | ppm | ppm | ppm | ppm | ppm | ppm | ppm | ppm | ppm | ppm | ppm | ppm | ppm |
| Group A | Average | 55.25 | 20.18 | 6.66 | 0.14 | 1.53 | 2.95 | 2.02 | 3.68 | 0.78 | 0.69 | 5.34 | 13 | 8 | 120 | 2292 | 2113 | 39 | 429 | 168 | 17 | 92 | 42 | 130 | 28 |
| | Minimum | 49.89 | 18.44 | 5.16 | 0.13 | 0.84 | 2.30 | 1.33 | 3.11 | 0.55 | 0.21 | 1.49 | 10 | 7 | 94 | 2045 | 1902 | 33 | 402 | 130 | 14 | 60 | 40 | 100 | 24 |
| n = 5 | Maximum | 61.08 | 21.41 | 7.32 | 0.15 | 2.19 | 3.47 | 2.93 | 4.27 | 0.87 | 1.19 | 11.15 | 14 | 10 | 146 | 2546 | 2331 | 42 | 450 | 240 | 20 | 130 | 50 | 150 | 31 |
| | Std.Dev. | 5.23 | 1.23 | 0.86 | 0.01 | 0.52 | 0.50 | 0.62 | 0.46 | 0.13 | 0.47 | 4.88 | 2 | 1 | 20 | 227 | 196 | 4 | 21 | 45 | 2 | 26 | 4 | 21 | 3 |
| Group B | Average | 53.72 | 18.69 | 6.81 | 0.13 | 2.04 | 4.75 | 1.76 | 3.50 | 0.77 | 0.65 | 6.91 | 13 | 6 | 121 | 1831 | 1510 | 30 | 329 | 223 | 18 | 120 | 42 | 82 | 24 |
| | Minimum | 47.70 | 15.29 | 5.82 | 0.10 | 1.17 | 2.44 | 1.27 | 3.07 | 0.61 | 0.25 | 1.86 | 11 | 4 | 93 | 1528 | 1343 | 24 | 235 | 130 | 15 | 70 | 40 | 30 | 19 |
| n = 6 | Maximum | 58.20 | 21.05 | 8.07 | 0.15 | 3.98 | 11.11 | 2.32 | 3.90 | 0.93 | 1.41 | 12.22 | 16 | 7 | 155 | 2144 | 1654 | 34 | 397 | 310 | 23 | 190 | 50 | 150 | 26 |
| | Std.Dev. | 4.86 | 1.95 | 0.83 | 0.02 | 1.14 | 3.39 | 0.42 | 0.32 | 0.11 | 0.55 | 4.12 | 2 | 1 | 20 | 215 | 105 | 4 | 58 | 70 | 3 | 44 | 4 | 48 | 3 |
| Group C | Average | 58.10 | 17.89 | 7.07 | 0.15 | 2.66 | 3.89 | 1.44 | 3.26 | 0.81 | 0.37 | 3.54 | 15 | 5 | 112 | 1384 | 1259 | 36 | 332 | 417 | 25 | 192 | 42 | 119 | 25 |
| | Minimum | 53.84 | 15.37 | 5.92 | 0.12 | 1.33 | 1.75 | 1.20 | 3.00 | 0.64 | 0.22 | 1.00 | 14 | 4 | 94 | 1243 | 995 | 27 | 239 | 210 | 21 | 130 | 30 | 80 | 20 |
| n = 9 | Maximum | 64.25 | 20.13 | 7.87 | 0.21 | 4.89 | 9.05 | 1.66 | 3.45 | 0.96 | 0.59 | 6.33 | 17 | 6 | 125 | 1571 | 1504 | 49 | 423 | 600 | 30 | 330 | 60 | 150 | 29 |
| | Std.Dev. | 2.79 | 1.67 | 0.61 | 0.03 | 1.13 | 2.59 | 0.18 | 0.16 | 0.10 | 0.13 | 1.58 | 1 | 1 | 10 | 105 | 169 | 8 | 66 | 126 | 3 | 63 | 8 | 21 | 3 |
| Group D | Average | 53.95 | 15.87 | 7.39 | 0.14 | 3.87 | 7.93 | 0.97 | 2.65 | 0.81 | 0.46 | 5.66 | 16 | 4 | 118 | 991 | 770 | 33 | 256 | 428 | 39 | 233 | 58 | 115 | 21 |
| | Minimum | 46.23 | 13.87 | 6.55 | 0.10 | 2.41 | 6.14 | 0.65 | 2.13 | 0.71 | 0.28 | 1.52 | 14 | 3 | 102 | 833 | 567 | 27 | 198 | 350 | 22 | 160 | 30 | 30 | 17 |
| n = 8 | Maximum | 59.13 | 18.86 | 9.03 | 0.19 | 5.79 | 11.26 | 1.51 | 3.15 | 0.96 | 1.28 | 11.54 | 19 | 5 | 147 | 1245 | 928 | 43 | 319 | 660 | 113 | 320 | 110 | 170 | 26 |
| | Std.Dev. | 4.40 | 1.81 | 0.88 | 0.03 | 1.37 | 1.94 | 0.26 | 0.34 | 0.10 | 0.34 | 3.03 | 2 | 1 | 15 | 153 | 124 | 5 | 41 | 102 | 30 | 59 | 28 | 41 | 3 |
| Group E1 / E2 | Average | 53.99 | 14.07 | 6.72 | 0.10 | 4.41 | 9.53 | 0.74 | 2.80 | 0.81 | 0.33 | 6.37 | 17 | 3 | 120 | 424 | 249 | 24 | 164 | 328 | 24 | 204 | 52 | 114 | 18 |
| | Minimum | 47.23 | 10.58 | 5.53 | 0.05 | 2.08 | 5.46 | 0.21 | 1.97 | 0.59 | 0.16 | 3.92 | 11 | 2 | 82 | 276 | 123 | 20 | 124 | 80 | 10 | 50 | 20 | 70 | 14 |
| n = 12 | Maximum | 58.43 | 17.98 | 8.21 | 0.15 | 7.03 | 15.62 | 1.26 | 3.40 | 0.91 | 0.60 | 10.40 | 21 | 4 | 158 | 586 | 444 | 28 | 209 | 620 | 35 | 320 | 90 | 150 | 20 |
| | Std.Dev. | 3.01 | 1.67 | 0.92 | 0.03 | 1.69 | 2.62 | 0.31 | 0.40 | 0.11 | 0.15 | 2.04 | 3 | 1 | 23 | 84 | 103 | 3 | 23 | 176 | 9 | 114 | 18 | 28 | 2 |
| Group E3 / E4 | Average | 56.61 | 19.05 | 7.78 | 0.09 | 3.47 | 3.14 | 0.89 | 3.82 | 0.94 | 0.22 | 2.96 | 19 | 4 | 137 | 526 | 197 | 32 | 213 | 178 | 21 | 91 | 48 | 110 | 26 |
| | Minimum | 51.24 | 16.58 | 6.95 | 0.06 | 1.19 | 1.25 | 0.50 | 2.40 | 0.73 | 0.14 | 1.84 | 15 | 3 | 110 | 337 | 118 | 25 | 166 | 110 | 11 | 50 | 30 | 30 | 23 |
| n = 10 | Maximum | 64.42 | 23.19 | 9.41 | 0.13 | 5.08 | 4.82 | 1.18 | 5.39 | 1.08 | 0.35 | 4.15 | 23 | 6 | 165 | 678 | 435 | 36 | 317 | 380 | 26 | 180 | 70 | 160 | 31 |
| | Std.Dev. | 3.48 | 2.22 | 0.78 | 0.03 | 1.32 | 1.23 | 0.25 | 0.89 | 0.12 | 0.06 | 0.79 | 3 | 1 | 17 | 110 | 91 | 3 | 41 | 79 | 5 | 40 | 11 | 45 | 3 |
| Group F | Average | 53.32 | 12.41 | 8.49 | 0.15 | 9.15 | 9.32 | 0.71 | 1.43 | 0.72 | 0.17 | 3.71 | 37 | 2 | 173 | 274 | 116 | 21 | 104 | 1740 | 46 | 295 | 68 | 98 | 15 |
| | Minimum | 52.37 | 10.64 | 6.32 | 0.11 | 7.90 | 6.47 | 0.42 | 0.60 | 0.53 | 0.11 | 1.83 | 29 | 1 | 108 | 179 | 94 | 18 | 83 | 1200 | 32 | 190 | 40 | 90 | 14 |
| n = 8 | Maximum | 54.89 | 14.38 | 10.67 | 0.20 | 10.50 | 11.09 | 1.00 | 2.40 | 0.93 | 0.27 | 4.99 | 49 | 2 | 222 | 330 | 146 | 25 | 117 | 2480 | 58 | 360 | 120 | 100 | 17 |
| | Std.Dev. | 1.17 | 1.70 | 2.50 | 0.04 | 1.34 | 2.02 | 0.27 | 0.90 | 0.19 | 0.07 | 1.43 | 9 | 1 | 57 | 66 | 24 | 3 | 15 | 576 | 13 | 73 | 36 | 5 | 1 |

**Table 3** - *continued*

| | | As | Rb | Nb | Sn | Cs | La | Ce | Pr | Nd | Sm | Eu | Gd | Tb | Dy | Ho | Er | Tm | Yb | Lu | Hf | Ta | Tl | Pb | Th | U |
|---|---|---|---|---|---|---|---|---|---|---|---|---|---|---|---|---|---|---|---|---|---|---|---|---|---|---|
| | | ppm | ppm | ppm | ppm | ppm | ppm | ppm | ppm | ppm | ppm | ppm | ppm | ppm | ppm | ppm | ppm | ppm | ppm | ppm | ppm | ppm | ppm | ppm | ppm | ppm |
| Group A | Average | 10 | 139 | 45 | 3 | 8 | 244 | 386 | 36 | 121 | 17 | 4 | 11 | 1 | 7 | 1 | 4 | 1 | 4 | 1 | 11 | 3 | 1 | 74 | 56 | 8 |
| | Minimum | 6 | 95 | 41 | 2 | 5 | 200 | 330 | 28 | 88 | 12 | 3 | 8 | 1 | 6 | 1 | 3 | 0 | 3 | 0 | 10 | 3 | 1 | 47 | 49 | 6 |
| n = 5 | Maximum | 16 | 176 | 53 | 3 | 9 | 279 | 452 | 44 | 136 | 20 | 5 | 13 | 2 | 8 | 2 | 4 | 1 | 4 | 1 | 12 | 3 | 1 | 114 | 64 | 10 |
| | Std.Dev. | 4 | 34 | 5 | 1 | 1 | 30 | 48 | 6 | 19 | 3 | 1 | 2 | 0 | 1 | 0 | 0 | 0 | 0 | 0 | 1 | 0 | 0 | 27 | 7 | 2 |
| Group B | Average | 7 | 123 | 32 | 2 | 6 | 163 | 275 | 26 | 87 | 13 | 3 | 8 | 1 | 6 | 1 | 3 | 0 | 3 | 0 | 7 | 2 | 1 | 49 | 41 | 7 |
| | Minimum | 5 | 99 | 25 | 1 | 6 | 138 | 233 | 21 | 66 | 10 | 3 | 6 | 1 | 5 | 1 | 3 | 0 | 3 | 0 | 3 | 2 | 0 | 26 | 28 | 4 |
| n = 6 | Maximum | 11 | 145 | 37 | 3 | 8 | 195 | 357 | 33 | 122 | 17 | 4 | 11 | 1 | 8 | 1 | 4 | 1 | 4 | 0 | 10 | 3 | 1 | 72 | 48 | 8 |
| | Std.Dev. | 2 | 22 | 5 | 1 | 1 | 24 | 46 | 5 | 20 | 3 | 1 | 2 | 0 | 1 | 0 | 1 | 0 | 0 | 0 | 3 | 0 | 0 | 19 | 8 | 1 |
| Group C | Average | 36 | 129 | 32 | 3 | 9 | 158 | 251 | 23 | 76 | 12 | 3 | 8 | 1 | 6 | 1 | 3 | 0 | 3 | 0 | 9 | 2 | 1 | 49 | 36 | 7 |
| | Minimum | 5 | 96 | 24 | 2 | 6 | 112 | 183 | 15 | 51 | 8 | 2 | 6 | 1 | 5 | 1 | 3 | 0 | 2 | 0 | 6 | 2 | 1 | 18 | 28 | 3 |
| n = 9 | Maximum | 260 | 147 | 38 | 3 | 25 | 225 | 362 | 30 | 100 | 15 | 4 | 10 | 2 | 8 | 2 | 4 | 1 | 4 | 1 | 10 | 3 | 1 | 81 | 43 | 10 |
| | Std.Dev. | 84 | 18 | 5 | 0 | 6 | 37 | 60 | 5 | 18 | 3 | 1 | 2 | 0 | 1 | 0 | 1 | 0 | 0 | 0 | 1 | 0 | 0 | 20 | 6 | 3 |
| Group D | Average | 19 | 104 | 24 | 2 | 8 | 97 | 163 | 17 | 60 | 10 | 2 | 7 | 1 | 6 | 1 | 3 | 0 | 3 | 0 | 7 | 2 | 1 | 25 | 24 | 4 |
| | Minimum | 7 | 69 | 19 | 1 | 4 | 77 | 127 | 14 | 43 | 7 | 2 | 5 | 1 | 5 | 1 | 3 | 0 | 2 | 0 | 5 | 2 | 0 | 8 | 21 | 3 |
| n = 8 | Maximum | 80 | 145 | 29 | 3 | 24 | 122 | 200 | 20 | 79 | 12 | 3 | 9 | 1 | 8 | 1 | 4 | 1 | 4 | 1 | 8 | 2 | 1 | 48 | 34 | 5 |
| | Std.Dev. | 25 | 24 | 4 | 1 | 7 | 15 | 27 | 2 | 13 | 2 | 0 | 1 | 0 | 1 | 0 | 1 | 0 | 0 | 0 | 1 | 0 | 0 | 13 | 5 | 1 |
| Group E1 / E2 | Average | 8 | 103 | 16 | 3 | 6 | 36 | 68 | 8 | 27 | 5 | 1 | 5 | 1 | 5 | 1 | 3 | 0 | 3 | 0 | 5 | 1 | 1 | 20 | 11 | 3 |
| | Minimum | 5 | 74 | 10 | 1 | 4 | 28 | 56 | 6 | 22 | 5 | 1 | 4 | 1 | 4 | 1 | 2 | 0 | 2 | 0 | 4 | 1 | 0 | 7 | 8 | 2 |
| n = 12 | Maximum | 28 | 148 | 24 | 6 | 10 | 49 | 97 | 9 | 35 | 7 | 1 | 5 | 1 | 6 | 1 | 3 | 0 | 3 | 0 | 6 | 2 | 2 | 36 | 15 | 3 |
| | Std.Dev. | 7 | 21 | 4 | 1 | 2 | 6 | 12 | 1 | 3 | 1 | 0 | 1 | 0 | 1 | 0 | 0 | 0 | 0 | 0 | 1 | 0 | 0 | 9 | 2 | 0 |
| Group E3 / E4 | Average | 11 | 152 | 23 | 5 | 9 | 54 | 106 | 11 | 39 | 7 | 2 | 6 | 1 | 6 | 1 | 3 | 1 | 3 | 0 | 6 | 2 | 1 | 32 | 16 | 4 |
| | Minimum | 7 | 96 | 15 | 2 | 5 | 36 | 70 | 8 | 28 | 6 | 1 | 5 | 1 | 5 | 1 | 3 | 0 | 3 | 0 | 5 | 1 | 0 | 11 | 12 | 3 |
| n = 10 | Maximum | 23 | 222 | 60 | 7 | 13 | 75 | 149 | 15 | 49 | 10 | 2 | 8 | 1 | 7 | 1 | 4 | 1 | 4 | 1 | 8 | 4 | 2 | 52 | 21 | 5 |
| | Std.Dev. | 5 | 40 | 13 | 2 | 3 | 11 | 22 | 2 | 6 | 1 | 0 | 1 | 0 | 1 | 0 | 1 | 0 | 0 | 0 | 1 | 1 | 1 | 14 | 3 | 1 |
| Group F | Average | 6 | 53 | 8 | 2 | 3 | 23 | 45 | 4 | 17 | 4 | 1 | 4 | 1 | 4 | 1 | 2 | 0 | 2 | 0 | 3 | 1 | 0 | 17 | 6 | 2 |
| | Minimum | 5 | 19 | 5 | 1 | 1 | 12 | 24 | 3 | 13 | 3 | 1 | 3 | 1 | 3 | 1 | 2 | 0 | 2 | 0 | 3 | 0 | 0 | 7 | 4 | 1 |
| n = 8 | Maximum | 7 | 89 | 11 | 2 | 5 | 33 | 64 | 6 | 22 | 5 | 1 | 4 | 1 | 4 | 1 | 3 | 0 | 3 | 0 | 4 | 1 | 1 | 25 | 9 | 2 |
| | Std.Dev. | 1 | 37 | 3 | 1 | 2 | 10 | 19 | 1 | 4 | 1 | 0 | 0 | 0 | 0 | 0 | 0 | 0 | 0 | 0 | 0 | 0 | 0 | 7 | 3 | 1 |

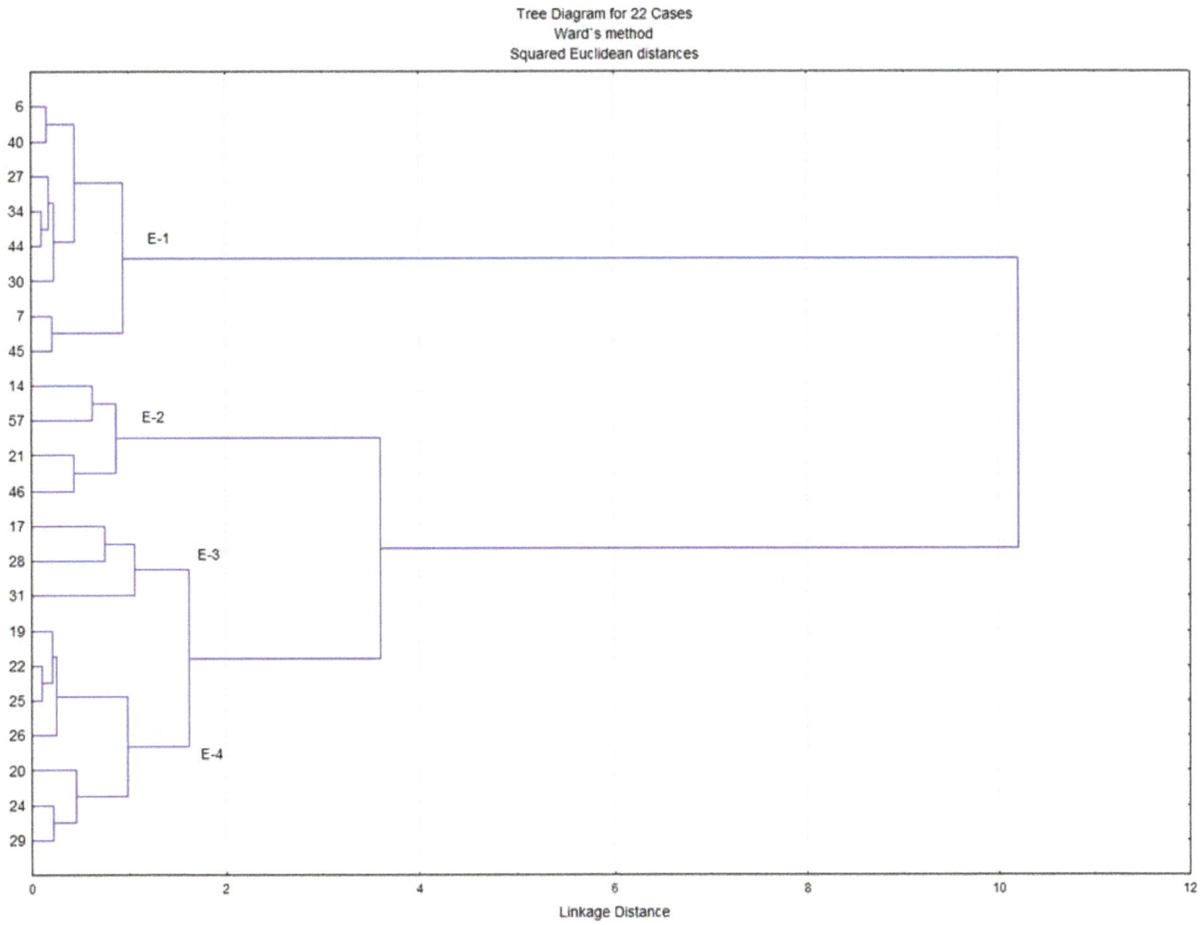


**Figure 4.** Hierarchical clustering dendrogram using Ward's method and squared Euclidean distances of cluster E.

in feldspars under weathering conditions, biotite, apatite, pyroxene, and precipitated in limestone (Wedepohl 1978), suggesting an association with the mineralogy of the immediate hinterland of the mountainous region in the eastern part (Rollinson 1993; Muchez *et al.* 2008).

No absolute technological or chronological patterning can be ascribed to any of the clusters. However, it becomes clear that cluster A encompasses mainly material characteristic for the site of Aykırıkça, associated with a well-defined Archaic 'fabric' group. This seems consistent when evaluating material attributed to cluster B by focusing on the same type of ware groups.

The extent of chemical variability remains consistent for cluster C, even though mostly 'orange' wares are attributed to this group. 'Matt-painted' sherds seem to concentrate within cluster D, along with other slipped ceramics. It is possible that the higher CaO in cluster D is significant as a provenance indicator, or at least as an indicator of some significant recipe pattern for these wares; this possibility remains to be tested. Cluster E shows a more complex pattern. Although statistically relatively close clusters, variability between the subgroups seems more elaborate than with previous clusters. In this respect, subgroup E-1 is composed of a wide variety of wares incorporating so-called 'matt-painted', 'buff', 'orange', 'white', 'grey-cored', and dark coloured Archaic ceramics. At this point, it is virtually unfeasible to offer interpretations for this group, except for its high circulation. Another explanation for this group composition is the statistical reflection of the mixing of resources as both high Cr, Ni, CaO and Sr values, and depletions of $Na_2O$, Th and REEs reflect. Cluster E-2 is similarly composed of 'orange', 'white', and Southwest Anatolian wares, while only characterised by low REE and Th values, together with strong variability in $Na_2O$, Sr, Cr, Co, and Cs. Clusters E-3 and E-4 are exclusively composed of Düver material, strongly suggesting that these ceramics were locally produced, given the presence of all ceramic categories ranging from the Archaic to the Hellenistic period. A conservative use of mineral resources seems manifest, as for example both 'black on red' ceramics and Hellenistic red slipped tableware have a nearly identical chemical pattern. Also, on a synchronic scale, local 'black on red', 'matt-painted' and technomic ceramics were produced from this quite homogenous resource, indicating a socially stratified community and a specialisation of the pottery craft. Chemically, the sherds attributed to cluster F show the most significant deviation from the other clusters, as mainly recognisable in REE pattering and the depletion of alkali metals. As such, these sherds appear to be derived from alternative resources not readily associated with the study area. Again, material from this cluster adds to the variability of wares and provenance at Düver. It is remarkable however that no sherds with a signature of

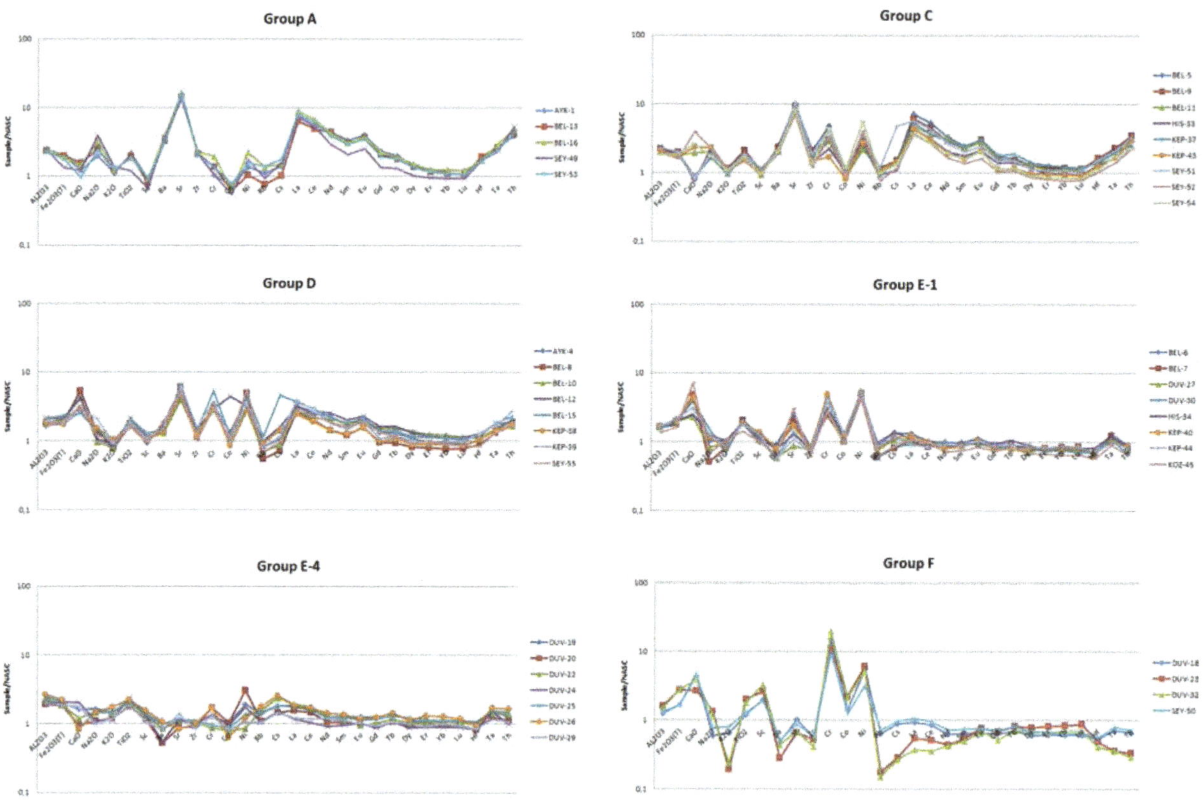


**Figure 5.** Multi-element plots of clusters A, C, D, E-1, E-4, and F, showing NASC-normalised chemical data.

the eastern mountainous region were found at Düver, while the opposite pattern is clearly visible in the dataset.

It can be concluded that the wide array of technomic wares attributed to the Archaic period, especially at Aykırıkça and Düver, are consistent in production. These present the picture of a more restricted local production, quite diversified internally, and with only limited circulation. Possibly, this could even be indicative of political or socio-economic differences. The category of 'matt-painted' wares seems a regional phenomenon covering a wider distribution. The 'orange' ware group also seems to represent a consistent widespread pottery practice in the region, which, although mainly attributed to the Classical-Hellenistic period, may already have been produced during the Archaic period. The ceramic assemblages suggest a strong continuity in the use of both sources and technologies, even though the variability documented suggests the presence of several producers.

## 4. Conclusions

Geochemical analysis proves to be a robust guide for describing the provenance and circulation of ceramic wares. The application of multi-element plots following statistical procedures is useful both for discerning provenance and/or 'recipe' signals of ceramic groups. Also, it provides the opportunity to examine in detail the effectiveness of the statistical method chosen. As such, some 'statistical' clusters are often of either more or less relevance archaeologically. In this respect, the post-depositional influence on chemical composition in this region remains a topic calling for future research. In this paper, geochemical analysis of a regional pottery assemblage revealed the presence of at least two largely different production regions, corresponding to their respective geological and geographical situations. However, only very limited exchange seems to have taken place between these two regions (the so-callled 'Lake District' and 'Çeltikçı valley system'). Nevertheless, especially when considering Kepez Kalesi, Belören, Hisar, Aykırıkça, and Seydiköy, circulation of wares among these sites seems to have occurred. No attribution to a specific production site can be undertaken, and therefore it is plausible — when taking into account the variation of the wares produced — that several production workshops were active in the region, the scales of which are yet to be determined. From a technological point of view, it is apparent that no distinction in terms of resources was made regarding table wares, cooking wares, or storage vessels. The consistent use of few resources seems to have been employed for all ceramic products. As such, both 'matt-painted' wares and large storage vessels display the same geochemical signal. One could also expect some differentiation or specialisation of the craft chronologically, relating to resource procurement. Instead, the obvious chronological disparity between the Archaic and Classical-Hellenistic material can be accounted for on a technological level rather than related to a substantial change in the organisation of production. This socio-economic pattern and its historical implications represent important aspects calling for future research in this area.

## References


Aydal S., Michell S., Robinson Th. and Vandeput L., 1997, The Pisidian survey 1995. *Anatolian Studies*, 47, 141-172.

Birmingham J., 1964, Surface finds from various sites. *Anatolian Studies*, 14, 29-34.

Cummer W.W., 1970, Phrygian rooftiles in the Burdur Museum. *Anadolu*, 14, 29-71.

Çokay-Kepçe S., 2006, *The Karaçalli Necropolis near Antalya*, Adalya Supplementary Series 4, AKMED/ Suna-Inan Kirac Akdeniz Medeniyetleri Arastirma Enstitusu, Antalya.

Degryse P. and Waelkens M., 2008, *Sagalassos VI. Geo- and Bio-Archaeology at Sagalassos and in its Territory*, Leuven University Press, Leuven.

Gromet L.P., Dymek R.F., Haskin L.A. and Korotev R.L., 1984, The "North American Shale Composite": its compilation, major and trace element characteristics. *Geochimica et Cosmochimica Acta*, 48, 2469-2482.

Johnson R.A. and Wichern D.W., 1992, *Applied Multivariate Statistical Analysis*, Prentice Hall, New Jersey.

Mee Ch.B., 1978, Aegean Trade and Settlement in Anatolia in the Second Millennium B.C. *Anatolian Studies*, 28, 121-155.

Mellaart J., 1954, Preliminary report on a survey of pre-classical remains in southern Turkey. *Anatolian Studies*, 4, 175-240.

Mellaart J., 1955, Some Prehistoric Sites in North-Western Anatolia. *Istanbuler Mitteilungen*, 6, 53-88.

Mellaart J. and Murray A., 1995, *Beycesultan 3.2: Late Bronze Age and Phrygian Pottery and Middle and Late Bronze Age Small Objects*, Occasional Publication of the British Institute of Archaeology at Ankara 12, British Institute of Archaeology at Ankara, London.

Muchez Ph., Lens S., Degryse P., Callebaut K., Dederen M., Hertogen J., Joachimski M., Keppens E., Ottenburgs R., Schroyen K. and Waelkens M., 2008, Petrography, mineralogy and geochemistry of the rocks in the area of the archaeological site of Sagalassos. In *Sagalassos VI. Geo- and Bio-Archaeology at Sagalassos and in its Territory*, (eds. P. Degryse and M. Waelkens), 25-52, Leuven University Press, Leuven.

Peacock D.P.S., 1977, *Pottery and Early Commerce: Characterization and Trade in Roman and Later Ceramics*, Academic Press, London.

Poblome J. and Degeest R., 2003, The ceramic evidence. In *Chora of Sagalassos. The Evolution of the Settlement Pattern from Prehistoric until Recent Times*, (eds. H. Vanhaverbeke and M. Waelkens), 138-142, Studies in Eastern Mediterranean Archaeology 5, Brepols, Turnh-out.

Robinson D.M., 1950, *Excavations at Olynthos 13: Vases found in 1934 and 1938*, Johns Hopkins Press, Baltimore.

Rollinson H.R., 1993, *Using Geochemical Data: Evaluation, Presentation, Interpretation*, Longman Publishing Group, Harlow.

Rotroff S.I., 1997, *The Athenian Agora. Hellenistic Pottery. Athenian and imported wheelmade table ware and related materials, Vol. XXIX, Part 1*, The American School of Classical Studies at Athens, Princeton.

Rotroff S.I., 2006, *The Athenian Agora. Hellenistic Pottery. The Plain Wares, Vol. XXXIII*, The American School of Classical Studies at Athens, Princeton.

Sparkes B.A. and Talcott L., 1970, *The Athenian Agora XII, Black and Plain Pottery of the 6th, 5th, and 4th Centuries B.C.*, American School of Classical Studies at Athens, New York.

Talloen P., Poblome J., Vanhaverbeke H. and Waelkens M., 2006, Matar in Pisidia. Phrygian Influences in Southwestern Anatolia. In *Pluralismus autochtoner in Anatolien. Vom Zusammenbruch der Hethitischen Großreiches bis zum Beginn der Hellenisiering (Bonn 19-20 May 2005)*, (ed. M. Hutter), 175-190, Ugarit-Verlag, Münster.

Vanhaverbeke H. and Waelkens M., 2003, *The Chora of Sagalassos. The Evolution of the settlement Pattern from Prehistoric until Recent Times*, Studies in Eastern Mediterranean Archaeology 5, Brepols, Turnhout.

Waelkens M. and Loots L., 2000, *Sagalassos V. Report on the survey and excavation campaigns of 1996 and 1997*, Acta Archaeologica Lovaniensia Monographiae 11, Leuven University Press, Leuven.

Waelkens M. and Poblome J., 1995, *Sagalassos III. Report on the fourth excavation campaign of 1993*, Acta Archaeologica Lovaniensia Monographiae 7, Leuven University Press, Leuven.

Waelkens M. and Poblome J., 1997, *Sagalassos IV. Report on the survey and excavation campaigns of 1994 and 1995*, Acta Archaeologica Lovaniensia Monographiae 9, Leuven University Press, Leuven.

Waelkens M., Paulissen E., Vanhaverbeke H., Öztürk I., De Cupere B., Ekıncı H.A., Vermeersch P.M., Poblome J. and Degeest R., 1997, The 1994 and 1995 Surveys on the Territory of Sagalassos. In *Sagalassos IV. Report on the Survey and Excavation Campaigns of 1994 and 1995 (Acta Archaeologica Lovaniensia Monographiae 9)*, (eds. M. Waelkens and J. Poblome), 11-102, Leuven University Press, Leuven.

Ward J.H., 1963, Hierarchical grouping to optimize an objective function. *Journal of the American Statistical Association*, 58, 236-244.

Wedepohl K.H., 1978, *Handbook of Geochemistry, Vol. II, Part 5*, Springer-Verlag, Berlin and New York.


## ACKNOWLEDGEMENTS


This research was supported by the Belgian Programme on Interuniversity Poles of Attraction (IAP 6/22), initiated by the Belgian Federal Science Policy Office - Belgian State and the Research Fund of the K.U. Leuven (BOF-GOA 07/02 - Methusalem 07/01), next to projects G.0788.09 and G.0421.06 of the Fund for Scientific Research, Flanders (FWO).The authors are grateful to Steven Luypaers for the preparation of the samples, Rebecca Scott for correcting the English text, and the anonymous referees who helped improve the paper.

# 23

# Experiments with double chamber sunken up-draught kilns

**Dragoş Gheorghiu**
Department of Master, National University of Arts, 19 Budisteanu, Bucharest, Romania

**Abstract** - Sunken up-draught kilns appeared in the 5th millennium BC in the context of the South Eastern European Chalcolithic Cucuteni-Tripolye tradition, as a mechanism for the mass production of ceramics fired at temperatures up to 900°C. Compared with the other pyro-instruments of the respective epoch, the sunken up-draught kiln presented a series of advantages, such as high temperatures, protection of vessels made of fine paste, and a higher output of vessels. Experiments allowed identifying and refining the stages of the *chaîne opératoire* of this type of kiln, as well as the discovery of new features, such as ergonomics or micro-physical processes, which will help archaeologists to understand the complex technology involved in Chalcolithic ceramic making.

## 1. Introduction

Kilns are a subject often encountered in the archaeological and experimental archaeology literature (Rhodes 1981; Dawson and Kent 1984; Kingery 1997), and they were among the first subjects of experimental archaeology in the last decades. Although this might appear to be a well-exhausted subject, experiments with this pyro-instrument have the potential of contributing new data.

This paper will discuss one of the types of prehistoric draught kilns, which the author believes to represent one of the most extraordinary prehistoric 'machines' due to its simplicity and efficiency, as demonstrated by experiments (Gheorghiu 2002; 2006; 2007). This proposition is also based on the fact that sunken up-draught kilns did not change from the 6th millennium BC until post-medieval times in some parts of South Eastern Europe.

Sunken up-draught kilns are attested in the archaeological record starting from the 6th millennium BC in the Near East (Simpson 1997, 39), but only from the 5th -4th millennia BC in South Eastern Europe (Coms¸a 1976; Markevic, 1981). Their invention occurred very early, probably before the emergence of ceramics in the Near East, and they could have been the source of the *vaiselle blanche*/white ware (de Contenson and Courtois, 1979), produced by plastering wickerwork containers with lime plaster (Gourdin and Kingery 1975), the latter material requiring a minimum temperature of 900°C (Rehder 2000, 47-8), which could have been reached in draught kilns.

## 2. The archaeological background

The emergence of efficient sunken up-draught kilns corresponds to a mastery of fire by means of the control of the ascending flow of hot gases that carried and fed the combustion through perforated surfaces conferring an air-absorbing character to the platform dividing the inner space of the kiln (Rye and Evans 1976, 164; Arnold 1997, 213). This technological innovation allowed the control of the thermal shock, and provided a clean firing for the decorated pottery, which was separated from the fuel, thus creating the premises of an increase in the volume of ceramic production and the emergence of craft specialisation in Chalcolithic South Eastern Europe (Gheorghiu 2001; 2007, 45).

Sunken up-draught kilns dating to the 5th millennium BC were excavated in this region, belonging to the Chalcolithic Cucuteni-Tripolye tradition (Dumitrescu *et al.* 1954; Markevic, 1981). Additional evidence of the existence of such kilns in the Cucuteni-Tripolye tradition is the vast quantity of fine ceramics, modelled from a high quality paste (resulting from levigated clays), with very thin walls (down to 3 mm), decorated before firing with complex patterns, and fired at high temperatures (Ellis 1984; Gâţa 2000; Marinescu-Bîlcu, 2000). A large quantity of pottery could be produced with the help of the slow potter's wheel or *tournette*, identified in the archaeological record of the middle and final phases of this tradition (Ellis 1984).

The replicas used for experimentation in this study were reproduced after the sunken up-draught kilns discovered at Costeşti (Markevic, 1981) and Glăvăneşti Vechi (Comşa 1976, 25), dated to the final phase of the Cucuteni-Tripolye tradition (5th-4th millennia BC). All the theoretical work presented in this paper is based on the experiments carried out with replicas of these two kilns during the last decade.

### *Open firings vs. controlled firings*

While in South Eastern Europe Chalcolithic container-like pyro-instruments (like ovens or pits) were successfully used for firing ceramics (Comşa 1976; Gheorghiu 2002; 2007), the level of a mass production of pottery was reached only after the introduction of sunken up-draught kilns. The differences between the open structures of combustion and the sunken up-draught kilns consist of the following features:

- the fuel is separated from the clay objects, which are not in direct contact with the flame;
- the inner space of the kiln is divided by a perforated platform into a firebox and a firing chamber;
- the thermal shock of the hot gases and flames is moderated by the tubes of the perforated platform;
- the amount of fuel in the firing box is regulated in order to control the interior temperature;
- the opening or closing of the apertures of the kiln (Rye and Evans 1976, 164) resulted in a control of the oxygen, an operation which could create an oxidising or reduced atmosphere.

All these features defined by the notion of 'control' (control of air flow and consequently control of the fire by changing the time and temperature of firing by means of manipulating the fuel, or control of the atmosphere by closing the apertures) render the sunken up-draught kiln a veritable 'machine', i.e., a complex instrument with inputs and outputs which can be regulated. The term 'machine' indicates in this context that the pyro-container was no longer a simple box, but a space where the fire was separated, fed and controlled throughout the firing.

### *The shape of sunken up-draught kilns*

Depending on the direction of the flow of air, up-draught and downdraught kilns can be distinguished, the former category being encountered in the archaeological record of the late Eastern European Chalcolithic. A sunken up-draught kiln was cut into a slopping bank, whose medium height allowed the operator, at the same time, to see the opening of the firebox and the flue of the firing chamber. The functional shape of a sunken up-draught kiln was composed of the following parts: a fire tunnel/firebox situated below a firing chamber, a perforated platform, a support for the platform, and ceramic waste or a movable dome (Dawson and Kent 1985) to cover the flue.

The shapes of the Cucuteni sunken up-draught kilns that were replicated for the purposes of this study could be reduced to the following basic volumes: a semi-cylindrical firebox (1.5 × 0.5 m), intersected with an ovoid volume (diameter at the base 1.5 m, depth 2 m) divided by a clay platform (height: 45-50 cm) with perforations (diameter 0.03-0.05 m) (Fig. 1).

It is evident that the small openings of the firebox and of the flue allowed an easy shutting down of the kiln at the end of the firing process to homogenise the temperature inside, a process which could generate a reduced atmosphere inside the vase chamber. This post-firing shuttering of the kiln was the method employed for the production of the light grey ceramics in the final phase Cucuteni B (Gheorghiu 2006, 38). The ovoid shape seems

**Figure 1.** A cross-section through the Cucuteni sunken up-draught kiln (after Markevic 1981).

to have been the most efficient for creating a good air-draught and for conserving thermal energy during the combustion process and after; the difficulty with it consists in the loading and unloading of the ceramic objects.

### *Functioning of sunken up-draught kilns*

There is general agreement that the key operations involved in the functioning of a draught kiln are the following: a gradual evacuation of capillary water (Arnold 1997, 61), heat containment (Shepard 1956, 75; Kingery 1997, 11) and heat transfer to wares, protection of ceramic objects contained by the firing chamber from thermal shocks during the processes of heating or cooling, protection against humidity (rain or moisture) (Arnold 1997, 213), and the production of higher temperatures (Shepard 1956, 75) compared to other pyro-instruments. The perforated platform has two important roles: to control the thermal shock (i.e., to transform the large flames originating from the firebox into a set of small flames which came in contact with the vessels long after the initiation of the fire), and to store heat, being the principal source for transferring heat (Fig. 2).

## 3. Experiments

### Chaîne opératoire. *Physical determinism and subjective choices of the potter*

It is not the purpose of this paper to present a detailed account of firing with a sunken up-draught kiln, but merely to emphasise those aspects that have not been mentioned in the existing literature hitherto. Thus, a significant number of details concerning the firing process, such as the time duration of the procedures, the temperatures reached, or the rate of waste in each firing are not mentioned here.

A first set of operations entailed the loading of the firing chamber through the flue, followed by the covering of the flue with shards, in order to form a ceramic strainer, and the structured positioning of the fuel in front of the firebox. Both the fuel and shard covering were positioned so as to allow the creation of an airflow between the firebox and the flue. Because the environment associated to the Cucuteni-Trypolye culture was deciduous forest, pinewood was used in these experiments only to initiate the fire, and beech wood to continue the firing.

**Figure 2.** The perforated platform of the sunken up-draught kiln with a thermocouple fitted; Vadastra, 2002. Photograph by Dragoş Gheorghiu.

After initiating the fire, the fired fuel was gradually introduced inside the firebox, the fire being stabilised at a temperature between 20°C and 120°C (Hamer 1975, 22) to allow 'water smoking', i.e., the slow elimination of capillary water (Arnold 1997, 61) (Fig. 3). There is a relationship between the duration of this stage and the dimensions of the sunken up-draught kiln, as well as of the pieces which form the load.

The temperature can be increased to over 350°C, the limit of combustion for the organic material contained into the clay's fabric, but stabilised under 700°C. This represents the second stage of firing, when the process of sintering of the internal particles is initiated (Kingery 1997, 12). At a temperature between 570°C and 600°C, the elimination of water from the crystal structure of the minerals begins (Shepard 1956, 72, 81; Arnold 1997, 610; Gâţa 2000, 124), which is associated with the beginning of the emission of red glow from the incandescent clay objects (Fig. 4). The positioning of the wasters on the flue allows the potter to observe this phenomenon, and consequently to control the firing.

The heating of the interior of the sunken up-draught kilns and of the ceramic load over 700°C produces a strong air draught, which allows the flames from the firebox to reach up to the flue. Frequently, when the temperature reaches 900°C at the middle of the load, coals already fill the 1.5 × 0.5 m firebox and the air draught diminish, consequently impeding the rise of the temperature.

One can infer that the volumetric structure and the dimensions of this type of kiln are optimal for temperatures under 900°C (as the analyses of most of the Cucuteni-Tripolye fine ceramics show). In order to raise the temperature over this thermal threshold, one should

**Figure 3.** The 'water smoking' of the sunken up-draught kiln; Vadastra, 2002. Photograph by Dragoş Gheorghiu.

**Figure 4.** Incandescent clay objects seen through the flue; Vadastra, 2004. Photograph by Dragoş Gheorghiu.

discharge the firebox and initiate a new fire, which would help increase the temperature to over 1000°C. As this is a procedure with high energy expenditure and a high level of difficulty, it is possible that high temperatures were not reached in draught sunken up-draught kilns in the East European prehistory, a hypothesis supported by the analysis of prehistoric ceramics (Gâţa 2000, 125).

It is important to mention that a sunken up-draught kiln does not produce a uniform temperature inside the vessel chamber, the difference between the temperatures (measured with the help of Platinum-Rhodium thermocouples placed on the upper side of the platform and at the middle of the load) sometimes being up to 200°C (Gheorghiu 2002).

Except for the stages related to water elimination, which require the observance of a well determined time span, the other stages of the firing process can be carried out using different time spans. Therefore, a complete firing process could last between 5 and 18 hours (Deca 1982, 214) depending on the type of load, fuel, and temperature of the environment, but also on the style of firing adopted by the potter.

## 4. Discussion and archaeological implications

The experiments carried out with sunken up-draught kilns brought to light a series of problems, not mentioned so far in the specialist literature. Some of these leave visible traces in the archaeological record, while some only help experimenters to refine the *chaînes opératoires* pertaining

**Figure 5.** A child unloading the kiln; Vadastra, 2003. Photograph by Dragoş Gheorghiu.

to the construction and use of this type of kilns. From the first category, four examples will be discussed.

The first example refers to the technology involved in digging the ovoid and cylindrical shapes, which is a difficult operation due to the narrow dimensions of these features. This is the reason why the author assumed that children performed this operation. During the experiments, children were very efficient in loading and unloading the vessels, with the platform supporting their weight. The dimensions of the opening of the sunken up-draught (Fig. 5) kiln only allowed children to excavate its shapes.

When the author built the platform, he asked a child to plait a twig structure inside the excavated fire chamber, which he fixed in the walls of the sunken up-draught kiln and on

**Figure 6.** The 'walls' of the kiln after 5 successive firings; Vadastra, 2004. Photograph by Dragoş Gheorghiu.

**Figure 7.** The partially collapsed platform at 1200°C. One can notice the filled perforations near the fracture line; Vadastra, 2003. Photograph by Dragoş Gheorghiu.

the upper part of a pedestal of clay left at the bottom of the fire chamber. One of the Cucuteni-Tripolye sunken up-draught kilns displays a profile left on the walls of the firing chamber (Markevic 1981) for a better fixing of the twig structure of the platform.

After finishing the plaiting of twigs, the child pressed wet clay on this structure and carried out some of the perforations sitting on the un-perforated area on top of the pedestal. Other perforations were made from outside the sunken up-draught kiln, with the help of a long pointed stick. As the Cucuteni-Tripolye excavated kilns show (Markevic 1981), the large dimensions of the pedestals left the central part of the platforms without perforations, and so the positioning of the vessels inside the firing chamber should have been done in such a manner as to allow the circulation of the air and flames in this area. One might also assume that a child (entering from the firebox aperture) finished the lower surface of the wet platform and enlarged the perforations.

The second example refers to the changes in the dimensions of the sunken up-draught kiln due to repeated firings. In a first stage, a sunken up-draught kiln is a negative shape dug into the ground, but consecutive firings transform into ceramic not only the load of vessels but also the soil, creating the 'walls' of the kiln, which are visible in the archaeological record. One can thus conclude that the kiln is a ceramic construction produced by itself (Fig. 6).

The third example refers to the behaviour of the platform at high temperatures. For a better understanding of the functioning of the sunken up-draught kilns, the perforated platform can be approached as a set of tubes for air absorption, with the role of inducing a distribution of the thermal shock. At temperatures over 1150°C, one can notice the beginning of a process of vitrification in the filling up of the air-draught tubes with glass and slag, with consequences on the lowering of the temperature in the vessel chamber. It is assumed that the tubes of the platforms were cleaned periodically, because the occurrence of glass is not mentioned in relation to the Cucuteni-Tripolye excavated sunken up-draught kilns. The experiment with the collapsed alluvial clay platform at temperatures over 1150°C led to the hypothesis that the respective feature was made from a refractory material, probably clay with a high proportion of sand. The repair of collapsed platforms (Fig. 7) in the Cucuteni-Tripolye tradition is not mentioned either in the existing literature, but only the repeated plastering of the interior of the firing chambers (Markevic 1981). This experiment could thus offer some hints for future analyses of the platform materials and for potential reconstructions.

The fourth aspect refers to the micro-processes of evaporation of capillary water. When a wet sunken up-draught kiln is fired for the first time, one can observe rhythmic fluctuations of the temperature inside the vessel chamber (Gheorghiu 2002, 92, diagram 3), due to the recurring evaporation of the water contained in the platform and the walls, which decreases the level of heat inside the sunken up-draught kiln on each occasion. One may infer that this phenomenon occurs (but at a micro-scale) also on the surface of the vessels fired during the 'water smoking' stage.

## 5. Conclusion

Although sunken up-draught kilns seem to have been studied in depth, one can conclude that every time experiments are conducted, they could bring new data and a new perspective that would lead to a better understanding of these complex ancient machines. Experiments demonstrate that the Cucuteni-Tripolye sunken up-draught kilns showed good visual ergonomics (since one operator could control the inputs and outputs of the kiln at every stage of the process), a good ergonomic dimension (both for adults and for the children who helped in the process) (Gheorghiu 2012), and that these dimensions, together with the shape of the vessel chamber, conferred to the machine the advantages of thermal efficiency. From the physical processes related to the growing in time of the efficiency of the vessel chamber, one can infer that sunken up-draught kilns were used for long periods of time, thereby constituting specialised places that lasted for years.

## References

Arnold E.D., 1997, *Ceramic theory and cultural process*, Cambridge University Press, Cambridge.

Comşa E., 1976, Caracteristicile şi însemna tatea cuptoarelor de ars din aria culturii Cucuteni-Ariuşd. *Studii şi cerceta ri de istorie veche şi arheologie*, 27, 23-33.

de Contenson H. and Courtois L.C., 1979, A propos des vases en chaux: Recherches sur leur fabrication et leur origine. *Paléorient*, 5, 177-182.

Dawson D. and Kent O., 1984, Methods of kiln reconstruction. *Bulletin of the Experimental Firing Group*, 2, 13-17.

Dawson D. and Kent O., 1985, Kiln superstructures - The Bickley experiment. *Bulletin of the Experimental Firing Group*, 3, 70-79.

Deca E., 1982, Centrul de ceramica din comuna Lungeşti, Jud. Vâlcea. *Buridava - Studii şi materiale*, 211-216, Râmnicu Vâlcea Museum, Râmnicu Vâlcea.

Ellis L., 1984, *The Cucuteni-Tripolye culture: A study in technology and the origins of complex society*, British Archaeological Reports International Series 217, Oxford.

Dumitrescu Vl., Dumitrescu H., Petrescu Dâmboviţa M. and Gostar N., 1954, *Ha ba şeşti. Monografie arheologica* , Editura Academiei, Bucharest.

Gâţa G., 2000, A technological survey of the pottery. In *Dra guşeni. A Cucutenian community*, (eds. S. Marinescu-Bîlcu and A. Bolomey), 111-129, Editura Enciclopedica , Bucharest; Wasmuth Verlag, Tübingen.

Gheorghiu D., 2001, South-Eastern European Early and Late Chalcolithic. In *Encyclopedia of Prehistory, Vol. IV*, (eds. P. Peregrine and M. Ember), 356-380, Kluwer Academic and Plenum, New York and London.

Gheorghiu D., 2002, Fire and air draught: Experimenting the Chalcolithic pyroinstruments. In *Fire in archaeology*, (ed. D. Gheorghiu), 83-94, British Archaeological Reports International Series 1089. Archaeopress, Oxford.

Gheorghiu D., 2006, On Chalcolithic ceramic technology: A study case from the Lower Danube traditions. In *Ceramic studies. Papers on social and cultural significance of ceramics in Europe and Eurasia from prehistoric and historic times*, (ed. D. Gheorghiu), 29-42, British Archaeological Reports International Series 1553. Archaeopress, Oxford.

Gheorghiu D., 2007, Chalcolithic pyroinstruments with air-draught - An outline. In *Fire as an instrument. The Archaeology of pyrotechnologies*, (ed. D. Gheorghiu), 41-51, British Archaeological Reports International Series 1619. Archaeopress, Oxford.

Gheorghiu D., 2012, The ergonomics of the chaines-operatoires. In *Estudos de tecnologia ceramica*, (eds. L. Oosterbeek and J.F. Cerezer), 59-62, Arkeos 31, Tomar.

Gourdin W.H. and Kingery W.D., 1975, The Beginnings of Pyrotechnology: Neolithic and Egyptian lime plaster. *Journal of Field Archaeology*, 2, 133-150.

Hamer F., 1975, *The Potter's Dictionary of materials ad technologies*. Pitman Publishing, London; Watson Guptill Publications, New York.

Kingery W.D., 1997, Operational Principles of Ceramic kilns. In *The Prehistory and History of Ceramic Kilns*, (ed. P.M. Rice), 11-20, Proceedings of the Prehistory and History of Ceramic Kilns, 98th Annual Meeting of the American Ceramic Society in Indianapolis, American Ceramic Society, Colombus, OH.

Marinescu-Bîlcu S., 2000, The pottery. Tradition and innovation. In *Dra guşeni. A Cucutenian community*, (eds. S. Marinescu-Bîlcu and A. Bolomey), 91-110, Editura Enciclopedica, Bucharest; Wasmuth Verlag, Tübingen.

Markevic V.I., 1981, *Pozne-Tripolskie plemena severnog moldovii*. Kishinev.

Rehder J.E., 2000, *The mastery and uses of fire in Antiquity*, McGill-Queen's University Press, Montreal.

Rhodes D., 1981, *Kilns. Design, construction and operation*, Chilton Book Company, Radnor, PA.

Rye O.S. and Evans C., 1976, *Traditional pottery techniques of Pakistan: field and laboratory studies*, Smithsonian Contributions to Anthropology 21, Smithsonian Institution Press, Washington DC.

Shepard A.O., 1956, *Ceramics for the archaeologist*, Carnegie Institution of Washington Publication 609, Washington DC.

Simpson J., 1997, Prehistoric ceramics in Mesopotamia. In *Pottery in the making. World Ceramic Traditions*, (eds. I. Freestone and D. Gaimster), 38-43, British Museum Press, London.

## ACKNOWLEDGEMENTS

I thank Dr. Michela Spataro for support and Professor George Nash for reviewing the English of the manuscript and for useful comments. I would also like to thank the two reviewers who improved the coherence of the text. The pyro-experiments were financed by three CNCSIS grants (Nos. 1612; 112; 945), and my first experiments with sunken up-draught kilns were carried out with Dr. Alex Gibson.

24

# Archaeometric investigation of Celtic graphitic pottery from two archaeological sites in Hungary

**Izabella Havancsák[a], Bernadett Bajnóczi[a], Mária Tóth[a], Attila Kreiter[b] and Szilvia Szöllősi[c]**

[a]Institute for Geological and Geochemical Research, Research Centre for Astronomy and Earth Sciences, Hungarian Academy of Sciences, H-1112 Budapest, Budaörsi út 45., Hungary, havancsaki@gmail.com, bajnoczi@geochem.hu, totyi@geochem.hu

[b]Hungarian National Museum National Heritage Protection Centre, H-1113 Budapest, Daróci út 3., Hungary, attila.kreiter@mnm-nok.gov.hu

[c]Institute of Archaeological Sciences, Eötvös Loránd University, H-1088 Budapest, Múzeum krt. 4/B., Hungary, szolloszilva@gmail.com

**Abstract** - Graphite-tempered ceramics from two Celtic archaeological sites in South Hungary, Dunaszentgyörgy (Highway No. 6) (LT B2-C1) and Bátaszék (Körtvélyes-dűlő) (LT D), were investigated in the context of the present project. The main aim of the research was to compare the graphitic temper of the vessels found at the archaeological sites and to provide a preliminary outline about the possible provenance of the graphite.

Ceramics from both Dunaszentgyörgy and Bátaszék contain similar graphitic temper: graphite fragments and clasts of graphitic paragneiss. Graphitic gneiss contains sillimanite and kyanite. Metamorphic rocks with similar medium- to high-grade minerals cannot be found in outcrops located on the territory of present-day Hungary. Based on the mineralogical composition, the potential source area of the graphitic temper can be restricted to the Variegated Unit of the Moldanubicum zone, in the Czech Republic.

In addition to the graphitic rock fragments found as archaeological artefacts at Bátaszék, the comparison of graphite-tempered ceramics to non-graphite-tempered vessels from the same archaeological sites also suggests that graphite was imported as raw material and that ceramics were produced 'locally'. Firing temperatures for both graphitic and non-graphitic wares were usually below 800°C, and for several non-graphitic vessels below 650°C.

## 1. Introduction

The Celtic 'graphitic ware' is a distinctive type of pottery known from most parts of the Central European Celtic world. The Celtic 'graphitic ware' category includes a range of pottery forms; the situla-like pot with decorations of vertically incised bundles of lines being the most common. Other vessel types, such as bottles, shallow and deep bowls, plates, cups, and mugs also occur (Kappel 1969). The use of graphite in pottery making had already started in Neolithic times, but the importance of graphite as temper increased considerably starting with the early La Tène period (von Carnap-Bornheim 1998). The use of ceramics containing graphite as non-plastic inclusion (temper) reached a peak in the territory of present-day Hungary during the period between the beginning of the 3rd century BC and the middle of the 1st century AD in the form of the classical Celtic 'graphitic ware', mostly situla-like pots (Szabó *et al.* 1999). These vessels occur both in settlements and cemeteries (Benadík 1961).

Graphitic ceramics can be found in a wide geographical area, not only around graphite sources, an aspect which suggests extended trade. Determination of the exact provenance (i.e., mine or mining district) of the graphite used for preparing vessels constitutes a significant topic within research on Celtic artefacts. Another aspect which is also not clear pertains to the form in which graphite was distributed. There is evidence for distribution of raw graphite (e.g., Meduna 1998; Sievers 2006), clay mixed with graphite (Kappel 1969), as well as of finished graphitic vessels (Waldhauser 1992). Graphite or graphitic raw materials are thought to be useful for objects which are constantly exposed to high temperature (Kappel 1969; Martinón-Torres and Rehren 2009). There are numerous open questions regarding the function of graphite in the vessels, the most widely accepted view being that these vessels were used for cooking and/or storing food and liquids (Trebsche 2003).

This study was performed on Celtic graphitic ceramics from the Dunaszentgyörgy (Highway No. 6) (LT B2-C1) and Bátaszék (Körtvélyes-dűlő) (LT D) settlement sites, which are located in South Hungary, on the flood plain of the River Danube, about 40 km from each other (Fig. 1). The main purpose of this study was to compare the graphitic temper, as well as the technological characteristics (firing conditions) of the vessels found at the two sites, and to assess the possible provenance of the graphite using petrographic and X-ray diffraction analyses. Non-graphitic ceramic artefacts from the two archaeological sites were also studied for comparison.

Graphitic rock fragments (graphitic gneiss) were found in pits and post holes from the Celtic layer in Bátaszék. These rock fragments were probably used as temper for the graphitic vessels; therefore, their mineral assemblage was compared with the temper of ceramics.

## 2. Sample description

In this work, 132 ceramic vessels from the two archaeological sites were studied. The examined sherds are both typologically and technologically representative of the ceramic assemblages of the studied sites.

From Dunaszentgyörgy, 19 graphitic and 21 non-graphitic ceramic sherds were analysed (Havancsák *et al.* 2009a; Kreiter *et al.* 2009). The group of graphitic ceramics contained fragments of 16 situla-like vessels (Fig. 2a), two dishes, and one bowl. Most of the sherds have grey colour and metallic lustre; however, two sherds have a light red colour. Several samples (11 sherds) exhibit a sandwich structure with 0.5 to 3 mm light surface rim and thick (4 to 8 mm) black core or a thin red surface on one side.

The non-graphitic ceramics from Dunaszentgyörgy belong to the following forms: biconical pots (4 pieces), bowls with s-profile (9), bowls (3), pot with swollen rim (1), cooking pot (1), storage vessel (1), bottle (1), and mug (1). Half of the non-graphitic ceramics from Dunaszentgyörgy exhibit a sandwich structure with 1.5 to 5 mm light red surface and black-grey core, while two sherds have black-grey surfaces and red cores. The non-graphitic sherds are mostly red coloured, fine and semi-fine ceramics.

A total of 35 graphitic and 57 non-graphitic ceramic sherds from the Bátaszék archaeological site was studied (Havancsák *et al.* 2009b). The graphitic sherds mostly belong to the categories of situla-like pots (Fig. 2b) and dishes. The graphitic sherds are grey-black in colour (24 pieces), and 11 sherds show a sandwich structure with red surface and black core.

The non-graphitic ceramics can be classified as bowls with s-profile (39 pieces) (Fig. 2b), bowls (5), deep conical bowls (4), dishes (4), cooking pots (3), hemispherical bowl (1), and biconical cooking pot (1). Among the non-graphitic ceramics, fine, semi-fine and coarse sherds can be found.

## 3. Analytical methods

The petrographic investigation was carried out using a Nikon Eclipse E600 polarising microscope. The texture

**Figure 1.** Location of the studied archaeological sites on a schematic map of Hungary.

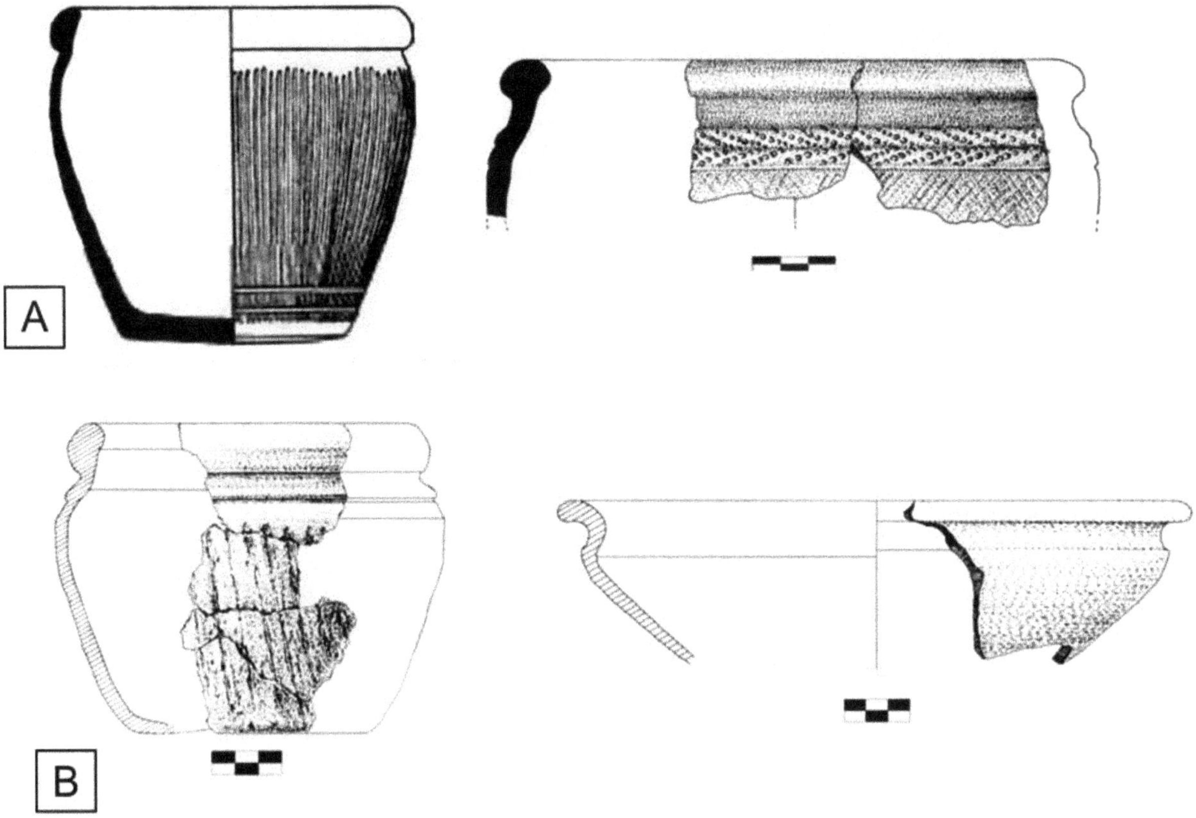


**Figure 2.** (a) Situla-like pots from Dunaszentgyörgy; (b) Situla-like pot and bowl with s-profile from Bátaszék.

recording suggested by the Prehistoric Ceramics Research Group (PCRG 1997) was used for these ceramics. The mineralogical composition of pottery was determined by X-ray diffraction analysis using a Philips PW 1730 diffractometer with a Bragg-Brentano alignment. The instrument parameters were: Cu Kα radiation, 45 kV tension, 35 mA intensity, 0.05°-0.01° 2θ step size, time constant 1 sec, and graphite monochromator. In the case of sandwich-structured ceramics, the core and the surface were analysed separately.

## 4. Results

### *Petrography*

Graphite-bearing sherds from Dunaszentgyörgy have black or brown colour with one polariser (1N), depending on the amount of the graphite. Graphite appears as an opaque fine powder colouring the matrix and may disguise the real isotropy of the groundmass. The original isotropy can be observed where the amount of graphite is subordinate. The isotropy of the graphitic ceramics from Dunaszentgyörgy is weak. The sherds display a hiatal texture with two characteristic grain size maxima. The first grain size maximum is between 0.1 and 0.3 mm. The dominant non-plastic components are graphite, quartz, K-feldspar, plagioclase, and muscovite. Occasionally, calcite, rutile, and opaque phases also occur. Quartz and feldspar often contain numerous apatite and zircon inclusions.

The second grain size maximum is between 0.8 and 5 mm. Graphite grains, angular graphitic lithoclasts (Figs. 3a, b), sharp crystal fragments (quartz, K-feldspar and plagioclase), and spherical argillaceous rock fragments (ARF) occur in this grain size interval. Graphite grains appear as opaque, individual fresh flakes, and in lithoclasts with silicate minerals (Figs. 3a, b). Graphitic lithoclasts consist of quartz, K-feldspar, sillimanite, kyanite, tourmaline, mica (muscovite and biotite), occasionally amphibole, calcite, and opaque minerals (Fig. 3b). Quartz and muscovite crystals in the lithoclasts often show petrographic signs of compression: quartz crystals have cataclastic texture, and muscovite flakes are kink-banded. Sherds show great variability in the amount and size of graphite. The size of graphite flakes varies between very fine and coarse (from 0.05 to 5 mm). The amount of graphite is rare or sparse in 8 sherds, moderate in 1, common in 2, and very common in 8 ceramic fragments.

The non-graphitic ware from the Dunaszentgyörgy archaeological site shows a light and dark brown groundmass with weak isotropy. The texture of the samples is dominantly serial, and fundamentally composed of very fine and fine (rarely medium), moderately and well-sorted non-plastic components (Fig. 3c). The dominant grain size varies between 0.1 and 0.3 mm in the serial samples, containing quartz, K-feldspar, plagioclase, muscovite, and opaque minerals. Numerous apatite and zircon crystals occur in quartz and feldspar grains (Fig. 3d). Four non-graphitic ceramics show a hiatal texture with another grain size maximum of 0.8 to 1 mm. Angular quartz, K-feldspar, spherical calcareous nodules, ARF, occasionally epidote, calcite, and biotite can be found in this grain size interval.

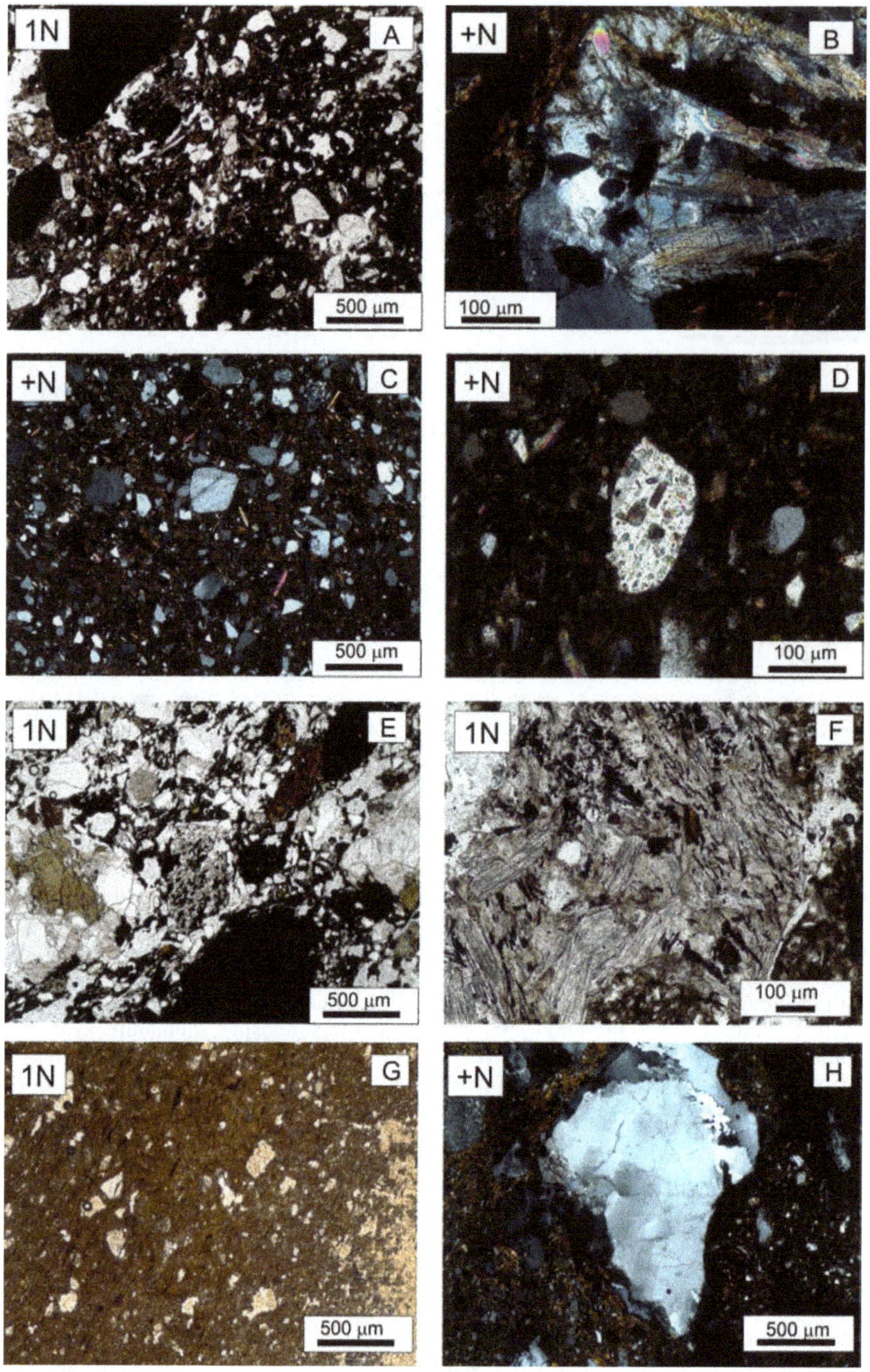


**Figure 3.** Texture and mineralogical assemblage of the studied ceramics from Dunaszentgyörgy (a, b, c, d) and Bátaszék (e, f, g, h). (a) texture of graphitic ceramic with graphitic lithoclast and graphite flakes; (b) graphitic lithoclast in graphitic sherd with sillimanite, quartz, feldspar, zircon, and graphite; (c) quartz, feldspar, and muscovite in non-graphitic ceramic; (d) feldspar-hosted apatite and zircon inclusions in a non-graphitic ceramic; (e) texture of graphitic ceramic with graphite, quartz, feldspar, and biotite; (f) graphitic lithoclast in graphitic ceramic with sillimanite pseudomorphs, quartz and feldspar; (g) texture of non-graphitic ceramic with quartz and feldspar; (h) granitic lithoclast in non-graphitic ceramic.

Graphitic vessels from Bátaszék are black and brown in colour with 1N, and only two samples are light brown-grey colour. The graphitic vessels show weak isotropy and a hiatal texture with two grain size maxima of 0.1 to 0.2 mm and 0.5 to 1.2 mm. Well-sorted grains of 0.1 to 0.2 mm size are composed of quartz, feldspar, biotite, muscovite, calcite, and fine-powdered graphite (Fig. 3e). Occasionally, tourmaline, calcite, garnet, epidote, and amphibole also occur. Some of the feldspar grains contain zircon inclusions.

Crystal fragments and lithoclasts occur in the second grain size interval. Quartz, K-feldspar, plagioclase, biotite, calcite,

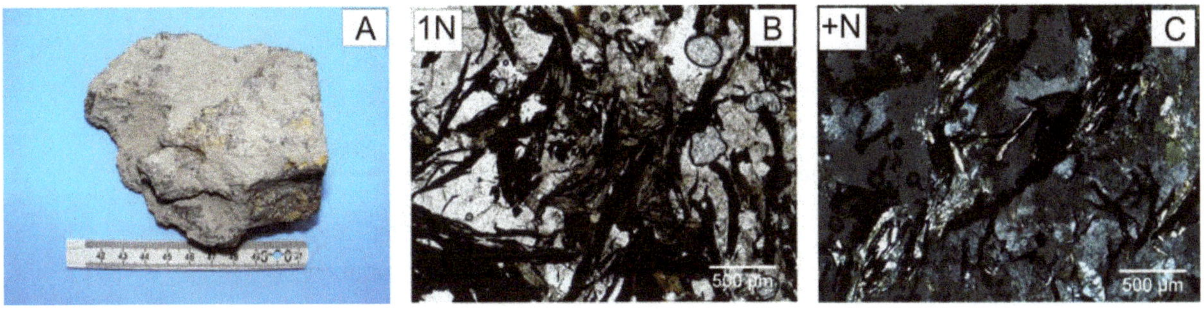


**Figure 4.** Graphitic rock found as an archaeological artefact at Bátaszék. (a) macroscopic view of the graphitic rock fragment; (b), (c) texture of the graphitic rock with graphite, quartz, feldspar, sillimanite, and biotite.

amphibole, and graphite occur as fragments; they are generally idiomorphic and fresh, except for the K-feldspar grains, which show slight alteration. Angular granitic and graphitic lithoclasts can be found in the studied ceramics (Figs. 3e, f). Granitic rock fragments consist of quartz, K-feldspar (often altered into clay minerals), plagioclase, biotite, and amphibole. Graphitic lithoclasts contain quartz, K-feldspar, plagioclase, sillimanite, kyanite, biotite, calcite, and also graphite. Sillimanite crystals are mostly pseudomorphs after biotite and kyanite (Fig. 3f). The size of graphite in the ceramics varies between 0.05 and 7 mm. The sherds have variable amounts of graphite: traces in 6 sherds, rare or sparse in 8, moderate in 14, common in 5, and very common in 2 ceramic fragments.

Non-graphitic ceramics from Bátaszék have a light brown-grey colour, and the isotropy of the groundmass is weak. The sherds are predominantly serial in texture (Fig. 3g), but hiatal texture occurs as well (Fig. 3h). Serial samples contain very fine to fine (0.1 to 0.4 mm), well-sorted non-plastic components, such as quartz, K-feldspar, plagioclase, muscovite, and, in some samples, calcite and shell fragments (Fig. 3g). Non-plastic components are altered and fractured. Sherds with hiatal texture have another characteristic grain size maximum at 0.5 to 1.5 mm. The grains are angular granitic lithoclasts (made of quartz, K-feldspar, plagioclase, muscovite, and biotite) (Fig. 3h), ARF, calcareous nodules, and occasional shell fragments.

The graphitic rock found at the Bátaszék archaeological site is a dark grey, fine- to medium-grained metamorphic gneiss with a slight foliation (Fig. 4a). It has a granoblastic-lepidoblastic texture and consists of idiomorphic K-feldspar crystals, isometric quartz grains, graphite, sillimanite and biotite flakes, and some calcite (Figs. 4b, c). The texture shows petrographic signs of compression and shearing stress: K-feldspar grains are often fractured, polycrystalline quartz grains show undulatory extinction, and biotite is kink-banded.

### *X-ray diffraction analysis*

The graphitic ceramics from Dunaszentgyörgy consist predominantly of graphite, quartz, plagioclase, K-feldspar and small amounts of calcite, 10 Å phyllosilicate (sericite-illite), magnetite and/or maghemite, and sillimanite. One sample contains chlorite (14 Å), and diopside is also present in another sample.

The non-graphitic ceramics from Dunaszentgyörgy are composed predominantly of quartz, plagioclase, and K-feldspar. Small amounts of calcite (and dolomite), 10 Å phyllosilicate, magnetite and/or maghemite, chlorite (14 Å in 9 samples), and amphibole (in 2 samples) also occur. Diopside and gehlenite were observed in two samples.

The graphite-bearing ceramics from Bátaszék show a similar mineralogical composition to the graphitic vessels from Dunaszentgyörgy: quartz, graphite, K-feldspar, plagioclase, 10 Å phyllosilicate and small amounts of magnetite and/or maghemite, sillimanite, and occasionally calcite. Three samples contain chlorite (14 Å), and one sample contains diopside.

The non-graphitic sherds from Bátaszék mainly contain quartz, plagioclase, K-feldspar, 10 Å phyllosilicate, calcite, dolomite, maghemite and/or magnetite, some dolomite, and amphibole. Small amounts of chlorite can be observed in 24 samples (14 Å). Three samples contain gehlenite and/or diopside.

Both graphitic and non-graphitic ceramics with a sandwich structure from Dunaszentgyörgy and Bátaszék contain trace amounts of hematite in their red surfaces.

## 5. Discussion

Based on the petrographic observations, both the graphite-tempered and non-graphitic ceramics are characterised by similar very fine to fine-grained non-plastic components. Apatite and zircon inclusions are frequently present in the quartz and feldspar grains. It follows, therefore, that the graphite-tempered and non-graphitic ceramics from Dunaszentgyörgy were most likely made from similar raw materials. The results of X-ray diffraction analysis correspond well with the petrographic observations. Except for the graphite and graphitic lithoclasts, the phases of both types of ceramics are similar. The hiatal texture of graphitic ceramics suggests that they were intentionally tempered with graphite-bearing rock fragments; however, the argillaceous rock fragments (ARF) found in the hiatal-textured ceramics were originally present in the clay raw material.

The non-graphitic ceramics from Bátaszék can be divided into two types based on the petrographic observations: serial- and hiatal-textured. Serial sherds contain common non-plastic components (quartz, feldspars, calcite, and shell fragments), while hiatal-textured ceramics contain granitic lithoclasts, as well as granitic phenocrysts.

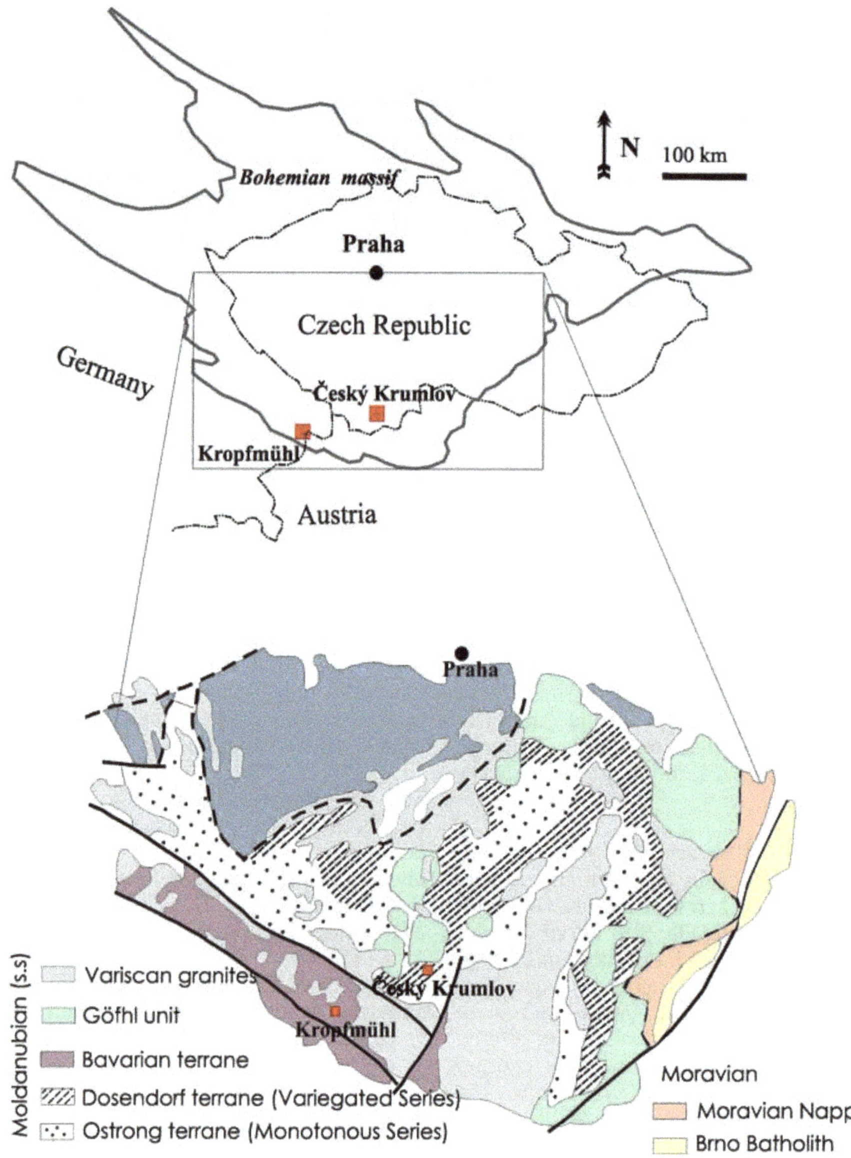


**Figure 5.** Location of the Bohemian Massif and the detailed geological map of the Moldanubicum (modified after Dallmeyer *et al.* 1995 and Finger *et al.* 2007).

Outcrops of granite (the Carboniferous Mórágy Granite) with the observed mineralogical composition can be found in the vicinity of the archaeological site, about 6 to 8 km from Bátaszék. The granitic rock fragments could originally have been present in the argillaceous raw material as eroded clasts, but it is more probable that they were added intentionally to the clay as temper. The clast and crystal fragments (mostly feldspar and biotite) do not show signs of transportation, and are mostly angular and idiomorphic, which suggest that granitic rock fragments were added to the raw material as temper.

Graphitic ceramics from the same archaeological site also contain granite-related lithoclasts and phenocrysts, as well as graphitic rock fragments added as temper. Raw graphite in the form of the graphitic rock artefact found at Bátaszék has a similar mineralogical composition to the graphitic lithoclasts found in the ceramics. The utilisation of granitic temper and the presence of raw graphite at the archaeological site confirm that graphite-bearing and (at least hiatal-textured) non-graphitic ceramics in Bátaszék were produced locally or in the vicinity of the site, and imported graphite temper was used for the graphitic ceramics.

### *Firing conditions*

The firing temperature of the ceramics was estimated on the basis of optical observations and X-ray diffraction analysis, with consideration of the results of former studies (e.g., Heimann 1979; Maggetti 1982; Duminuco *et al.* 1998; Cultrone *et al.* 2001).

Ca silicate minerals (gehlenite and diopside) formed at high-temperature firing were detected only in three sherds from Dunaszentgyörgy (one graphitic, two non-graphitic). The absence of Ca silicate minerals in most of the graphitic and non-graphitic vessels, as well as the presence of unreacted carbonate indicate firing below a range of 800 to 850°C. The 14 Å reflection of chlorite in nine non-graphitic samples and in one graphitic sherd indicates that the firing temperature was below 650°C for these ceramics.

Only a few ceramics (three non-graphitic and one graphitic) from Bátaszék contain Ca silicates (gehlenite and diopside). Absence of Ca silicates in the majority of graphitic and non-graphitic sherds suggests that the maximum firing temperature was usually between 800 and 850°C. The 14 Å peak of chlorite indicating firing temperature below 650°C is detected in almost half of the non-graphitic ceramics (24 samples) and in three graphitic vessels.

The mineralogical composition of ceramics indicates that the vessels from both archaeological sites were fired at similar maximum firing temperatures: ~800°C for graphitic vessels and ~800°C or ~650°C for non-graphitic vessels.

The occurrence of magnetite and/or maghemite as iron-bearing compound(s) in graphitic ceramics indicates firing under reducing conditions. Reducing firing allowed the survival of graphite. The presence of hematite in the surface of sandwich-structured ceramics (both graphitic and non-graphitic) suggests a locally oxidising atmosphere. The thin red surfaces indicate ceramic firing with high heating rate and short duration. However, the thin red outer layers of the ceramics could also have developed as a result of the use of pots for cooking on fire.

Earlier studies on the firing conditions of graphitic ceramics sometimes indicated lower firing temperatures. Gregor and Čambal (2009) observed firing temperatures of 500 to 600°C in an oxidising atmosphere, and in some cases of 900°C in a reducing atmosphere for Celtic graphitic ceramics from Bratislava (Slovakia). Jerem and Kardos (1985) inferred a firing temperature range of 600 to 650°C for the Celtic graphitic ceramics of the Sopron-Krautacher (Hungary) archaeological site. However, Friederich *et al.* (2010) determined that Celtic graphitic and non-graphitic ceramics found at Bopfingen (Germany) were fired in a reducing atmosphere at 800 to 850°C, followed by re-oxidation during cooling. According to Kappel (1969), the firing temperature for Celtic graphitic ceramics in general did not exceed 900°C.

### Provenance of graphite

The studied ceramics were tempered with graphite and graphitic lithoclasts. The best way to determine the provenance of the graphitic temper is to study the mineral assemblage of the graphitic lithoclasts in the ceramics. Graphitic lithoclasts in the vessels of both archaeological sites show similar mineral assemblages, suggesting that the graphitic raw materials presumably originated from similar geological environments. The mineralogical composition (quartz, K-feldspar, muscovite, biotite, sillimanite, kyanite, and calcite) and petrographic characteristics of the rock fragments indicate that the temper was derived from metamorphic gneiss, which was formed during the metamorphism of organic matter-bearing sediments (paragneiss). The presence of kyanite and sillimanite suggests medium- to high-grade (amphibolite facies) metamorphism (Bucher and Frey 2002).

Low-grade metamorphic graphitic phyllite outcrops adjacent to Velemszentvid in West Hungary; however, based on the mineral assemblage of the graphitic lithoclasts, this phyllite was not used for the tempering of the studied ceramics. Medium- to high-grade metamorphic graphite-bearing rocks cannot be found in outcrops located in the surroundings of the archaeological sites, or even in Hungary. However, there are large graphite resources hosted by paragneiss, located in the Moldanubian zone of the southern Bohemian Massif in the area of Czech Republic, Austria, and Germany, which could represent potential sources of the graphitic temper (Fig. 5). In all of these countries there are localities where graphite was mined and also used for vessels in historical times, already present in the Celtic period. Graphite from Český Krumlov (Czech Republic) was used for ceramic production in the Celtic oppidum of Třísov (Břeň 1987). Graphite was transported from Kropfmühl (Germany) to the Manching oppidum (Kappel 1969; Gebhard *et al.* 2004), and graphite from Obernzell (Austria) was mixed with argillaceous raw material for the production of triangular crucibles in the late Middle Ages (Martinón-Torres and Rehren 2005; 2009).

The possible provenance territory can be further restricted based on geological analogies. In Kropfmühl (Germany), graphite occurs in micaceous cordierite-bearing gneiss (Wayland 1951). Cordierite was not identified in the studied ceramics. The presence of cordierite in metamorphic rocks suggests lower grade metamorphism (Bucher and Frey 2002) than that of the observed index minerals (sillimanite, kyanite) of the graphitic lithoclasts of ceramics. Calc-silicate, gneiss and marble host the graphite in the Austrian part of Moldanubicum (Schrauder *et al.* 1993). The mineral assemblage of the gneiss in Austria is quartz, graphite, K-feldspar, plagioclase, calcite, muscovite, kyanite, and sulphide minerals. Sillimanite is not yet known in the metamorphics of the territory; however, sillimanite was observed in most cases in the graphitic ceramics of Dunaszentgyörgy and Bátaszék. Following a comparison of the geology of the main tectonic units and the mineralogical composition of graphite-bearing rocks with the mineralogical composition of graphitic lithoclasts in the ceramics, the most likely potential source for the graphite seems to be located in the Variegated (Varied) Series of the Moldanubicum, in the Czech Republic. In the Variegated Series, mainly biotite-sillimanite paragneiss, graphitic paragneiss, graphitic quartzite, marble, and amphibolite occur (Kachlík 1999; Finger *et al.* 2007). Graphite-bearing paragneiss is composed of quartz, feldspar, mica (biotite, muscovite), sillimanite, and occasionally kyanite (Kachlík 1999; Houzar and Novák 2002; Janousek *et al.* 2008). The graphitic rock temper in the Dunaszentgyörgy and Bátaszék ceramics contains the above-mentioned minerals.

From the archaeological point of view, graphitic temper is not the only link between Celtic cultures inhabiting the geographical areas of the present-day Czech Republic and Hungary. For example, shale used as raw material for the LT B2-C1 age 'black bracelets' commonly found in Hungary, originated from the Kounov area of the Czech Republic. The bracelets are likely to have been manufactured in the area of the Czech Republic, and were transported to the area of present-day Hungary through trade networks (Venclová 2001).

## 6. Conclusions

In this study, graphitic and non-graphitic ceramics from two Celtic archaeological sites in Hungary were examined. The graphitic and non-graphitic ceramics from Dunaszentgyörgy have similar raw materials except for the graphitic temper. Similar conclusions can be drawn from the results corresponding to the Bátaszék ceramics: both graphitic and some of the non-graphitic ceramics were tempered with locally available granitic lithoclasts, and, in addition, the graphitic ceramics were tempered with graphitic raw material.

Based on the mineralogical composition, the graphitic and non-graphitic ceramics at both archaeological sites were fired at a temperature below 800°C, and several non-graphitic vessels below ~650°C.

Graphitic lithoclasts in the ceramics of Dunaszentgyörgy and Bátaszék show similar mineral assemblages, which suggest that the graphitic raw materials originate from similar geological environments. Graphite-bearing rock fragments derive from medium- to high-grade graphitic (para)gneiss - an aspect that can be inferred from the presence of sillimanite and kyanite minerals. The nearest locality of the graphitic gneiss with these index minerals is the Moldanubicum zone of the Bohemian Massif. The potential source for the graphite was thus presumably located in the Variegated Series, in the area of the present-day Czech Republic.

## References


Benadík B., 1961, Grafitová keramika v laténskych hroboch na Slovensku [Die Graphitkeramik in latènezeitlichen Gräbern in der Slowakei]. *Slovenská Archeologia*, 9, 175-208.

Břeň J., 1987, K výrobě tuhové keramiky na keltském oppidu Třísově, okres Český Krumlov [Zur Herstellung der Graphittonkeramik aus dem keltischen Oppidom Třísov-Bezirk Český Krumlov (Süd-Böhmen)]. Časopis Národniho Muzea v Praze, 156, 1-9.

Bucher K. and Frey M., 2002, *Petrogenesis of metamorphic rocks*, Springer, Berlin and London.

Carnap-Bornheim C. von, 1998, Graphit und Graphittonware. *Reallexikon des Germanischen Altertumskunde*, 12, 593-598.

Cultrone G., Rodriguez-Navarro C., Sebastian E., Cazalla O. and De La Torre M.J., 2001, Carbonate and silicate phase reactions during ceramic firing. *European Journal of Mineralogy*, 13, 621-634.

Dallmeyer R.D., Franke W., Weber K. (eds.), 1995, *Pre-Permian Geology of Central and Eastern Europe*, Springer, Berlin.

Duminuco P., Messiga B. and Riccardi M.P., 1998, Firing process of natural clays. Some microtextures and related phase compositions. *Thermochimica Acta*, 321, 185-190.

Finger F., Gerdes A., Janousek V., René M. and Riegler G., 2007, Resolving the Variscan evolution of the Moldanubian sector of the Bohemain massif: the significance of the Bavarian and the Moravo-Moldanubian tectonometamorphic phases. *Journal of Geosciences*, 52, 9-28.

Friederich J., Gebhard R., Krause R., Riederer J., Wagner F.E. and Wagner U., 2010, Mössbauer study of Celtic pottery from Bopfingen, Baden-Württemberg, Germany. *Journal of Physics: Conference Series*, 217, 012064.

Gebhard R., Bott R., Distler N., Michálek J., Riederer J., Wagner F.E. and Wagner U., 2004, Ceramics from the Celtic oppidum of Manching and its influence in Central Europe. *Hyperfine Interactions*, 154, 199-214.

Gregor M. and Čambal R., 2009, Preliminary mineralogical and petrographic study of La Tène household ceramics from Bratislava's oppidum (Slovakia). In *Proceedings of 9th European Meeting on Ancient Ceramics (EMAC'07) - Vessels: Inside and Outside*, (eds. K.T. Biró, V. Szilágyi and A. Kreiter), 255-264, Hungarian National Museum, Budapest.

Havancsák I., Bajnóczi B., Tóth M., Kreiter A. and Szöllő si Sz., 2009a, Kelta grafitos kerámia: elmélet és gyakorlat dunaszentgyörgyi kerámiák ásványtani, petrográfiai és geokémiai vizsgálatának tükrében [Celtic graphitic pottery: theory and practice in the light of mineralogical, petrographic and geochemical study of ceramics from Dunaszentgyörgy (S. Hungary)]. *Archeometriai Műhely*, VI/1, 39-52.

Havancsák I., Bajnóczi B., Szakmány Gy., Kreiter A., Szöllősi Sz. and Gáti Cs., 2009b, A petrográfiai vizsgálatok jelentősége a kelta kerámiák grafitos sorványítóanyagának proveniencia meghatározásában [Significance of petrographic investigations in the determination of provenance of graphitic temper in Celtic ceramics]. *Archeometriai Műhely*, VI/4, 1-14.

Heimann R.B., 1979, Archäometrie: Methoden der Brenntemperaturbestimmung von antiker Keramik, Fridericana. *Zeitschrift der Universität Karlsruhe*, H.24, 17-34.

Houzar S. and Novák M., 2002, Marbles with carbonatite-like geochemical signature from variegated units of the Bohemian Massif, Czech Republic, and their geological significance. *Journal of the Czech Geological Society*, 47, 103-110.

Janousek V., Vrána S., Erban V., Vokurka K. and Drábek M., 2008, Metabasic rocks in the Varied Group of the Moldanubian Zone, southern Bohemia - their petrology, geochemical character and possible petrogenesis. *Journal of Geosciences*, 53, 31-46.

Jerem E. and Kardos J., 1985, Entwicklung und Charakter der eisenzeitlichen Graphittonware. *Mitteilungen der Österreichischen Arbeitsgemeinschaft für Ur- und Frühgeschichte (MUAG)*, 35, 65-75.

Kachlík V., 1999, Relationship between Moldanubicum, the Kutná Hora Crystalline Unit and Bohemicum (Central Bohemia, Czech Republic): A result of the polyphase Variscan nappe tectonics. *Journal of the Czech Geological Society*, 44, 201-292.

Kappel I., 1969, *Die Graphittonkeramik von Manching*, Ausgrabungen in Manching Band 2, F. Steiner, Wiesbaden.

Kreiter A., Bajnóczi B., Havancsák I., Szakmány Gy. and Tóth M., 2009, Macroscopic and petrographic analysis of Celtic ceramics. In *Settlement and cemetery excavations at the borders of Dunaszentgyörgy,* (ed. J. Kvassay), 157-199, VIA - Monographia Minor in Cultural Heritage 1, Field Service of Cultural Heritage, Budapest.

Maggetti M., 1982, Phase analysis and its significance for technology and origin. In *Archaeological ceramics*, (eds. J.S. Olin and A.D. Franklin), 121-131, Smithsonian Institution Press, Washington DC.

Martinón-Torres M. and Rehren Th., 2005, Ceramic materials in fire assay practices: a case study of 16th century laboratory equipment. In *Archaeometric and Archaeological Approaches to Ceramics. Papers presented at EMAC'05, 8th European Meeting on Ancient Ceramics, Lyon 2005*, (ed. S.Y. Waksman), 139-149, British Archaeological Reports International Series 1691. Archaeopress, Oxford.

Martinón-Torres M. and Rehren Th., 2009, Post-medieval crucible production and distribution: a study of materials and materialities. *Archaeometry*, 51, 49-74.

Meduna J., 1998, Zur Problematik der latènezeitlichen Graphittonkeramik. In *Frühmittelalterliche Graphittonkeramik in Mitteleuropa - Naturwissenschaftliche Keramikuntersuchungen*, (eds. T. Kempke, P. Kouřil, L. Poláček and Č. Staňa), 11-15, Internationale Tagungen in Mikulčice Band 4, Archeologický ústav AVČR, Brno.

Prehistoric Ceramics Research Group (PCRG), 1997, *The study of later prehistoric pottery: general policies and guidelines for analysis and publication*, Prehistoric Ceramics Research Group Occasional Papers 1-2, Oxford.

Schrauder M., Beran A., Hoernes S. and Richter W., 1993, Constraints on the origin and genesis of graphite-bearing rocks from the Variegated sequence of the Bohemian Massif (Austria). *Mineralogy and Petrology*, 49, 175-188.

Sievers S., 2006, Der Fernhandel am Ende der Latènezeit. In *Celtes et Gaulois. L'Archéologie face á l'Historie. Les mutations de la fin de l'âge du Fer*, (ed. C. Haselgrove), 67-81, Collection Bibracte 12, Centre archéologique européen, Glux-en-Glenne.

Szabó M., Guillaumet J.-P. and Kriveczky B., 1999, Polgár-Király-érpart vaskori település a Kr.e. IV-III. évszázadban. *A debreceni Déri Múzeum Évkönyve* 1997/1998, 177-181.

Trebsche P., 2003, *Keramik mit Feinkammstrich aus keltischen Siedlungen im Grossraum Linz: Untersuchungen zu Werkstätten, Funktion, Verbreitung und Datierung*, Linzer Archäologische Forschungen Band 35, Museen der Stadt Linz, Linz.

Venclová N., 2001, External contacts: visible and invisible. In *Fernkontakte in der Eisenzeit. Konferenz Liblice 2000*, (eds. A. Lang and V. Salač), 72-82, Archäologisches Institut der Akademie der Wissenschaften der Tschechischen Republik, Praha.

Waldhauser J., 1992, Keltische Distributionssysteme von Graphittonkeramik und die Ausbeutung der Graphitlagerstätten während der fortgeschrittenen Latènezeit. *Archäologisches Korrespondenzblatt*, 22, 377-392.

Wayland R.G., 1951, The graphite of the Passau area, Bavaria. *Transactions of the American Institute of Mining and Metallurgical Engineers*, 190, 166-172.

# 25

# Inferring provenance, manufacturing technique, and firing temperatures of the Monagrillo ware (3520–1300 cal BC), Panama's first pottery

**Fumie Iizuka[a,b], Richard Cooke[b], Lesley Frame[a,c] and Pamela Vandiver[a,c]**
[a]School of Anthropology, The University of Arizona
[b]Smithsonian Tropical Research Institute, Panama
[c]Department of Materials Science and Engineering, The University of Arizona

**Abstract** - Monagrillo (3520–1300 cal BC) is Panama's oldest pottery. Archaeologists assumed it was a low-fired expedient ware made with any available clay. We studied 1) clay sources (thin sections; DTA; shrinkage, porosity, and plasticity tests), 2) manufacturing techniques (xeroradiography; thin sections; visual inspection), and 3) firing temperature (SEM-EDS; porosity tests). We identified two clay types, one restricted to the Pacific coast, one widely distributed. Vessels were made by layering slabs and occasionally lumps. Rim- and lip-finishing is variable. Firing temperature (>800–950°C) is relatively high for open firing. Porosity is quite low. These aspects indicate that Monagrillo is not an experimental or expedient ware.

## 1. Monagrillo, Panama's first pottery

In the Americas, the earliest known pottery becomes progressively younger along the Central American isthmus and into Mexico. Therefore, some archaeologists have reasonably assumed that diffusion was the primary mechanism for the dispersal of ceramics northwards from South America (Ford 1969; Fonseca Zamora 1997; Meggers 1997). Others have argued that independent invention is as plausible an explanation, as there are geographical gaps in the distribution of the earliest known wares in Central and Mesoamerica which differ stylistically and technologically from each other and from contemporary South American wares (Cooke 1995; 2005; Hoopes 1995). Clark and Gosser (1995) propose that invention was 'dependent' – i.e., that people adopted the technology from other groups but manipulated style and function for their own purposes.

The Monagrillo ware (Willey and McGimsey 1954) is the earliest known pottery in Panama. It was produced between 2600 and 1200 BC (3520–1300 cal BC) over an area of 5600 km$^2$ in the central part of the country, between Parita Bay on the Pacific coast and the Coclé del Norte drainage in the Caribbean foothills (Fig. 1; Cooke 1995, Fig. 14.1; Griggs 2005). It has not been reported elsewhere in Panama. It appears stratified directly above Preceramic layers at two rock-shelters (Cueva de los Ladrones (Cl1) and the Aguadulce Shelter (Ag13)). This fact supports the hypothesis that it was the first pottery made in central Panama.

Current evaluations suggest that the Monagrillo pottery is typologically coherent, showing little evidence for diachronic change during its long period of manufacture. Vessel shapes are limited to bowls and restricted collarless vessels (Cooke 1995, Fig. 14.2). No handles, lugs, or feet have been found. Decoration consists of red- painted bands and daubs, and rare incised decoration using lines and volutes. Towards the end of the tradition, short collars are added to jars, and the variety of incised, impressed and modelled motifs increases. These developments, however, are poorly dated (Cooke 1995, Fig. 14.3; Cooke and Sánchez-Herrera 2004).

Previous assessments of Monagrillo pottery have relied on comparing data obtained from visual (usually non-instrumental) inspection of manufacturing processes, intuitive typological studies and regional site surveys supported by radiocarbon chronologies. Our study of the Monagrillo ware uses rigorous analytical methods derived from materials science and geology, focusing on the production process, with particular attention to raw material selection and firing temperatures.

## 2. Subsistence economy, settlement patterns and social interactions of the Monagrillo potters

Data about the subsistence economy, settlement patterns, and social interactions of the Monagrillo potters have been provided by (1) archaeological surveys (Cooke and Ranere 1992),

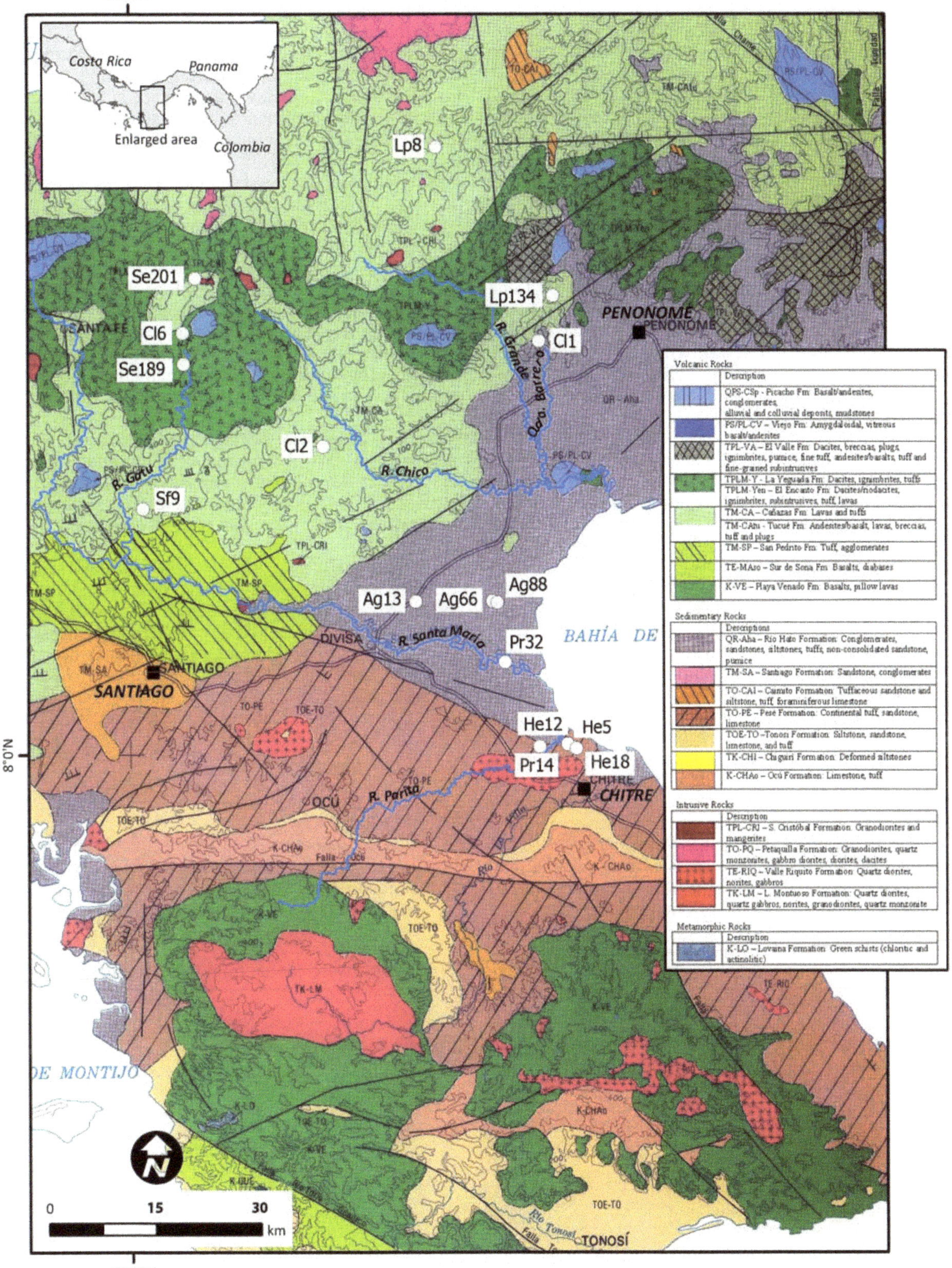


**Figure 1.** Geological map of Panama with Monagrillo sites The map, modified from del Giudice and Recchi (1969), was produced with the assistance of Natalia Hoyos.

(2) excavations at stratified sites (Willey and McGimsey 1954; Bird and Cooke 1978; Ranere and Hansell 1978; Cooke 1984), (3) archaeobotany (Dickau 2005, 2010; Piperno and Pearsall 1998; Perry *et al.* 2007), (4) zooarchaeology (Hansell 1979; Carvajal Contreras and Hansell 2008; Cooke and Jiménez 2008; Cooke *et al.* 2008), (5) geomorphology (Clary *et al.* 1984), and (6) vegetation history derived from a lake sediment record at Laguna La Yeguada, located within the Monagrillo pottery production zone (Piperno and Pearsall 1998; Piperno 2006).

Like their Preceramic predecessors in central Panama, Monagrillo potters were farmers who by 3500 cal BC had cleared extensive tracts of forest on the Pacific watershed for planting early forms of New World staple crops, e.g., maize *(Zea mays),* manioc *(Manihot esculenta)*, sweet potatoes *(Ipomoea batatas)*, squash (*Cucurbita* spp.) and peppers (*Capsicum* spp.). They were not hunter-gatherers and 'incipient' agriculturalists bound to the marine littoral – as formerly believed (Willey 1971). However, the degree and duration of sedentism at individual sites, and the seasonality of their occupations, are difficult to gauge from the existing record.

On the Pacific watershed of Panama under present-day climatic conditions, rain-fed farming is only possible between May and December. Domestic refuse at the largest Monagrillo site (Zapotal [Pr32]) covers ca. 1 ha. This site, located near prime agricultural soils, produced a large number of plant-processing tools. The remains of a small dwelling were also found here. It is likely that it was a village (Cooke and Ranere 1992; Perry *et al.* 2007). Ag13 and Cl1 – two rock-shelters located 18 and 25 km from the present-day shore of Parita Bay – were probably occupied for long periods at a time (i.e., during the farming season). Game was regularly brought back to each site. Shellfish, crabs, and fish from the marine littoral, also regularly consumed, would have been obtained by making trips to the coast or through exchange. Palm nuts were processed intensively at the Ag13. The primary species – the Neotropical oil palm, *Elaeis oleifera* – requires swampy habitats, and is most productive in rainy season months. The processing of palm fruits (*Atallea alenii*) and balsam seeds (*Humirastrum diguense*) is in evidence at the only Monagrillo site known on the Caribbean watershed: Calaveras (Lp8). This shelter's inhabitants consumed maize and may well have lived year round in this perennially humid habitat.

Hansell's (1979) preliminary growth-ring study of marine shells found in middens at the Monagrillo (He5) type site suggests that this settlement, dated between 2400 and 1300 BC (2800–1400 cal BC) and located then on the active marine shore of Parita Bay, was occupied mostly in the dry season (non-farming) months by people who spent the rest of the year elsewhere. Fishing in in-shore marine habitats was an important activity here and at Pr32. The high number of small shoaling fish in the middens, e.g, thread-herrings (*Opithonema* spp.), small jacks (*Carangidae* spp.), and sea catfish (*Ariidae* spp.), suggests the use of gill-nets and/or inter-tidal weirs. Inland-coast seasonal transhumance is still practised in this region in the dry season. He5 and Pr32 may have provided the inland shelters with dried and salted fish – another practice that persists in the region (Zohar and Cooke 1997). Smaller rock-shelters in the foothills and mountains where Monagrillo pottery is scarce are likely to have been occupied intermittently or irregularly as hunting-and-gathering camps or as rest stations on paths.

These data allow us to infer that Monagrillo communities interacted regularly, exchanging or transporting foodstuffs and other produce from many different habitats in an environmentally heterogeneous interaction zone. It can be assumed that pottery was one of these products.

The primary goal of this study is to use a detailed examination of the production processes and circulation of the Monagrillo pottery in order to improve knowledge about residential and interactive behaviour and pottery use before the appearance of those better -known regional societies that archaeologists consider to be chiefdoms with well-defined social classes and extensive trade connections (e.g., Linares, 1977; Helms, 1979; Drennan 1996; Isaza-Aizuprúa, 2007; Haller, 2008).

## 3. Materials and methods

### *Sample inventory*

Analyses were conducted on 110 Monagrillo sherds obtained through excavations conducted at sites in four environmental zones, and also on 12 raw clay samples collected near Cl1 and He5. Sample details are provided in Table 1.

**Table 1.** Inventory of Monagrillo sherd samples (samples for each analysis were chosen from this inventory).

| Environmental Zone | Site | River Drainage | Abbrev. | Number of sherds | Number of clay samples |
|---|---|---|---|---|---|
| Pacific Coast | Monagrillo | Parita | He5 | 30 | 3 |
| | Zapotal | Santa María | Pr32 | 20 | |
| | La Mula-Sarigua | Parita | Pr14 | 3 | |
| Pacific Plains | Aguadulce | Santa María | Ag13 | 20 | |
| Pacific Foothills | Cueva de los Ladrones | Grande | Cl1 | 31 | 9 |
| | Carabalí | Santa María | Sf9 | 2 | |
| Caribbean Foothills | Calaveras | Coclé del Norte | Lp8 | 4 | |

### *Pottery provenance*

In order to ascertain the locations where Monagrillo pottery was made, we studied the physical properties of archaeological sherds and raw clays deemed suitable for pottery production according to the results of Iizuka and Vandiver (2006). The mineralogical analysis used petrography (He5, 9 sherds and 2 modern clay samples; Pr32, 5 sherds; Pr14, 3 sherds; Ag13, 4 sherds; Sf9, 2 sherds; Cl1, 7 sherds and 4 modern clay samples). The manufacturing techniques of archaeological sherds were identified using visual and petrographic analysis in conjunction with xeroradiography.

Physical clay properties including shrinkage, plasticity, and porosity were tested using three samples of clays gathered from He5 and nine clay samples from Cl1. The plasticity test was carried out by determining the minimum radius of curvature that could be made with 1 cm diameter coils of hydrated clay before the clay began to crack. For the shrinkage test, we measured the linear drying shrinkage comparing wet and dry states for three days. The porosity test was conducted by firing the clays in 100°C increments from 500°C to 800°C using a rapid ramp and 15 minute soak sequentially at each temperature. After each firing, the sample was weighed and then boiled in distilled water for 1 hour. The samples then sat in water for 24 hours before re-weighing. The percentage of increased weight was directly related to the amount of water taken up by the pores of the ceramic. This method is useful for measuring open porosity, but does not work for closed porosity.

The composition of the ceramics and raw clay samples was determined using a variety of techniques. Because illitic and smectitic clays have very distinctive thermal decomposition characteristics, differential thermal analysis (DTA) on one sample from Cl1 and one sample from He5 provides a good indication of the clay type (Mackenzie 1970). Petrographic examination of the sherds and sample clays fired to 750°C for 15 minutes in an electric furnace was useful for identifying the mineral constituents of the ceramics. The mineral determinations were compared to the geological map of Central Panama. Based on mineralogy and morphology, we were able to identify both intentionally added temper and natural inclusions present in the clays. Natural mineral inclusions tend to be rounded, well integrated with the clay matrix, and often weathered. Intentionally added inclusions are often angular and have fresh (non-weathered) minerals, including feldspars.

### *Pottery manufacturing techniques*

Detailed visual analyses allowed us to identify butt and bevel joints, cracks, thickened body parts, and indentations. Petrography and xeroradiography were used to examine elongation and direction of pores, and ultimately identify manufacturing lines in the sherds.

### *Firing temperature*

Microstructural changes in pottery occur with the progressive sintering and vitrification of the clay matrix due to increased temperature (Tite and Maniatis 1975; Kingery 1987; Tite 1995, 37). Original firing temperatures were estimated by examining the microstructural features of sherds using scanning electron microscopy (SEM) and energy dispersive spectroscopy (EDS) on a Hitachi S-2460. Archaeological sherds were reheated at 100°C increments, and the original microstructure was compared to the reheated specimens. The point at which the microstructure changes marks the upper boundary of the original firing temperature range. A requisite of this experiment is that the sherd composition undergoes no changes during burial. For example, additions of soluble salts can cause changes to the thermo-physical properties of a pottery sample, e.g., melting at a lower temperature. In order to meet this requirement, the sherds selected for this experiment were screened so that no salts were present on their surfaces. EDS measurements ensured that the composition of the examined area was representative of the clay and not nearby mineral inclusions or salt accretions. Sherd fragments were imaged at x10,000 magnification. In addition to the microstructural analysis, the porosity of Monagrillo sherds was measured and compared to the porosity of clays fired at 100°C intervals, as described above.

## 4. Results

### *Geological setting and the provenance of Monagrillo pottery*

A principal objective of this study is to relate petrographic thin sections of Monagrillo pottery to the extremely complex geology of Central Panama. Ongoing research by others (e.g., geologists of the Smithsonian Tropical Research Institute Geology Program; Wörner *et al.* 2009) is rapidly modifying the current knowledge of regional geology. The following summary highlights salient features that should enable us to differentiate clay sources across the region.

The clearest petrographic differences – and therefore the ones most relevant to identifying the sources of clays used in the Monagrillo pottery – exist between the geologically older Azuero Peninsula to the south and the younger Cordillera Central to the north (Fig. 1). The Azuero Peninsula is itself divided into four zones: (1) the *Azuero Plateau*, which contains the oldest rocks in Western Panama, e.g., Mesozoic pillow basalts, diabase, and occasional gabbroic intrusives (Okaya and Ben-Avraham 1987, 794; Buchs *et al.* 2010), (2) the *Ocú Formation* (Fig. 1, K-CHAo) including Late Cretaceous formations of deep ocean hemipelagic limestones, (3) *Proto-Arc* igneous rocks including basaltic to basaltic trachyandesitic lava flows and Late Cretaceous dykes (Buchs *et al.* 2010, 20, 23, 28), and (4) the mature arc of the *Azuero Arc Group*, which includes igneous rocks with felsic intrusives and mafic to felsic volcanic rocks of Late Cretaceous-Middle Eocene age (del Giudice and Recchi 1969; Fig. 1, TK-LM, K-VE; Buchs *et al.* 2010, 22–4).

Ceramics which used clays and geological tempering materials from the *Proto-Arc* and *Azuero Arc* should be distinguishable using petrography. The subophitic, intersertal and porphyritic igneous rocks of the *Proto-Arc* contain minerals such as clinopyroxene, plagioclase, orthopyroxene, amphibole, alkali-feldspar, and glass (Buchs *et al.* 2010, 20, 24). On the other hand, the *Azuero Arc* outcrops contain granodiorites with zircon and amphibole minerals. While the lavas have inter-granular to porphyritic texture, the porphyritic rocks contain zoned plagioclase, alkali-feldspar, greenish clinopyroxene, amphibole, and quartz (Buchs *et al.* 2010, 21). The rocks of these zones can also be distinguished geochemically (Buchs *et al.* 2010, 16).

The Cordillera Central consists mostly of Miocene and Pleistocene-Holocene rocks (de Boer *et al.* 1988, 278; Lissinna 2005, 51–2, 73; Wörner *et al.* 2009, 192). The subduction-related calc-alkaline volcanism from the Late Tertiary comprises: (1) the *Cañazas group* (Fig. 1, TM-CA) containing basaltic to andesitic lavas and tuffs, (2) the *La Yeguada Formation* (Fig. 1, TPLM-Y) containing dacitic to rhyolitic volcanic rocks, and (3) the *El Valle Formation* (Fig. 1, TPL-VA) containing dacitic lavas, breccias, ignimbrites, tuffs, and some basalts/andesites (Lissinna 2005, 51). In the Cordillera Central, eruptions occurring between the El Barú Volcano in the west and the El Valle Volcano in the east resulted in the presence of orthopyroxene phenocrysts, amphibole, and clino-pyroxene, but at El Valle, *not olivine*. In addition, $K_2O$ values relative to $SiO_2$ decrease eastwards from El Barú to El Valle (de Boer *et al.* 1988, 280–1). The numerous recent eruptions – up to about 13,000 years ago (Bush and Colinvaux 1990; Defant et al. 1991; Knutsen 2010) – complicate ceramic sourcing by petrography within the Cordillera Central. We believe, however, that combining petrography with geochemistry could potentially be effective.

Rivers deposit sediments from entire drainages. Therefore, the mineral signatures of clays obtained in the lower stretches of rivers will reflect the basal geology from nearby areas and from farther upstream, where different kinds of rocks may be found. The River Parita rises in the mountains of the Azuero Peninsula. Thus, we predict that Monagrillo pottery produced at He5 and Pr14, in the lower stretches of this river, will contain clays derived from *igneous intrusive* and *sedimentary* rocks found along the course of the river. On the other hand, pottery made near Cl1 in the River Grande drainage is likely to use materials derived from the Cordillera Central, where igneous *extrusive* rocks predominate.

To sum up, the geology of the production zone of Monagrillo pottery is sufficiently differentiated to enable us to distinguish pottery made with raw materials from either the Azuero Peninsula or the Cordillera Central. As our study progresses, it is likely that we will be able to identify the provenance of some clays and tempers within these two areas. Sites located along the lower stretches of rivers on the Pacific side will provide greater challenges because of the mixed geological signatures of the alluvial zone.

## Petrographic analysis

The results of our petrographic analyses of Monagrillo sherds (Iizuka, 2013) show that some sherds contain only natural mineral inclusions, while others have both natural *and* added inclusions. The added inclusions are monomineralic cracked igneous intrusive rocks or monomineralic angular sands from a similar geological context. He5 sherds are classified into six types, Pr32 sherds into two types, Pr14 sherds into three types, Ag13 sherds into three types, Sf9 sherds into two types, CL1 sherds into four types, and Lp8 sherds into one type. In addition, raw clays collected near He5 and Cl1 were classified into two types. We grouped together the pottery paste types with the same mineralogical characteristics, regardless of site provenance. These groupings are summarised in Table 2.

Group 1 consists of clays with extrusive felsic natural inclusions and intentionally added temper of cracked intrusive igneous felsic rock or monomineralic angular sand with igneous felsic rock constituency. Pottery types belonging to Group 1 are found at six sites (He5, Sf9, Pr32, Ag13, Cl1 and Lp8) located near the Parita, Santa María, Grande, and Coclé del Norte rivers. The next most frequent type (Group 2) includes clays whose natural inclusions are igneous felsic extrusive and intermediate extrusive rock fragments. Igneous felsic intrusive cracked rock fragments or monomineralic angular sand with igneous felsic constituency have been added as temper. Sherds with these characteristics were found at five sites (He5, Pr32, Ag13, Sf9 and Pr14) located near the Parita and Santa María rivers. They were not identified at sites in other drainages, Cl1 along the Grande river, or Lp8 along the Coclé del Norte river.

Granitic rock natural inclusions in clays were only found at He5 and Pr32, which are both located near the Pacific coast, in the Parita and Santa María drainages, respectively. However, He5 granitic clays (Groups 9, 10, 11) have granitic rocks as the major natural inclusions, whereas Pr32 clays (Group 5) have granite in addition to igneous extrusive felsic and intermediate rocks as the natural inclusions. Some He5 granitic clays (He5-Type 1 and 2) contain tourmaline, a distinctive mineral which was not found at Pr32. One He5-Type 1 sherd has a sandstone inclusion.

To sum up, our petrographic analyses revealed that all river drainages in which Monagrillo sites are found produced sherds with clays derived from igneous felsic extrusive rocks, regardless of whether mineral temper was added intentionally to the clay. Raw clay samples gathered from both Cl1 and He5 (and fired to 750°C) contained clay with similar igneous felsic extrusive rock inclusions. Archaeological sherds found at He5 were the only ones that had granitic rock-based clays, indicating a local origin. A combination of both felsic and intermediate igneous extrusive rock fragments was only present in sherds from sites in the Santa María drainage basin, including He5, located by the Parita river but close to the mouth of the Santa María river. None of the Cl1 sherds contained intermediate extrusive rock inclusions.

We do not know whether the absence of intermediate extrusive rock inclusions in the Cl1 clays is due to: (1) the weathering away of previously existing potassium-rich feldspars in intermediate extrusive rocks in the clays; or (2) people gathering clays from areas where there were no intermediate extrusive rocks, although the vicinity of Cl1 can geologically contain these rock types. At this point, it is difficult to conclude whether the discrepancy between the geology surrounding Cl1 and Lp8 and the ceramic pastes of the sherds from these sites – that is, the likely presence of intermediate extrusive rocks in the area, with the pottery not containing intermediate extrusive rock inclusions – is due to the movement of pottery or the movement of raw materials into these regions, or to the preferential selection of weathered clays that are local to these sites. Bulk compositional analyses of the ceramic pastes and raw materials using a portable X-ray fluorescence (XRF) unit, inductively coupled plasma mass spectrometry (ICP-MS), and electron probe micro-analysis (EPMA) may allow us to distinguish between these regions chemically.

**Table 2.** Pottery groups according to added inclusions and clay types.

| Added inclusion types and clay types | |
|---|---|
| Group 1 | [Igneous felsic intrusive temper] + [igneous extrusive felsic natural inclusions] Lp8-Type 1, Cl1-Type 1, Sf9-Type 1, Pr32-Type 1, He5-Type 7, Ag13-Type 3, He5-Type 3 |
| Group 2 | [Igneous felsic intrusive temper] + [igneous intermediate extrusive + felsic extrusive natural inclusions] He5-Type 5, Pr32-Type 1, Ag13-Type 2, Sf9-Type 2, Pr14-Type 2 |
| Group 3 | [Igneous felsic intrusive temper] + [igneous felsic extrusive + mafic natural inclusions] He5-Type 6 |
| Group 4 | [Igneous intermediate intrusive temper] + [igneous felsic extrusive natural inclusions] Cl1-Type 2, Cl1-Type 3 |
| Group 5 | [Igneous intermediate intrusive temper] + [igneous felsic extrusive + igneous intermediate extrusive natural inclusions + igneous felsic intrusive natural inclusions] Pr32-Type 2 |
| Group 6 | [Igneous intermediate intrusive temper] + [igneous felsic extrusive + intermediate extrusive + mafic extrusive natural inclusions] Pr14-Type 1 |
| Group 7 | [Cracked igneous mafic intrusive rock temper or sand] + [igneous felsic extrusive natural inclusions] Pr14-Type 3 |
| Group 8 | [Igneous mafic intrusive temper] +[igneous felsic extrusive + intermediate extrusive and/or mafic extrusive natural inclusions] Ag13-Type 1 ((?) extrusive inclusions are heavily weathered) |
| Group 9 | [Natural inclusions with granitic rock fragments with tourmaline and minor amounts of igneous extrusive rocks and sedimentary sandstone without added inclusions in clay] He5-Type 1 |
| Group 10 | [Natural inclusions with granitic rock fragments with tourmaline and minor amounts of igneous extrusive rocks without added inclusions in clay] He5-Type 2 |
| Group 11 | [Natural inclusions with granitic rock fragments without tourmaline and with minor amounts of igneous felsic extrusive rocks without added inclusions] He5-Type 4 |
| Group 12 | [Clays with major inclusions that are igneous felsic extrusive rocks]<br>– He5-Clay 1, Cl1-Clay 1, Cl1-Type 4<br>– Clays (excluding added inclusions) of Lp8-Type 1, CL1-Type 1, SF9-Type 1, Pr32-Type 1, He5-Type 7, Ag13-Type 3, Cl1-Type 2, Cl1-Type 3, Pr14-Type 3, Ag13-Type1 |
| Group 13 | [Clays with igneous felsic extrusive and intermediate extrusive rock inclusions]<br>– He5-Clay 2<br>– Clays (excluding added inclusions) of He5-Type 5, Pr32-Type 1, Ag13-Type 2, Sf9-Type 2, Pr32-Type 2, Ag13-Type 1 (?) |
| Group 14 | [Clays with biotite and felsic extrusive rock inclusions] Cl1-Clay 2 |

### *Manufacturing technique*

Iizuka and Vandiver's (2006) clay suitability test for pottery production of clays collected near Cl1 and He5 indicated that, at Cl1, clays have to be carefully selected, while at He5, clays are workable and can be readily used for pottery production. DTA showed that both areas contained illitic clays.

Visual inspection, xeroradiography, and petrography showed that most Monagrillo sherds were constructed by overlapping slabs of clay. However, a few sherds at Pr32 (n=5) and Ag13 (n=3) combined slabs at their seams with small clay lumps. Lips and rims are finished in many different ways. Some have lips folded on top of slabs or consist of two clay pieces attached to a slab. The radiography in Figure 2 (top image) shows elongated pores in sherd Cl1-45-F13, indicating joints in the ceramic fabric. The bottom image in Figure 2 shows the thin section of He5-76-F1, with elongated pores aligned parallel to the vessel surface.

### *Porosity test*

The porosities of clays suitable for pottery production (5 samples) were all above 27% (the minimum porosity was 26.84% [He5-DO-1, fired at 800°C]). On the other hand, the porosity of Monagrillo pottery (8 samples) ranged between 26.19% (He5-63-F12) and 9.77% (Cl1-3-F120). The porosities of the archaeological sherds were much lower than porosities obtained for the clay samples collected near Cl1 and He5 after they were fired to 800°C. We interpret this characteristic to signify that all archaeological sherds must have been fired above 800°C if the clays were gathered locally.

### *Firing temperature*

The re-firing experiments corroborate the conclusion reached from the porosity tests. The results of our microstructural study using SEM-EDS (Fig. 3) are the following:

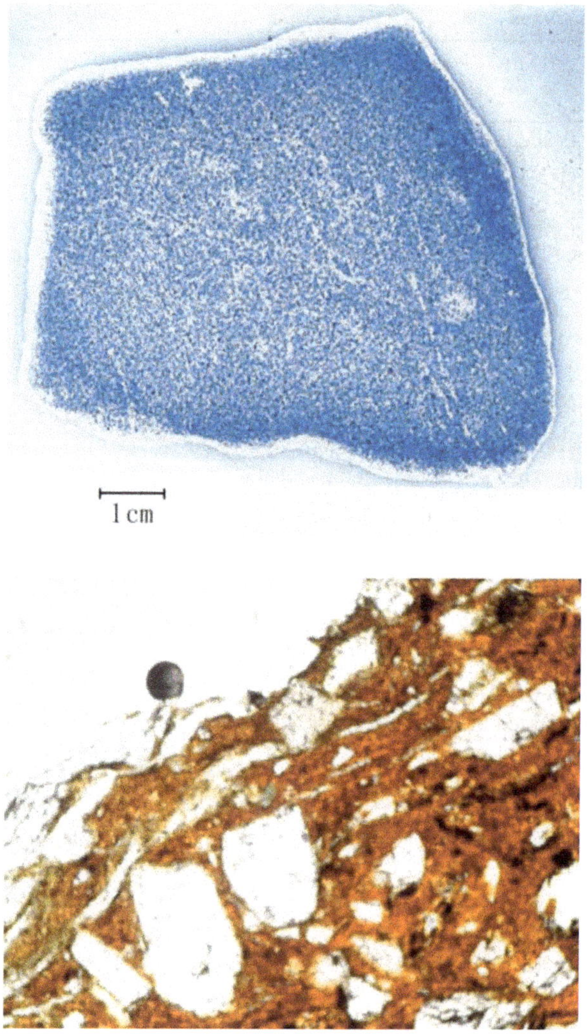


**Figure 2.** (top) Radiograph of a Monagirllo sherd (Cl1-45-F13) from xeroradiography. The radiograph shows locations and alignment of pores, indicators of joints created during manufacturing. (bottom) Image of a ceramic thin section under plane polarized light. The thin section shows alignment of elongated pores parallel to the vessel align with the left margin.

*Cl1-3-F137*
Microstructure remains unchanged up to 900°C. Slight rounding of clay particles becomes visible at 1000°C and continues up to 1100°C. We conclude that the firing temperature was slightly lower than 1000°C, probably 950°C.

*Cl1-45-F26*
No change occurs in this re-fired sherd up to 900°C, but a drastic change takes place at 1000°C. We infer that the original firing temperature was a little over 900°C. Figure 3 shows the SEM-EDS images and data of Cl1-45-F26.

*He5-63-F12*
The microstructural analysis of this re-fired sherd revealed that there is no detectable change between 500°C and 800°C. Some rounding of the pores in the clay matrix and smoothing of the pore walls is visible at 900°C. A similar sintering stage continues at 1000°C. A major microstructural change occurs at 1100°C as the sherd deforms plastically, losing its original shape. We infer that this vessel was originally fired at approximately 850°C.

*He5-76-F1*
The clay matrix of this sherd shows intensified rounding of the pore edges at 900°C as compared to the images taken at 800°C. We infer a firing temperature slightly above 800°C, although, unlike for He5-63-F12, microstructural changes at 1000°C are more pronounced.

## 5. Conclusions

### *Raw material selection*

Archaeologists have assumed that Monagrillo pottery was crude and expedient, made with any available local clay, and fired at low temperatures (500°C). This opinion was not based on instrumental or microscopic analyses (Cooke and Ranere 1992; Cooke 1995). A petrographer (Hill 2002) who analysed 20 samples sent to him by John Griggs from four sites (He5, Ag13, Cl1, and Lp8) deduced that local clays were used and that no temper was added. However, in contradiction to Hill's (2002) results, our study suggests that a temper of freshly cracked monomineralic coarse rocks or angular monomineralic sands were added to some Monagrillo vessels. In the future, we plan to compare the petrography and geochemistry of the intentionally added mineral temper with those of rocks gathered from zones of igneous intrusive rocks in Central Panama.

Geological studies of Central Panama show that the Azuero Peninsula and isthmus proper have different formation histories. Azuero is older than the Cordillera Central. Our project analysed Monagrillo sherds obtained from seven sites located in four distinct drainages (Coclé del Norte, Grande, Santa María, and Parita) and in four different environmental zones (Pacific coast, Pacific plains, Pacific foothills, and Caribbean foothills). Our results to date suggest that, mineralogically, the most ubiquitous pottery fabric (from six sites) incorporates natural inclusions of igneous felsic extrusive rock and temper of igneous felsic intrusive rocks. This type is found at sites from every major river drainage. Sites located in the Santa María and Parita drainage basins (He5, Pr32, Ag13, Sf9, Pr14) all have pottery with igneous felsic intrusive characteristics, and clays based on igneous extrusive felsic *and* intermediate rocks. However, this fabric type was not found in Cl1 or Lp8, located in the Grande and Coclé del Norte drainages, respectively. Geochemical analyses are necessary to further classify the above clay types. Pottery made with clay containing natural granitic inclusions that appear to contain tourmaline were only found at He5. In addition, a sherd with sandstone inclusion was only found at He5. He5 is located on the coast of the Azuero Peninsula, at the mouth of the Parita River and near the La Villa River, both of which receive alluvial deposits from igneous intrusive and sedimentary (limestone and sandstone) host rocks. This suggests that the pottery found at He5 likely has a local signature that is distinct from that of the pottery found in Cl1 and Lp8.

### *Manufacturing technique*

Willey and McGimsey (1954) proposed that Monagrillo vessels were made using the coil building technique.

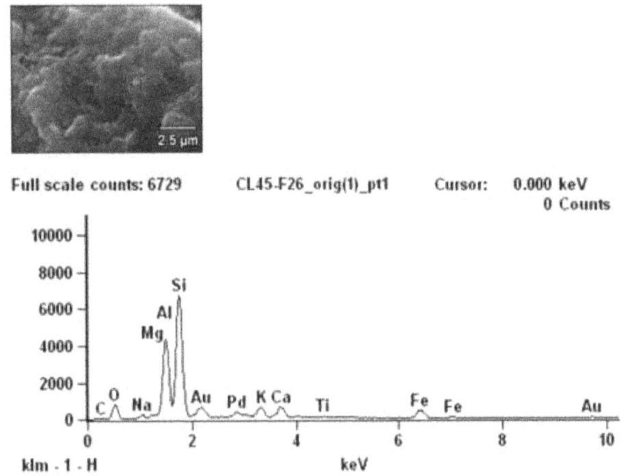


CL45-F26 (Original), Accelerating Voltage: 15.0 kV,

Magnification: 10,000X

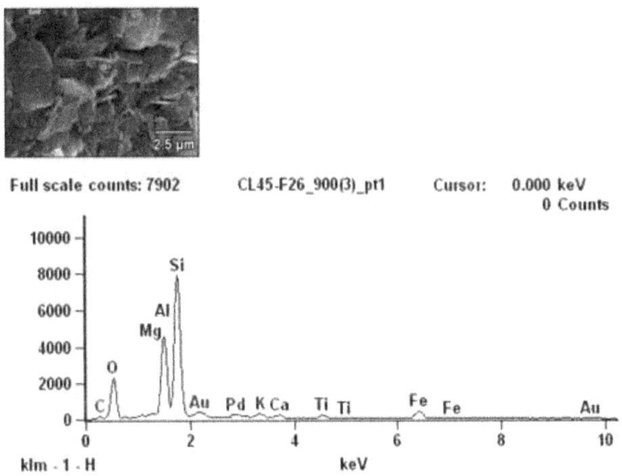


CL45-F26 (900 °C), Accelerating Voltage: 15.0 kV,

Magnification: 10,000 X

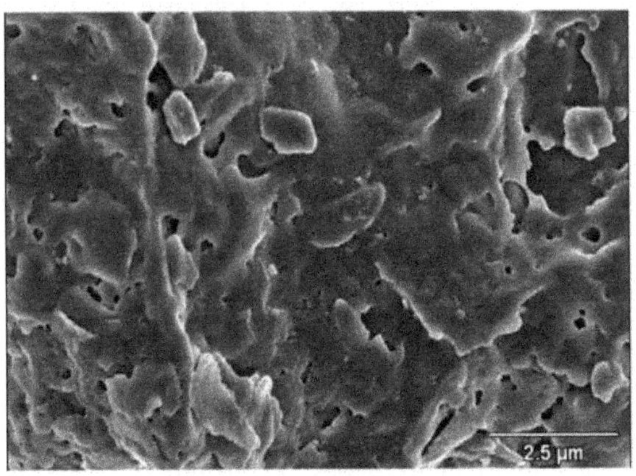


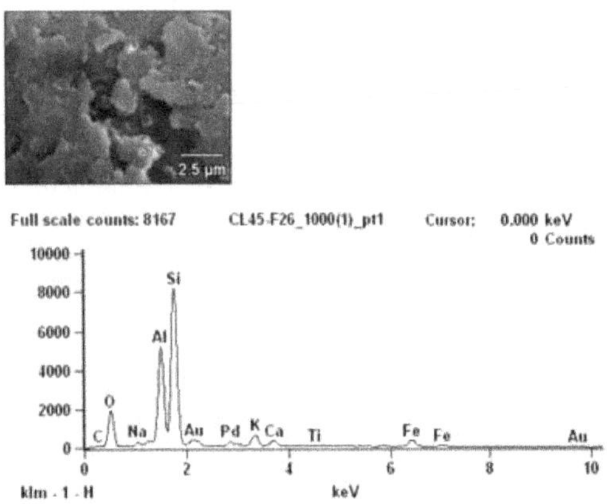


CL45-F26 (1,000 °C), Accelerating Voltage: 5.0 kV

Magnification: 10,000 X

**Figure 3.** SEM examination of sherd CL45-F26 after refiring experiments. The ceramic microstructure remains uchanged at 900°C but appears more sintered at 1000°C, indicating an original firing temperature between those two thresholds.

Our research indicates that most Monagrillo pottery did not resort to this technique, but rather used layered slabs. Some vessel bodies from Pr32 and Ag13 were made by combining slabs and small lumps of clay. In general terms, there is very little variation in the techniques employed for building vessels. The finishing of rims and lips exhibits the greatest variability.

### Porosity

Our porosity tests suggest that the porosity index of Monagrillo vessels was always lower than that of clays gathered near Cl1 and He5, fired at 100°C intervals between 500°C and 800°C. This means that if potters from these sites gathered nearby clays and fired them, they would have fired the pottery above 800°C to reach the porosity of the archaeological sherds.

### Firing temperatures

It has been proposed that Monagrillo pottery was crude and fired at very low temperatures (Cooke and Ranere 1992; Cooke 1995). Although our sample size is still small, our results contradict this hypothesis, indicating that pottery from Cl1 and He5 was fired at 950°C and between 800°C and 850°C, respectively. Experimental and ethno-archaeological studies suggest that temperatures attained with the open firing method normally range between 500 and 900°C (Tite 1995), and occasionally reach 950°C (Gosselain 1992) to 1000°C (Colton 1953; Shepard 1980; Rye 1981). Thus, it is likely that the Monagrillo pottery was fired at the higher end of the open firing temperature range. We thus conclude that Monagrillo pottery is more expertly made than it was previously suggested.

### Regional interpretations

*Mobility of the Monagrillo population*

The techniques we have employed so far have only identified two clay types within the samples of Monagrillo sherds: (1) clays with mostly granitic rock-based natural inclusions that contain tourmaline, and (2) clays with natural inclusions that consist of igneous extrusive rocks. The fact that the former is found only at He5 suggests that some of the pottery used by this site's inhabitants used nearby sources of raw materials. Since the second clay type is so ubiquitous, we cannot identify its precise provenance. The granitic-based ceramic paste with tourmaline has not yet been found outside He5. Therefore, it is reasonable to infer that Monagrillo people did not carry or exchange locally made pottery far from this site. It is hoped that further investigations will enable us to identify additional evidence for local production, and thus improve our ability to determine how Monagrillo pottery was circulating.

*Manufacturing methods*

Despite previous arguments to the contrary, this study has shown that the manufacturing methods of the Monagrillo ware were relatively homogenous. Regardless of the archaeological sites from which they originated, all of the sherds were produced in similar ways, using layered slabs or slabs joined with lumps of clay. Firing temperatures were relatively uniform for the entire collection and clustered at the upper range of temperatures attainable by open-firing. The main difference in manufacturing methods seen with this collection is that some ceramics show additions of mineral temper, while others contain only natural temper. Such skill and attentiveness reflect a well-established and carefully performed craft.

We propose that the Monagrillo potters prioritised impermeability and resistance to breakage. High firing temperatures enhanced their vessels' suitability for holding liquids and for being transported over long distances. Vessels whose primary functions are serving, cooking, and storage can perform these functions even if they are fired at low temperatures (500–800°C). However, the addition of temper does increase thermal shock resistance and therefore optimality for cooking.

### Future directions

Our research has shed new light on the Monagrillo pottery of the Central American isthmus, providing support for the argument that this was not an experimental or hastily produced ware. Raw materials were carefully selected, and vessels were manufactured to fit specific performance criteria. This interpretation contradicts the previously held notion that the Monagrillo ware was crude and roughly made. Future research will address the outstanding issue of trace element compositional differences in Monagrillo pottery, focusing on identifying additional meaningful geochemical signatures of clays and firing tempers across the entire production zone.

## References

Bird J.B. and Cooke R.G., 1978, La Cueva de los Ladrones: datos preliminares sobre la ocupación Formativa. *Actas del V Symposium Nacional de Antropología, Arqueología y Etnohistoria de Panamá*, 283-304, Instituto Nacional de Cultura, Panamá.

de Boer J.Z., Defant M.J., Stewart R.H., Restrepo J.F., Clark L.F. and Ramirez A.H., 1988, Quaternary calc-alkaline volcanisms in western Panama: regional variation and implication for the plate tectonic framework. *Journal of South American Earth Science*, 1, 275-293.

Buchs D.M., Arculus J., Baumgartner P., Baumgartner-Mora C. and Ulianov A., 2010, Late Cretaceous Arc development on the SW margin of the Caribbean Plate: insights from the Golfito (Costa Rica) and Azuero (Panama) complex. *Geochemistry, Geophysics, Geosystems*, 11, 1-35.

Bush M.B. and Colinvaux P.A., 1990, A pollen record of a complete glacial cycle from lowland Panama. *Journal of Vegetation Science*, 1, 105-118.

Carvajal Contreras D.R. and Hansell P., 2008, Molluscs in Central Panama: a review. *Archaeofauna*, 17, 157-74.

Clary J., Hansell P., Ranere A., Buggey T. and Lange F.W., 1984, The Holocene geology of the Western Parita Bay coastline of Central Panama. *Recent Developments in Isthmian Archaeology*, 55-83, British Archaeological Reports, International Series 212, Oxford.

Clark J. and Gosser D., 1995, Reinventing Mesoamerica's first pottery. In *The emergence of pottery: technology and innovation in ancient societies*, (eds. W. Barnett and J. Hoopes), 209-221, Smithsonian Institution Press, Washington DC.

Colton H.S., 1953, *Potsherds*, Museum of Northern Arizona Bulletin 25, Northern Arizona Society of Science and Art, Flagstaff.

Cooke R.G., 1984, Archaeological research in central and eastern Panama: a review of some problems. In *The archaeology of Lower Central America*, (eds. F.W. Lange and D.Z. Stone), 263-302, University of New Mexico Press (School for American Research), Albuquerque.

Cooke R.G., 1995, Monagrillo, Panama's first pottery: summary of research, with new interpretations. In *The emergence of pottery: technology and innovation in ancient societies*, (eds. W. Barnett and J. Hoopes), 169-184, Smithsonian Institution Press, Washington DC.

Cooke R.G., 2005, Prehistory of Native Americans on the Central American land bridge: Colonization, dispersal, and divergence. *Journal of Archaeological Research*, 13, 129-187.

Cooke R.G. and Ranere A.J., 1992, The origins of wealth and hierarchy in the central region of Panama, with observations on its relevance to the phylogeny of Chibchan-speaking polities in Panama and elsewhere. In *Wealth and hierarchy in the Intermediate Area*, (ed. F. Lange), 243-316, Dumbarton Oaks, Washington DC.

Cooke R.G. and Sánchez-Herrera L.A., 2004, Arqueología en Panama (1888-2002). In *Panama: cien años de República*, (ed. A. Figueroa), 111-149, Editorial Universitaria, Panamá.

Cooke R.G. and Jiménez M., 2008, Pre-Columbian use of freshwater fish in the Santa Maria biogeographical province, Panama. *Quaternary International*, 185, 46-58.

Cooke R.G., Jiménez M. and Ranere A.J., 2008, Archaeozoology, art, documents, and the life assemblage. In *Case studies in environmental archaeology*, (eds. E.J. Reitz, C.M. Scarry and S.J. Scudder), 95-121, Springer, New York.

Defant M.J., Richerson P.M., de Boer J.Z., Stewart R.H., Maury R.C., Bellon H., Drummond M.S., Feigenson M.D. and Jackson T.E., 1991, Dacite genesis via both slab melting and differentiation: petrogenesis of La Yeguada volcanic complex, Panama. *Journal of Petrology*, 32, 1101-1142.

Dickau R., 2005, *Resource use, crop dispersals, and the transition to agriculture in prehistoric Panama: evidence from starch grains and macroremains*, PhD Dissertation, Temple University, Philadelphia.

Dickau R., 2010, Microbotanical and macrobotanical evidence of plant use and the transition to agriculture in Panama. In *Integrating zooarchaeology and paleoethnobotany: a consideration of issues, methods and cases*, (eds. A.M. VanDerwacker and T.M. Peres), 99-134, Springer, New York.

Drennan R.D., 1996, Betwixt and between in the Intermediate Area. *Journal of Archaeological Research*, 4, 95-131.

Fonseca Zamora O., 1997, Cerámica temprana de Costa Rica en el contexto del área histórica Chibchoide (400-2500 A.P.). *Revista de Arqueología Americana*, 13, 41-68.

Ford J.A., 1969, A comparison of formative cultures in the Americas: diffusion or the psychic unity of man? *Smithsonian Contributions to Anthropology*, 11, Smithsonian Institution, Washington DC.

del Giudice D. and Recchi G., 1969, *Geología del area del proyecto minero de Azuero, informe técnico preparado para el gobierno de la Republica de Panama: Panama, programa para el desarrollo de las Naciones Unidas*, 48 pp.

Gosselain O.P., 1992, Bonfire of the enquiries: pottery firing temperatures in archaeology: What for? *Journal of Archaeological Science*, 19, 243-259.

Griggs J.C., 2005, *The archaeology of Central Caribbean Panama*, PhD Dissertation, The University of Texas at Austin, Austin.

Haller M.J., 2008, *Asiento Viejo and the development of the Río Parita chiefdom, Panama*, University of Pittsburgh Memoirs in Latin American Archaeology 19, University of Pittsburgh Latin American Archaeology Publications, Pittsburgh.

Hansell P., 1979, *Shell analyis: a case study from Panama*, MA Thesis, Temple University, Philadelphia.

Helms M.W., 1979, *Ancient Panama: chiefs in search of power*, University of Texas Press, Austin.

Hill D.V., 2002, *Petrographic analysis of ceramics from five pre-Columbian sites in central Panama*, Unpublished Report.

Hoopes J., 1995, Interaction in hunting and gathering societies as a context for the emergence of pottery in Central American isthmus. In *The emergence of pottery: technology and innovation in ancient societies*, (eds. W. Barnett and J. Hoopes), 185-198, Smithsonian Institution Press, Washington DC.

Iizuka F., 2013, *Early pottery in the tropics of Panama (ca. 4,500-3,200 B.P.): production processes, circulation, and diagenesis*. PhD Dissertation, The University of Arizona, Tucson.

Iizuka F. and Vandiver P., 2006, *Clay suitability tests for pottery production on clay gathered around Monagrillo Pottery sites (ca. 4,800 – 3,200 B.P.) from central Panama*, [Presentation] Tinker Symposium on Graduate Research in Latin America. The University of Arizona, Tucson, 3 November 2006.

Isaza-Aizuprúa I., 2007, *The ancestors of Parita: pre-Columbian settlement patterns in the lower La Villa River Valley, Azuero Peninsula*, PhD Dissertation, Boston University, Boston.

Kingery D., 1987, Microstructure analysis as part of a holistic interpretation of ceramic art and archaeological artifacts. *Archeomaterials*, 1, 91-99.

Knutsen K.L., 2010, *La Yeguada volcanic complex, Western Panama: an assessment of the geologic hazards using new 40AR/39Ar ages*, MA Thesis, Michigan Technological University, Houghton, Michigan.

Linares O.F., 1977, *Ecology and the arts in ancient Panama: On the Development of Rank and Symbolism in the Central Provinces*, Studies in Precolumbian Art and Archaeology 17, Dumbarton Oaks, Washington DC.

Lissinna B., 2005, *A profile through the Central American landbridge in Western Panama: 115 Ma interplay between the Galápagos Hotspot and the Central American subduction zone*, 47-102, PhD Dissertation, Christian-Albrechts-Universität zu Kiel, Kiel.

Mackenzie R.C. (ed.), 1970, *Differential Thermal Analysis, Vol. 1, Fundamental Aspects*, Academic Press, London. and New York.

Meggers B., 1997, La cerámica temprana en América del Sur: ¿Invención independiente o difusión? *Revista de Arqueología Americana*, 13, 7-40.

Okaya D.A. and Ben-Avraham Z., 1987, Structure of the continental margin of southwestern Panama. *Geological Society of America Bulletin*, 99, 792-802.

Ranere A. and Hansell P., 1978, Early subsistence patterns along the Pacific coast of Central Panama. In *Prehistoric Coastal Adaptations*, (eds. B.L. Stark and B. Voorhies), 31-48, Academic Press, New York.

Perry L., Dickau R., Zarrillo S., Holst I., Pearsall D.M., Piperno D.R., Berman M.J., Cooke R.G., Rademaker K., Ranere A.J., Raymod J.S., Sandweiss D.H., Scaramelli F., Tarble K. and Zeidler J.A., 2007, Starch fossils and the domestication and dispersal of chili peppers (Capsicum spp. L.) in the Americas. *Science*, 315, 986-988.

Piperno D.R. and Pearsall D.M., 1998, *The origins of agriculture in the lowland tropics*, Academic Press, San Diego.

Piperno D.R., 2006, Quaternary environmental history and agricultural impact on vegetation in Central America. *Annals of the Missouri Botanical Garden*, 93, 274-296.

Rye O.S., 1981, *Pottery technology: principles and reconstruction*, Taraxacum, Washington DC.

Shepard A.O., 1980, *Ceramics for the archaeologists*, Carnegie Institution of Washington, Washington DC.

Tite M., 1995, Firing temperature determinations: how and why? *KVHAA Konferenser*, 34, 37-42.

Tite M. and Maniatis Y., 1975, Examination of ancient pottery using the scanning electron microscope. *Nature*, 257, 122-123.

Willey G.R., 1971, *An introduction to American Archaeology, Vol. 2, South America*, Prentice-Hall, Englewood. Cliffs NJ.

Willey G. and McGimsey C.R.III, 1954, *Monagrillo culture of Panama*, Papers of the Peabody Museum of Archaeology and Ethnology 49(2), Harvard University Press, Cambridge.

Wörner G., Harmon R.S. and Wegner W., 2009, Geochemical evolution of igneous rocks and changing magma sources during the formation and closure of Central American land bridge of Panama. In *The backbone of the Americas: shallow subduction, plateau uplift, and ridge and terrane collision*, (eds. S.M. Kay, V.A. Ramos and W.R. Dickinson), 183-196, Geological Society of America Memoir 204. The Geological Society of America, Boulder.

Zohar I. and Cooke R.G., 1997, The impact of salting and drying on fish skeletons: preliminary observations from Parita Bay, Panama. *Archaeofauna*, 6, 59-66.

## ACKNOWLEDGEMENTS

The authors would like to acknowledge the invaluable and generous support provided by the following people: David Killick, Camilo Montes, Natalia Hoyos, Stewart Redwood, Agustín Cardona, Nikki Strong, Rory McFadden, Diego Ramírez, Cesar Silva, Ruth Siddall, Carlos Jaramillo, Sara Moron, and James Wilson. We also acknowledge the financial and logistical support provided by the Smithsonian Tropical Research Institute, National Science Foundation Integrative Graduate Education and Research Traineeship, and the School of Anthropology of the University of Arizona.

# 26

# Molding the 'collapse': Technological analysis of the Terminal Classic molded-carved vases from Altun Ha, Belize

**Carmen Ting, Elizabeth Graham and Marcos Martinón-Torres**
UCL Institute of Archaeology, London, United Kingdom, carmen.k.ting@gmail.com

**Abstract** - Technological analyses by visual examination, thin-section petrography, INAA and SEM-EDS of an assemblage of elite serving vessels from the site of Altun Ha, Belize, provide important data on technology and organization of production. According to the manner in which they were decorated, these vessels are referred to in the literature as 'molded-carved', but prior research has also shown that they share a distinctive iconographic program. Evidence so far indicates that they constitute a ceramic tradition that is reflective of social and political changes that characterized the Terminal Classic period in the Maya lowlands. The results of the technological analyses described here combined with prior research on contexts, iconography and glyphic texts suggest that alterations in political and social systems during the Terminal Classic stimulated changes not only in the type of elite pottery being produced but also in the manufacturing technology and concomitantly in the organization of production.

## 1. Introduction

Despite more than a century of excavations and research conducted in the Maya Lowlands, the nature of the Classic Maya collapse remains a source of speculation and debate (Aimers 2007; Rice *et al.* 2004). What Mayanists call the 'Maya collapse' is the end of the Classic period, or the Terminal Classic (A.D. 750-1050), in which drastic changes in the sociopolitical order took place in the Maya Lowlands. Among these changes were the disintegration of divine kingship and the discontinuation of cultural traits that had been widespread in the Southern Lowlands in Classic times. One of the more noticeable changes that takes place during the Terminal Classic period is the virtual disappearance in the archaeological record of polychrome vessels circulated by elites and their apparent replacement, at least in some areas, by vases decorated by molding and carving rather than by polychromatic painting (Aimers 2004, 80; Chase and Chase 2004, 343-345; Forsyth 2005, 11; Rice and Forsyth 2004; Smith 1955, 37-48; Willey *et al.* 1967, 311). The research described below comprises a technological analysis of the molded-carved assemblage from the site of Altun Ha, Belize. The aims are threefold: to distinguish the potential production groups; to reconstruct the manufacturing technology; and to characterize the organization of production. Based on these data, we are able not only to assess the impact of changes in the sociopolitical order but also to piece together the kinds of changes that occurred in the Maya Lowlands during the Terminal Classic period.

## 2. Classification

Molded-carved vases, which occur in various forms and iconographic designs, were widely distributed in the Maya Lowlands during the Terminal Classic. All are considered part of what is known as the Pabellon System, a loose but useful category based on commonalities in decorative technique and molded-carved execution but not on details of design or pastes (Aimers 2009; Henderson and Agurcia 1987). At present, three types or traditions of molded-carved vases have been named in the literature, although the bases for the names are not the same in the three cases. In only one type, Ahk'utu' Molded-carved (Helmke and Reents-Budet 2008), is the name based on a distinctive iconographic program and dedicatory glyphic phrases which indicate that the vases might be produced for a single patron (Helmke and Reents-Budet 2008, 41-43).

Pabellon Molded-carved and Sahcaba Molded carved — both originally described as 'modeled-carved' before the evidence accumulated for the use of molds — were established as 'types' in the assemblage from Uaxactun, a site in the central Peten region of Guatemala (Smith 1955, 34, 43-45, 95, 192, 194, 195; Smith and Gifford 1966, 160, 162). Pabellon Molded-carved as a type is characterized not only by a consistent set of iconographic imagery but also by a highly distinctive fine-orange paste (see Helmke and Reents-Budet 2008, 37 for references, e.g., Sabloff and Willey 1967; Smith 1958). The paste of Sahcaba as described in the literature is neither fine nor orange (Helmke and Reents-Budet 2008, 39) but its fabric is distinctive enough to have been included by Smith and Gifford (1966, 162) in the Teabo Ceramic Group as Puuc Red Ware; indeed Sahcaba was originally described by Smith as 'Carved Ferruginous Ware' (Smith 1955, Fig. 86), which suggests a paste dark red in color. Sahcaba has been identified at Caracol by Arlen Chase and Diane Chase, but apparently only on the basis of the absence of fine paste (Chase 1994, 173, 175, Figs. 13.11d and m; Chase and Chase 2001, Figs. 16b and 16m). Thus far, no distinguishing iconographic or glyphic characteristics have been defined as particular to Sahcaba at Caracol that are not shared by Ahk'utu' (Helmke and Reents-Budet 2008, 39). The Sahcaba example from Uaxactun, illustrated in Smith (1955, Fig. 86 i,m,o), has a scene that appears to depict interacting elite individuals in a style roughly comparable to that of Pabellon and Ahk'utu', but scene details and the style of execution are different.

We have assigned the sherds analysed in this research to the Ahk'utu' Molded-carved type as defined by Helmke and Reents-Budet (2008). We base this assignment on the fact that where moulded-carved decoration is preserved on Altun Ha vessels, the iconographic program and/or the glyphs can clearly be seen to fall within the Ahk'utu tradition (Graham *et al.* 1980, 164-165, Figs. 7, 8; Helmke and Reents-Budet 2008, Fig. 6). We emphasize nonetheless that this attribution is provisional, owing to the possibility that pastes of some of the sherds without decoration may turn out to connect them with a different tradition.

At one level, owing to the fact that courtly scenes continued to be represented on the molded-carved vase exteriors as they were on polychrome vases, these vases can be seen to be a continuation of the great Classic polychrome tradition. The molded-carved vases have been said, in fact, to take over "at least some of the functions of the polychromes during the Terminal Classic" (Forsyth 2005, 18). Nonetheless, although the vase form and the presence of courtly scenes are common features, the mode of execution changes from painting to the use of gouge-incision. The appearance of the molded-carved ceramic tradition therefore marks the beginning of a new era in the Maya Lowlands, and the period of transition remains little known.

## 3. Background

Previously known as 'imitation Pabellon' or described as 'modeled-carved' (Graham *et al.* 1980, 164-165; Graham 1987, 79), the molded-carved vases from Altun Ha display a characteristic cylindrical or barrel shape with either flat or concave bases supported by hollow tripod oven-shaped feet. Each tripod support is perforated and contains a ceramic rattler (Fig. 1a). The ceramic pastes are tempered with white inclusions that are visible to the naked eye. Vessel walls are covered with a thin layer of slip that ranges from dark red to light orange in color. Where the molded-carving design is in evidence, the exterior surfaces of the vases are decorated with three stacked horizontal moldings which frame the primary glyphic text and the iconographic scenes (Fig. 1b). Interpretation of the glyphic texts suggests that these vases were manufactured for an elite woman named Lady Olom,

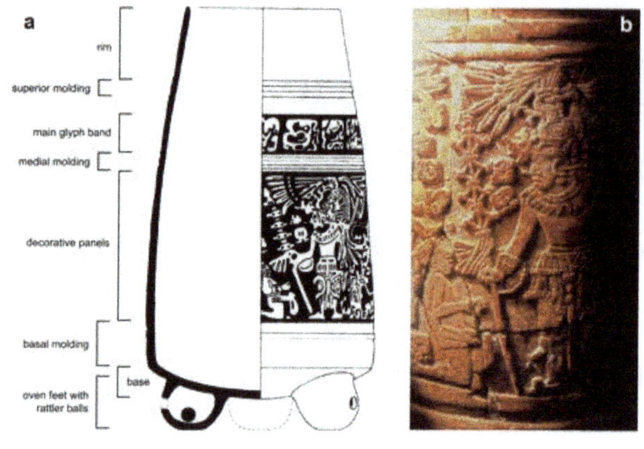


**Figure 1.** An Ahk'utu' molded-carved vase: a) vessel profile; b) panel showing the iconographic design. Illustration by Christophe Helmke and photograph by Reents-Budet (Helmke and Reents-Budet 2008, 38-41).

who lived during the first half of the ninth century in the eastern Peten region (Helmke and Reents-Budet 2008, 41-43). Interestingly, the distribution of these molded-carved vases is restricted to an area that stretches from the eastern Peten lowlands of Guatemala to the Caribbean coast of central Belize, where their presence is reported from 23 sites located along the Belize River and its tributaries. The site of Altun Ha and the Marco Gonzalez site on Ambergris Caye seem to be outliers, but their inclusion may reflect interaction between the inland sites and the coast.

Some archaeologists propose that the molded-carved vases were used in competitive feasting during the Terminal Classic period, a time of intensive jockeying among Maya elites for positions, power, and tribute rights (Helmke 2001; LeCount 1999; 2005; Pohl and Pohl 1994, 140). Feasting here not only refers to the sharing of an elaborately prepared meal but also includes acts of giving and receiving prestige goods, such as the molded-carved vases, which would have served to symbolize and maintain social ties (Clark and Blake 1994, 26-27; LeCount 2001). Within a local context, the liberal distribution of gifts served the purpose of binding leaders through indebtedness; on the regional level, gift exchange aided the building of alliances (Brumfiel 1994, 10). The molded-carved vases, therefore, functioned as both social and political currency. They served as symbols in validating hierarchies, consolidating support, and amassing tribute, but as a result they also encouraged the development of factionalism during the Terminal Classic period in the Maya Lowlands (Helmke 2001, 74-75).

## 4. Geological and archaeological settings of Altun Ha

Altun Ha yielded one of largest molded-carved assemblages, with approximately 200 sherds. The site is situated on the north-central coastal plain of Belize and was excavated from 1964 to 1970 (Fig. 2) (Pendergast 1979, 1982, 1990). Altun Ha is underlain by flinty siliceous Eocene and Miocene limestone, which serves as the parent material for clay loam and sandy loam in some areas (Pendergast 1979, 7; Wright *et al.* 1959, 77). The leaching and eluviation of limestone leaves the topsoil mottled black, yellow, and dull red color, with frequent flints.

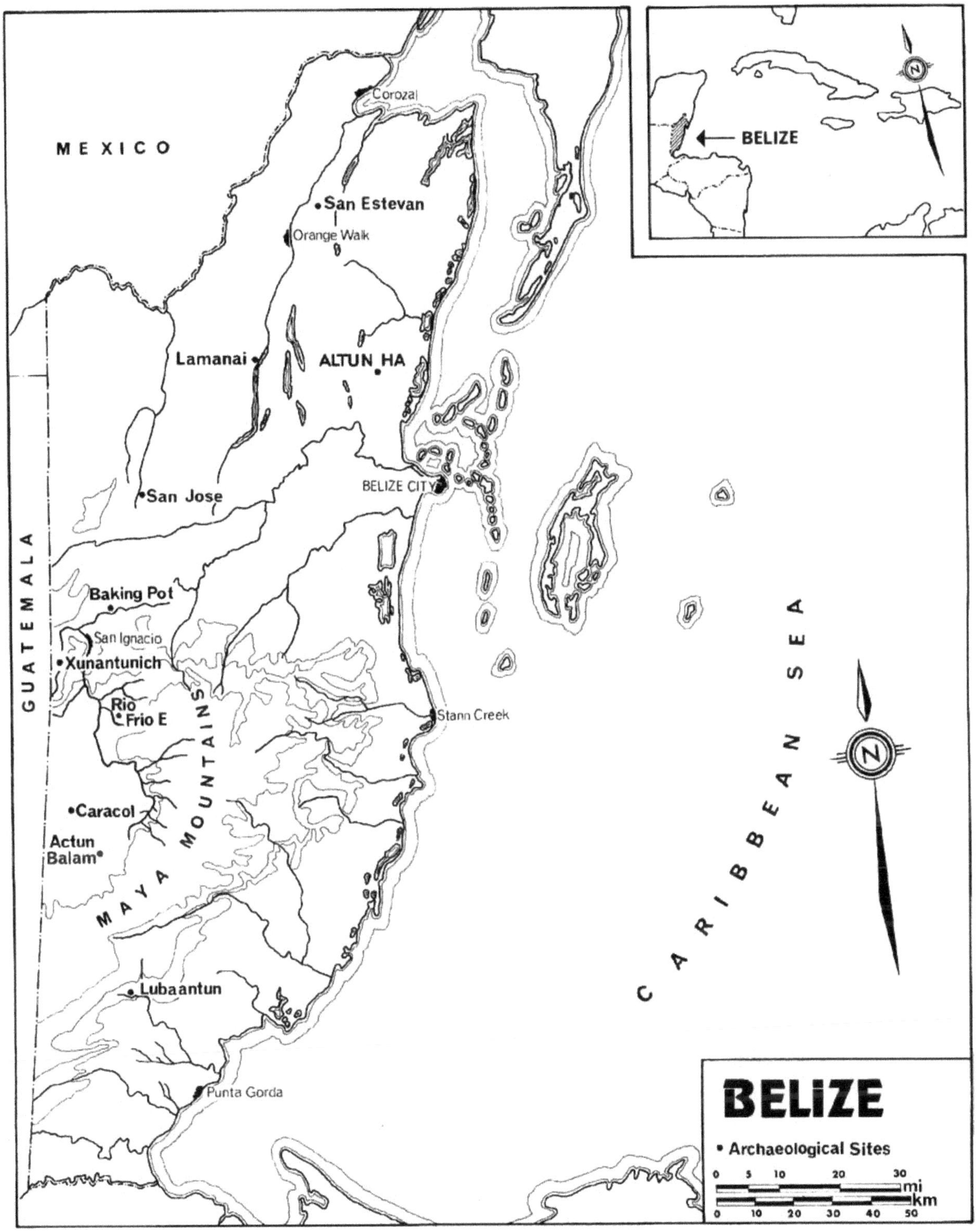


**Figure 2.** Map of Belize showing the site of Altun Ha (Pendergast 1979, 6)

In fact, the presence of limestone as underlying bedrock dominates the geology of Northern Belize which is described as a low-lying shelf, as opposed to the predominance of igneous (e.g. granite) and metamorphic rocks (e.g. gneiss, shales) as underlying bedrock in the mountainous southern region (Howie 2005, 120-135; King *et al.* 1992, 26).

As a small but important, and in fact staggeringly wealthy, urban centre in the central Maya Lowlands, Altun Ha had a long history of development. Early occupation dates to the Preclassic period, but the community reached its pinnacle during the Late Classic. Like many sites in the Southern Lowlands, Altun Ha experienced decline and eventually was to "descend into the dust" (Pendergast 1992, 71) toward the end of the 9$^{th}$ century. Yet the process of 'collapse' was unlikely to have been either sudden or massive, as is evident in the considerable quantity of Terminal Classic materials, including the molded-carved vases, retrieved through excavation. The majority of the molded-carved sherds were found in terminal occupation debris of the vaulted masonry (Structures C-6, C-10,E-7, E-14, E-44, E-51) and plazuela group residential structures (Structures J-1, K-32, K-34). The C- and E-Group structures are believed to have been residences of non-royal lineage elites whereas the J- and K-Group structures may have been the homes of extended families of the royal lineages.

## 5. Methods

A variety of archaeometric techniques, coupled with systematic sampling and statistical methods were employed to examine the molded-carved assemblage. Twenty-one molded-carved sherds were chosen for analyses employing a stratified sampling framework. The strata were created by visual examination based on variations in color, texture or 'feel', size and frequency of tempering materials, and the presence or absence of slip. Although the sherd abundances are not equal across these sampling strata, the selection procedure ensured a more balanced representation of paste variability within the assemblage (Drennan 1996, 237-41; Orton 2000, 26-30). Macroscopic examination of sherd color with reference to the Munsell color system was also useful in the preliminary assessment of firing conditions (Rice 2005, 343-345; Shepard 1976, 107-113). With the aid of a stereomicroscope, visual examination also helped to determine the decorative forming and surface finishing techniques.

Instrumental neutron activation analysis (INAA) was chosen because it allows the detection of a large number of elements, in particular trace elements, which are useful in discriminating production-related compositional groups with a high degree of precision and accuracy (Blackman and Bishop 2007, 321; Glascock 1992, 12; Pollard *et al.* 2007, 132; Speakman and Glascock 2007, 180). The samples were prepared according to the analytical protocols of the Smithsonian Institution (Blackman and Bishop 2007); samples were then activated by the 20MW research reactor housed at the National Institute of Standards and Technology (NIST) Centre for Neutron Research. 29 elements were detected: Na, K, Ca, Sc, Cr, Fe, Co, Zn, As, Br, Rb, Zr, Mo, Sb, Cs, Ba, La, Ce, Nd, Sm, Eu, Tb, Yb, Lu, Hf, Ta, Th, U and W. After experimentation with numerous bivariate plots to identify those elements with clearest grouping tendencies, 23 elements were selected and processed statistically using hierarchical cluster and principal component analysis (PCA) (Baxter 2003; Bishop and Neff 1989; Shennan 1997). These elements were those named above except for Co, Mo, Sb, Hf, U and W. An additional PCA was conducted on all minor and trace elements (excluding Na, K, Ca, and Fe) which are useful in discriminating the differences in the clay sources themselves.

Thin-section petrography was used to identify mineralogical constituents. This procedure helped in assigning sherds to fabric groups in the absence of access to information on raw material sources (Freestone 1991). Petrography is particularly useful in this case because the predominance of limestone as parent material in northern Belize, Yucatan and Peten homogeneous the chemical composition of disparate clay sources and thus increases the difficulty in discriminating among local sources (Howie 2005, 128-136). The application of thin-section petrography, therefore, serves to confirm and even refine the compositional groupings identified by INAA. The samples were studied by using Whitbread's ceramic thin-section descriptive system (1995). Whitbread's system allows the qualitative and semi-quantitative analyses, and its focus on the textural characteristics of the samples is useful in identifying manufacturing technology. For instance, the shape and orientation of voids are indicative of forming techniques whereas the optical states and color of the clay matrix are influenced by the general firing condition.

Scanning electron microscopy energy dispersive spectrometry (SEM-EDS) was chosen because this microanalytical technique allows the analysis of a specific spot or small area on the sample by directing the electron beam precisely on the area to be analyzed. High-resolution images can also be produced with the use of the secondary electron and backscattered electron detectors that are attached to the system. In this way, the SEM-EDS enables detection of the presence of a slip layer and the determination of slip composition. Firing temperature can also be estimated by focusing on the micromorphology (Maniatis and Tite 1981; Tite and Maniatis 1975; Tite *et al.* 1982; Wolf 2002). Analyses were conducted on polished cross sections using a Hitachi S-3400N SEM with an Oxford Instruments EDS, housed at the Wolfson Laboratories at the UCL Institute of Archaeology. Each sample was analyzed at a fixed working distance of 10 mm with the accelerating voltage set at 20kV. The process time of each analysis was set to 5 with deadtime ranging from 26% to 34%. Calibration with cobalt was conducted at the beginning and at regular intervals throughout the analysis of each sample. The results presented here are presented as stoichiometric oxides and have been normalized to 100% to account for sample porosity and small drifts in beam intensity.

## 6. Results

### *6.1. Compositional variability*

Three chemically distinct groups were recognized and are shown in the plot of the first three principal components of the compositional data by INAA (Fig. 3; Table 1). These three groups can also be recognized by plotting the first, second and third components of the data of the minor and trace elements only on the bivariate plot, which suggests that the clay sources of these three groups are different. Group 1 is the core group which comprises the majority of the samples. These samples contain a significantly high bulk calcium (Ca) concentration that ranges from 11.6% to 22.1%. Such high Ca concentration is clearly caused by the addition of calcite as a tempering material, as the ceramic matrices themselves, analyzed by SEM-EDS, revealed relatively low CaO levels (Table 2). Thin-section petrography shows that polycrystalline calcite inclusions of medium- to fine-grained size predominate (Fig. 4a). Monocrystalline and polycrystalline quartz inclusions are common. Textural concentration features are also

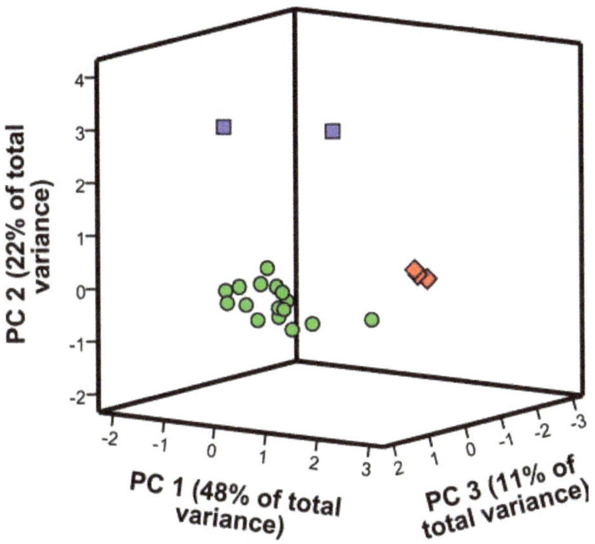


**Figure 3.** Scatterplot based on the principal component analysis of the INAA data on the overall composition (excluding Co, Mo, Sb, Hf, U and W) of the Ahk'utu' samples from Altun Ha. Circles represent Group 1 samples, diamonds represent Group 2 samples, and squares represent Group 3 samples.

common. In terms of the microstructure, planar voids, which display a preferred orientation parallel to the margin, are common in all samples. In terms of groundmass, all samples exhibit fine matrices with poorly sorted inclusions. The matrices are homogeneous throughout, varying from light yellowish brown to brown under PPL, and from dark brown to olive brown under XP, and are optically moderately active to inactive. Thin layers of slip, which appears to be red (2.5YR 5/8) in PPL and changes from dark brown (7.5YR 3/4) to reddish yellow (7.5YR 7/8) upon rotation in XP, can be identified in three samples.

Group 2 comprises only three samples, which contain bulk Ca concentrations comparable to those in Group 1, with an average of 17%. Group 2 is distinguished from Group 1 by its relatively higher bulk concentrations of Fe, Cr, La, Nd and Sm. Subtle variation between the two groups is also noticeable under thin-section petrography. In particular, although the mineralogy and groundmass of the samples of the two groups are largely similar, the size and frequency of inclusions vary (Fig. 4b). The polycrystalline calcite inclusions of the samples in Group 2 are much coarser-grained than those in the previous group. Textural concentration features are more frequent, which explains the higher Fe content because these concentrations are rich in Fe as shown in the SEM-EDS data.

Group 3 consists of only two samples. This group is completely different from the previous two groups in terms of chemical composition and mineralogical constituents. Bulk Ca concentrations are low and range from only 0.9% to 4.5%, but the samples have higher bulk Na and K concentrations. Thin-section petrography further confirms that these samples belong to an entirely different fabric group in which crystalline calcite is lacking. Instead, volcanic ash is the predominant type of inclusion, recognizable in thin section but also under the SEM due to their large and abundant round pores (Fig. 4c & d). Polycrystalline and monocrystalline quartz inclusions are common. Feldspar, biotite, muscovite and chert are present but very rare. Only a few textural concentration features are present. Planar voids are common. All samples exhibit very fine matrices with moderately to well-sorted inclusions. The matrices are homogenous throughout and appear light brown to brown under PPL to dark brown under XP, and are optically inactive. A thin layer of slip, which appear to be yellowish red (5YR 4/6) in PPL and brownish yellow (10YR 6/8) in XP, can be identified in one sample.

### 6.2. *Technological variability*

Macroscopically, the samples of Group 1 and Group 2 share similar physical attributes. The presence of a dark core in most samples indicates that the vessels were fired in an incomplete oxidizing atmosphere. The firing temperature of the samples is estimated to have been below 800°C, as evident in the presence of flaky structure of clay minerals in the secondary electron images on SEM (Fig. 5a) (Maniatis and Tite 1981; Tite and Maniatis 1975; Tite *et al.* 1982;

**Table 1.** Average chemical composition by INAA for the three groups of Ahk'utu' vases from Altun Ha.

| Group | | Na | K | Ca | Sc | Cr | Fe | Zn | As | Br | Rb | Zr | Cs | Ba | La | Ce | Nd | Sm | Eu | Tb | Yb | Lu | Ta | Th |
|---|---|---|---|---|---|---|---|---|---|---|---|---|---|---|---|---|---|---|---|---|---|---|---|---|
| 1 | Mean | 588 | 5891 | 168500 | 11 | 52 | 24350 | 34 | 11 | 13 | 87 | 149 | 6 | 347 | 25 | 47 | 18 | 4 | <1 | <1 | 3 | <1 | 1 | 11 |
| | Max. | 859 | 7460 | 221000 | 13 | 58 | 39700 | 47 | 17 | 23 | 118 | 261 | 8 | 457 | 33 | 64 | 32 | 6 | <1 | <1 | 3 | <1 | 2 | 13 |
| | Min. | 344 | 3710 | 116000 | 9 | 41 | 20800 | 24 | 7 | 4 | 66 | 0 | 4 | 272 | 18 | 36 | 0 | 3 | <1 | <1 | 2 | <1 | <1 | 9 |
| 2 | Mean | 617 | 4060 | 176000 | 14 | 64 | 37333 | 73 | 16 | 6 | 66 | 55 | 4 | 357 | 55 | 75 | 54 | 12 | 2 | 2 | 6 | <1 | <1 | 11 |
| | Max. | 671 | 4380 | 188000 | 14 | 66 | 38800 | 78 | 18 | 9 | 82 | 166 | 5 | 384 | 56 | 90 | 57 | 12 | 2 | 2 | 6 | <1 | 1 | 11 |
| | Min. | 570 | 3850 | 165000 | 14 | 63 | 35800 | 70 | 15 | 5 | 51 | 0 | 4 | 318 | 54 | 58 | 49 | 11 | 2 | 1 | 6 | <1 | <1 | 11 |
| 3 | Mean | 13700 | 11500 | 27505 | 9 | 28 | 23850 | 73 | 9 | 7 | 123 | 187 | 6 | 595 | 24 | 52 | 21 | 4 | <1 | <1 | 2 | <1 | <1 | 15 |
| | Max. | 14500 | 12600 | 45300 | 10 | 29 | 24300 | 80 | 11 | 10 | 167 | 216 | 8 | 700 | 28 | 62 | 25 | 5 | <1 | <1 | 3 | <1 | 1 | 18 |
| | Min. | 12900 | 10400 | 9710 | 8 | 26 | 23400 | 67 | 6 | 4 | 80 | 157 | 4 | 489 | 21 | 42 | 16 | 3 | <1 | <1 | 2 | <1 | <1 | 11 |

**Table 2.** Average composition by SEM-EDS for the ceramic matrices of the three groups of Ahk'utu' vases from Altun Ha. All figures are normalized to 100%.

| Group | | $Na_2O$ | MgO | $Al_2O_3$ | $SiO_2$ | $P_2O_5$ | $K_2O$ | CaO | $TiO_2$ | FeO |
|---|---|---|---|---|---|---|---|---|---|---|
| 1 | Average | 0.2 | 0.4 | 26.7 | 60.9 | 0.3 | 1.6 | 3.8 | 1.0 | 5.0 |
| | Max. | 0.4 | 0.7 | 35.0 | 66.4 | 0.6 | 2.1 | 5.2 | 1.2 | 6.7 |
| | Min. | 0.1 | 0.2 | 22.9 | 51.1 | 0.1 | 1.1 | 2.8 | 0.6 | 3.7 |
| 2 | Average | 0.2 | 0.5 | 31.1 | 54.4 | 0.5 | 1.0 | 4.2 | 1.1 | 7.1 |
| | Max. | 0.2 | 0.5 | 32.1 | 56.0 | 0.5 | 1.1 | 4.5 | 1.2 | 7.8 |
| | Min | 0.1 | 0.4 | 29.1 | 53.1 | 0.5 | 0.9 | 3.7 | 1.0 | 6.4 |
| 3 | Average | 1.0 | 0.8 | 21.4 | 60.9 | 5.5 | 0.8 | 4.3 | 0.8 | 4.6 |
| | Max. | 1.6 | 0.9 | 22.3 | 69.1 | 10.8 | 1.0 | 6.0 | 0.5 | 5.6 |
| | Min. | 0.4 | 0.7 | 20.5 | 52.7 | 0.2 | 0.5 | 2.6 | 1.0 | 3.6 |

Wolf 2002). In contrast, signs of incomplete oxidation are lacking in the samples of Group 3, which exhibit homogeneous colour throughout the cross section with no dark core; this suggests that the sherds are from vessels fired in a relatively complete oxidizing atmosphere. Again, the firing temperature of the samples cannot have been too high judging from the flaky structure of the clay minerals visible in the secondary electron images (Fig. 5b). The flaky structure of clay minerals seems to be contradicting with the observation of optical inactivity of the samples under

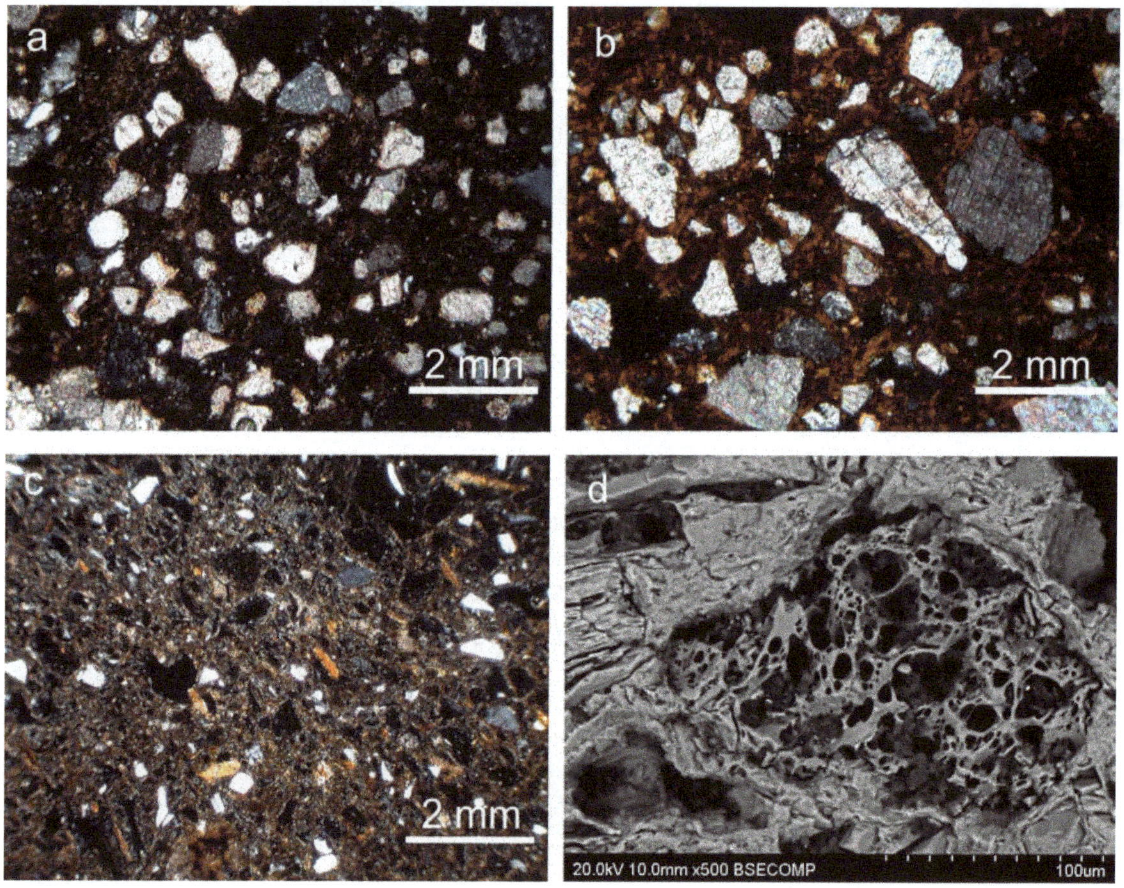


**Figure 4.** Photomicrographs of samples from a) Group 1; b) Group 2; and c) Group 3 under thin section petrography; and d) the presence of volcanic ash in a backscattered electron image in the SEM. All thin-section photomicrographs were taken at x40 in XP.

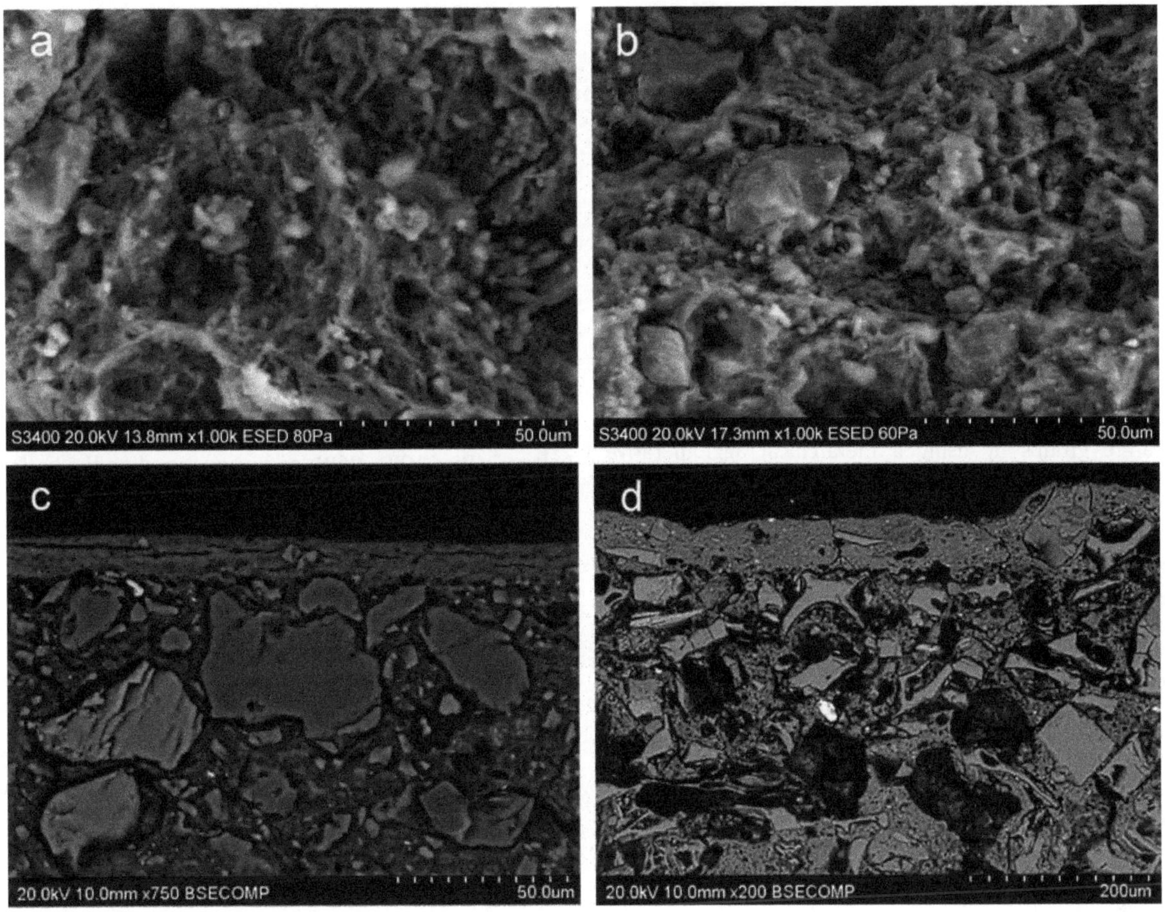


**Figure 5.** Secondary electron images showing the flaky structure of a) a calcite-tempered sample from Group 1; and b) a volcanic ash-tempered sample from Group 3. Backscattered electron images showing the slip layers (as indicated by arrows) of samples from c) Group 1; and d) Group 3.

petrography. One possible explanation is that the firing atmosphere of the vessels from which these samples are derived may have reached a high enough temperature to initiate the process of vitrification, but the temperature at which they were fired at was not high enough to effect complete vitrification. It is also possible that the volcanic ash temper may act locally as a flux, promoting vitrification and then the overall hardness and homogeneity, but this hypothesis requires further investigation.

Although, based on the appearance of the larger and better preserved sherds, we believe that the exterior surfaces of all molded-carved vases were once slipped, a slip layer was preserved on only three samples from Group 1 and one from Group 3. The slips on the samples from Group 1 were applied before carving as evident in the absence of slip in the troughs under the stereomicroscope, whereas the slip on the sample from Group 3 was applied to the surface after carving, as shown in the presence of slip remnants in both the ridges and troughs of the carving lines. In all cases, slip could be easily detected as a homogeneous layer, with partially vitrified microstructure, attached to the surface of the sherd in the backscattered electron images on the SEM (Fig 5c & d). The slip layers do not exceed 20$\mu$m in thickness. No apparent variations in brightness can be distinguished between the slip layers and the rest of the matrices in the backscattered electron images, a fact which suggests that both slip and vessel body were made of clay of similar composition. SEM-EDS analyses of the slips and their associated matrices confirm that only slight variations in composition exist between the two, although a general trend of higher iron oxide (FeO) and relatively lower potassium oxide ($K_2O$) can be observed in the slip layers of three out of four samples; such variation may be caused by the removal of impurities from the clay during preparation of the slip.

## 7. Discussion

### *7.1. Discrimination of potential production groups*

The rationale behind the physiochemical analyses of ceramic materials is that potters' recipes for combination of clay and non-plastic tempers are unique to these production groups (Harry 2005; Reents-Budet *et al.* 1994; Reents-Budet *et al.* 2000, 101; Rice 2009, 128). Therefore, based on the results, we propose that the three distinct compositional groups potentially represent three production groups, each of which produced molded-carved vases according to its own ceramic recipe. Group 1 vases were tempered with a mixture of medium- and fine-grained calcite; Group 2 vases were tempered with coarse-grained calcite; and Group 3 vases were tempered with volcanic

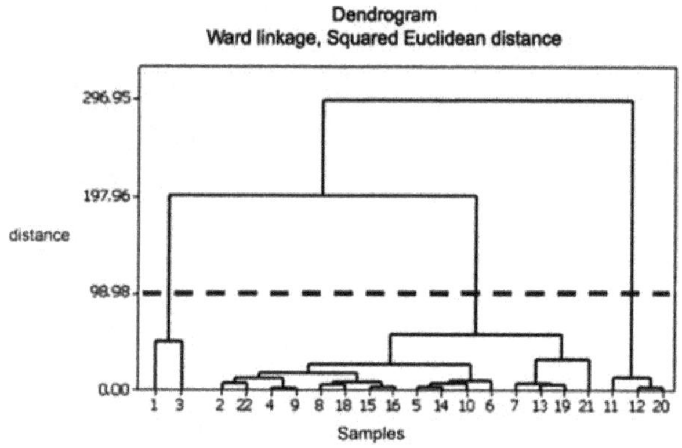


**Figure 6.** Dendrogram showing the presence of three separate groups by using a hierarchical agglomerative cluster analysis on the Ahk'utu' samples from Altun Ha, Ward's method and squared Euclidean distance.

ash. Although more detailed clay sampling is required to determine the provenance of the production groups, the mineralogy of the samples in Group 1 and 2 is consistent with the local geology which is underlain by limestone, thus suggesting that these two groups were local to the area. On the other hand, the mineralogy of the samples in Group 3 is consistent with the metamorphic and granitic nature of the Maya Mountain, thus suggesting that Group 3 may have its origin in Maya Mountain area. Nevertheless, the co-existence of three compositional groups within the same assemblage from Altun Ha suggests the existence of exchange mechanisms, either with regard to the pots themselves or involving the materials used in manufacture.

## *7.2. Reconstruction of manufacturing technology*

The manufacturing technology exhibited by the molded-carved vases is largely the same across the three production groups. Molding was the primary forming technique in which the clay was pressed against concave molds. This hypothesis is supported by the identification of planar voids aligning with the margins of the thin sections at microscopic level, as well as the recovery of concave ceramic molds and matching sherds from the Terminal Classic site of Cerro Palenque in Honduras (Lopiparo *et al.* 2005, 112).

For Groups 1 and 2, we suggest that thin layers of slip were then applied to the exterior surface of the vessels, which were dried to leather-hard before carving, as shown in the presence of smooth beds and sharp margins along the carving lines under the stereomicroscope in most samples. Once the clay and the slip were dried, a plano-relief carving method was used to produce a raised design, which created a three-dimensional effect (Rice 2005, 146). This order of slip application and carving was reversed in the case of Group 3, with slip applied after plano-relief carving.

All groups reflect firing at temperatures below 800°C. The Group 1 and 2 vases were fired in an incomplete oxidizing atmosphere as indicated by the presence of dark core and the moderately active matrix upon rotation in nearly all thin sections. Group 3 vases are likely to have been fired in an atmosphere that resulted in complete oxidation. Low firing temperatures, varying degrees of oxidation and uneven access to air suggest that the molded-carved vases were fired in a non-kiln, open firing structure. Non-kiln open firing method does not involve the construction of permanent firing structure, thus making the identification of the pottery production location or workshop in the Maya areas more difficult (Rice 1997; Rye 1981, 96-98).

## *7.3. Characterization of the organization of production*

In the absence of archaeological evidence of production or firing locales, the organization of production of the molded-carved vases from Altun Ha can be assessed on the basis of indirect evidence as proposed by Costin (1991). Costin's parameters include standardization, skill and efficiency.

Standardization refers to "the relative degree of homogeneity or reduction in variability in the characteristics of the pottery" (Rice 1991, 268), and is measured on the basis of the compositional and technological variability in this study. Judging from the distance of coefficient between samples on the dendrogram of INAA data by agglomerative hierarchical analysis (Fig. 6), Group 3 is obviously separated from the other two groups whereas the difference between Group 1 and 2, noticeable in the chemistry but also in the petrographic structure, is sufficient enough to suggest the involvement of different production groups. The small size of Groups 2 and 3 makes it difficult to make any reliable assessment of the internal fabric standardization. For Group 3 in particular (the ash-tempered ware), there is a significant chemical difference between the two samples, perhaps suggesting that they represent different subfabrics. However, it should be noted that SEM-EDS of one of these sherds showed unusually high $P_2O_5$ contents (around 10%); given that this oxide is easily subject to post-depositional alteration, the divergent chemistries between these two sherds may just reflect different degrees of contamination.

It is worth highlighting that, in spite of the technical differences noted between the three groups (in temper, order of slipping/carving, and firing regime), they all broadly conform to the same technology in manufacturing; thus resulting in similar style of the end-products. These similarities indicate that some technological information may have been shared by all producers, but more importantly, they also suggest that all potting groups aimed to manufacture pottery that would be recognized by consumers over a broad area on the basis of color, shape, and pictorial representation. Consumer demands can therefore be seen to have influenced what could be said to be the 'type' of the vessel. The co-existence of three production groups reflects, however, localized community traditions of resource use and manufacture.

Turning to the other two parameters to characterize the organization of production, skill and efficiency, these are influenced by the political, economic and social conditions under which production is organized, and remain even harder to access archaeologically (Costin 1991; Costin and Hagstrum 1995, 622-623). It is generally assumed that the levels of skill and efficiency are positively correlated with degree of craft specialization, with the exception of objects that have significant social or political value in which case extra energy, time and resources are required in their production. Although archaeological evidence suggests that the molded-carved vases had significant social and political value, they were efficiently produced as evidenced by the use of molds. The use of molds in the manufacture of the Ahk'utu' vases has the advantage of producing standardized products (at least in terms of their physical appearance) efficiently; reducing the level of skills required; and thus reducing the risk and cost involved (Arnold 1999). The apparent discrepancy between the value of goods and the costs involved in their production can perhaps be explained by political developments in the Maya Lowlands during the Terminal Classic. We believe that competing elites required the efficient production of molded-carved vases as political and social currency to build alliances and legitimize their precarious positions during the relative political instability and reorganization characteristic of Terminal Classic times.

Based on the above information, and with reference to the eight-part typology of craft specialization created by Costin (1991), we propose that the production of the molded-carved vases from Altun Ha was organised as dispersed corvée, which entailed the involvement of part-time labor producing for elite or government institutions within a household or community setting.

## 8. Conclusion

The results of this study in conjunction with earlier research on molded-carved vases, cited above, suggest that these vases were circulated among elites who were responding to changes in the sociopolitical order during the Terminal Classic period (Graham 1985, 227-228). Change was clearly occurring both in the kinds of elite pottery being produced and in methods of manufacture. Like in the Classic period, consumer demand could be satisfied regionally or inter-regionally depending on the type or style of vase desired, a fact that suggests continued lowland integration at the elite level (Chase and Chase 2005, 88). Unlike the Classic period, however, the Terminal Classic saw the disappearance of polychrome vases and their replacement by the molded-carved vases.

As in the Classic period, there was a wide range of groups producing pottery in the Terminal Classic throughout the lowlands (Rice 1987, 79-80), but the case of the Ahk'utu' vases as reflected in the Altun Ha samples demonstrates greater variety in local production. One possible explanation is that elites throughout the Southern Lowlands continued to share a value structure but there existed greater flexibility and fluidity in the way the material that reinforced this structure could be produced. With regard to social hierarchy, this situation may reflect a lower pyramid in which more individuals had access to means of accumulating status and wealth than had existed in Classic times. Helmke and Reents-Budet (2008, 47) suggest that molded-carved special service wares were not associated with the highest royal segment of Maya society. They propose that what we may be seeing is a restructuring of social negotiations from paramount royalty during the Late Classic to lesser nobles during Terminal Classic times (Helmke and Reents-Budet 2008, 47; Graham 1985, 228). With regard to the distribution zone of Ahk'utu' vases in particular, they suggest that this reflects the spatial extent of a particular Terminal Classic social network. We concur but add that the social network may have had an economic basis, in which case it may also map out tribute relationships.

Although further analyses of a greater number of vessels are essential, our results nonetheless show clearly the distinctive contribution that can be made from technological studies. From a study of style, iconography and epigraphy, Helmke and Reents-Budet (2008) were able to distinguish the limits of an area encompassing the eastern Peten and parts of Belize in which the vessels with the Ahk'utu' scene and reference to a particular patron occurred. This commonality might at first suggest the existence of a single production source. The Altun Ha technological evidence alone, however, indicates that the efforts of different production groups were marshalled to produce such vessels, which in turn suggests that the relationship between consumers (elites) and producers was different from the Classic period when polychromes were circulated and producers are believed to have been attached to elite households. We might even ask if molds were traded to stimulate production over a wide area in response to elite demands. As analysis proceeds, we hope to have answers to such questions, and to shed more light on the dynamics of the changing relationships of the period of 'collapse' and the Classic to Postclassic transition.

## References


Aimers J.J., 2004, *Cultural Change on a Temporal and Spatial Frontier: Ceramics of the Terminal Classic to Postclassic transition in the Upper Belize River Valley*, BAR International Series 1325, Archaeopress, Oxford.

Aimers J.J., 2007, What Maya Collapse? Terminal Classic Variation in the Maya Lowlands. *Journal of Archaeological Research*, 17, 329-377.

Aimers J.J., 2009, Bring It On: Using Ceramic Systems at Lamanai. Research Reports in Belizean Archaeology 6:245-252. Papers from the 2008 Belize Archaeology Symposium, Institute of Archaeology, NICH, Belmopan, Belize.

Arnold D.E., 1999, Advantages and Disadvantages of Vertical-Half Molding Technology: Implications for Production Organization. In *Pottery and People: A Dynamic Approach*, (eds. J.M. Skibo and G.M. Feinman), 59-80, University of Utah Press, Salt Lake City.

Baxter M., 2003, *Statistics in Archaeology*, Oxford University Press, New York.

Bishop R.L. and Neff H., 1989, Compositional Data Analysis in Archaeology. In *Archaeological Chemistry*, (ed. R.O. Allen), 57-86, American Chemical Society, Washington D.C.

Blackman M.J. and Bishop R.L., 2007, The Smithsonian-NIST Partnership: the application of instrumental neutron activation analysis to archaeology. *Archaeometry*, 49(2), 321-341.

Brumfiel E.M., 1994, Factional Competition and Political Development in the New World: an introduction. In *Factional Competition and Political Development in the New World*, (eds. E.M. Brumfiel and J.W. Fox), 1-14, Cambridge University Press, Cambridge.

Chase A.F., 1994, A Contextual Approach to the Ceramics of Caracol, Belize. In *Studies in the Archaeology of Caracol, Belize*, (eds. D.Z. Chase and A.F. Chase),

157-182, Pre-Columbian Art Research Institute, Monograph 7, San Francisco.

Chase A.F., 2005, Contextualizing the Collapse Hegemony and Terminal Classic Ceramics from Caracol, Belize. In *Geographies of Power: understanding the nature of Terminal Classic Pottery n the Maya Lowlands*, (eds. S.L. Lopez and A.E. Foias), 73-92, BAR International Series 1447, Archaeopress, Oxford.

Chase A.F. and Chase D.Z., 2001, The Royal Court of Caracol, Belize: Its Palaces and People. In *Royal Courts of the Ancient Maya*, (eds. T. Inomata and S.D. Houston), 102-137, Westview Press, Boulder, Colorado.

Chase A.F. and Chase D.Z., 2004, Terminal Classic Status-Linked Ceramics and the Maya "Collapse": de facto refuse at Caracol, Belize. In *The Terminal Classic in the Maya Lowlands: collapse, transition, and transformation*, (eds. A.A. Demarest, P.M. Rice and D.S. Rice), 342-366, University Press of Colorado, Colorado.

Clark J.E. and Blake M., 1994, The Power of Prestige: competitive generosity and the emergence of rank societies in lowland Mesoamerica. In *Factional Competition and Political Development in the New World*, (eds. E. M. Brumfiel and J.W. Fox), 17-30, Cambridge University Press, Cambridge.

Costin C.L., 1991, Craft Specialization: issues in defining, documenting, and explaining the organization of production. *Archaeological Method and Theory*, 3, 1-56.

Costin C.L. and Hagstrum M.B., 1995, Standardization, Labour Investment, Skill, and the Organization of Ceramic Production in Late Prehispanic Highland Peru. *American Antiquity*, 60(4), 619-639.

Drennan R.D., 1996, *Statistics for Archaeologists: a common sense approach*, Plenum Press, New York.

Forsyth D.W., 2005, A Survey of Terminal Classic Ceramic Complexes. In *Geographies of Power: understanding the nature of Terminal Classic pottery in the Maya Lowlands*, (eds. S.L López Varela and A.E. Foias), 23-40, BAR International Series 1447, Archaeopress, Oxford.

Freestone I.C., 1991, Extending Ceramic Petrology. In *Recent Developments in Ceramic Petrology*, (eds. A. Middleton and I.C. Freestone), 399-410, British Museum Press, London.

Glascock M.D., 1992, Characterization of Archaeological Ceramics at MURR by Neutron Activation Analysis and Multivariate Statistics. In *Chemical Characterization of Ceramic Pastes in Archaeology*, (ed. H. Neff), 11-26, Prehistory Press, Madison, Wisconsin.

Graham E., 1985, Facets of Terminal to Postclassic Activity in the Stann Creek District, Belize. In *The Lowland Maya Postclassic*, (eds. A.F. Chase and P.M. Rice), 215-229, University of Texas, Austin.

Graham E., 1987, Terminal Classic to Early Historic Period Vessel Forms from Belize. In *Maya Ceramics: papers from the 1985 Maya Ceramic Conference*, (eds. P.M. Rice and R.J. Sharer), 73-98, BAR. International Series 345(i), BAR, Oxford.

Graham E., McNatt L. and Gutchen M.A., 1980, Excavations in Footprint Cave, Caves Branch, Belize. *Journal of Field Archaeology*, 7, 153-172.

Harry K.G., 2005, Ceramic Specialization and Agricultural Marginality: do ethnographic models explain the development of specialized pottery production in the prehistory American Southwest? *American Antiquity*, 70 (2), 295-319.

Helmke C.G.B., 2001, The Last Supper: competitive feasting and the Terminal Classic Moulded-carved ceramic tradition in the Central Maya Lowlands. Unpublished MA Thesis. Institute of Archaeology, University College London.

Helmke C. and Reents-Budet D., 2008, A Terminal Classic Molded-Carved Ceramic Type of the Eastern Maya Lowlands. In *Research Reports in Belizean Archaeology* 5, (eds. J. Morris, S. Jones, J. Awe and C. Helmke), 37-49, Papers from the 2007 Belize Archaeology Symposium, Institute of Archaeology, NICH, Belmopan, Belize.

Henderson J.S. and Agurcia R.F., 1987, Ceramic Systems: Facilitating Comparison in Type-Variety Analysis. In *Maya Ceramics: Papers from the 1985 Maya Ceramic Conference*, (eds. P.M. Rice and R.J. Sharer), 431-438, BAR International Series 345(ii), Oxford.

Howie L.A., 2005, Ceramic Production and Consumption in the Maya Lowlands During the Classic to Postclassic Transition: a technological study of ceramics at Lamanai, Belize. Unpublished PhD Dissertation, University of Sheffield, Sheffield.

King R.B., Bailie I.C., Abell T.M.B., Dunsmore J.R., Gray D.A., Pratt J.H., Versey H.R., Wright A.C.S. and Zisman S.A., 1992, *Land Resource Assessment of Northern Belize*, Volume I., Overseas Development Natural Resources Institute, Natural Resources Institute, Kent.

LeCount L.J., 1999, Polychrome Pottery and Political Strategies in late and Terminal Classic Lowland Maya Society. *Latin American Antiquity*, 10(3), 239-258.

LeCount L.J., 2001, Like Water for Chocolate: feasting and political ritual among the Late Classic Maya at Xunantunich, Belize. *American Anthropologist*, 103(4), 935-953.

LeCount L.J., 2005, Continuity and Change in the Ceramic Complex of Xunantunich during the Late and Terminal Classic Periods. In *Geographies of Power: understanding the nature of Terminal Classic pottery in the Maya Lowlands*, (eds. S.L. Lopez Varela and A.E. Foias), 93-106, BAR International Series 1447, Archaeopress, Oxford.

Lopiparo J.L., Joyce R.A. and Hendon J.A., 2005, Terminal Classic Pottery Production in the Ulua Valley, Honduras. In *Geographies of Power: understanding the nature of Terminal Classic pottery in the Maya Lowlands*, (eds. S.L. Lopez Varela and A.E. Foias), 107-120, BAR International Series 1447, Archaeopress, Oxford.

Maniatis Y. and Tite M.S., 1981, Technological Examination of Neolithic-Bronze Age Pottery from Central and Southeast Europe and from the Near East. *Journal of Archaeological Science*, 8, 59-76.

Orton C., 2000, *Sampling in Archaeology*, Cambridge University Press, Cambridge.

Pendergast D.M., 1979, *Excavations at Altun Ha, Belize, 1964-1970 Volume 1*, Royal Ontario Museum, Toronto.

Pendergast D.M., 1982, *Excavations at Altun Ha, Belize, 1964-1970 Volume 2*, Royal Ontario Museum, Toronto.

Pendergast D.M., 1990, *Excavations at Altun Ha, Belize, 1964-1970 Volume 3*, Royal Ontario Museum, Toronto.

Pendergast D.M., 1992, Noblesse Oblige: the elites of Altun Ha and Lamanai, Belize. In *Mesoamerican Elites: an archaeological assessment*, (eds. D.Z. Chase and A.F. Chase), 61-79, University of Oklahoma Press, Norman.

Pollard M., Batt C., Stern B. and Young S.M.M., 2007, *Analytical Chemistry in Archaeology*, Cambridge University Press, Cambridge.

Pohl M.E.D. and Pohl J.M.D., 1994, Cycles of conflict: political factionalism in the Maya Lowlands. In *Factional Competition and Political Development in the New World*, (eds. E.M. Brumfiel and J.W. Fox), 138-157, Cambridge University Press, Cambridge.

Reents-Budet D., Bishop R.L. and MacLeod B., 1994, Painting Styles, Workshop Locations and Pottery. In *Painting the Maya Universe: Royal Ceramics at the Classic Period*, (ed. D. Reents-Budet), 164-233, Duke University Press, Durham.

Reents-Budet D., Bishop R.L., Taschek J.T. and Ball J.W., 2000, Out of Palace Dumps: ceramic production and use at Buenavista del Cayo. *Ancient Mesoamerica*, 11, 99-121.

Rice P.M., 1987, Lowland Maya Pottery Production in the Late Classic Period. In *Maya Ceramic: Papers from the 1985 Maya Ceramic Conference*, (eds. P.M. Rice and R.J. Sharer), 525-544, BAR International Series 345(i), Oxford.

Rice P.M., 1991, Specialization, Standardization, and Diversity: A Retrospective. In *The Ceramic Legacy of Anna O. Shepard*, (eds. R.L. Bishop and F.W. Lange), 257-279, University of Colorado, Colorado.

Rice P.M., 1997, *The Prehistory and History of Ceramic Kilns*, The American Ceramic Society, Westerville, Ohio.

Rice P.M., 2005, *Pottery Analysis: a sourcebook*, University of Chicago Press, Chicago.

Rice P.M., 2009, Late Classic Maya Pottery Production: review and synthesis. *Journal of Archaeological Method and Theory*, 16, 117-156.

Rice P.M. and Forsyth D.W., 2004, Terminal Classic-Period Lowland Ceramics. In *The Terminal Classic in the Maya Lowlands: collapse, transition, and transformation*, (eds. A.A. Demarest, P.M. Rice and D.S. Rice), 28-59, University Press of Colorado, Colorado.

Rice P.M., Demarest A.A. and Rice D.S., 2004, The Terminal Classic and The "Classic Maya Collapse" in Perspective. In *The Terminal Classic in the Maya Lowlands: collapse, transition, and transformation*, (eds. A.A. Demarest, P.M. Rice and D.S. Rice), 1-11, University Press of Colorado, Colorado.

Rye O.S., 1981, *Pottery Technology: principles and reconstruction*, Taraxacum, Washington D.C..

Sabloff J.A. and Willey G.R., 1967, The Collapse of Maya Civilization in the Southern Lowlands: A Consideration of History and Process. *Southwestern Journal of Anthropology*, 23(4), 311-336.

Shepard A.O., 1976, *Ceramics for the Archaeologist*, Carnegie Institution of Washington, Washington D.C.

Smith R.E., 1955, *Ceramic Sequence at Uaxactun, Guatemala, Volumes 1, II*, Middle American Research Institute Publication 20, Tulane University, New Orleans.

Smith R.E., 1958, The Place of Fine Orange Pottery in Mesoamerican Archaeology. *American Antiquity*, 24(2), 151-160.

Smith R.E. and Gifford J.C., 1966, *Maya Ceramic Varieties, Types, and Wares at Uaxactun: Supplement to 'Ceramic Sequence at Uaxactun, Guatemala'*, Middle American Research Institute, Tulane University, New Orleans.

Speakman R.J. and Glascock M.D., 2007, Acknowledging Fifty Years of Neutron Activation Analysis in Archaeology. *Archaeometry*, 49(2), 179-183.

Shennan S., 1997, *Quantifying Archaeology*, University of Iowa Press, Iowa City.

Tite M.S., Freestone I.C., Meeks N.D. and Bimson M., 1982, The Use of Scanning Electron Microscopy in the Technological Examination of Ancient Ceramics. In *Archaeological Ceramics*, (eds. J.S. Olin and A.D. Franklin), 109-120, Smithsonian Institution Press, Washington D.C..

Tite M.S. and Maniatis Y., 1975, Examination of Ancient Pottery Using the Scanning Electron Microscope. *Nature*, 257, 122-123.

Wright A.C.S., Romney D.H., Arbuckle R.H. and Vial V.E., 1959, *Land in British Honduras*, Report of the British Honduras Land Use Survey Team, Colonial Research Publications No. 24, Her Majesty's Stationery Office.

Whitbread I.K., 1995, Appendix 3: The Collection, Processing and Interpretation of Petrographic Data. In *Greek Transport Amphorae: a petrological and archaeological study*, (ed. I.K. Whitbread), 365-395, British School at Athens, London.

Willey G.R., Culbert T.P. and Adams R.E.W., 1967, Maya Lowland Ceramics: a report from the 1965 Guatemala City Conference. *American Antiquity*, 32, 289-315.

Wolf S., 2002, Estimation of the Production Parameters of Very Large Medieval Bricks from St. Urban, Switzerland. *Archaeometry*, 44(1), 37-65.

## ACKNOWLEDGEMENTS

The authors would like to acknowledge the valuable contributions of a number of individuals who have helped with this research. We would like to thank Dr. David Pendergast for granting access to the Altun Ha collections and reading the manuscript of this paper. We would like to express our gratitude to Drs. Ronald Bishop and James Blackman of the Smithsonian Institution on conducting the neutron activation analysis of the samples. We would also like to thank Dr. Christophe Helmke for conducting the epigraphic analyses on the molded-carved sherds. Thanks should be extended to Dr. Jim Aimers for his insightful comments on the type-variety system and ceramic system approach which have contributed positively to the final version of this paper. We also appreciate the technical support provided by the staff of the UCL Wolfson Archaeological Science Laboratory.

# Permissions

All chapters in this book were first published in CSIPAC, by Hamad Bin Khalifa University Press; hereby published with permission under the Creative Commons Attribution License or equivalent. Every chapter published in this book has been scrutinized by our experts. Their significance has been extensively debated. The topics covered herein carry significant findings which will fuel the growth of the discipline. They may even be implemented as practical applications or may be referred to as a beginning point for another development.

The contributors of this book come from diverse backgrounds, making this book a truly international effort. This book will bring forth new frontiers with its revolutionizing research information and detailed analysis of the nascent developments around the world.

We would like to thank all the contributing authors for lending their expertise to make the book truly unique. They have played a crucial role in the development of this book. Without their invaluable contributions this book wouldn't have been possible. They have made vital efforts to compile up to date information on the varied aspects of this subject to make this book a valuable addition to the collection of many professionals and students.

This book was conceptualized with the vision of imparting up-to-date information and advanced data in this field. To ensure the same, a matchless editorial board was set up. Every individual on the board went through rigorous rounds of assessment to prove their worth. After which they invested a large part of their time researching and compiling the most relevant data for our readers.

The editorial board has been involved in producing this book since its inception. They have spent rigorous hours researching and exploring the diverse topics which have resulted in the successful publishing of this book. They have passed on their knowledge of decades through this book. To expedite this challenging task, the publisher supported the team at every step. A small team of assistant editors was also appointed to further simplify the editing procedure and attain best results for the readers.

Apart from the editorial board, the designing team has also invested a significant amount of their time in understanding the subject and creating the most relevant covers. They scrutinized every image to scout for the most suitable representation of the subject and create an appropriate cover for the book.

The publishing team has been an ardent support to the editorial, designing and production team. Their endless efforts to recruit the best for this project, has resulted in the accomplishment of this book. They are a veteran in the field of academics and their pool of knowledge is as vast as their experience in printing. Their expertise and guidance has proved useful at every step. Their uncompromising quality standards have made this book an exceptional effort. Their encouragement from time to time has been an inspiration for everyone.

The publisher and the editorial board hope that this book will prove to be a valuable piece of knowledge for researchers, students, practitioners and scholars across the globe.

# List of Contributors

**P. Guibert and A. Bouviera**
IRAMAT-CRP2A - UMR 5060, CNRS - Université de Bordeaux 3, Pessac, France

**I.K. Bailiff**
Luminescence Laboratory, Dept of Archaeology, University of Durham, South Road, Durham, DH1 3LE, England

**C. Sapin**
Laboratoire Artehis - UMR 5594, CNRS -Université de Bourgogne, Dijon, France

**D. Prigent**
Service archéologique départemental de Maine-et-Loire, France

**S. Blain**
IRAMAT-CRP2A - UMR 5060, CNRS - Université de Bordeaux 3, Pessac, France
FRS-FNRS, Centre Européen d'Archéométrie, Université de Liège, Belgium

**Alice M.W. Hunt**
Center for Applied Isotope Studies, University of Georgia

**Bruno Fabbri and Sabrina Gualtieri**
CNR, Institute of Science and Technology for Ceramics, Faenza, Italy

**Enrico Acquaro**
University of Bologna, Ravenna, Italy

**Yona Waksman**
CNRS UMR5138, Lyon, France

**Claudio Capelli**
DIP.TE.RIS, Università degli Studi di Genova, Italy

**Trinitat Pradell**
Dpt. FEN. Universitat Politècnica de Catalunya, Barcelona, Spain

**Judit Molera**
GRMAiA, EPS, Universitat de Vic, Spain

**Noémi S. Müller, Vassilis Kilikoglou and George Vekinis**
Institute of Materials Science, N.C.S.R. "Demokritos", Aghia Paraskevi, 15310 Athens, Greece

**Peter M. Day**
Department of Archaeology, University of Sheffield, Northgate House, West Street, Sheffield S1 4ET, UK

**Noémi S. Müller**
Fitch laboratory, British School at Athens, Souedias 52, 10676 Athens, Greece

**P.C. Gutierrez and A. Climent-Font**
Centro de Micro-Análisis de Materiales and Departamento de Física Aplicada, C-XII, Universidad Autónoma de Madrid, Campus de Cantoblanco, 28049 Madrid, Spain

**T. Pradell**
Departament de Física i Enginyeria Nuclear, Universitat Politècnica de Catalunya, Campus Baix Llobregat, ESAB, Av. Canal Olímpic 15, 08860 Castelldefels, Spain

**J. Molera**
GR-MECAMAT. Escola Politècnica Superior. Universitat de Vic-Universitat Central de Catalunya . c) de la Laura, 13, 08500 Vic, Spain

**A.D. Smith**
dSTFC, Daresbury Laboratory, Sci-Tech Daresbury, Daresbury, Warrington WA4 4AD UK
Dalton Cumbrian Facility, University of Manchester, Westlakes Science & Technology Park, Moor Row, Cumbria CA24 3HA, UK

**M.S. Tite**
Research Laboratory for Archaeology and the History of Art, Dyson Perrins Building, South Parks Road, Oxford OX13QY, UK

**Clive R. Orton**
UCL Institute of Archaeology, 31–34 Gordon Square, London, WC1H 0PY, UK

**Hector Neff**
Department of Anthropology and IIRMES, California State University Long Beach, 1250 Bellflower Blvd. Long Beach, CA90815

**Bruno Fabbri, Sabrina Gualtieri and Francesca Amato**
CNR-ISTEC, National Research Council, Institute of Science and Technology for Ceramic Materials, Faenza, Italy

**Anna Gómez, Walter Cruells and Miquel Molist**
Department of Prehistory, Universitat Autònoma de Barcelona, Seminar on Prehistoric Archaeology of the Near East (SAAPO)

**Rob Ixer, Sara Lunt and Bill Sillar**
UCL Institute of Archaeology, 31-34 Gordon Square, London WC1H 0PY, United Kingdom

**Nigel Wood**
Research Laboratory for Archaeology and the History of Art, University of Oxford, UK, OX1 3QY

**Sabrina Rastelli**
Ca' Foscari University, Department of East Asian Studies, Dorsoduro 3462, 30123 Venezia, Italy

**Ioannis Karatasiosa, Noémi S. Müller and Vassilis Kilikoglou**
Institute of Materials Science, NCSR "Demokritos", Greece

**Konstantinos Alexiou**
21st Archaeological Ephorate of Prehistorical and Classical Antiquities, Ministry of Culture, Greece

**Noémi S. Müller**
Science and Technology in Archaeology Research Center (STARC), The Cyprus Institute, Cyprus

**Baoqiang Kang, Hongying Duan, Yinzhong Ding, He Li and Jianmin Miao**
Palace Museum, Beijing, 100009, PRC

**Simon Groom**
UCL Institute of Archaeology, 31-34 Gordon Square, London WC1H 0PY, United Kingdom

**Guanglie Lu**
Central Laboratory, Zhejiang University, Hangzhou, 310028, PRC

**K.T. Biró**
Hungarian National Museum, Budapest, Hungary

**P. Csengeri and J.S. Koós**
Herman Ottó Museum, Miskolc, Hungary

**Gy. Szakmány**
Eötvös Loránd University, Budapest, Hungary

**H. Taubalde, C. Berthold and J. Zöldföldi**
University of Tübingen, Germany

**J. Mihály**
Research Centre for Natural Sciences, HAS, Budapest, Hungary

**Javier G. Iñañez**
Grupo Investigación en Patrimonio Construido (GPAC), University of the Basque Country (UPV/EHU), CIEA Lascaray, Avda.Miguel de Unamuno, 3, 01006, Vitoria-Gasteiz, Álava, Spain
Grupo Investigación en Patrimonio Construido (GPAC), University of the Basque Country (UPV/EHU), CIEA Lascaray, Avda. Miguel de Unamuno, 3, 01006, Vitoria-Gasteiz, Álava, Spain

**V. Szilágyi**
Centre for Energy Research, Budapest, HAS, Hungary

**T.Á. Rácz**
Tragor Ignác Museum, Directorate of Pest County Museums

**V. Gál-Mlakár**
Herman Ottó Museum, Directorate of Borsod-Abaúj-Zemplén County Museum

**E. Simonyi**
Hungarian National Museum

**I.E. Sajó**
Research Centre for Natural Sciences, HAS, Budapest, Hungary

**Massimo Saracino**
Independent researcher, Verona, Italy

**Lara Maritan and Claudio Mazzoli**
Department of Geosciences, University of Padova, Via G. Gradenigo 6, 35131 Padova, Italy

**David Ben-Shlomo, Avshalom Karasik and Uzy Smilansky**
Department of Physics of Complex Systems, The Weizmann Institute of Science, Rehovot 76100, Israel

**Avshalom Karasik**
The Institute of Archaeology, The Hebrew University, Mount Scopus, Jerusalem 91904, Israel

**Stephan Karl**
Department of Archaeology, University Graz, Universitätsplatz 3/II, 8010 Graz, Austria

**Daniel Jungblut and Gabriel Wittum**
Goethe Center for Scientific Computing (G-CSC), University Frankfurt, Kettenhofweg 139, 60325 Frankfurt am Main, Germany

**Dragoş Gheorghiu**
Department of Master, National University of Arts, 19 Budisteanu, Bucharest, Romania

**Hubert Mara and Susanne Krömker**
Interdisciplinary Center for Scientific Computing (IWR), University Heidelberg, Im Neuenheimer Feld 368, 69120 Heidelberg, Germany

**Yuval Goren**
Laboratory for Comparative Microarchaeology, Department of Archaeology and Ancient Near Eastern Cultures, Tel Aviv University

**Shira Gurwin**
The Eretz-Israel Museum, Tel Aviv, and Department of Archaeology and Ancient Near Eastern Cultures, Tel Aviv University

**Eran Arie**
The Israel Museum, Jerusalem, and Department of Archaeology and Ancient Near Eastern Cultures, Tel Aviv University

**Dennis Braekmans**
Laboratory for Ceramic Studies, Faculty of Archaeology, Leiden University, 2300 RA Leiden, The Netherlands
Materials in Art and Archaeology, Laboratory of Materials Science, Delft University of Technology, Mekelweg 2, 2628 CD Delft, The Netherlands

**Patrick Degryse and Bert Neyt**
Section Geology, Earth and Environmental Sciences, Katholieke Universiteit Leuven, Celestijnenlaan 200 E - bus 2408, B-3001 Leuven, Belgium

**Jeroen Poblome and Marc Waelkens**
Sagalassos Archaeological Research Project, Katholieke Universiteit Leuven, Blijde-Inkomststraat 21 - bus 3314, B-3000, Leuven, Belgium

**Izabella Havancsák, Bernadett Bajnóczi and Mária Tóth**
Institute for Geological and Geochemical Research, Research Centre for Astronomy and Earth Sciences, Hungarian Academy of Sciences, H-1112 Budapest, Budaörsi út 45., Hungary

**Attila Kreiter**
Hungarian National Museum National Heritage Protection Centre, H-1113 Budapest, Daróci út 3., HungVary

**Szilvia Szöllősi**
Institute of Archaeological Sciences, Eötvös Loránd University, H-1088 Budapest, Múzeum krt. 4/B., Hungary

**Fumie Iizuka**
School of Anthropology, The University of Arizona
Smithsonian Tropical Research Institute, Panama

**Richard Cooke**
Smithsonian Tropical Research Institute, Panama

**Lesley Frame and Pamela Vandiver**
School of Anthropology, The University of Arizona
Department of Materials Science and Engineering, The University of Arizona

**Carmen Ting, Elizabeth Graham and Marcos Martinón-Torres**
UCL Institute of Archaeology, London, United Kingdom

# Index

CPSIA information can be obtained
at www.ICGtesting.com
Printed in the USA
LVHW060852030122
707669LV00002B/111

9 781639 870554